D1286635

Asymptotics and Mellin-Barnes Integrals is a comprehensive account of the properties of Mellin-Barnes integrals and their application to problems involving special functions, primarily the determination of asymptotic expansions. An account of the basic analytical properties of Mellin-Barnes integrals and Mellin transforms and their use in applications ranging from number theory to differential and difference equations is followed by a systematic analysis of the asymptotics of Mellin-Barnes representations of many important special functions, including hypergeometric, Bessel and parabolic cylinder functions. An account of the recent developments in the understanding of the Stokes phenomenon and of hyperasymptotics in the setting of Mellin-Barnes integrals ensues. The book concludes with the application of ideas set forth in the earlier parts of the book to higher-dimensional Laplace-type integrals and sophisticated treatments of Euler-Jacobi series, the Riemann zeta function and the Pearcey integral. Detailed numerical illustrations accompany many of the results developed in the text.

Richard Paris is a Reader in Mathematics at the University of Abertay Dundee. He is the author of numerous papers that have appeared in *Proceedings A of the Royal Society of London*, *Methods and Applications of Analysis* and the *Journal of Computational and Applied Mathematics* and is the co-author of *Asymptotics of High Order Differential Equations*.

David Kaminski is an Associate Professor of Mathematics at the University of Lethbridge. He is the author of numerous papers that have appeared in such publications as *SIAM Journal on Mathematical Analysis*, the *Journal of Computational and Applied Mathematics*, *Utilitas Mathematica* and *Methods and Applications of Analysis*.

ENCYCLOPEDIA OF MATHEMATICS AND ITS APPLICATIONS

Asymptotics and Mellin-Barnes Integrals

R. B. PARIS   D. KAMINSKI

CAMBRIDGE
UNIVERSITY PRESS

PUBLISHED BY THE PRESS SYNDICATE OF THE UNIVERSITY OF CAMBRIDGE
The Pitt Building, Trumpington Street, Cambridge, United Kingdom

CAMBRIDGE UNIVERSITY PRESS
The Edinburgh Building, Cambridge CB2 2RU, UK
40 West 20th Street, New York, NY 10011-4211, USA
10 Stamford Road, Oakleigh, VIC 3166, Australia
Ruiz de Alarcón 13, 28014 Madrid, Spain
Dock House, The Waterfront, Cape Town 8001, South Africa

http://www.cambridge.org

First published 2001

Printed in the United Kingdom at the University Press, Cambridge

Typeface Times Roman 10/13 pt.   *System* LaTeX 2_ε [KW]

A catalogue record for this book is available from the British Library.

Library of Congress Cataloging in Publication Data
Paris, R.B.
  Asymptotics and Mellin-Barnes integrals/R. B. Paris, D. Kaminski.
    p.   cm. – (Encyclopedia of mathematics and its applications; v. 85)
  Includes bibliographical references and index
  ISBN 0-521-79001-8
    1. Mellin transform.   2. Asymptotic expansions.   I. Kaminski, D. (David), 1960–   II. Title
  III. Series.
  QA432.P34 2001
  515′.723 – dc21   00-067492

ISBN 0 521 79001 8 hardback

Contents

Preface

Mellin-Barnes integrals are characterised by integrands involving one or more gamma functions (and possibly simple trigonometric or other functions) with integration contours that thread their way around sequences of poles of the integrands. They are a powerful tool in the development of asymptotic expansions of functions defined by integrals, sums or differential equations and, combined with the closely related Mellin transform, form an important part of the toolkit of any practising analyst. The great utility of these integrals resides in the facts that the asymptotic behaviour near the origin and at infinity of the function being represented is related to the singularity structure in the complex plane of the resulting integrand and to the inherent flexibility associated with deformation of the contour of integration over subsets of these singularities.

It is a principal aim of this book to describe the theory of these integrals and to illustrate their power and usefulness in asymptotic analysis. Mellin-Barnes integrals have their early history bound up in the study of hypergeometric functions of the late nineteenth and early twentieth centuries. This association has lent a classical feel to their use and in the domain of asymptotic analysis, the account of their utility in other works has largely been restricted to the analysis of special sums or their role in inversion of Mellin transforms. For their part, Mellin transforms have appeared in several settings within mathematics, as far back as Riemann's memoir on the distribution of primes, and continue to see application through to the present day.

This work gathers a detailed account of the asymptotic analysis of Mellin-Barnes integrals and, conversely, the use of Mellin-Barnes integral representations to problems in asymptotics, from basic results involving their early application to hypergeometric functions, to work that is still appearing at the beginning of this new century. Our account differs from earlier work in the latter half of the twentieth century. For example, texts such as those by Sneddon, *The Use of Integral*

Transforms (1972), and by Davies, *Integral Transforms and Their Application* (1978), are primarily concerned with Mellin transforms and their use in the construction of solutions to differential equations. The well-known monograph by Copson, *Asymptotic Expansions* (1965), barely mentions Mellin-Barnes integrals, and that by Olver, *Asymptotics and Special Functions* (1974), makes little use of them outside of the problem of determining the asymptotics of sums of special type. Mellin-Barnes integrals are more prominently employed in the accounts of Bleistein and Handelsman, *Asymptotic Expansion of Integrals* (1975), of Wong, *Asymptotic Approximation of Integrals* (1989) and of Marichev, *Handbook of Integral Transforms and Higher Transcendental Functions: Theory and Algorithmic Tables* (1982), but there the roles are primarily confined to consequences of the Parseval formula for Mellin transforms. The classic texts on analysis by Whittaker and Watson, *Modern Analysis* (1965), and Copson, *Theory of Functions of a Complex Variable* (1935), and *Generalised Hypergeometric Functions* by Slater (1960), include important sections describing the development of asymptotic expansions of functions represented by Mellin-Barnes integrals. The monograph by Paris and Wood, *Asymptotics of High Order Differential Equations* (1986), contains much of the foundations of the analysis found here, but in more limited scope, and restricted to solutions of differential equations of a particular type. We believe that this present volume, then, is to date the most comprehensive account of Mellin-Barnes integrals and their interactions with asymptotics.

Additionally, this work liberally employs numerical studies to better display the calibre of the asymptotic approximations obtained, a strategy we feel gives the non-expert practitioner a good sense of the concept or method being showcased. A wide-ranging collection of special functions is used to illustrate the ideas under discussion in the fine tradition of the texts mentioned earlier, from Bessel and parabolic cylinder functions, to more exotic functions such as the Mittag-Leffler function and a Riemann-Siegel type of expansion of the zeta function. This book should be accessible to anyone with a solid undergraduate background in functions of a single complex variable.

The book begins with a brief foray into general notions common in asymptotic analysis, and illustrated with the asymptotic behaviour of some classical (and more recent) special functions. The main tools employed in the asymptotics of integrals are found here, including Watson's lemma and the method of steepest descent. Also present is a description of the notion of optimal truncation, which plays a significant role later in the discussion of hyperasymptotics. Brief historical sketches of the namesakes of the type of integrals under examination round out the introductory chapter.

Basic results pertaining to Mellin-Barnes integrals and Mellin transforms are detailed in the next three chapters. Since rational functions of the gamma function are to be found in almost every Mellin-Barnes integral, a thorough account of the behaviour of these rational functions is provided, along with convergence rules for Mellin-Barnes integrals and error estimates for expansions of ratios of gamma functions that occur throughout the remainder of the monograph. Mellin transforms and their properties follow, with applications to the evaluation of slowly convergent sums, number-theoretic sums, and also to differential, integral and difference equations. While these latter applications are not strictly speaking necessarily concerned with asymptotics, they add to the value of the volume and, hopefully, render it more useful as a reference.

The theme of asymptotics comes to the fore in the remaining chapters. In the fifth chapter, a careful and systematic analysis is undertaken which extracts both algebraic and exponential behaviours of Mellin-Barnes-type integrals, with attention paid to the errors committed in the approximation process. These methods are illustrated in the settings of several classical special functions, and the calibre of the approximations illuminated with numerical comparisons. An account of the Stokes phenomenon ensues in the setting of Bessel functions, and the reader is drawn into a detailed account of the recent theory of hyperasymptotics applied to the confluent hypergeometric functions (which incorporate many of the commonly used special functions). An illustration of this theory is made to the exponentially-improved asymptotics of the gamma function and amplified by a study of the numerics of this new expansion.

The penultimate chapter illustrates the manner in which Mellin-Barnes-type integrals may be successfully deployed to extract the algebraic asymptotic behaviour of multidimensional Laplace-type integrals in a systematic manner, and further, interpret the results geometrically. The monograph closes with sophisticated applications of the ideas developed in the text to three particular problems: the determination of the asymptotics of the generalised Euler-Jacobi series, expansions for the zeta function on the critical line and the Pearcey integral, a two-variable generalisation of the classical Airy function.

There is much in this book that is encyclopaedic, but much also is of recent vintage – a good deal of the mathematics present is less than a decade old, and continues to develop apace. We feel we have captured the most important tools and techniques surrounding the analysis and asymptotics of Mellin-Barnes integrals, and by gathering them in a single source, have made the task of their continued application to both mathematics and physical science a more tractable and, we hope, interesting affair.

The authors gratefully acknowledge the long-suffering forbearance of their respective wives, Jocelyne and Laurie, during the lengthy duration of this project. The authors also acknowledge the support of their institutions, the Universities of Abertay Dundee and of Lethbridge, and in the case of the second author, the research funding made available by the Natural Sciences and Engineering Research Council of Canada, which underwrote some of the investigations reported on in this volume. We also owe a considerable debt of gratitude to J. Boersma of Eindhoven University of Technology for the meticulous care with which he studied the text and for his many critical comments on all but two of the chapters, which have improved the calibre of the volume before you. In spite of our best efforts, however, it is certain that some errors and misprints are bound to have crept into the text, and we ask for the reader's forgiveness for those that prove to be vexing.

R. B. Paris and D. Kaminski

1

Introduction

1.1 Introduction to Asymptotics

Before venturing into our examination of Mellin-Barnes integrals, we present an overview of some of the basic definitions and ideas found in asymptotic analysis. The treatment provided here is not intended to be comprehensive, and several high quality references exist which can provide a more complete treatment than is given here: in particular, we recommend the tracts by Olver (1974), Bleistein & Handelsman (1975) and Wong (1989) as particularly good treatments of asymptotic analysis, each with their own strengths.†

1.1.1 Order Relations

Let us begin our survey by defining the *Landau symbols* O and o and the notion of asymptotic equality.

Let f and g be two functions defined in a neighbourhood of x_0. We say that $f(x) = O(g(x))$ as $x \to x_0$ if there is a constant M for which

$$|f(x)| \leq M |g(x)|$$

for x sufficiently close to x_0. The constant M depends only on how close to x_0 we wish the bound to hold. The notation $O(g)$ is read as 'big-oh of g', and the constant M, which is often not explicitly calculated, is termed the *implied constant*.

In a similar fashion, we define $f(x) = o(g(x))$ as $x \to x_0$ to mean that

$$|f(x)/g(x)| \to 0$$

---

† Olver provides a good balance between techniques used in both integrals and differential equations; Bleistein & Handelsman present a relatively unified treatment of integrals through the use of Mellin convolutions; and Wong develops the theory and application of (Schwartz) distributions in the setting of developing expansions of integrals.

as $x \to x_0$, subject to the proviso that $g(x)$ be nonzero in a neighbourhood of x_0. The expression $o(g)$ is read as 'little-oh of g', and from the preceding definition, it is immediate that $f = o(g)$ implies that $f = O(g)$ (merely take the implied constant to be any (arbitrarily small) positive number).

The last primitive asymptotic notion required is that of asymptotic equality. We write

$$f(x) \sim g(x)$$

as $x \to x_0$ to mean that

$$\lim_{x \to x_0} \frac{f(x)}{g(x)} = 1,$$

provided, of course, that g is nonzero sufficiently close to x_0. The tilde here is read 'is asymptotically equal to'. An equivalent formulation of asymptotic equality is readily available: for $x \to x_0$,

$$f(x) \sim g(x) \quad \text{iff} \quad f(x) = g(x)\{1 + o(1)\}.$$

Example 1. The function $\log x$ satisfies the order relation $\log x = O(x - 1)$ as $x \to \infty$, since the ratio $(\log x)/(x - 1)$ is bounded for all large x. In fact, it is also true that $\log x = o(x - 1)$ for large x, and for $x \to 1$, $\log x \sim x - 1$.

Example 2. Stirling's formula is a well-known asymptotic equality. For large n, we have

$$n! \sim (2\pi)^{\frac{1}{2}} e^{-n} n^{n+\frac{1}{2}}.$$

This result follows from the asymptotic expansion of the gamma function, a result carefully developed in §2.1.

Example 3. The celebrated Prime Number Theorem is an asymptotic equality. If we denote by $\pi(x)$ the number of primes less than or equal to x, then for large positive x we have the well-known result

$$\pi(x) \sim \frac{x}{\log x}.$$

With the aid of Gauss' logarithmic integral,†

$$\text{li}(x) = \int_2^x \frac{dt}{\log t}$$

we also have the somewhat more accurate form

$$\pi(x) \sim \text{li}(x) \quad (x \to \infty).$$

---

† We note here that $\text{li}(x)$ is also used to denote the same integral, but taken over the interval $(0, x)$, with $x > 1$. With this larger interval, the integral is a Cauchy principal value integral. The notation in this example appears to be in use by some number theorists, and is also sometimes written $\text{Li}(x)$.

That both forms hold can be seen from a simple integration by parts:

$$\mathrm{li}(x) = \frac{x}{\log x} - \frac{2}{\log 2} + \int_2^x \frac{dt}{(\log t)^2}.$$

An application of l'Hôpital's rule reveals that the resulting integral on the right-hand side is $o(x/\log x)$, from which the $x/\log x$ form of the Prime Number Theorem follows.

A number of useful relationships exist for manipulating the Landau symbols. The following selections are all easily obtained from the above definitions, and are not established here:

(a) $O(O(f)) = O(f)$ (e) $O(f) + O(f) = O(f)$

(b) $o(o(f)) = o(f)$ (f) $o(f) + o(f) = o(f)$

(c) $O(fg) = O(f) \cdot O(g)$ (g) $o(f) + O(f) = O(f)$ (1.1.1)

(d) $O(f) \cdot o(g) = o(fg)$ (h) $O(o(f)) = o(O(f)) = o(f)$.

It is easy to deduce linearity of Landau symbols using these properties, and it is a simple matter to establish asymptotic equality as an equivalence relation. In the transition to calculus, however, some difficulties surface.

A moment's consideration reveals that differentiation is, in general, often badly behaved in the sense that if $f = O(g)$, then it does not necessarily follow that $f' = O(g')$, as the example $f(x) = x + \sin e^x$ aptly illustrates: for large, real x, we have $f = O(x)$, but the derivative of f is not bounded (i.e., not $O(1)$).

The situation for integration is a good deal better. It is possible to formulate many results concerning integrals of order estimates, but we content ourselves with just two.

Example 4. For functions f and g of a real variable x satisfying $f = O(g)$ as $x \to x_0$ on the real line, we have

$$\int_{x_0}^x f(t)\, dt = O\left(\int_{x_0}^x |g(t)|\, dt \right) \qquad (x \to x_0).$$

A proof can be fashioned along the following lines: for $f(t) = O(g(t))$, let M be the implied constant so that $|f(t)| \le M\,|g(t)|$ for t sufficiently close to x_0, say $|t - x_0| \le \eta$. (For $x_0 = \infty$, a suitable interval would be $t \ge N$ for some large positive N.) Then

$$-M\,|g(t)| \le f(t) \le M\,|g(t)| \qquad (|t - x_0| \le \eta),$$

whence the result follows upon integration.

Example 5. If f is an integrable function of a real variable x, and $f(x) \sim x^\nu$, $\mathrm{Re}(\nu) < -1$ as $x \to \infty$, then

$$\int_x^\infty f(t)\, dt \sim -\frac{x^{\nu+1}}{\nu + 1} \qquad (x \to \infty).$$

A proof of this claim follows from $f(x) = x^v\{1 + \psi(x)\}$ where $\psi(x) = o(1)$ as $x \to \infty$, for then

$$\int_x^\infty f(t)\,dt = -\frac{x^{v+1}}{v+1} + \int_x^\infty t^v \psi(t)\,dt.$$

But $\psi(t) = o(1)$ implies that for $\epsilon > 0$ arbitrarily small, there is an $x_0 > 0$ for which $|\psi(t)| < \epsilon$ whenever $t > x_0$. Thus, the remaining integral may be bounded as

$$\left| \int_x^\infty t^v \psi(t)\,dt \right| < \epsilon \int_x^\infty |t^v|\,dt \qquad (x > x_0).$$

Accordingly, we find

$$\int_x^\infty f(t)\,dt = -\frac{x^{v+1}}{v+1} + o\left(\frac{x^{v+1}}{v+1} \right) = -\frac{x^{v+1}}{v+1}\{1 + o(1)\},$$

from which the asymptotic equality is immediate.   □

It is in the complex plane that we find differentiation of order estimates becomes better behaved. This is due, in part, to the fact that the Cauchy integral theorem allows us to represent holomorphic functions as integrals which, as we have noted, are better behaved in the setting of Landau symbols. A standard result in this direction is the following:

Lemma 1.1. *Let f be holomorphic in a region containing the closed annular sector $S = \{z : \alpha \le \arg(z - z_0) \le \beta, |z - z_0| \ge R \ge 0\}$, and suppose $f(z) = O(z^v)$ (resp. $f(z) = o(z^v)$) as $z \to \infty$ in the sector, for fixed real v. Then $f^{(n)}(z) = O(z^{v-n})$ (resp. $f^{(n)} = o(z^{v-n})$) as $z \to \infty$ in any closed annular sector properly interior to S with common vertex z_0.*

The proof of this result follows from the Cauchy integral formula for $f^{(n)}$, and is available in Olver (1974, p. 9).

1.1.2 Asymptotic Expansions

Let a sequence of continuous functions $\{\phi_n\}$, $n = 0, 1, 2, \ldots$, be defined on some domain, and let x_0 be a (possibly infinite) limit point of this domain. The sequence $\{\phi_n\}$ is termed an *asymptotic scale* if it happens that $\phi_{n+1}(x) = o(\phi_n(x))$ as $x \to x_0$, for every n. If f is some continuous function on the common domain of the asymptotic scale, then by an (infinite) *asymptotic expansion* of f with respect to the asymptotic scale $\{\phi_n\}$ is meant the formal series $\sum_{n=0}^\infty a_n \phi_n(x)$, provided the coefficients a_n, independent of x, are chosen so that for any nonnegative integer N,

$$f(x) = \sum_{n=0}^N a_n \phi_n(x) + O(\phi_{N+1}(x)) \qquad (x \to x_0). \tag{1.1.2}$$

In this case we write

$$f(x) \sim \sum_{n=0}^{\infty} a_n \phi_n(x) \qquad (x \to x_0).$$

Such a formal series is uniquely determined in view of the fact that the coefficients a_n can be computed from

$$a_N = \lim_{x \to x_0} \frac{1}{\phi_N(x)} \left\{ f(x) - \sum_{n=0}^{N-1} a_n \phi_n(x) \right\} \qquad (N = 0, 1, 2, \dots).$$

The formal series so obtained is also referred to as an asymptotic expansion of *Poincaré type*, or an asymptotic expansion in the sense of Poincaré or, more simply, a Poincaré expansion. Examples of asymptotic scales and asymptotic expansions built with them are easy to come by. The most commonplace is the asymptotic power series: an *asymptotic power series* is a formal series

$$\sum_{n=0}^{\infty} a_n (x - x_0)^{\nu_n},$$

where the appropriate asymptotic scale is the sequence $\{(x - x_0)^{\nu_n}\}$, $n = 0, 1, 2, \dots$, and the ν_n are constants for which $(x - x_0)^{\nu_{n+1}} = o((x - x_0)^{\nu_n})$ as $x \to x_0$. Any convergent Taylor series expansion of an analytic function f serves as an example of an asymptotic power series, with x_0 a point in the domain of analyticity of f, $\nu_n = n$ for any nonnegative integer n, and the coefficients in the expansion are the familiar Taylor coefficients $a_n = f^{(n)}(x_0)/n!$.

Asymptotic expansions, however, need not be convergent, as the next two examples illustrate.

Example 1. WATSON'S LEMMA. A well-known result of Laplace transform theory is that the Laplace transform of a piecewise continuous function on the interval $[0, +\infty)$ is $o(1)$ as the transform variable grows without bound. By imposing more structure on the small parameter behaviour of the function being transformed, a good deal more can be said about the growth at infinity of the transform.

Lemma 1.2. *Let $g(t)$ be an integrable function of the variable $t > 0$ with asymptotic expansion*

$$g(t) \sim \sum_{n=0}^{\infty} a_n t^{(n+\lambda-\mu)/\mu} \quad (t \to 0+)$$

for some constants $\lambda > 0$, $\mu > 0$. Then, provided the integral converges for all sufficiently large x, the Laplace transform of g, $\mathcal{L}[g; x]$, has the asymptotic behaviour

$$\mathcal{L}[g; x] \equiv \int_0^{\infty} e^{-xt} g(t) \, dt \sim \sum_{n=0}^{\infty} \Gamma\left(\frac{n+\lambda}{\mu}\right) \frac{a_n}{x^{(n+\lambda)/\mu}} \quad (x \to \infty).$$

Proof. To see this, let us put, for positive integer N and $t > 0$,

$$g_N(t) = g(t) - \sum_{n=0}^{N-1} a_n t^{(n+\lambda-\mu)/\mu}$$

so that the Laplace transform has a finite expansion with remainder given by

$$\mathcal{L}[g; x] = \sum_{n=0}^{N-1} \Gamma\left(\frac{n+\lambda}{\mu}\right) \frac{a_n}{x^{(n+\lambda)/\mu}} + \int_0^\infty e^{-xt} g_N(t) \, dt. \qquad (1.1.3)$$

Since $g_N(t) = O(t^{(N+\lambda-\mu)/\mu})$, there are constants K_N and t_N for which

$$|g_N(t)| \le K_N \, t^{(N+\lambda-\mu)/\mu} \quad (0 < t \le t_N).$$

Use of this in the remainder term in our finite expansion (1.1.3) allows us to write

$$\left| \int_0^{t_N} e^{-xt} g_N(t) \, dt \right| \le K_N \int_0^{t_N} e^{-xt} t^{(N+\lambda-\mu)/\mu} dt$$

$$< \Gamma\left(\frac{N+\lambda}{\mu}\right) \frac{K_N}{x^{(N+\lambda)/\mu}}. \qquad (1.1.4)$$

By hypothesis, $\mathcal{L}[g; x]$ exists for all sufficiently large x, so the Laplace transform of g_N must also exist for all sufficiently large x, by virtue of (1.1.3). Let X be such that $\mathcal{L}[g_N; x]$ exists for all $x \ge X$, and put

$$G_N(t) = \int_{t_N}^t e^{-Xv} g_N(v) \, dv.$$

The function G_N so defined is a bounded continuous function on $[t_N, \infty)$, whence the bound

$$L_N = \sup_{[t_N, \infty)} |G_N(t)|$$

exists. Then for $x > X$, we have

$$\int_{t_N}^\infty e^{-xt} g_N(t) \, dt = \int_{t_N}^\infty e^{-(x-X)t} e^{-Xt} g_N(t) \, dt$$

$$= (x - X) \int_{t_N}^\infty e^{-(x-X)t} G_N(t) \, dt$$

after one integration by parts. After applying the uniform bound L_N to the integral that remains, we arrive at

$$\left| \int_{t_N}^\infty e^{-xt} g_N(t) \, dt \right| \le (x - X) L_N \int_{t_N}^\infty e^{-(x-X)t} \, dt = L_N e^{-(x-X)t_N} \qquad (1.1.5)$$

for $x > X$.

Together, (1.1.4) and (1.1.5) yield

$$\left| \int_0^\infty e^{-xt} g_N(t)\, dt \right| < \Gamma\left(\frac{N+\lambda}{\mu}\right) \frac{K_N}{x^{(N+\lambda)/\mu}} + L_N e^{-(x-X)t_N}$$

which, since $L_N e^{-(x-X)t_N}$ is $o(x^{-\nu})$ for any positive ν, establishes the asymptotic expansion for $\mathcal{L}[g; x]$. □

As a simple illustration of the use of Watson's lemma, consider the Laplace transform of $(1+t)^{\frac{1}{2}}$. From the binomial theorem, we have the convergent expansion as $t \to 0$

$$(1+t)^{\frac{1}{2}} = 1 + \tfrac{1}{2}t - \tfrac{1}{8}t^2 + \sum_{n=3}^\infty (-)^{n-1} \frac{1 \cdot 3 \cdot 5 \cdots (2n-3)}{2^n n!} t^n.$$

Since $(1+t)^{\frac{1}{2}}$ is of algebraic growth, its Laplace transform clearly exists for $x > 0$, and Watson's lemma produces the asymptotic expansion

$$\mathcal{L}[(1+t)^{\frac{1}{2}}; x] \sim \frac{1}{x} + \frac{1}{2x^2} - \frac{1}{4x^3} + \sum_{n=3}^\infty (-)^{n-1} \frac{1 \cdot 3 \cdot 5 \cdots (2n-3)}{2^n x^{n+1}}$$

as $x \to \infty$. The resulting asymptotic series is divergent, since the ratio of the $(n+1)$th to nth terms in absolute value is $(2n-1)/(2x)$ which, for fixed x, tends to ∞ with n. The reason for this divergence is a simple consequence of our applying the binomial expansion for $(1+t)^{\frac{1}{2}}$ (valid in $0 \le t \le 1$) in the Laplace integral beyond its interval of convergence.

Example 2. The confluent hypergeometric function† $U(1; 1; z)$ (which equals the exponential integral $e^z E_1(z)$) has the integral representation

$$U(1; 1; z) = \int_0^\infty \frac{e^{-t} dt}{t+z} \tag{1.1.6}$$

for z not a negative number or zero. In fact, it is relatively easy to show that this integral representation converges uniformly in the closed annular sector $\mathcal{S}_{\epsilon,\delta} = \{z : |z| \ge \epsilon, |\arg z| \le \pi - \delta\}$ for every positive ϵ and every positive $\delta < \pi$. Such a demonstration can proceed along the following lines.

Put $\theta = \arg z$ for $z \in \mathcal{S}_{\epsilon,\delta}$ and observe that for any nonnegative t, $|t+z|^2 = t^2 + |z|^2 + 2|z|t \cos\theta \ge t^2 + |z|^2 - 2|z|t \cos\delta \ge |z|^2 \sin^2\delta$. Thus, the integrand of (1.1.6) admits the simple bound

$$e^{-t}|t+z|^{-1} \le e^{-t}|z|^{-1} \operatorname{cosec}\delta$$

whence we have, upon integrating the bound,

$$|U(1; 1; z)| \le |z|^{-1} \operatorname{cosec}\delta$$

---

† An alternative notation for this function is $\Psi(1; 1; z)$.

for $z \in S_{\epsilon,\delta}$. The uniform convergence of the integral follows, from which we see that $U(1; 1; z)$ is holomorphic in the z plane cut along the negative real axis.

Through repeated integration by parts, differentiating in each case the factor $(t + z)^{-k}$ appearing at each step, we arrive at

$$U(1; 1; z) = \sum_{k=1}^{n} (-)^{k-1}(k-1)! z^{-k} + R_n(z), \tag{1.1.7}$$

where the remainder term $R_n(z)$ is

$$R_n(z) = (-)^n n! \int_0^{\infty} \frac{e^{-t}}{(t+z)^{n+1}} dt. \tag{1.1.8}$$

Evidently, each term produced in the series in (1.1.7) is a term from the asymptotic scale $\{z^{-j}\}$, $j = 1, 2, \ldots$, so that if we can show that for any n, $R_n(z) = O(z^{-n-1})$, we will have established the asymptotic expansion

$$U(1; 1; z) \sim \sum_{k=1}^{\infty} (-)^{k-1}(k-1)! z^{-k}, \tag{1.1.9}$$

for $z \to \infty$ in the sector $|\arg z| \le \pi - \delta < \pi$.

To this end, we observe that the bound used in establishing the uniform convergence of the integral (1.1.6), namely $1/|t + z| \le 1/|z| \sin \delta$, can be brought to bear on (1.1.8) to yield

$$|R_n(z)| \le \frac{n!}{(|z| \sin \delta)^{n+1}}.$$

The expansion (1.1.9) is therefore an asymptotic expansion in the sense of Poincaré. It is, however, quite clearly a divergent series, as ratios of consecutive terms in the asymptotic series diverge to ∞ as $(n!/|z|^{n+1})/((n-1)!/|z|^n) = n/|z|$, as $n \to \infty$, irrespective of the value of z. Nevertheless, the divergent character of this asymptotic series does not detract from its computational utility.   □

In Tables† 1.1 and 1.2, we have gathered together computed and approximate values of $U(1; 1; z)$, with approximate values derived from the finite series approximation

$$S_n(z) = \sum_{k=1}^{n} (-)^{k-1}(k-1)! z^{-k},$$

obtained by truncating the asymptotic expansion (1.1.9) after n terms. It is apparent from the tables that the calibre of even modest approximations to $U(1; 1; z)$ becomes quite good once $|z|$ is of the order of 100, and is good to two or more significant digits for values of $|z|$ as small as 10. This naturally leads one to

---

† In Tables 1.1 and 1.2 we have adopted the convention of writing $x(y)$ in lieu of the more cumbersome $x \times 10^y$.

Table 1.1. *Computed and approximate values of*
U(1; 1; z) for real values of z

z	$U(1;1;z)$	$S_5(z)$	$S_{10}(z)$
10	0.915633(−1)	0.916400(−1)	0.915456(−1)
50	0.196151(−1)	0.196151(−1)	0.196151(−1)
100	0.990194(−2)	0.990194(−2)	0.990194(−2)

Table 1.2. *Computed and approximate values of*
U(1; 1; z) for imaginary values of z

z	$U(1;1;z)$
10i	0.948854(−2) − 0.981910(−1)i
50i	0.399048(−3) − 0.199841(−1)i
100i	0.999401(−4) − 0.999800(−2)i

z	$S_5(z)$
10i	0.940000(−2) − 0.982400(−1)i
50i	0.399040(−3) − 0.199841(−1)i
100i	0.999400(−4) − 0.999800(−2)i

z	$S_{10}(z)$
10i	0.950589(−2) − 0.982083(−1)i
50i	0.399048(−3) − 0.199841(−1)i
100i	0.999401(−4) − 0.999800(−2)i

wonder how the best approximation can be obtained, in view of the utility of these finite approximations and the divergence of the full asymptotic expansion: how can we select n so that the approximation furnished by $S_n(z)$ is the best possible?

The strategy we detail here, called *optimal truncation*, is easily stated: for a fixed z, the successive terms in the asymptotic expansion will reach a minimum in absolute value, after which the terms must necessarily increase without bound given the divergent character of the full expansion; see Fig. 1.1. It is readily shown that the terms in $S_n(z)$ attain their smallest absolute value when $k \sim |z|$ (except when $|z|$ is an integer, in which case there are two equally small terms corresponding to $k = |z| - 1$ and $k = |z|$). If the full series is truncated just before this minimum modulus term is reached, then the finite series that results is the optimally truncated series, and will yield the best approximation to the original function, in the present case, $U(1; 1; z)$.

To see that this is so, observe for $U(1; 1; z)$ that for $z > 0$ the remainder in the approximation after n terms of the asymptotic series,

$$R_n(z) = U(1;1;z) - S_n(z) = (-)^n n! \int_0^\infty \frac{e^{-t}dt}{(t+z)^{n+1}},$$

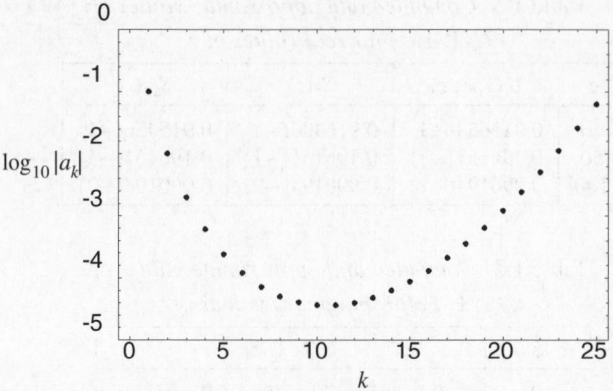

Fig. 1.1. Magnitude of the terms $a_k = (-)^{k-1}\Gamma(k)z^{-k}$ in the expansion $S_n(z)$ against ordinal number k when $z = 10$.

is of the sign opposite to that in the last term in $S_n(z)$ and further, is of the same sign as the first term left in the full asymptotic series after excising $S_n(z)$. In absolute value, we also have

$$|R_n(z)| = \frac{n!}{z^{n+1}} \int_0^\infty \frac{e^{-t}dt}{(1+t/z)^{n+1}} < \frac{n!}{z^{n+1}},$$

so the remainder term is numerically smaller in absolute value than the modulus of the first neglected term. Since the series $S_n(z)$ is an alternating series, it follows that $S_n(z)$ is alternately bigger than $U(1; 1; z)$ and less than $U(1; 1; z)$ as n increases. The sum $S_n(z)$ will therefore be closest in value to $U(1; 1; z)$ precisely when we truncate the full expansion just before the numerically smallest term (in absolute value) in the full expansion. From the preceding inequality, it is easy to note that the remainder term will then be bounded by this minimal term.

To see the order of the remainder term at optimal truncation, we substitute $n \sim z$ ($\gg 1$) in the above bound for $R_n(z)$, and employ Stirling's formula to approximate the factorial, to find

$$|R_n(z)| < \frac{n!}{z^{n+1}} \simeq (2\pi)^{\frac{1}{2}} \frac{e^{-n} n^{n+\frac{1}{2}}}{z^{n+1}} \simeq \left(\frac{2\pi}{z}\right)^{\frac{1}{2}} e^{-z}.$$

This shows that at optimal truncation the remainder term for $U(1; 1; z)$ is of order $z^{-\frac{1}{2}} e^{-z}$ as $z \to +\infty$ and consequently that evaluation of the function by this scheme will result in an error that is *exponentially small* in z; these results can be extended to deal with complex values of z – see Olver (1974, p. 523) for a more detailed treatment. We remark that this principle is found to apply to a wide range of asymptotic series yielding in each case an error term at optimal truncation that is typically exponentially small in the asymptotic variable.

We observe that not all asymptotic series present the regular behaviour of the coefficients depicted in Fig. 1.1. In certain compound expansions, with coefficients containing gamma functions in the numerator, it is possible to find situations where some of the arguments of the gamma functions approach a nonpositive integer value. This gives rise to a series of 'peaks' superimposed on the basic structure of Fig. 1.1. A specific example is provided by the compound expansion

$$z^{-2/\mu}(I_1 + I_2), \tag{1.1.10}$$

where $I_r = \sum_{k=0}^{\infty} a_k^{(r)}$ $(r = 1, 2)$ and, for positive parameters m_1, m_2 and μ,

$$a_k^{(1)} = \frac{(-)^k}{k!} \Gamma\left(\frac{1 + \mu k}{m_1}\right) \Gamma\left(\frac{m_1 - m_2(1 + \mu k)}{m_1 \mu}\right) z^{-(1+\mu k)/m_1}$$

with a similar expression for $a_k^{(2)}$ with m_1 and m_2 interchanged. Expansions of this type arise in the treatment of certain Laplace-type integrals discussed in Chapter 7. If the parameters m_1, m_2 and μ are chosen such that the arguments of the second gamma function in $a_k^{(1)}$ and $a_k^{(2)}$ are not close to zero or a negative integer, then the variation of the modulus of the coefficients with ordinal number k will be similar to that shown in Fig. 1.1. If, however, the parameter values are chosen so that these arguments become close to a nonpositive integer† for subsets of k values, then we find that the variation of the coefficients becomes irregular with a sequence of peaks of variable height. Such a situation for the coefficients $a_k^{(1)}$ is shown in Fig. 1.2 for two sets of parameter values. The truncation of such series has been investigated in Liakhovetski & Paris (1998), where it is found that even if the series I_1 is truncated at a peak (provided that the corresponding peak associated with the coefficients $a_k^{(2)}$ is included) increasingly accurate asymptotic approximations are obtained by steadily increasing the truncation indices in the series I_1 and I_2 until they correspond roughly to the global minimum of each curve. An inspection of Fig. 1.2, however, would indicate that these optimal points are not as easily distinguished as in the case of Fig. 1.1.

The notion of optimal truncation will surface in a significant way in the subject matter of the Stokes phenomenon and hyperasymptotics, and so we defer further discussion of it until Chapter 6, where a detailed analysis of remainder terms is undertaken. We do mention, however, that apart from optimally truncating an asymptotic series, one can sometimes obtain dramatic improvements in the numerical utility of an asymptotic expansion if one is able to extract exponentially small (measured against the scale being used) terms prior to developing an asymptotic expansion. This particular situation can be seen in the following example.

---

† If the parameter values are such that the second gamma-function argument equals a nonpositive integer for a subset of k values, then the expansion (1.1.10) becomes nugatory. In the derivation of (1.1.10) by a Mellin-Barnes approach this would result in a sequence of double poles and the formation of logarithmic terms.

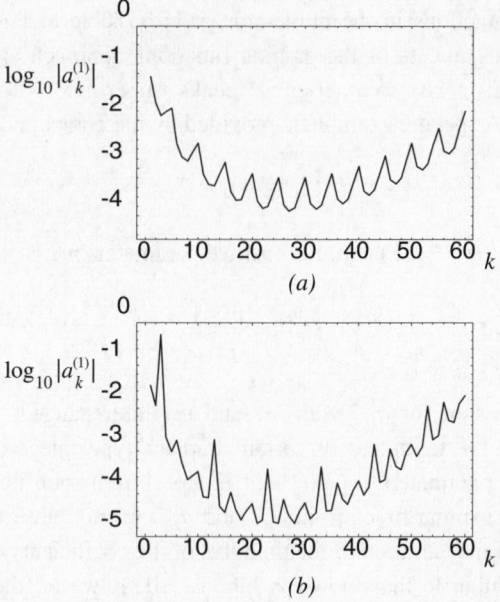

Fig. 1.2. Magnitude of the coefficients $a_k^{(1)}$ against ordinal number k for $\mu = 3$, $m_1 = 1.5$ when (a) $m_2 = 1.2$, $z = 3.0$ and (b) $m_2 = 1.049$, $z = 3.6$. For clarity the points have been joined.

Example 3. Let us consider the finite Fourier integral

$$J(\lambda) = \int_{-1}^{1} e^{i\lambda(x^3/3 + x)} dx$$

with λ large and positive. Introduce the change of variable $u = \frac{1}{3}x^3 + x$ and observe that over the interval of integration, the change of variable is one-to-one, fixes the origin and maps ± 1 to $\pm \frac{4}{3}$ respectively, resulting in

$$J(\lambda) = \int_{-4/3}^{4/3} e^{i\lambda u} x'(u) du,$$

where $x(u)$ is the function inverse to the $x \mapsto u$ change of variable. An explicit formula for $x(u)$ is available to us from the classical theory of equations, resulting from the trigonometric solution to the cubic equation, and takes the form

$$x = 2\sinh\theta, \qquad \text{where} \quad 3\theta = \operatorname{arcsinh}\left(\tfrac{3}{2}u\right),$$

or

$$x = \left(\tfrac{3}{2}u + \sqrt{\tfrac{9}{4}u^2 + 1}\right)^{1/3} - \left(\tfrac{3}{2}u + \sqrt{\tfrac{9}{4}u^2 + 1}\right)^{-1/3}.$$

It is a straightforward matter to deduce that $x^{(k)}(-u) = (-)^{k-1} x^{(k)}(u)$, where $x^{(n)}(u)$ as usual indicates the nth derivative of the inverse function.

By repeatedly applying integration by parts, the latter representation for $J(\lambda)$ can be seen to yield a finite asymptotic expansion with remainder,

$$J(\lambda) = \sum_{n=1}^{N} \left\{ e^{4i\lambda/3} x^{(n)}\left(\tfrac{4}{3}\right) - e^{-4i\lambda/3} x^{(n)}\left(-\tfrac{4}{3}\right) \right\} \frac{(-)^{n-1}}{(i\lambda)^n}$$

$$+ \frac{(-)^N}{(i\lambda)^N} \int_{-4/3}^{4/3} e^{i\lambda u} x^{(N+1)}(u)\, du. \tag{1.1.11}$$

In view of the Riemann-Lebesgue lemma, the remainder term is seen to be $o(\lambda^{-N})$, so the finite expansion (1.1.11) leads, after exploiting $x^{(k)}(-\tfrac{4}{3}) = (-)^{k-1} x^{(k)}(\tfrac{4}{3})$, to the large-$\lambda$ expansion†

$$J(\lambda) \sim 2 \sin\left(\tfrac{4}{3}\lambda\right) \sum_{n=0}^{\infty} \frac{(-)^n}{\lambda^{2n+1}} x^{(2n+1)}\left(\tfrac{4}{3}\right) - 2 \cos\left(\tfrac{4}{3}\lambda\right) \sum_{n=1}^{\infty} \frac{(-)^n}{\lambda^{2n}} x^{(2n)}\left(\tfrac{4}{3}\right).$$

If we evaluate the first few derivatives $x^{(n)}(\tfrac{4}{3})$ and employ optimal truncation for modest values of λ, say $\lambda = 4, 5, 6, 7$, we obtain the approximate values shown in the fourth column of Table 1.3. The columns labelled N_s and N_c show respectively, for each value of λ, the number of terms of the sine and cosine series in the expansion of $J(\lambda)$ retained after optimally truncating each series. As comparison with the last column of Table 1.3 reveals, the asymptotic approximations obtained for these modest values of λ are of poor calibre.

However, an improvement in the numerical utility of the expansion can be obtained by rewriting the integral representation of $J(\lambda)$ in the following manner. Because of the exponential decay in the integrand, we can, by Cauchy's theorem, write

$$J(\lambda) = \left\{ -\int_{1}^{\infty e^{\pi i/6}} + \int_{-1}^{\infty e^{5\pi i/6}} + \int_{\infty e^{5\pi i/6}}^{\infty e^{\pi i/6}} \right\} e^{i\lambda(x^3/3+x)}\, dx. \tag{1.1.12}$$

The third integral in this sum can be expressed in terms of the Airy function

$$\text{Ai}(z) = \frac{1}{2\pi i} \int_{\infty e^{-\pi i/3}}^{\infty e^{\pi i/3}} \exp\left(\tfrac{1}{3}t^3 - zt\right) dt,$$

namely,

$$2\pi \lambda^{-1/3} \text{Ai}(\lambda^{2/3}) = \int_{\infty e^{5\pi i/6}}^{\infty e^{\pi i/6}} e^{i\lambda(x^3/3+x)}\, dx$$

upon making the substitution $x = it\lambda^{-1/3}$. From this, and integration by parts applied to each of the remaining integrals in (1.1.12), we arrive at the same expansion and approximation for $J(\lambda)$ that we found earlier, only now the expansion

---

† This expansion does not fit the form of a Poincaré-type expansion as we have defined it previously, but rather is an example (after separating sine and cosine terms) of a compound asymptotic expansion, discussed in the next section.

Table 1.3. *Comparison of optimally truncated asymptotic approximation, asymptotic approximation and exponentially decaying correction and computed values of the Fourier integral $J(\lambda)$*

λ	N_s	N_c	Optimally truncated series	Optimally truncated series with Airy term	$J(\lambda)$
4	2	2	-0.213739	-0.153525	-0.154260
5	3	2	0.055788	0.083551	0.083545
6	6	5	0.164661	0.177709	0.177703
7	6	5	0.022816	0.029031	0.029034

includes the term involving the Airy function:

$$J(\lambda) \sim \frac{2\pi}{\lambda^{1/3}} \mathrm{Ai}(\lambda^{2/3}) + 2\sin\left(\tfrac{4}{3}\lambda\right) \sum_{n=0}^{\infty} \frac{(-)^n}{\lambda^{2n+1}} x^{(2n+1)}\left(\tfrac{4}{3}\right)$$

$$- 2\cos\left(\tfrac{4}{3}\lambda\right) \sum_{n=1}^{\infty} \frac{(-)^n}{\lambda^{2n}} x^{(2n)}\left(\tfrac{4}{3}\right).$$

The Airy function of positive argument can be shown to exhibit exponential decay as the argument increases, so the additional Airy function term in the above expression is $o(\lambda^{-k})$ for any nonnegative integer k and can be eliminated entirely from the asymptotic expansion in view of the definition of asymptotic expansions of Poincaré type. If it is instead retained, the resulting approximations for the same modest values of λ used in Table 1.3 show dramatic improvement, giving several significant figures of the computed values of $J(\lambda)$ as a comparison of the last two columns of Table 1.3 reveals.   □

Another interesting fact concerning asymptotic power series stems from the observation that given an arbitrary sequence of complex numbers $\{a_n\}_{n=0}^{\infty}$, there is a function $f(z)$ holomorphic in a region containing a closed annular sector which has the formal series $\sum_{n=0}^{\infty} a_n z^{-n}$ as its asymptotic expansion.

One such construction† proceeds by taking the closed annular sector to be $S = \{z : |\arg z| \le \theta, |z| \ge R > 0\}$ – other sectors can be used by translating and rotating this initial choice. Then set

$$f(z) = \sum_{n=0}^{\infty} \frac{a_n e_n(z)}{z^n}$$

where for nonzero a_n,

$$e_n(z) = 1 - \exp\left(-z^{\phi} r^n / |a_n|\right),$$

† This account is drawn from Olver (1974, § I.9). Other examples along this line are also to be found there.

for numbers ϕ and r chosen to satisfy $0 < \phi < \pi/(2\theta)$ and $0 < r < R$. Should an a_n vanish, the corresponding e_n is taken to be the zero function, so that the corresponding term in the sum defining f is effectively excised.

With these terms so defined, in the sector of interest we have $|\arg(z^\phi)| < \frac{1}{2}\pi$ and

$$\left| \frac{a_n e_n(z)}{z^n} \right| \leq r^n |z|^{\phi-n} \leq |z|^\phi \left(\frac{r}{R} \right)^n, \tag{1.1.13}$$

since $|1 - e^{-\zeta}| \leq |\zeta|$ when $|\arg \zeta| \leq \frac{1}{2}\pi$. The series defining f therefore converges uniformly on compact subsets of our sector, and so defines a holomorphic function there.

That f has the desired asymptotic expansion can be seen from

$$f(z) - \sum_{n=0}^{N-1} \frac{a_n}{z^n} = -\sum_{n=0}^{N-1} \frac{a_n}{z^n} \exp\left(-\frac{z^\phi r^n}{|a_n|} \right) + \sum_{n=N}^{\infty} \frac{a_n e_n(z)}{z^n}, \tag{1.1.14}$$

where it bears noting that the infinite series here is uniformly convergent. Because of the exponential decay of each term in the finite sum on the right, the entire sum is $o(z^{-n})$ for any n as $z \to \infty$ in our sector. The remaining series on the right-hand side is easily bounded using (1.1.13) to give

$$\left| \sum_{n=N}^{\infty} \frac{a_n e_n(z)}{z^n} \right| \leq |z|^\phi \sum_{n=N}^{\infty} \left(\frac{r}{|z|} \right)^n = |z|^\phi \left(\frac{r}{|z|} \right)^N \frac{|z|}{|z|-r} = O(z^{\phi-N}).$$

Upon replacing N by $N + \lfloor \phi \rfloor + 1$, we obtain a similar expression to that in (1.1.14), for which the right-hand side is $O(z^{-N})$ but for which there are "extra" terms on the left-hand side. These additional terms, $a_n z^{-n}$ for $n \geq N$, are also $O(z^{-N})$ and so can be absorbed into the order estimate that results on the right-hand side.

1.1.3 Other Expansions

Expansions other than Poincaré-type also have currency in asymptotic analysis. Here, we mention but three types.

To begin, let $\{\phi_n\}$ be an asymptotic scale as $x \to x_0$. A formal series $\sum f_n(x)$ is a *generalised asymptotic expansion* of a function $f(x)$ with respect to the asymptotic scale $\{\phi_n\}$ if

$$f(x) = \sum_{n=0}^{N} f_n(x) + o(\phi_N(x)) \qquad (x \to x_0, \ N = 0, 1, 2, \ldots).$$

In this event, we write, as we have for Poincaré-type expansions,

$$f(x) \sim \sum_{n=0}^{\infty} f_n(x) \qquad (x \to x_0, \ \{\phi_n\}),$$

indicating with the formal series the asymptotic scale used to define the expansion.

The important difference between Poincaré and generalised asymptotic expansions is that the functions f_n appearing in the formal series expansion for f need not, themselves, form an asymptotic scale.

Example 1. Define the sequence of functions $\{f_n\}$, for nonnegative integer n and nonzero x, by

$$f_n(x) = \frac{\cos nx}{x^n}.$$

For $x \to \infty$, it is apparent that each $f_n(x) = O(x^{-n})$, and that $\{\phi_n(x)\} = \{x^{-n}\}$ is an asymptotic scale. However, the sequence $\{f_n(x)\}$ fails to be an asymptotic scale, as a ratio of consecutive elements in the sequence gives

$$\frac{f_{n+1}(x)}{f_n(x)} = \frac{\cos(n+1)x}{x \cos nx},$$

which fails to be $o(1)$ for all x sufficiently large.

Generalised asymptotic expansions are less commonplace than expansions of Poincaré type, and are not used in our development of asymptotic expansions of Mellin-Barnes integrals.

A different mechanism for extending Poincaré-type expansions presents itself naturally in the setting of the method of stationary phase or steepest descent, and in the domain of expansions of solutions of differential equations. The idea here is to replace the series expansion of a function, as in (1.1.2), by several different series, each with different scales.

Put more precisely, by a *compound asymptotic expansion* of a function f, we mean a finite sum of Poincaré-type series expansions

$$f(x) \sim A_1(x) \sum_{n=0}^{\infty} a_{1n}\phi_{1n}(x) + A_2(x) \sum_{n=0}^{\infty} a_{2n}\phi_{2n}(x)$$

$$+ \cdots + A_k(x) \sum_{n=0}^{\infty} a_{kn}\phi_{kn}(x) \qquad (x \to x_0),$$

where, for $1 \le m \le k$, the sequences $\{\phi_{mn}\}$ are asymptotic scales, the coefficient functions $A_m(x)$ are continuous, and for $N_1, N_2, \ldots, N_k \ge 0$, we have

$$f(x) = A_1(x) \left\{ \sum_{n=0}^{N_1} a_{1n}\phi_{1n}(x) + O(\phi_{1,N_1+1}(x)) \right\}$$

$$+ A_2(x) \left\{ \sum_{n=0}^{N_2} a_{2n}\phi_{2n}(x) + O(\phi_{2,N_2+1}(x)) \right\}$$

$$+ \cdots + A_k(x) \left\{ \sum_{n=0}^{N_k} a_{kn}\phi_{kn}(x) + O(\phi_{k,N_k+1}(x)) \right\} \qquad (x \to x_0).$$

It is entirely possible that some of the series $A_j(x) \sum a_{jn}\phi_{jn}(x)$ could, by virtue of the coefficient function $A_j(x)$, or choice of scale $\{\phi_{jn}\}$, be $o(\phi_{mn})$ for some $m \neq j$, and so be absorbed into the error terms implied in other series in the compound expansion. However, in some numerical work, the retention of such negligible terms, when measured against the other scales in the expansion, can add to the numerical accuracy of asymptotic approximations of f, especially for values of x that are at some distance from x_0. This, in turn, extends the utility of such expansions.

In some circumstances, it may be possible to embed the scales $\{\phi_{mn}\}$ in a larger scale, say $\{\psi_\nu\}$, and so collapse the sum of Poincaré expansions into a single-series expansion involving this larger scale $\{\psi_\nu\}$. Success in this direction depends in part on the coefficient functions $A_j(x)$.

Example 2. STEEPEST DESCENT METHOD. An integral of the form

$$I(\lambda) = \int_C g(z)e^{\lambda f(z)}dz,$$

is said to be of *Laplace type* if the functions f and g are holomorphic in a region containing the contour C, and the integral converges for some λ. In the most common setting, C is an infinite contour, and the parameter λ is large in modulus. Thus, we require that the integral $I(\lambda)$ exist for all λ sufficiently large in some sector.

The idea behind the steepest descent method is deceptively simple: deform the integration contour C into a sum of contours, $C_1, C_2, \ldots, C_k$, so that along each of the contours C_n, the *phase* function $f(z)$ has a single point z_n – a *saddle* or *saddle point*†– at which $f'(z_n)$ vanishes, and as z varies along the contour C_n, $\lambda[f(z) - f(z_n)] \leq 0$, with this difference tending to $-\infty$ as $|z| \to \infty$ along the contour. If this deformation is possible, the contours $C_1, C_2, \ldots, C_k$ are termed *steepest descent contours*, and the integral can be recast as

$$I(\lambda) = \sum_{n=1}^{k} e^{\lambda f(z_n)} \int_{C_n} g(z)e^{\lambda[f(z)-f(z_n)]}dz.$$

In the case where $f''(z_n) \neq 0$ for all saddle points z_n, each integral in the sum can be represented as a Gaussian integral, namely

$$e^{\lambda f(z_n)} \int_{C_n} g(z)e^{\lambda[f(z)-f(z_n)]}dz = e^{\lambda f(z_n)} \int_{-\infty}^{\infty} g(z(t))e^{-|\lambda|t^2}z'(t)\,dt.$$

The transformation $z \mapsto t$ will map one branch of the steepest descent curve from z_n to ∞ into the positive real t axis, and the remainder of the steepest descent curve will be mapped into the negative real t axis. By splitting the integral into integrals

---

† Saddle points of Fourier-type integrals are often referred to as stationary points.

taken over negative and positive real t axes separately, a further reduction to a sum of two Laplace transforms can be achieved, to each of which Watson's lemma can then be applied.

For a concrete example, we consider the Pearcey integral

$$P(x, y) = \int_{-\infty}^{\infty} \exp\{i(t^4 + xt^2 + yt)\}\, dt,$$

where, for the purpose of illustration, we will assume $|x|$ and $|y|$ are both large, with $x < 0$ and $y > 0$. We will also replace x by $-x$ and take $x > 0$. Thus, we consider

$$P(-x, y) = x^{\frac{1}{2}} \int_{-\infty}^{\infty} \exp\{ix^2(u^4 - u^2 + yx^{-3/2}u)\}\, du, \qquad (1.1.15)$$

where we have applied the simple change of variable $t = x^{\frac{1}{2}}u$. Denoting the phase function of this integral by

$$\psi(u) = u^4 - u^2 + yx^{-3/2}u,$$

we have

$$\psi'(u) = 4(u^3 - \tfrac{1}{2}u + \tfrac{1}{4}yx^{-3/2})$$
$$= 4(u^3 - (u_1 + u_2 + u_3)u^2 + (u_1u_2 + u_1u_3 + u_2u_3)u - u_1u_2u_3),$$

where the roots of $\psi'(u) = 0$ are indicated by u_1, u_2 and u_3. Because $\psi'(u)$ is a real cubic polynomial, we always have one real zero. If x is sufficiently large compared to y, we can ensure that the other two zeros of $\psi'(u)$ are also real, and that all three are distinct. Additionally, the elementary theory of equations furnishes us with

$$\sum u_i = 0, \qquad \sum_{i<j} u_i u_j = -\tfrac{1}{2}, \qquad u_1 u_2 u_3 = -\tfrac{1}{4}yx^{-3/2},$$

from which we deduce that one $u_i < 0$, and the other two are positive. Let us label these so that $u_1 < 0 < u_2 < u_3$.

We mention here that the theory of equations also provides a trigonometric form for the roots u_i, namely,

$$\begin{aligned} u_1 &= -\sqrt{2/3} \cdot \sin(\phi + \tfrac{1}{3}\pi), \\ u_2 &= \sqrt{2/3} \cdot \sin\phi, \\ u_3 &= \sqrt{2/3} \cdot \sin(\tfrac{1}{3}\pi - \phi), \end{aligned} \qquad (1.1.16)$$

where the angle ϕ is given by

$$\sin(3\phi) = y\left(\tfrac{2}{3}x\right)^{3/2} \qquad (1.1.17)$$

which, under the hypothesis of $y(\tfrac{2}{3}x)^{-3/2} < 1$, can be guaranteed to be real. The zeros displayed in (1.1.16) undergo a confluence when the angle ϕ tends to $\tfrac{1}{6}\pi$. The curve this value of ϕ defines is the so-called caustic in the real plane: $y = (\frac{2}{3}x)^{3/2}$. The saddles u_i are therefore, successively, the locations of a local minimum, a local maximum and a local minimum of $\psi(u)$.

For real x and y, we may rotate the contour of integration in (1.1.15) onto the line from $\infty e^{9\pi i/8}$ to $\infty e^{\pi i/8}$ through an application of Jordan's lemma. Since there are three real saddle points for $(-x, y)$ satisfying $\phi < \frac{1}{6}\pi$, we may further represent $P(-x, y)$ as a sum of three contour integrals,

$$P(-x, y) = x^{\frac{1}{2}} \sum_{j=1}^{3} \int_{\Gamma_j} e^{ix^2\psi(u)} du, \qquad (1.1.18)$$

where the contours Γ_j are the steepest descent curves: Γ_1, beginning at $\infty e^{9\pi i/8}$, ending at $\infty e^{5\pi i/8}$ and passing through $u_1 < 0$; Γ_2, beginning at $\infty e^{5\pi i/8}$, ending at $\infty e^{-3\pi i/8}$ and passing through $u_2 > 0$; and Γ_3, beginning at $\infty e^{-3\pi i/8}$, ending at $\infty e^{\pi i/8}$ and passing through $u_3 > u_2$. Along these contours, the phase $i\psi(u)$ is real and decreases to $-\infty$ as we move along the Γ_j away from the saddle points so that each integral is effectively a Gaussian integral. The general situation is depicted in Fig. 1.3.

Let us set

$$d_j = \{(-)^j(1 - 6u_j^2)\}^{\frac{1}{2}} \qquad (j = 1, 2, 3).$$

In accordance with the steepest descent methodology mentioned previously, we set $\psi(u) - \psi(u_j) = (-)^{j+1}d_j^2 v^2$, to find at each saddle point u_j,

$$v = (u - u_j)\left\{1 + \frac{4u_j(u - u_j)}{6u_j^2 - 1} + \frac{(u - u_j)^2}{6u_j^2 - 1}\right\}^{1/2}$$

whence reversion yields the expansion, for each j,

$$u - u_j = \sum_{k=1}^{\infty} b_{k,j} v^k,$$

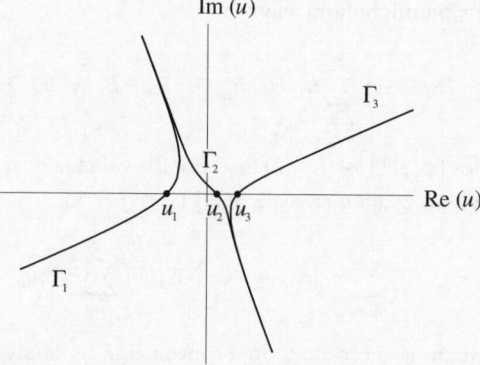

Fig. 1.3. Steepest descent curves through the saddles u_1, u_2 and u_3.

convergent in a neighbourhood of $v = 0$. We observe that $b_{1,j} = 1$ for each $j = 1, 2, 3$. Substitution into each term in (1.1.18) followed by termwise integration will furnish

$$\int_{\Gamma_j} e^{ix^2\psi(u)}du \sim e^{ix^2\psi(u_j)+(-)^{j+1}\pi i/4} \frac{\pi^{\frac{1}{2}}}{xd_j} S_j(x, \phi),$$

where $S_j(x, \phi)$ denotes the formal asymptotic sum

$$S_j(x, \phi) = \sum_{k=0}^{\infty}(2k+1)b_{2k+1,j}\frac{\Gamma(k+\frac{1}{2})}{\Gamma(\frac{1}{2})}\left((-)^{j+1}i\right)^k(d_jx)^{-2k}.$$

It then follows that

$$P(-x, y) \sim \sqrt{\frac{\pi}{x}} \sum_{j=1}^{3} \frac{e^{ix^2\psi(u_j)+(-)^{j+1}\pi i/4}}{d_j} S_j(x, \phi)$$

for large x. This is evidently a compound asymptotic expansion with each constituent asymptotic series corresponding to a single saddle point of $P(-x, y)$. We shall meet the Pearcey integral again in Chapter 8, in a less restricted setting. $\square$

There also arise situations in which functions depending on parameters other than the asymptotic one may possess asymptotic expansions which not only depend on such auxiliary parameters, but may also undergo discontinuous changes of scale as these parameters vary. Such a discontinuity in the scale can occur, even if the function involved is holomorphic in the control parameter. In more specific terms, let us suppose that a function $F(\lambda; \mu)$ has asymptotic parameter λ and control parameter μ. For $\lambda \to \lambda_0$, and $\mu < \mu_0$, say, one might have an asymptotic form

$$F(\lambda; \mu) \sim A_1(\lambda; \mu) \sum_{n=0}^{\infty} a_{1n}(\mu)\phi_{1n}^{-} + \cdots + A_k(\lambda; \mu) \sum_{n=0}^{\infty} a_{kn}(\mu)\phi_{kn}^{-},$$

where $\{\phi_{jn}^{-}\}$ $(1 \le j \le k)$ are asymptotic scales in the variable λ, while for $\mu > \mu_0$, a different expansion might hold, say

$$F(\lambda; \mu) \sim B_1(\lambda; \mu) \sum_{n=0}^{\infty} b_{1n}(\mu)\phi_{1n}^{+} + \cdots + B_r(\lambda; \mu) \sum_{n=0}^{\infty} b_{rn}(\mu)\phi_{rn}^{+},$$

for different scales $\{\phi_{jn}^{+}\}$ $(1 \le j \le r)$ in λ. For the value $\mu = \mu_0$, a third expansion may hold, involving yet another scale $\{\phi_{jn}\}$ $(1 \le j \le s)$,

$$F(\lambda; \mu_0) \sim C_1(\lambda) \sum_{n=0}^{\infty} c_{1n}\phi_{1n} + \cdots + C_s(\lambda) \sum_{n=0}^{\infty} c_{sn}\phi_{sn} \qquad (\lambda \to \lambda_0).$$

Distinct forms such as these may apply, even if F is analytic in a neighbourhood of μ_0, and the limiting forms of the expansions may not exist as $\mu \to \mu_0^{\pm}$, compounding the difficulty of using such expansions in a neighbourhood of $\mu = \mu_0$.

This setting can be dealt with through the use of a *uniform asymptotic expansion*, a (usually) compound expansion

$$F(\lambda; \mu) \sim D_1(\lambda; \mu) \sum_{n=0}^{\infty} d_{1n}\psi_{1n} + \cdots + D_k(\lambda; \mu) \sum_{n=0}^{\infty} d_{kn}\psi_{kn},$$

where the asymptotic scale $\{\psi_{jn}\}$ $(1 \le j \le k)$ is a sequence of functions of the asymptotic parameter, which retains its character as an asymptotic scale for all values of the control parameter in a neighbourhood of $\mu = \mu_0$, i.e., $\psi_{j,n+1} = o(\psi_{jn})$ for $\lambda \to \lambda_0$, for every μ in some neighbourhood of $\mu = \mu_0$.

On first glance, it may appear there is little that is new captured in this account. The essential difference is that the coefficient functions D_j must be continuous in a neighbourhood of $\mu = \mu_0$ for all λ in a neighbourhood of λ_0. Furthermore, for $\mu \ne \mu_0$, each D_j must have expansions for $\lambda \to \lambda_0$ which, when combined with the associated Poincaré expansion $\sum d_{jn}\psi_{jn}$, allows the recovery of either the A- or B-coefficient expansions, and for $\mu = \mu_0$, the recovery of the C-coefficient series. Because the D-series is continuous in μ in a neighbourhood of μ_0, the D-coefficient expansion interpolates continuously from the A-series to the C-series to the B-series as μ varies. This continuous interpolation is possible only through additional complexity in the form of the coefficients D_j.

Example 3. BESSEL FUNCTIONS OF LARGE ORDER.   As an illustration, we cite the asymptotic expansion of the Bessel function $J_\nu(\nu x)$ for large positive order and argument in the form

$$J_\nu(\nu x) \sim \begin{cases} \dfrac{e^{\nu(\tanh \alpha - \alpha)}}{(2\pi \nu \tanh \alpha)^{\frac{1}{2}}} \displaystyle\sum_{k=0}^{\infty} \dfrac{c_k(\coth \alpha)}{\nu^k} \\[1.5em] \left(\dfrac{2}{\pi \nu \tan \beta}\right)^{\frac{1}{2}} \left\{ \cos \Psi \displaystyle\sum_{k=0}^{\infty} \dfrac{c_{2k}(i \cot \beta)}{\nu^{2k}} \right. \\[1.5em] \qquad \left. -i \sin \Psi \displaystyle\sum_{k=0}^{\infty} \dfrac{c_{2k+1}(i \cot \beta)}{\nu^{2k+1}} \right\}, \end{cases} \tag{1.1.19}$$

where in the first expansion $0 < x < 1$ with $x = \operatorname{sech} \alpha$ and in the second expansion $x > 1$ with $x = \sec \beta$. The coefficients $c_k(t)$ are polynomials in t of degree $3k$, with $c_0(t) = 1$, $c_1(t) = \frac{1}{24}(3t - 5t^3), \ldots$ and $\Psi = \nu(\tan \beta - \beta) - \frac{1}{4}\pi$; see Abramowitz & Stegun (1965, p. 366). These expansions describe the asymptotic structure of $J_\nu(\nu x)$ on either side of the transition point $x = 1$. When $0 < x < 1$, $J_\nu(\nu x)$ decays exponentially *away* from the point $x = 1$ while when $x > 1$, $J_\nu(\nu x)$ changes to an oscillatory form with an amplitude that eventually decays like $x^{-\frac{1}{2}}$ as $x \to +\infty$. Both these expansions break down in the neighbourhood of $x = 1$ and so cannot describe uniformly the behaviour of $J_\nu(\nu x)$ for $x > 0$.

A uniformly valid expansion which incorporates both the expansions in (1.1.19) is given by [Abramowitz & Stegun (1965, p. 368)]

$$J_\nu(\nu x) \sim \left(\frac{4\zeta}{1-x^2}\right)^{\frac{1}{4}} \left\{ \frac{\text{Ai}(\nu^{2/3}\zeta)}{\nu^{1/3}} \sum_{k=0}^{\infty} \frac{a_k(\zeta)}{\nu^{2k}} + \frac{\text{Ai}'(\nu^{2/3}\zeta)}{\nu^{5/3}} \sum_{k=0}^{\infty} \frac{b_k(\zeta)}{\nu^{2k}} \right\},$$

$$(1.1.20)$$

where $\text{Ai}(z)$ denotes the Airy function. This expansion holds for $\nu \to +\infty$ uniformly with respect to x in the sector $|\arg x| \leq \pi - \epsilon$, $\epsilon > 0$. The variable ζ is defined by

$$\tfrac{2}{3}\zeta^{3/2} = \log\left\{ (1 + \sqrt{1-x^2})/x \right\} - \sqrt{1-x^2}$$

the branches being chosen so that ζ is real when $x > 0$. The coefficients $a_k(\zeta)$, $b_k(\zeta)$ are complicated functions of ζ and are expressed in terms of finite sums of the coefficients $c_k((1-x^2)^{-\frac{1}{2}})$, with

$$a_0(\zeta) = 1, \qquad b_0(\zeta) = -\tfrac{5}{48}\zeta^{-2} + \zeta^{-\frac{1}{2}}\left\{ \tfrac{5}{24}(1-x^2)^{-\frac{3}{2}} - \tfrac{1}{8}(1-x^2)^{-\frac{1}{2}} \right\}.$$

Although the coefficient functions $a_k(\zeta)$ ($k \geq 1$) and $b_k(\zeta)$ ($k \geq 0$) are analytic in the neighbourhood of the transition point $x = 1$ ($\zeta = 0$), they are, in common with many uniform expansions, expressed in a form that possesses a removable singularity at this point.

The asymptotic forms (1.1.19) can be obtained from (1.1.20) by insertion of the expansion for the Airy function and its derivative; see below for the leading-order terms. When $0 < x < 1$, ζ is bounded away from zero and the arguments of the Airy functions in (1.1.20) are large and positive. These functions are therefore exponential in character and the expansion (1.1.20) reduces to the first form in (1.1.19). On the other hand, when $x > 1$, $\zeta < 0$ and is bounded away from zero, so that the arguments of the Airy functions are large and negative and consequently produce oscillatory terms. In this case the expansion (1.1.20) reduces to the second form in (1.1.19).

At the transition point $x = 1$ ($\zeta = 0$), we employ the evaluations $\text{Ai}(0) = \Gamma(\tfrac{1}{3})/(2 \cdot 3^{1/6}\pi)$, $\text{Ai}'(0) = -3^{1/6}\Gamma(\tfrac{2}{3})/(2\pi)$ together with the limiting value

$\{4\zeta/(1-x^2)\}^{\frac{1}{4}} = 2^{\frac{1}{3}}$ to find the expansion†

$$J_\nu(\nu) \sim \frac{2^{1/3}\nu^{-1/3}}{3^{2/3}\Gamma\left(\frac{2}{3}\right)} \sum_{k=0}^{\infty} \frac{a_k(0)}{\nu^{2k}} - \frac{2^{1/3}\nu^{-5/3}}{3^{1/3}\Gamma\left(\frac{1}{3}\right)} \sum_{k=0}^{\infty} \frac{b_k(0)}{\nu^{2k}}.$$

Example 4. THE PEARCEY INTEGRAL REVISITED. As a further illustration, let us again consider the Pearcey integral. Because of the additional complexity involved, we shall only consider asymptotic behaviour to leading order; the character of the uniform expansion will still be apparent in our terse account.

To leading order, for $y(\frac{2}{3}x)^{-3/2} < 1$ to ensure that the angle ϕ in (1.1.17) satisfies $\phi < \frac{1}{6}\pi$, the Pearcey integral has the asymptotic form

$$P(-x, \mu x^{3/2}) \sim \sum_{j=1}^{3} \frac{e^{ix^2\psi(u_j)+(-)^{j+1}\pi i/4}}{d_j} \sqrt{\frac{\pi}{x}} \qquad (1.1.21)$$

where we have set $\mu = y/x^{3/2}$. When $\mu = (2/3)^{3/2}$, so that $\phi = \frac{1}{6}\pi$, the saddle points u_2 and u_3 coalesce into a single saddle of order 2, i.e., a saddle point at which, additionally, the phase function $\psi(u)$ has a vanishing second derivative. A modification of the steepest descent method then allows us to deduce the approximation [Bleistein & Handelsman (1986, pp. 263–265)]

$$P(-x, (2/3)^{3/2}x^{3/2}) \sim \frac{e^{ix^2\psi(u_1)+\pi i/4}}{d_1} \sqrt{\frac{\pi}{x}}$$

$$+ \frac{e^{ix^2/12}}{2^{1/2}3^{1/3}x^{1/6}} \left\{ \Gamma\left(\tfrac{1}{3}\right) - \frac{i\Gamma\left(\tfrac{2}{3}\right)}{2 \cdot 3^{1/3}x^{2/3}} \right\} \qquad (1.1.22)$$

for $x \to \infty$; observe that $\psi(u_1) = -\frac{2}{3}$ when $\mu = (2/3)^{3/2}$, and that $d_2 = d_3 = 0$ of Example 2. For $\mu > (2/3)^{3/2}$, the asymptotic behaviour of the Pearcey integral is dominated by the contribution from the saddle point u_1, for which we find

$$P(-x, \mu x^{3/2}) \sim \frac{e^{ix^2\psi(u_1)+\pi i/4}}{d_1} \sqrt{\frac{\pi}{x}} \qquad (x \to \infty). \qquad (1.1.23)$$

As μ increases from below $(2/3)^{3/2}$, to $(2/3)^{3/2}$ and then beyond, we see a discontinuous change in the asymptotic scales used in (1.1.21) and in (1.1.22), with the complete disappearance of the last two terms in (1.1.22) as we move to $\mu > (2/3)^{3/2}$. The uniform asymptotic approximation that interpolates continuously between these disparate forms in a neighbourhood of $\mu = (2/3)^{3/2}$ results from an application of the cubic transformation introduced by Chester et al. (1957). This transformation captures the essential features of the circumstance of a

---

† We note that the leading term of this expansion yields the well-known approximation due to Cauchy given by

$$J_\nu(\nu) \sim \frac{2^{1/3}\nu^{-1/3}}{3^{2/3}\Gamma\left(\frac{2}{3}\right)} \qquad (\nu \to +\infty).$$

Laplace-type integral undergoing a confluence of two neighbouring simple saddle points. Applied to the Pearcey integral, this method yields an approximation of the form [Kaminski (1989)]

$$P(-x, \mu x^{3/2}) \sim \frac{e^{ix^2\psi(u_1)+\pi i/4}}{d_1}\sqrt{\frac{\pi}{x}}$$
$$+ \frac{2\pi e^{ix^2\eta}}{x^{1/6}}\left\{ p_0(\mu)\mathrm{Ai}(-x^{4/3}\zeta) + \frac{iq_0(\mu)}{x^{2/3}}\mathrm{Ai}'(-x^{4/3}\zeta)\right\},$$

$$(1.1.24)$$

where the quantities η and ζ are given by $\eta = \frac{1}{2}\{\psi(u_2) + \psi(u_3)\}$ and $\zeta^{3/2} = \frac{3}{4}\{\psi(u_2) - \psi(u_3)\}$. The coefficients $p_0(\mu)$ and $q_0(\mu)$ are continuous functions in a neighbourhood of $\mu = (2/3)^{3/2}$ and satisfy $p_0((2/3)^{3/2}) = 2^{-1/2}3^{-1/6}$ and $q_0((2/3)^{3/2}) = 2^{-3/2}3^{-5/6}$. We note that $\zeta < 0$ for $\mu > (2/3)^{3/2}$ and vice versa, and that $\psi(u_2) = \psi(u_3) = \frac{1}{12}$ when $\mu = (2/3)^{3/2}$, at which point $u_2 = u_3$.

The original asymptotic forms can be recovered from (1.1.24) by applying the asymptotic forms of the Airy function and its derivative for large $|z|$, namely [Abramowitz & Stegun (1965, pp. 448–449)]

$$\mathrm{Ai}(z) \sim \frac{e^{-2z^{3/2}/3}}{2\sqrt{\pi}\, z^{1/4}} \quad (|\arg z| < \pi),$$

$$\mathrm{Ai}(-z) \sim \frac{1}{\sqrt{\pi}\, z^{1/4}}\left\{\sin\left(\tfrac{1}{4}\pi + \tfrac{2}{3}z^{3/2}\right) - \tfrac{5}{48}z^{-3/2}\cos\left(\tfrac{1}{4}\pi + \tfrac{2}{3}z^{3/2}\right)\right\}$$

$$(|\arg z| < \tfrac{2}{3}\pi),$$

$$\mathrm{Ai}'(z) \sim -\frac{z^{1/4}e^{-2z^{3/2}/3}}{2\sqrt{\pi}} \quad (|\arg z| < \pi),$$

$$\mathrm{Ai}'(-z) \sim -\frac{z^{1/4}}{\sqrt{\pi}}\left\{\cos\left(\tfrac{1}{4}\pi + \tfrac{2}{3}z^{3/2}\right) - \tfrac{7}{48}z^{-3/2}\sin\left(\tfrac{1}{4}\pi + \tfrac{2}{3}z^{3/2}\right)\right\}$$

$$(|\arg z| < \tfrac{2}{3}\pi).$$

For $\mu < (2/3)^{3/2}$ and bounded away from $(2/3)^{3/2}$, the arguments of Ai and Ai$'$ in (1.1.24) are negative, so the preceding asymptotic forms for the Airy function and its derivative produce oscillatory terms, whence the approximation (1.1.24) reduces to (1.1.21). Conversely, if $\mu > (2/3)^{3/2}$, we find $\zeta^{3/2}$ is pure imaginary, in which case the exponentially decaying asymptotic forms for the Airy function and its derivative apply. In this event, the leading term in the asymptotic approximation of $P(-x, \mu x^{3/2})$ is the one arising from the saddle u_1, evident in (1.1.23).

Finally, at the point of confluence when $\mu = (2/3)^{3/2}$, we employ the values of Ai(0) and Ai$'$(0) given earlier along with the values $u_2 = u_3 = 1/\sqrt{6}$ and $u_1 = -2/\sqrt{6}$ to recover (1.1.22) from (1.1.24).

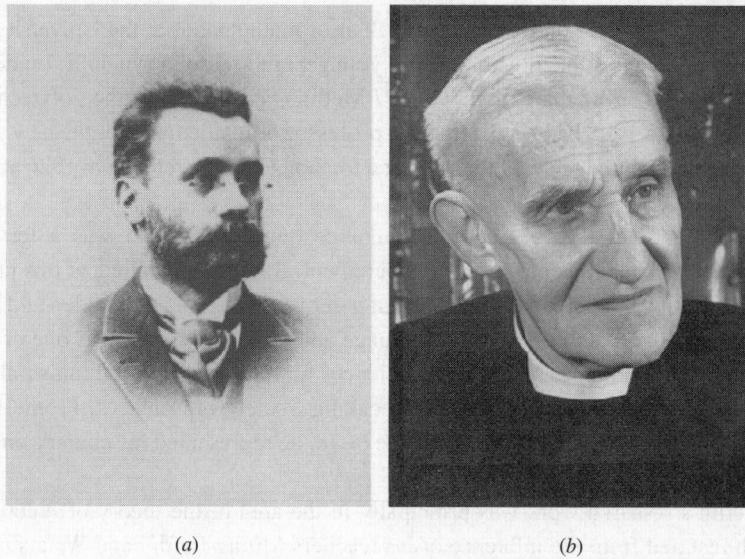

(a)   (b)

Fig. 1.4. Portraits of (a) R.H. Mellin (1854–1933) and (b) E.W. Barnes (1874–1953) (*reproduced with permission*).

1.2 Biographies of Mellin and Barnes

The names of Mellin and Barnes (Fig. 1.4) are intimately linked with, and were the main exponents of, the asymptotic procedure discussed in this book. We give below a brief biographical account of these two eponymous mathematicians, together with a description of their main mathematical contributions. These accounts are based on Lindelöf (1933) and Elfving (1981) (for Mellin) and Whittaker (1954), Rawlinson (1954) and the Obituary Notices of *The Times* in November 1953 (for Barnes).

Robert Hjalmar Mellin

Robert Hjalmar Mellin, the son of a clergyman, was born in Liminka, northern Ostrobothnia, in Finland on 19 June 1854. He grew up and received his schooling in Hämeenlinna (about 100 km north of Helsinki) and undertook his university studies in Helsinki, where his teacher was the Swedish mathematician G. Mittag-Leffler. In the autumn of 1881 Mellin defended his doctoral dissertation on algebraic functions of a single complex variable. He made two sojourns in Berlin in 1881 and 1882 to study under K. Weierstrass and in 1883–84 he returned to continue his studies with Mittag-Leffler in Stockholm.

Mellin was appointed as a docent at the University of Stockholm from 1884–91 but never actually gave any lectures. Also in 1884 he was appointed a senior lecturer in mathematics at the recently founded Polytechnic Institute which was later (in 1908) to become the Technical University of Finland. In 1901 Mellin withdrew his application for the vacant chair of mathematics at the University of Helsinki in favour of his illustrious (and younger) fellow countryman E. Lindelöf (1870–1946). During the period 1904–07 Mellin was Director of the Polytechnic Institute and in 1908 he became the first professor of mathematics at the new university. He remained at the university for a total of 42 years, retiring in 1926 at the age of 72.

With regard to the ever-burning language question, Mellin was a fervent fennoman with an apparently fiery temperament. It must be recalled, at this juncture, that Finland had for a long time been part of the kingdom of Sweden and had consequently been subjected to its language and culture.† Mellin was one of the founders of the Finnish Academy of Sciences in 1908 as a purely Finnish alternative to the predominantly Swedish-speaking Society of Sciences. From 1908 until his death on 5 April 1933, at the age of 78, he represented his country on the editorial board of *Acta Mathematica*.

Mellin's research work was principally in the area of the theory of functions which resulted from the influence of his teachers Mittag-Leffler and Weierstrass. He studied the transform which now bears his name‡ and established its reciprocal properties. He applied this technique systematically in a long series of papers to the study of the gamma function, hypergeometric functions, Dirichlet series, the Riemann zeta function and related number-theoretic functions. He also extended his transform to several variables and applied it to the solution of partial differential equations. The use of the inverse form of the transform, expressed as an integral along a path parallel to the imaginary axis of the complex plane of integration, was developed by Mellin as a powerful tool for the generation of asymptotic expansions. In this theory, he included the possibility of higher-order poles (thereby leading to the inclusion of logarithmic terms in the expansion) and to several sequences of poles yielding sums of asymptotic expansions of very general form.

During the last decade of his life Mellin was, rather curiously for an analyst, preoccupied by Einstein's theory of relativity and he wrote no less than 10 papers on this topic. In these papers, where he was largely concerned with general philosophical problems of time and space, he adopted a quixotic standpoint in his attempt to refute the theory as being logically untenable.

Ernest William Barnes

Ernest William Barnes was born in Birmingham on 1 April 1874, the eldest of four sons of John Starkie Barnes and Jane Elizabeth Kerry, both elementary

---

† After the Napoleonic wars Finland became an autonomous Grand Duchy under Russia, to finally emerge as an independent republic in the aftermath of the First World War.

‡ We point out that similar studies of an incomplete nature had been carried out earlier by Pincherle; for references, see Watson (1966, p. 190).

school head-teachers. In 1883 Barnes' father was appointed Inspector of Schools in Birmingham, a position that he occupied throughout the rest of his working life. Barnes was educated at King Edward's School, Birmingham and in 1893 went up to Cambridge as a Scholar of Trinity College. He was bracketed Second Wrangler in 1896 and was placed in the first division of the first class in Part II of the Mathematical Tripos in 1897. In the following year he was awarded the first Smith's Prize and was duly elected to a Trinity Fellowship. He was appointed a lecturer in mathematics in 1902, junior dean in 1906–08 and a tutor in 1908. He graduated Sc.D. of the University of Cambridge in 1907 and was elected a Fellow of the Royal Society in 1909.

In the same year he became a lecturer in mathematics, Barnes was ordained deacon by the Bishop of London and from 1906 to 1908 was Junior Dean of Trinity. In 1915, Barnes left Cambridge, and his career as a professional mathematician, upon his appointment as Master of the Temple in London. This was followed in 1918 to a Canonry of Westminster and finally, in 1924, to the Bishopric of Birmingham, an office he held until 1952 when he had to retire on account of ill-health. He died on 29 November 1953 at his home in Sussex at the age of 79, survived by his wife and two sons.

Barnes' episcopate was marked by a series of controversies stemming from his outspoken views and, rather surprisingly for someone who held such high office in the Church, often unorthodox religious beliefs. In 1940 he lost a libel case in which he had attacked the Cement Makers' Federation for allegedly holding up the supply of cement, for their own profit, at a time of great national need in the construction of air-raid shelters. Undaunted by this set-back, Barnes returned to his accusations on the cement ring in a speech he delivered in the House of Lords the following year, in which he claimed that powerful business concerns were using libel and slander action to suppress criticism. As a theological author, Barnes' book in 1947, entitled *The Rise of Christianity*, aroused such fierce opposition and criticism from more orthodox members of the Church, that it was strongly suggested he should renounce his episcopal office, a hint which Barnes did not take.

In all, Barnes wrote 29 mathematical papers during the years 1897–1910. His early work was concerned with various aspects of the gamma function, including generalisations of this function given by the so-called Barnes G function, which satisfies the equation $G(z + 1) = \Gamma(z)G(z)$, and to the double gamma function. Barnes next turned his attention to the theory of integral functions, where he investigated their asymptotic structure in a series of papers. He also considered second-order linear difference equations connected with the hypergeometric functions. In the last five of his papers dealing with the hypergeometric functions, Barnes made extensive use of the integrals studied by Mellin in which the integral involves gamma functions of the variable of integration. It was in these papers that he brought to the attention of British mathematicians the power and simplicity associated with these integrals, and which now bear the name Mellin-Barnes integrals. His last mathematical paper, published in 1910, was a short and elegant demonstration of a previously known result of Thomæ concerning a transformation of a $_3F_2$ generalised hypergeometric function of unit argument into a more rapidly convergent function of the same kind.

2

Fundamental Results

2.1 The Gamma Function $\Gamma(z)$

Mellin-Barnes integral representations involve integrands which contain one or more gamma functions with a path of integration that is typically parallel to the imaginary axis of the integration variable. The manipulation of these integrals to produce asymptotic expansions, or the determination of their domains of convergence, requires some basic properties of the gamma function defined by Euler's integral

$$\Gamma(z) = \int_0^\infty e^{-t} t^{z-1} dt \qquad (\text{Re}(z) > 0),$$

where the path of integration is the real axis and t^{z-1} has its principal value. In this opening section we accordingly collect together some standard results on the asymptotic expansion of $\Gamma(z)$ and discuss some useful bounds when z is a complex variable. This collection of results will be exploited in later sections of the book. Standard texts discussing various properties of $\Gamma(z)$ can be found in Olver (1974, pp. 31–40), Temme (1996, Ch. 3) and Whittaker & Watson (1965, Ch. 13).

2.1.1 The Asymptotic Expansion of $\Gamma(z)$

The well-known representation of the logarithm of $\Gamma(z)$ is given by [Whittaker & Watson (1965, p. 251)]

$$\log \Gamma(z) = \left(z - \tfrac{1}{2}\right) \log z - z + \tfrac{1}{2} \log 2\pi + \Omega(z), \qquad (2.1.1)$$

where $\Omega(z)$ is an analytic function of z possessing simple poles at $z = -k$, $k = 0, 1, 2, \ldots$. Several different functional forms for $\Omega(z)$ are known. Two such results (the first being due to Binet) are

$$\Omega(z) = 2 \int_0^\infty \frac{\arctan(t/z)}{e^{2\pi t} - 1} dt \qquad (\mathrm{Re}(z) > 0), \qquad (2.1.2)$$

where the arctan takes its principal value, and [Olver (1974, p. 294)]

$$\Omega(z) = \int_0^\infty \frac{\Delta_2(t)}{2(t + z)^2} dt \qquad (|\arg z| < \pi). \qquad (2.1.3)$$

The quantity $\Delta_{2r}(t) = B_{2r} - B_{2r}(t - [t])$, where $B_{2r}(t)$ denotes the Bernoulli polynomial, $B_{2r} \equiv B_{2r}(0)$ are the Bernoulli numbers and $[t]$ denotes the integer part.

Yet another representation (also due to Binet) is the *convergent* inverse factorial expansion [Whittaker & Watson (1965, p. 253); see also §4.7]

$$\Omega(z) = \frac{c_1}{z + 1} + \frac{c_2}{(z + 1)(z + 2)} + \frac{c_3}{(z + 1)(z + 2)(z + 3)} + \cdots \qquad (2.1.4)$$

valid in $\mathrm{Re}(z) > 0$, where the coefficients c_n are

$$c_1 = c_2 = \tfrac{1}{12}, \quad c_3 = \tfrac{59}{360}, \quad c_4 = \tfrac{29}{60}, \quad c_5 = \tfrac{533}{280}, \quad c_6 = \tfrac{1577}{168}, \ldots,$$

and generally

$$c_n = \frac{1}{n} \int_0^1 (x)_n \left(x - \tfrac{1}{2}\right) dx \qquad (n \geq 1),$$

with $(x)_n$ denoting the Pochhammer symbol given by

$$(x)_n = \frac{\Gamma(x + n)}{\Gamma(x)} = x(x + 1)(x + 2) \cdots (x + n - 1) \quad (n = 0, 1, 2, \ldots).$$

The convergence of this expansion in $\mathrm{Re}(z) > 0$ can be established from the fact that $c_n \leq \Gamma(n) \int_0^1 x(x - \tfrac{1}{2}) dx = \Gamma(n)/12$, so that the behaviour of the nth term in the sum is then controlled by the ratio $\Gamma(n)/\Gamma(n + z + 1) \sim n^{-z-1}$ as $n \to \infty$.

The expansion of the function $\Omega(z)$ for $|z| \to \infty$ in $|\arg z| < \pi$ can be obtained by repeated integration by parts of (2.1.3) to yield

$$\Omega(z) = \sum_{r=1}^{n-1} \frac{B_{2r}}{2r(2r - 1)z^{2r-1}} + R_n(z). \qquad (2.1.5)$$

The first few Bernoulli numbers† B_{2r} are

$$B_2 = \tfrac{1}{6}, \quad B_4 = -\tfrac{1}{30}, \quad B_6 = \tfrac{1}{42}, \quad B_8 = -\tfrac{1}{30}, \quad B_{10} = \tfrac{5}{66},$$

$$B_{12} = -\tfrac{691}{2730}, \quad B_{14} = \tfrac{7}{6}, \quad B_{16} = -\tfrac{3617}{510}, \ldots$$

---

† The expansion in Whittaker & Watson (1965, p. 252) involves the older notation for the Bernoulli numbers, where the present B_{2r} is denoted by $(-)^{r-1} B_r$.

and the remainder $R_n(z)$ is given by [Olver (1974, p. 294)]

$$R_n(z) = \int_0^\infty \frac{\Delta_{2n}(t)}{2n(t+z)^{2n}}\, dt \qquad (|\arg z| < \pi).$$

When $z (= x)$ is real and positive, we have the well-known bound [Whittaker & Watson (1965, p. 253)]

$$R_n(x) = \frac{B_{2n}\Theta_n}{2n(2n-1)\, x^{2n-1}},$$

where Θ_n is a number in the interval $(0, 1)$. For complex $z = xe^{i\theta}$ in the sector $|\arg z| < \pi$, we employ the inequality

$$|t+z|^2 = t^2 + 2xt\cos\theta + x^2 = (t+x)^2 - 4xt\sin^2 \tfrac{1}{2}\theta \geq (t+x)^2 \cos^2 \tfrac{1}{2}\theta$$

to find the bound [Olver (1974, p. 294); Riekstiņš (1986, p. 112)]

$$|R_n(z)| \leq \sec^{2n}\left(\tfrac{1}{2}\theta\right) \int_0^\infty \frac{|\Delta_{2n}(t)|}{2n(t+x)^{2n}}\, dt = \sec^{2n}\left(\tfrac{1}{2}\theta\right)|R_n(x)|$$

$$\leq \frac{|B_{2n}|}{2n(2n-1)} \frac{\sec^{2n}\left(\tfrac{1}{2}\theta\right)}{|z|^{2n-1}} \qquad (|\arg z| < \pi) \tag{2.1.6}$$

upon using the fact that $\Delta_{2n}(t)$ does not change sign in $[0, \infty)$. A different form of bound for the remainder† can be derived by insertion of the Maclaurin series expansion for $\arctan(t/z)$ in (2.1.2), but the resulting bound is less sharp than (2.1.6) and is valid only in $|\arg z| < \tfrac{1}{2}\pi$; see Whittaker & Watson (1965, p. 252).

From (2.1.5) we then obtain the familiar asymptotic expansion, known as the *Stirling series*,

$$\Omega(z) \sim \sum_{r=1}^\infty \frac{B_{2r}}{2r(2r-1)z^{2r-1}} \tag{2.1.7}$$

valid in the Poincaré sense as $|z| \to \infty$ in $|\arg z| < \pi$; for an exponentially-improved version see §6.4. The explicit representation of the first few terms in this important expansion is

$$\Omega(z) \sim \frac{1}{12z} - \frac{1}{360z^3} + \frac{1}{1260z^5} - \frac{1}{1680z^7} + \frac{1}{1188z^9} - \frac{691}{360360z^{11}} + \cdots.$$

We remark that (2.1.7) involves only negative odd powers of z. Combination of (2.1.1) and (2.1.7) then yields *Stirling's formula*

$$\Gamma(z) = (2\pi)^{\frac{1}{2}} e^{-z} z^{z-\frac{1}{2}}\{1 + O(z^{-1})\} \tag{2.1.8}$$

for $|z| \to \infty$ in the sector $|\arg z| < \pi$.

† See also Spira (1971).

Exponentiation of (2.1.1) produces the expansion for $\Gamma(z)$, which we write in the form

$$\Gamma^*(z) \sim \sum_{k=0}^{\infty} (-)^k \gamma_k z^{-k} \qquad (|z| \to \infty, \ |\arg z| < \pi), \qquad (2.1.9)$$

where $\Gamma^*(z) = e^{\Omega(z)}$ denotes the *scaled* gamma function defined by

$$\Gamma^*(z) = (2\pi)^{-\frac{1}{2}} z^{\frac{1}{2}-z} e^z \Gamma(z) \qquad (2.1.10)$$

and γ_k are the *Stirling coefficients*. We note at this point that the expansion for the reciprocal of the scaled gamma function is given by [Temme (1996, p. 71)]

$$\frac{1}{\Gamma^*(z)} \sim \sum_{k=0}^{\infty} \gamma_k z^{-k} \qquad (|z| \to \infty, \ |\arg z| < \pi) \qquad (2.1.11)$$

and involves the same coefficients as (2.1.9) but with different signs of the coefficients with odd index.

2.1.2 The Stirling Coefficients

There is no known closed-form representation for the Stirling coefficients. Their values can be generated numerically by means of the following recurrence relation [Temme (1996, p. 70)]

$$\gamma_k = (-2)^k \frac{\Gamma\left(k + \frac{1}{2}\right)}{\sqrt{\pi}} d_{2k},$$

$$d_n = \frac{n+1}{n+2} \left\{ \frac{d_{n-1}}{n} - \sum_{j=1}^{n-1} \frac{d_j d_{n-j}}{j+1} \right\} \qquad (n \geq 1), \qquad (2.1.12)$$

where $d_0 = 1$ and an empty sum is to be interpreted as zero. The values of γ_k for $1 \leq k \leq 10$ (with $\gamma_0 = 1$) are presented in Table 2.1 [see also Wrench (1968)].

Table 2.1. *The Stirling coefficients γ_k for $1 \leq k \leq 10$*
(with $\gamma_0 = 1$)

k	γ_k	k	γ_k
1	$-\dfrac{1}{12}$	2	$\dfrac{1}{288}$
3	$\dfrac{139}{51,840}$	4	$-\dfrac{571}{2,488,320}$
5	$-\dfrac{163,879}{209,018,880}$	6	$\dfrac{5,246,819}{75,246,796,800}$
7	$\dfrac{534,703,531}{902,961,561,600}$	8	$-\dfrac{4,483,131,259}{86,684,309,913,600}$
9	$-\dfrac{432,261,921,612,371}{514,904,800,886,784,000}$	10	$\dfrac{6,232,523,202,521,089}{86,504,006,548,979,712,000}$

An interesting identity satisfied by the γ_k is given by

$$\sum_{j=0}^{m} (-)^j \gamma_j \gamma_{m-j} = 0 \qquad (m \geq 1).$$ (2.1.13)

This follows from the expansions for $\Gamma^*(z)$ and $1/\Gamma^*(z)$ in (2.1.9) and (2.1.11): the result (2.1.13) is the statement that the coefficient of z^{-m} in the product of these expansions must vanish for $m \geq 1$. For odd m, we note that (2.1.13) is a simple identity.

For large k, the form of the asymptotic behaviour of γ_k is found to depend on the parity of the index k and is given by [Dingle (1973, p. 159); Boyd (1994)]

$$\gamma_k = \begin{cases} \dfrac{1}{\pi}(-)^{(k+1)/2} \dfrac{\Gamma(k)}{(2\pi)^k} \left\{ 1 + O(k^{-1}) \right\} & (k \text{ odd}) \\[4mm] -\dfrac{1}{6}(-)^{k/2} \dfrac{\Gamma(k-1)}{(2\pi)^k} \left\{ 1 + O(k^{-1}) \right\} & (k \text{ even}). \end{cases}$$ (2.1.14)

From this last result, we can see that the nature of the divergence of the expansions (2.1.9) and (2.1.11) is controlled by the familiar 'factorial divided by a power' dependence given by $\Gamma(k)/(2\pi z)^k$, for odd k, and $\Gamma(k-1)/(2\pi z)^k$, for even k. Indeed, the same divergence is present in the expansion for $\Omega(z)$ in (2.1.5), since the large-order behaviour of the Bernoulli numbers can be obtained from the identity involving the Riemann zeta function

$$B_{2r} = 2(-)^{r-1} \frac{(2r)!}{(2\pi)^{2r}} \zeta(2r),$$ (2.1.15)

by using the fact that $\zeta(2r) \sim 1$ as $r \to \infty$.

2.1.3 Bounds for $\Gamma(z)$

We conclude this section with the derivation of certain bounds satisfied by the gamma function. Let $z = |z|e^{i\theta}$ be a complex variable and a, b be real parameters. Then we have the result expressing the quotient of two gamma functions

$$\begin{aligned} \frac{\Gamma(z+a)}{\Gamma(z+b)} &= \frac{1}{\Gamma(b-a)} \int_0^1 t^{z+a-1}(1-t)^{b-a-1} dt \\[3mm] &= \frac{1}{\Gamma(b-a)} \int_0^\infty e^{-(z+a)\tau}(1-e^{-\tau})^{b-a-1} d\tau \\[3mm] &= \frac{e^{-i\theta}}{\Gamma(b-a)} \int_0^\infty e^{-|z|u-aue^{-i\theta}}(1-e^{-ue^{-i\theta}})^{b-a-1} du, \end{aligned}$$

where $b > a \geq 0$ and $\text{Re}(z) > 0$. The last integral is obtained via rotation of the path of integration to the ray $\arg \tau = -\theta$, followed by the substitution $\tau = ue^{-i\theta}$

and use of Cauchy's theorem. Since $|1 - e^{-\zeta}| \leq |\zeta|$ when $|\arg \zeta| \leq \frac{1}{2}\pi$, we then obtain the bound when $b - a \geq 1$, $a \geq 0$ and $\mathrm{Re}(z) > 0$

$$\left|\frac{\Gamma(z+a)}{\Gamma(z+b)}\right| \leq \frac{1}{\Gamma(b-a)} \int_0^\infty e^{-(|z|+a\cos\theta)u} u^{b-a-1} du$$

$$\leq \frac{1}{(|z| + a\cos\theta)^{b-a}} \leq \frac{1}{|z|^{b-a}}. \tag{2.1.16}$$

When $0 < b - a < 1$, use of the result $\Gamma(z+1) = z\Gamma(z)$ shows that

$$\left|\frac{\Gamma(z+a)}{\Gamma(z+b)}\right| = \left|\frac{\Gamma(z+a)(z+b)}{\Gamma(z+b+1)}\right| < \frac{1+b/|z|}{|z|^{b-a}}. \tag{2.1.17}$$

A similar procedure can be used to deal with negative ranges of $b - a$.

A bound for $\Gamma(z)$ in $\mathrm{Re}(z) \geq 0$ can be derived from (2.1.1) expressed in the form

$$\Gamma(z) = (2\pi)^{\frac{1}{2}} z^{z-\frac{1}{2}} e^{-z+\Omega(z)},$$

where, from (2.1.5) and (2.1.6), we have

$$|\Omega(z)| \equiv |R_1(z)| \leq \frac{\sec^2\left(\frac{1}{2}\theta\right)}{12\,|z|} \qquad (|\arg z| < \pi).$$

When $\mathrm{Re}(z) \geq 0$ this yields the bound $|\Omega(z)| \leq \frac{1}{6}|z|^{-1}$, so that

$$|\Gamma(z)| \leq (2\pi)^{\frac{1}{2}} |z^{z-\frac{1}{2}} e^{-z}| \exp\left\{\tfrac{1}{6}|z|^{-1}\right\} \tag{2.1.18}$$

valid when $|\arg z| \leq \frac{1}{2}\pi$. If $z = x + iy$, $x \geq 0$ and $\phi(y) = \arctan(|y|/x)$ we have

$$\left|z^{z-\frac{1}{2}} e^{-z}\right| = |z|^{x-\frac{1}{2}} e^{-x-|y|\phi(y)}$$

$$= |z|^{x-\frac{1}{2}} e^{-\frac{1}{2}\pi|y|} \exp\left\{-x - |y|\left(\phi(y) - \tfrac{1}{2}\pi\right)\right\}$$

$$\leq |z|^{x-\frac{1}{2}} e^{-\frac{1}{2}\pi|y|},$$

since

$$\frac{x}{|y|} + \phi(y) - \tfrac{1}{2}\pi = \frac{x}{|y|} - \arctan(x/|y|) \geq 0.$$

Hence we obtain the upper bound given by

$$|\Gamma(z)| \leq (2\pi)^{\frac{1}{2}} |z|^{x-\frac{1}{2}} e^{-\frac{1}{2}\pi|y|} \exp\left\{\tfrac{1}{6}|z|^{-1}\right\}, \tag{2.1.19}$$

$$z = x + iy, \quad x \geq 0.$$

The reflection formula

$$\Gamma(z)\Gamma(1-z) = \frac{\pi}{\sin \pi z} \tag{2.1.20}$$

should be used when dealing with situations with $\mathrm{Re}(z) < 0$. The bounds in (2.1.16)–(2.1.19) are given in Riekstiņš (1983; 1986, pp. 32–35).

A sharper bound for $|\Gamma(z)|$, valid when $z = x + iy$ with $x > 0$, can be obtained from the fact that

$$|z^{z-\frac{1}{2}}e^{-z}| = x^{x-\frac{1}{2}}e^{-x}(|z|/x)^{x-\frac{1}{2}}e^{-|y|\phi(y)}$$

$$< (2\pi)^{-\frac{1}{2}}\Gamma(x)(|z|/x)^{x-\frac{1}{2}}e^{-|y|\phi(y)},$$

where we have used the well-known inequality† $\Gamma(x) > (2\pi)^{\frac{1}{2}}x^{x-\frac{1}{2}}e^{-x}$. This yields the alternative bound

$$|\Gamma(z)| < \Gamma(x)(|z|/x)^{x-\frac{1}{2}}e^{-|y|\phi(y)}\exp\left\{\tfrac{1}{6}|z|^{-1}\right\} \tag{2.1.21}$$

valid when $|\arg z| < \tfrac{1}{2}\pi$.

2.2 Expansion of Quotients of Gamma Functions

In this section we give two general lemmas on the expansion of quotients of gamma functions as inverse factorial expansions. This type of expansion will be found to play a significant role in the development of asymptotic expansions of Mellin-Barnes integrals.

Consider the quotient of gamma functions given by

$$P(s) = \frac{\prod_{r=1}^{p}\Gamma(\alpha_r s + a_r)}{\prod_{r=1}^{q}\Gamma(\beta_r s + b_r)}, \tag{2.2.1}$$

where p and q are nonnegative integers. The parameters α_r $(1 \leq r \leq p)$ and β_r $(1 \leq r \leq q)$ are assumed to be positive while a_r $(1 \leq r \leq p)$ and b_r $(1 \leq r \leq q)$ are arbitrary complex numbers. The function $P(s)$ is a single-valued meromorphic function of s possessing infinite sequences of poles situated at the points

$$s = -(a_r + k)/\alpha_r \qquad (k = 0, 1, 2, \ldots; 1 \leq r \leq p)$$

which result from the poles of the gamma functions $\Gamma(\alpha_r s + a_r)$. In general there will be p such sequences of simple poles though, depending upon the values of α_r and β_r, it could happen that some of these poles are multiple or that some are ordinary points if any of the $\Gamma(\beta_r s + b_r)$ are singular there.

---

† See Whittaker & Watson (1965, p. 253).

We define the following parameters

$$
\left.
\begin{aligned}
h &= \prod_{r=1}^{p} \alpha_r^{\alpha_r} \prod_{r=1}^{q} \beta_r^{-\beta_r}, \\
\vartheta &= \sum_{r=1}^{p} a_r - \sum_{r=1}^{q} b_r + \tfrac{1}{2}(q - p + 1), \quad \vartheta' = 1 - \vartheta, \\
\kappa &= \sum_{r=1}^{q} \beta_r - \sum_{r=1}^{p} \alpha_r,
\end{aligned}
\right\}
\tag{2.2.2}
$$

where, as usual, an empty sum and product are to be interpreted as 0 or 1, respectively. We remark that, by hypothesis, $h > 0$ but that ϑ is in general complex. We shall refer to the coefficients α_r and β_r as being the *multiplicities* of the arguments of the associated gamma functions. The significance of the parameter κ is that it supplies a measure of the 'imbalance' between the sum of the multiplicities of the denominatorial gamma functions and that of the numeratorial gamma functions. We say that $\kappa > 0$ (resp. $\kappa < 0$) corresponds to a positive (resp. negative) imbalance and that when $\kappa = 0$, the function $P(s)$ is 'balanced'.

2.2.1 Inverse Factorial Expansions

Use of the expansion in (2.1.1) and (2.1.7) shows that for positive integer M as $|s| \to \infty$ uniformly on the sector $|\arg s| \le \pi - \epsilon, \epsilon > 0$

$$
\begin{aligned}
\log P(s) &= \sum_{r=1}^{p} \left(\alpha_r s + a_r - \tfrac{1}{2}\right) \log(\alpha_r s) - \sum_{r=1}^{q} \left(\beta_r s + b_r - \tfrac{1}{2}\right) \log(\beta_r s) \\
&\quad + \kappa s + \tfrac{1}{2}(p - q) \log 2\pi + \sum_{j=1}^{M-1} C_j s^{-j} + O(s^{-M}) \\
&= -\kappa s \log s + s(\kappa + \log h) + \left(\vartheta - \tfrac{1}{2}\right) \log s \\
&\quad + \log \mathcal{A}_0 + \sum_{j=1}^{M-1} C_j s^{-j} + O(s^{-M}),
\end{aligned}
$$

where the coefficients C_j are independent of s and

$$
\mathcal{A}_0 = (2\pi)^{\frac{1}{2}(p-q)} \prod_{r=1}^{p} \alpha_r^{a_r - \frac{1}{2}} \prod_{r=1}^{q} \beta_r^{\frac{1}{2} - b_r}.
$$

Then $P(s)$ has the asymptotic expansion for $|s| \to \infty$ in $|\arg s| \le \pi - \epsilon$ given by

$$
P(s) = \mathcal{A}_0 h^s e^{\kappa s} s^{\vartheta - \kappa s - \frac{1}{2}} \left\{ 1 + \sum_{j=1}^{M-1} D_j s^{-j} + O(s^{-M}) \right\},
\tag{2.2.3}
$$

where the coefficients D_j depend on C_j. When $P(s)$ is 'balanced' (i.e., when $\kappa = 0$), we see that

$$P(s) = h^s s^{\vartheta - \frac{1}{2}} A_0 \{1 + O(s^{-1})\} \qquad (|s| \to \infty; \ |\arg s| \le \pi - \epsilon). \qquad (2.2.4)$$

When $\kappa > 0$, we proceed to obtain an expansion of $P(s)$ as a series of inverse factorials (cf. §4.7). If $a \ (> 0)$ and b denote constants, so that $|\arg(as + b)| < \pi$, then from (2.2.3) and the fact that $\Gamma(as + b) \sim (2\pi)^{\frac{1}{2}} e^{-as} (as)^{as+b-\frac{1}{2}}$ for large $|s|$, we obtain

$$P(s)\Gamma(as + b) \sim A_0 (h\kappa^\kappa)^s (\kappa s)^{-\kappa s + \vartheta - \frac{1}{2}} e^{(\kappa - a)s} (as)^{as+b-\frac{1}{2}}$$

for $|s| \to \infty$ in $|\arg s| \le \pi - \epsilon, \epsilon > 0$, where

$$A_0 = (2\pi)^{\frac{1}{2}} \kappa^{\frac{1}{2} - \vartheta} A_0. \qquad (2.2.5)$$

The choice $a = \kappa$ and $b = 1 - \vartheta$ then removes the last three terms in the above asymptotic expression for $P(s)\Gamma(as + b)$ and we accordingly find the result

$$P(s) = \frac{A_0 (h\kappa^\kappa)^s}{\Gamma(\kappa s + \vartheta')} \left\{ 1 + \sum_{j=1}^{M-1} E_j s^{-j} + O(s^{-M}) \right\}$$

as $|s| \to \infty$ in $|\arg s| \le \pi - \epsilon$, where the coefficients E_j depend on D_j, κ and ϑ. If we now introduce the coefficients $A_j \ (j \ge 1)$ by means of the expansion for large $|s|$

$$A_0 \sum_{j=1}^{M-1} E_j s^{-j} = \frac{A_1}{\kappa s + \vartheta'} + \frac{A_2}{(\kappa s + \vartheta')_2} + \cdots + \frac{A_{M-1}}{(\kappa s + \vartheta')_M} + O(s^{-M}),$$

we finally obtain the expansion of $P(s)$ when $\kappa > 0$ as a series of inverse factorials in the form

$$P(s) = (h\kappa^\kappa)^s \left\{ \sum_{j=0}^{M-1} \frac{A_j}{\Gamma(\kappa s + \vartheta' + j)} + \frac{O(1)}{\Gamma(\kappa s + \vartheta' + M)} \right\} \qquad (2.2.6)$$

valid uniformly as $|s| \to \infty$ in $|\arg s| \le \pi - \epsilon$.

To deal with the case when the quotient of gamma functions involves a negative imbalance between the multiplicities of the denominatorial and numeratorial gamma functions, we consider the related quotient given by

$$Q(s) = \frac{\prod_{r=1}^{q} \Gamma(1 - b_r + \beta_r s)}{\prod_{r=1}^{p} \Gamma(1 - a_r + \alpha_r s)}. \qquad (2.2.7)$$

The parameters h, ϑ and κ are as defined in (2.2.2). A negative imbalance in $Q(s)$ is seen to correspond again to $\kappa > 0$, where κ is defined in (2.2.2). A similar application of (2.1.1) shows that for $|s| \to \infty$ in $|\arg s| \le \pi - \epsilon$

$$\frac{Q(s)}{\Gamma(as + b)} \sim \frac{A_0}{(2\pi)^{p-q+1}} (h\kappa^\kappa)^{-s} e^{(a-\kappa)s} (\kappa s)^{\kappa s + \vartheta - \frac{1}{2}} (as)^{-as-b+\frac{1}{2}}$$

since $|\arg(as+b)| < \pi$, where A_0 is defined in (2.2.5). If we now make the choice $a = \kappa$, $b = \vartheta$, in order to remove the last three terms of the above expression, we obtain the expansion of $Q(s)$ in the form

$$Q(s) = \frac{A_0}{(2\pi)^{p-q+1}} (h\kappa^\kappa)^{-s} \Gamma(\kappa s + \vartheta) \left\{ 1 + \sum_{j=1}^{M-1} E'_j s^{-j} + O(s^{-M}) \right\}$$

as $|s| \to \infty$ in $|\arg s| \le \pi - \epsilon$. Then, in a similar manner as for $P(s)$ in (2.2.6), we obtain the inverse factorial expansion

$$Q(s) = (h\kappa^\kappa)^{-s} \Gamma(\kappa s + \vartheta) \left\{ \sum_{j=0}^{M-1} \frac{A'_j}{(-\kappa s - \vartheta + 1)_j} + O(s^{-M}) \right\}$$

$$= (h\kappa^\kappa)^{-s} \left\{ \sum_{j=0}^{M-1} (-)^j A'_j \Gamma(\kappa s + \vartheta - j) + O(1)\Gamma(\kappa s + \vartheta - M) \right\}$$

$$(2.2.8)$$

valid uniformly as $|s| \to \infty$ in $|\arg s| \le \pi - \epsilon$. The coefficients A'_j are independent of s and, in particular, $A'_0 = A_0/(2\pi)^{p-q+1}$.

To see how the coefficients A'_j are related to the A_j appearing in (2.2.6), we use the reflection formula for the gamma function in (2.1.20) to find

$$Q(-s) = P(s)\Xi(s), \qquad \Xi(s) = \pi^{q-p} \frac{\prod_{r=1}^{p} \sin \pi(\alpha_r s + a_r)}{\prod_{r=1}^{q} \sin \pi(\beta_r s + b_r)}. \qquad (2.2.9)$$

We now let $|s| \to \infty$ with $\arg s = \frac{1}{2}\pi$ and compare the coefficients in the expansion† of $Q(-s)$ with those on the right-hand side of (2.2.9). Since

$$\Xi(s) \sim i(2\pi)^{q-p} e^{-\pi\kappa|s|} e^{-\pi i \vartheta} \qquad \left(|s| \to \infty; \ \arg s = \tfrac{1}{2}\pi \right)$$

and

$$(-)^j A'_j \Gamma(-\kappa s + \vartheta - j) = \frac{\pi}{\sin \pi(\vartheta - \kappa s)} \frac{A'_j}{\Gamma(\kappa s + \vartheta' + j)},$$

where $\pi \operatorname{cosec} \pi(\vartheta - \kappa s) \sim 2\pi i e^{-\pi\kappa|s|} e^{-\pi i \vartheta}$ as $|s| \to \infty$ with $\arg s = \frac{1}{2}\pi$, we find from (2.2.6) and (2.2.8) that $A'_j = A_j/(2\pi)^{p-q+1}$ $(0 \le j \le M-1)$.

The results of this section can now be stated as two lemmas. For the functions $P(s)$ and $Q(s)$ defined in (2.2.1) and (2.2.7) and the parameters h, ϑ and κ defined

---

† This is given by (2.2.8) with s replaced by $-s$ and valid as $|s| \to \infty$ in $|\arg(-s)| \le \pi - \epsilon$.

in (2.2.2), we have the following inverse factorial expansions [Wright (1940); Braaksma (1963, §3)]:

Lemma 2.1. *Let M be a positive integer and suppose that $\kappa > 0$. Then there exist numbers A_j ($0 \le j \le M-1$) independent of s and M such that $P(s)$ defined by*

$$P(s) = \frac{\prod_{r=1}^{p} \Gamma(\alpha_r s + a_r)}{\prod_{r=1}^{q} \Gamma(\beta_r s + b_r)}$$

possesses the inverse factorial expansion given by

$$P(s) = (h\kappa^\kappa)^s \left\{ \sum_{j=0}^{M-1} \frac{A_j}{\Gamma(\kappa s + \vartheta' + j)} + \frac{\sigma_M(s)}{\Gamma(\kappa s + \vartheta' + M)} \right\},$$

where the parameters h, κ and ϑ' are defined in (2.2.2). In particular, the coefficient A_0 has the value

$$A_0 = (2\pi)^{\frac{1}{2}(p-q+1)} \kappa^{\frac{1}{2}-\vartheta} \prod_{r=1}^{p} \alpha_r^{a_r - \frac{1}{2}} \prod_{r=1}^{q} \beta_r^{\frac{1}{2} - b_r}.$$

The remainder function $\sigma_M(s)$ is analytic in s except at the points $s = -(a_r + k)/\alpha_r$, $k = 0, 1, 2, \ldots$ ($1 \le r \le p$), where $P(s)$ has poles, and is such that

$$\sigma_M(s) = O(1)$$

for $|s| \to \infty$ uniformly in $|\arg s| \le \pi - \epsilon$, $\epsilon > 0$.
In the 'balanced' case $\kappa = 0$, the behaviour of $P(s)$ is described by (2.2.4).

Lemma 2.2. *Let M be a positive integer and suppose that $\kappa > 0$. Then, with the parameters h, ϑ and κ defined by (2.2.2), the function $Q(s)$ defined by*

$$Q(s) = \frac{\prod_{r=1}^{q} \Gamma(1 - b_r + \beta_r s)}{\prod_{r=1}^{p} \Gamma(1 - a_r + \alpha_r s)}$$

possesses the inverse factorial expansion given by

$$Q(s) = \frac{(h\kappa^\kappa)^{-s}}{(2\pi)^{p-q+1}} \left\{ \sum_{j=0}^{M-1} (-)^j A_j \Gamma(\kappa s + \vartheta - j) + \rho_M(s) \Gamma(\kappa s + \vartheta - M) \right\},$$

where the coefficients A_j are those appearing in Lemma 2.1. The remainder function $\rho_M(s)$ is analytic in s except at the points $s = (b_r - 1 - k)/\beta_r$, $k = 0, 1, 2, \ldots$ ($1 \le r \le q$), where $Q(s)$ has poles, and is such that

$$\rho_M(s) = O(1)$$

for $|s| \to \infty$ uniformly in $|\arg s| \le \pi - \epsilon$, $\epsilon > 0$.

The above two lemmas are more in the way of existence theorems: they say nothing about the values of the coefficients A_j, apart from A_0. The actual computation of these coefficients in general turns out to be quite difficult. Riney (1956, 1958) has elaborated two algorithms for determining the A_j when the parameters $\alpha_r = \beta_r = 1$. Another approach arises from the theory of integral functions of hypergeometric type (see §2.3), where the Maclaurin coefficients are given by $P(n)$ $(n = 0, 1, 2, \ldots)$. The exponential part of the asymptotic expansion (see §2.3) of such functions is known to involve the coefficients A_j. If the differential equation satisfied by these functions is known, the exponential expansion can be substituted into the differential equation to produce a recurrence relation for the coefficients A_j [Wright (1958) for the case $\alpha_r = \beta_r = 1$; see also Paris & Wood (1986, §3.4) for an example of a case where $\alpha_r \neq 1$, $\beta_r = 1$].

Yet another method is to exploit the computer algebra in packages such as *Mathematica* or *Maple*. We illustrate this procedure in §2.2.4 for the case of quotients involving two and three gamma functions when the parameters $\alpha_r \neq 1$.

2.2.2 A Recursion Formula when $\alpha_r = \beta_r = 1$

In the special case $\alpha_r = \beta_r = 1$, a recursion formula for the coefficients in the inverse factorial expansion of

$$P(s) = \frac{\prod_{r=1}^{p} \Gamma(s + a_r)}{\prod_{r=1}^{q} \Gamma(s + b_r)} \qquad (q \geq p + 1) \qquad (2.2.10)$$

has been given by Riney (1956). The parameters in (2.2.2) associated with this quotient have the values

$$h = 1, \quad \kappa = q - p \ (\geq 1), \quad \vartheta = \sum_{r=1}^{p} a_r - \sum_{r=1}^{q} b_r + \tfrac{1}{2}(q - p + 1),$$

$$\vartheta' = 1 - \vartheta, \quad A_0 = (2\pi)^{\frac{1}{2}(p-q+1)} \kappa^{\frac{1}{2}-\vartheta}. \qquad (2.2.11)$$

Then, from Lemma 2.1, we have the inverse factorial expansion

$$P(s) = A_0 \kappa^{\kappa s} \left\{ \sum_{j=0}^{M-1} \frac{c_j}{\Gamma(\kappa s + \vartheta' + j)} + \frac{O(1)}{\Gamma(\kappa s + \vartheta' + M)} \right\} \qquad (2.2.12)$$

as $|s| \to \infty$ in $|\arg s| \leq \pi - \epsilon$, where $M = 1, 2, \ldots$ and $c_j = A_j/A_0$ denote the normalised coefficients, with $c_0 = 1$.

A recursion formula for the coefficients c_j can be derived from the difference equation satisfied by $P(s)$. Use of the functional equation for the gamma function readily shows that

$$P(s + 1) = T(s)P(s), \qquad (2.2.13)$$

where

$$T(s) = \prod_{r=1}^{p}(s + a_r)/\prod_{r=1}^{q}(s + b_r). \tag{2.2.14}$$

Provided $T(s)$ has only simple poles (that is, $b_r \neq b_k$ for $r \neq k$), then $T(s)$ has the partial fraction decomposition

$$T(s) = \sum_{r=1}^{q}\frac{D_r}{s + b_r}, \tag{2.2.15}$$

where

$$D_j = \prod_{r=1}^{p}(a_r - b_j)/\prod_{r=1}^{q}{}'(b_r - b_j) \quad (1 \le j \le q) \tag{2.2.16}$$

and the prime denotes the omission of the term corresponding to $r = j$.

A recursion formula for c_j is then given by the following lemma:

Lemma 2.3. *Let c_j be the normalised coefficients ($c_j = A_j/A_0$) in the inverse factorial expansion (2.2.12) of the function $P(s)$ defined in (2.2.10). Provided $b_r \neq b_k$ for $r \neq k$, we have the recursion formula*

$$c_j = -\frac{1}{j\kappa^\kappa}\sum_{k=0}^{j-1}c_k\,e(j,k) \quad (j \ge 1),$$

where $c_0 = 1$ and

$$e(j,k) = \sum_{r=1}^{q}D_r\frac{\Gamma(\vartheta' - \kappa b_r + \kappa + j)}{\Gamma(\vartheta' - \kappa b_r + k)}.$$

The partial fraction coefficients D_r are defined in (2.2.16) and the parameters ϑ' and κ are given in (2.2.11).

Proof. We now establish this result using Riney's method of proof. We shall make use of the following expansion involving a single gamma function. Let $\beta > 0$ and let a and b denote arbitrary complex parameters. Then the inverse factorial expansion

$$\frac{1}{(s + a)\Gamma(\beta s + b)} = \frac{\beta}{\Gamma(b - a\beta)}\sum_{n=0}^{\infty}\frac{\Gamma(b - a\beta + n)}{\Gamma(\beta s + b + n + 1)} \tag{2.2.17}$$

is *convergent* in the right half-plane defined by $\mathrm{Re}(s + a) > 0$. The proof of this well-known expansion is given in Ford (1936).

Then, from (2.2.12), (2.2.13) and (2.2.15), we find

$$
P(s+1) = A_0 \kappa^{\kappa s} T(s) \left\{ \sum_{j=0}^{M-1} \frac{c_j}{\Gamma(\kappa s + \vartheta' + j)} + \frac{O(1)}{\Gamma(\kappa s + \vartheta' + M)} \right\}
$$

$$
= A_0 \kappa^{\kappa s} \left\{ \sum_{r=1}^{q} \sum_{j=0}^{M-1} \frac{c_j D_r}{(s + b_r)\Gamma(\kappa s + \vartheta' + j)} \right.
$$

$$
\left. + \frac{O(1)}{\Gamma(\kappa s + \vartheta' + \kappa + M)} \right\}, \tag{2.2.18}
$$

where we have used the facts that $T(s) \sim s^{-\kappa}$ and $s^\kappa \Gamma(\kappa s + \vartheta' + M) = O(\Gamma(\kappa s + \vartheta' + \kappa + M))$ as $|s| \to \infty$. Let $B = \min_r \operatorname{Re}(b_r)$ and let $\mathcal{R}$ denote the half-plane $\operatorname{Re}(s + B) > 0$. It then follows from (2.2.17) that the double sum appearing in the above expansion can be written as

$$
\kappa \sum_{j=0}^{M-1} c_j \sum_{n=0}^{\infty} \frac{e(n + j - \kappa, j)}{\Gamma(\kappa s + \vartheta' + j + n + 1)} \tag{2.2.19}
$$

provided $s \in \mathcal{R}$.

At this point, we note the special values of $e(j, k)$ for $j \le k$ given by

$$
e(j, k) = \begin{cases} 0 & j \le k - 2 \\ \kappa^{\kappa-1} & j = k - 1 \\ k\kappa^\kappa & j = k. \end{cases} \tag{2.2.20}
$$

These values are established at the end of this proof. It follows that in the sum over n in (2.2.19) we may restrict $n \ge \kappa - 1$. If we set $k = n - \kappa + 1 + j$, the above double sum then becomes

$$
\kappa \sum_{j=0}^{M-1} c_j \sum_{k=j}^{\infty} \frac{e(k - 1, j)}{\Gamma(\kappa s + \vartheta' + \kappa + k)}.
$$

We further note that the sum over k may be restricted to $k \le M - 1$, since the remainder may be absorbed into the order term in (2.2.18). Then, upon interchanging the order of summation, we obtain the expansion of $P(s + 1)$ in the form

$$
P(s+1) = A_0 \kappa^{\kappa s + 1} \left\{ \sum_{k=0}^{M-1} \frac{1}{\Gamma(\kappa s + \vartheta' + \kappa + k)} \sum_{j=0}^{k} c_j e(k - 1, j) \right.
$$

$$
\left. + \frac{O(1)}{\Gamma(\kappa s + \vartheta' + \kappa + M)} \right\} \tag{2.2.21}
$$

as $|s| \to \infty$ when $s \in \mathcal{R}$.

Application of Lemma 2.1 to $P(s + 1)$ yields the inverse factorial expansion

$$P(s+1) = A_0 \kappa^{\kappa s + \kappa} \left\{ \sum_{j=0}^{M-1} \frac{c_j}{\Gamma(\kappa s + \vartheta' + \kappa + j)} + \frac{O(1)}{\Gamma(\kappa s + \vartheta' + \kappa + M)} \right\}$$

as $|s| \to \infty$ in $|\arg s| \le \pi - \epsilon$. Comparison with the coefficients in (2.2.21) (after interchange of j and k) then shows that

$$\kappa^{\kappa-1} c_j = \sum_{k=0}^{j} c_k e(j-1, k) = \sum_{k=0}^{j-1} c_k e(j-1, k) + \kappa^{\kappa-1} c_j.$$

Therefore, upon replacing j by $j + 1$, we find

$$\sum_{k=0}^{j} c_k e(j, k) = 0,$$

whence, since $e(j, j) = j\kappa^\kappa$, the result stated in Lemma 2.3 follows. The case when two or more of the b_r are equal is more complicated. In this case Lemma 2.3 is still applicable but the evaluation of $e(j, k)$ has to be combined with a limiting procedure; see §2.2.3, Example 2.

To conclude this demonstration it remains to establish (2.2.20). If we compare the coefficients of the expansion in powers of s^{-1} of $T(s)$ obtained from (2.2.14) with those obtained from (2.2.15), we readily obtain

$$\sum_{r=1}^{q} D_r (-b_r)^k = \begin{cases} 0 & (k \le \kappa - 2) \\ 1 & (k = \kappa - 1) \\ \vartheta - \frac{1}{2}(\kappa + 1) & (k = \kappa). \end{cases}$$

For positive integer n and arbitrary complex x, let the function $F_n(x)$ be defined by

$$F_n(x) = \sum_{r=1}^{q} D_r \frac{\Gamma(x - \kappa b_r + n)}{\Gamma(x - \kappa b_r)}$$

$$= \sum_{r=1}^{q} D_r \sum_{k=0}^{n} b_{n,k} (-b_r)^k = \sum_{k=0}^{n} b_{n,k} \sum_{r=1}^{q} D_r (-b_r)^k,$$

where $b_{n,k}$ are the coefficients in the expansion of $\Gamma(x - \kappa b_r + n)/\Gamma(x - \kappa b_r)$ in powers of $-b_r$. Straightforward algebra shows that

$$b_{n,n} = \kappa^n, \qquad b_{n,n-1} = \kappa^{n-1} \{ nx + \frac{1}{2} n(n-1) \},$$

and hence we find that

$$F_n(x) = \begin{cases} 0 & (n \le \kappa - 2) \\ \kappa^{\kappa-1} & (n = \kappa - 1) \\ \kappa^\kappa (x - \vartheta') & (n = \kappa). \end{cases}$$

The statement in (2.2.20) then follows upon letting $n = j - k + \kappa$ and $x = \vartheta' + k$ in the definition of $e(j, k)$ in the statement of the Lemma. □

We remark that the associated 'reciprocal' quotient

$$Q(s) = \frac{\prod_{r=1}^{q} \Gamma(1 - b_r + s)}{\prod_{r=1}^{p} \Gamma(1 - a_r + s)} \qquad (q \geq p + 1)$$

has, from Lemma 2.2, the inverse factorial expansion given by

$$Q(s) = \frac{A_0 \kappa^{-\kappa s}}{(2\pi)^{p-q+1}} \left\{ \sum_{j=0}^{M-1} (-)^j c_j \Gamma(\kappa s + \vartheta - j) + O(1)\Gamma(\kappa s + \vartheta - M) \right\}$$

$$(2.2.22)$$

as $|s| \to \infty$ in $|\arg s| \leq \pi - \epsilon$. The coefficients c_j in this expansion are the same as those appearing in the expansion of $P(s)$ in (2.2.12).

2.2.3 Examples

We now give two examples to illustrate the application of Lemma 2.3.

Example 1. Consider the quotient

$$\hat{P}(s) = \frac{\Gamma(s - 1)}{\Gamma(2s)\Gamma\left(s - \frac{1}{2}\right)}.$$

As it stands this is not in the form (2.2.10). However, application of the duplication formula for the gamma function

$$\Gamma(2z) = 2^{2z-1}\pi^{-\frac{1}{2}}\Gamma(z)\Gamma(z + \tfrac{1}{2}) \qquad (2.2.23)$$

shows that the above quotient can be written as

$$\hat{P}(s) = 2^{1-2s}\pi^{\frac{1}{2}} \frac{\Gamma(s - 1)}{\Gamma\left(s - \frac{1}{2}\right)\Gamma(s)\Gamma\left(s + \frac{1}{2}\right)}.$$

This is now in the required form and corresponds to the parameter values $\kappa = 2$, $\vartheta = \vartheta' = \frac{1}{2}$ with $A_0 = (2\pi)^{-\frac{1}{2}}$. Then we find from (2.2.12) that

$$\hat{P}(s) = 2^{\frac{1}{2}} \left\{ \sum_{j=0}^{M-1} \frac{c_j}{\Gamma\left(2s + \frac{1}{2} + j\right)} + \frac{O(1)}{\Gamma\left(2s + \frac{1}{2} + M\right)} \right\} \qquad (2.2.24)$$

as $|s| \to \infty$ in $|\arg s| \leq \pi - \epsilon$.

The coefficients D_r in the partial fraction expansion (2.2.15) have the values $D_1 = -1$, $D_2 = 4$ and $D_3 = -3$, whereupon

$$e(j, k) = -\left(k + \tfrac{3}{2}\right)_N + 4\left(k + \tfrac{1}{2}\right)_N - 3\left(k - \tfrac{1}{2}\right)_N,$$

where $N = j + 2 - k$. The recursion formula in Lemma 2.3 becomes

$$c_j = -\frac{1}{4j} \sum_{k=0}^{j-1} c_k \, e(j, k) \quad (j \geq 1),$$

and we find the values of the first few coefficients given by

$$c_0 = 1, \quad c_1 = \tfrac{9}{8}, \quad c_2 = \tfrac{345}{128}, \quad c_3 = \tfrac{9555}{1024}, \quad c_4 = \tfrac{1371195}{32768},$$

$$c_5 = \tfrac{60259815}{262144}, \quad c_6 = \tfrac{6264182925}{4194304}, \dots . \tag{2.2.25}$$

The associated 'reciprocal' quotient

$$\hat{Q}(s) = \frac{\Gamma(2s+1)\Gamma\left(s+\frac{3}{2}\right)}{\Gamma(s+2)} = 2^{2s}\pi^{-\frac{1}{2}}\frac{\Gamma\left(s+\frac{1}{2}\right)\Gamma(s+1)\Gamma\left(s+\frac{3}{2}\right)}{\Gamma(s+2)}$$

has the same values of the parameters κ, ϑ and A_0 and, from (2.2.22), has the expansion

$$\hat{Q}(s) = 2^{\frac{1}{2}}\left\{\sum_{j=0}^{M-1}(-)^j c_j \Gamma\left(2s+\tfrac{1}{2}-j\right) + O(1)\Gamma\left(2s+\tfrac{1}{2}-M\right)\right\}, \tag{2.2.26}$$

valid as $|s| \to \infty$ in $|\arg s| \leq \pi - \epsilon$. The coefficients c_j are those in (2.2.25).

Example 2. We consider an example of $P(s)$ in which two of the b_r are equal. Let

$$P(s) = \frac{\Gamma\left(s+\frac{1}{2}\right)}{\Gamma^2\left(s+\frac{1}{4}\right)},$$

so that $\kappa = 1$, $\vartheta' = 0$ and $A_0 = 1$. From (2.2.12) we therefore have

$$P(s) = \sum_{j=0}^{M-1}\frac{c_j}{\Gamma(s+j)} + \frac{O(1)}{\Gamma(s+M)}$$

as $|s| \to \infty$ in $|\arg s| \leq \pi - \epsilon$.

If we let

$$P(s) = \lim_{\delta \to 0}\frac{\Gamma\left(s+\frac{1}{2}\right)}{\Gamma\left(s+\frac{1}{4}\right)\Gamma\left(s+\frac{1}{4}+\delta\right)},$$

then, for finite δ, we now have $\vartheta = 1 - \delta$, $\vartheta' = \delta$ and the coefficients D_r given by $D_1 = 1/(4\delta)$, $D_2 = 1 - (1/4\delta)$. Then

$$e(j, k) = \lim_{\delta \to 0}\left\{\frac{\Gamma\left(j+\frac{3}{4}\right)}{\Gamma\left(k-\frac{1}{4}\right)} + \frac{1}{4\delta}\left[\frac{\Gamma\left(\delta+j+\frac{3}{4}\right)}{\Gamma\left(\delta+k-\frac{1}{4}\right)} - \frac{\Gamma\left(j+\frac{3}{4}\right)}{\Gamma\left(k-\frac{1}{4}\right)}\right]\right\}$$

$$= \frac{\Gamma\left(j+\frac{3}{4}\right)}{\Gamma\left(k-\frac{1}{4}\right)}\left\{1 + \tfrac{1}{4}\left[\psi\left(j+\tfrac{3}{4}\right) - \psi\left(k-\tfrac{1}{4}\right)\right]\right\},$$

where ψ denotes the logarithmic derivative of the gamma function.

The recursion formula in Lemma 2.3 accordingly yields the values

$$c_0 = 1, \quad c_1 = 0.0625000, \quad c_2 = 0.0175781, \quad c_3 = 0.0179443,$$
$$c_4 = 0.0339260, \quad c_5 = 0.0954169, \dots .$$

A closed-form representation for the coefficients c_j in this case is given by $c_j = [(-\frac{1}{4})_j]^2/j!$; see (2.2.39) and (2.2.40).

2.2.4 An Algebraic Method for the Determination of the A_j

In this section we discuss an algebraic method for the determination of the coefficients A_j in Lemmas 2.1 and 2.2. We illustrate this approach by considering two examples of the quotient of gamma functions in (2.2.1) with $p = q = 1$ and $p = 2, q = 1$.

Example 1. As a first example, let us consider the ratio of two gamma functions

$$\Gamma(a + bs)/\Gamma(1 + s),$$

where a is arbitrary and $b > 0$. When $0 < b < 1$, we have from Lemma 2.1 the expansion

$$\frac{\Gamma(a + bs)}{\Gamma(1 + s)} = (h\kappa^\kappa)^s \left\{ \sum_{j=0}^{M-1} \frac{A_j}{\Gamma(\kappa s + \vartheta' + j)} + \frac{\sigma_M(s)}{\Gamma(\kappa s + \vartheta' + M)} \right\}, \quad (2.2.27)$$

where $\sigma_M(s) = O(1)$ as $|s| \to \infty$ uniformly in $|\arg s| \leq \pi - \epsilon, \epsilon > 0$. The parameters are given by

$$\kappa = 1 - b \; (> 0), \quad \vartheta = a - \frac{1}{2}, \quad \vartheta' = \frac{3}{2} - a, \quad h = b^b$$

and $A_0 = (2\pi)^{\frac{1}{2}} \kappa^{\frac{1}{2} - \vartheta} b^\vartheta$. When $b > 1$, we find from Lemma 2.2 the expansion

$$\frac{\Gamma(a + bs)}{\Gamma(1 + s)} = \frac{(h\kappa^\kappa)^{-s}}{2\pi} \left\{ \sum_{j=0}^{M-1} (-)^j A_j \Gamma(\kappa s + \vartheta - j) + \rho_M(s) \Gamma(\kappa s + \vartheta - M) \right\}$$

$$(2.2.28)$$

with the same parameter ϑ and coefficients A_j as in (2.2.27) but with κ now given by $\kappa = b - 1 > 0$ and $h = b^{-b}$. When $b = 1$ we have $\kappa = 0$; this corresponds to the 'balanced' case with expansion given by (2.2.30) in inverse powers of s.

Let us consider the case $0 < b < 1$. If we introduce the scaled gamma function defined in (2.1.10) we find after some routine algebra that

$$\frac{\Gamma(a + bs)\Gamma(\kappa s + \vartheta')}{\Gamma(1 + s)} = A_0 (h\kappa^\kappa)^s R(s) \Upsilon(s),$$

where

$$R(s) = \frac{e^{-\frac{1}{2}}\left(1 + a(bs)^{-1}\right)^{\vartheta + bs}\left(1 + \vartheta'(\kappa s)^{-1}\right)^{\kappa s + \frac{1}{2} - \vartheta}}{\left(1 + s^{-1}\right)^{s + \frac{1}{2}}},$$

$$\Upsilon(s) = \frac{\Gamma^*(a + bs)\Gamma^*(\kappa s + \vartheta')}{\Gamma^*(1 + s)}.$$

Then, with the normalised coefficients $c_j = A_j/A_0$, the expansion (2.2.27) can be written in the form

$$R(s)\Upsilon(s) = \sum_{j=0}^{M-1} \frac{c_j}{(\kappa s + \vartheta')_j} + \frac{\sigma_M(s)}{A_0(\kappa s + \vartheta')_M}, \qquad (2.2.29)$$

where $c_0 = 1$.

The procedure now consists of expanding both sides of (2.2.29) in inverse powers of κs and making use of the expansions in (2.1.9) and (2.1.11) for the scaled gamma function. Straightforward algebra shows that

$$R(s) = 1 + \frac{1}{2\kappa s}[\kappa(a-1)a/b - \vartheta\vartheta'] + O[(\kappa s)^{-2}]$$

$$\Upsilon(s) = 1 - \frac{\gamma_1}{\kappa s}(b + \kappa/b) + O[(\kappa s)^{-2}],$$

where the Stirling coefficient $\gamma_1 = -\frac{1}{12}$, so that upon equating coefficients of $(\kappa s)^{-1}$ in (2.2.29) we obtain

$$c_1 = \tfrac{1}{2}\{\kappa(a-1)a/b - \vartheta\vartheta'\} - \gamma_1(b + \kappa/b).$$

The higher coefficients can be obtained by continuation of this process with the help of *Mathematica* to find†

$$c_0 = 1, \quad c_1 = (2 - 12a + 12a^2 + 7b - 12ab + 2b^2)/24b,$$

$$\begin{aligned}
c_2 =\ & (4 - 144a + 480a^2 - 480a^3 + 144a^4 + 172b - 984ab \\
& + 1320a^2b - 480a^3b + 417b^2 - 984ab^2 + 480a^2b^2 + 172b^3 \\
& - 144ab^3 + 4b^4)/1152b^2,
\end{aligned}$$

$$\begin{aligned}
c_3 =\ & (-1112 - 3600a + 65520a^2 - 161280a^3 + 151200a^4 - 60480a^5 \\
& + 8640a^6 + 9636b - 220320ab + 715680a^2b - 816480a^3b \\
& + 378000a^4b - 60480a^5b + 163734b^2 - 929700ab^2 + 1440180a^2b^2 \\
& - 816480a^3b^2 + 151200a^4b^2 + 336347b^3 - 929700ab^3 + 715680a^2b^3
\end{aligned}$$

---

† It is helpful in this expansion process to write terms such as $(1+a(bs)^{-1})^{\vartheta+bs}$ appearing in $R(s)$ in the form $\exp\{(\vartheta + b(\kappa\zeta)^{-1})\log(1 + a\kappa\zeta/b)\}$, where $\zeta = (\kappa s)^{-1}$.

$$- 161280a^3b^3 + 163734b^4 - 220320ab^4 + 65520a^2b^4 + 9636b^5$$
$$- 3600ab^5 - 1112b^6)/414720b^3, \ldots .$$

A considerable simplification in the form of these coefficients results when $a = b$ to yield

$$c_0 = 1, \quad c_1 = (2a - 1)(a - 2)/24a,$$

$$c_2 = \frac{(2a - 1)(a - 2)}{1152a^2}(2a^2 + 19a + 2),$$

$$c_3 = -\frac{(2a - 1)(a - 2)}{414720a^3}(556a^4 - 1628a^3 - 9093a^2 - 1628a + 556), \ldots .$$

When $a = b = \frac{1}{2}$ (or $a = b = 2$ in (2.2.28)), the quotient of gamma functions reduces to a single gamma function by virtue of the duplication formula. In these cases we see that the coefficients correctly reduce to $c_k = 0$ $(k \geq 1)$.

We remark that the inverse factorial expansion of the associated 'reciprocal' quotient

$$\frac{\Gamma(s)}{\Gamma(1 - a + bs)}$$

is given by the right-hand side of (2.2.27) when $b > 1$ and of (2.2.28) when $0 < b < 1$. An example of the use of this expansion when $a = b$, with $0 < b < 1$, is given in §8.1.6, where expressions for c_k for $0 \leq k \leq 8$ are derived.

Example 2. Consider the quotient of three gamma functions given by

$$P(s) = \frac{1}{s!}\Gamma\left(\frac{1 + ms}{\mu}\right)\Gamma\left(\frac{1 + ns}{\nu}\right),$$

where we suppose that $\mu > m > 0$ and $\nu > n > 0$. The parameters associated with this quotient are

$$\kappa = 1 - \frac{m}{\mu} - \frac{n}{\nu}, \quad \vartheta = \frac{1}{\mu} + \frac{1}{\nu} - 1, \quad h = (m/\mu)^{m/\mu}(n/\nu)^{n/\nu},$$

$$A_0 = 2\pi\kappa^{\frac{1}{2} - \vartheta}(m/\mu)^{(1/\mu) - \frac{1}{2}}(n/\nu)^{(1/\nu) - \frac{1}{2}}.$$

In addition, we shall suppose the constants μ, ν, m and n to be further restricted so that $0 < \kappa < 1$. Then, from Lemma 2.1, we have the inverse factorial expansion

$$P(s) = (h\kappa^\kappa)^s\left\{\sum_{j=0}^{M-1}\frac{A_j}{\Gamma(\kappa s + \vartheta' + j)} + \frac{\sigma_M(s)}{\Gamma(\kappa s + \vartheta' + M)}\right\},$$

where $\sigma_M(s) = O(1)$ as $|s| \to \infty$ uniformly in $|\arg s| \leq \pi - \epsilon, \epsilon > 0$.

Introduction of the scaled gamma function then leads to the expansion in (2.2.29) for the normalised coefficients $c_j = A_j/A_0$, where now

$$\Upsilon(s) = \frac{\Gamma^*(\kappa s + \vartheta')}{\Gamma^*(1+s)} \Gamma^*\left(\frac{1+ms}{\mu}\right) \Gamma^*\left(\frac{1+ns}{\nu}\right),$$

$$R(s) = \frac{(1+\vartheta'\zeta)^{\kappa s - \vartheta + \frac{1}{2}}}{e(1+\kappa\zeta)^{s+\frac{1}{2}}} \left(1+\frac{\kappa\zeta}{m}\right)^{(1+ms)/\mu - \frac{1}{2}} \left(1+\frac{\kappa\zeta}{n}\right)^{(1+ns)/\nu - \frac{1}{2}},$$

with $\zeta = (\kappa s)^{-1}$. The procedure is as described in the previous example and consists of expanding both sides of (2.2.29) in inverse powers of κs and making use of the expansion for the scaled gamma function.

Straightforward algebra shows that (with $\gamma_1 = -\frac{1}{12}$)

$$\Upsilon(s) = 1 - \frac{\gamma_1}{\kappa s}\left\{1 - \kappa + \kappa\left(\frac{\mu}{m} + \frac{\nu}{n}\right)\right\} + O[(\kappa s)^{-2}],$$

$$R(s) = 1 + \frac{1}{2\kappa s}\{\kappa\Lambda + \vartheta(\vartheta - 1)\} + O[(\kappa s)^{-2}],$$

where $\Lambda = (1 - \mu)/(m\mu) + (1 - \nu)/(n\nu)$. Upon equating coefficients of $(\kappa s)^{-1}$ in (2.2.29) we then obtain

$$c_1 = \tfrac{1}{2}\kappa\Lambda + \tfrac{1}{2}\vartheta(\vartheta - 1) - \gamma_1\left\{1 - \kappa + \kappa\left(\frac{\mu}{m} + \frac{\nu}{n}\right)\right\}.$$

The higher coefficients can be obtained by continuation of this process with the help of *Mathematica*. For example, in the case $\mu = 3$, $\nu = 6$, $m = 1$ and $n = 2$ we find the set of coefficients presented in Table 2.2.

2.2.5 Special Cases

To conclude our discussion on the expansion of quotients of gamma functions we present cases involving two and three gamma fuctions for which a closed-form representation for the coefficients can be given.

Table 2.2. *The coefficients c_j for $1 \leq j \leq 10$ (with $c_0 = 1$) when $\mu = 3$, $\nu = 6$, $m = 1$ and $n = 2$*

j	c_j	j	c_j
1	$\frac{5}{12}$	2	$\frac{55}{96}$
3	$\frac{4675}{3456}$	4	$\frac{752,675}{165,888}$
5	$\frac{4,365,515}{221,184}$	6	$\frac{1,680,723,275}{15,925,248}$
7	$\frac{127,975,072,225}{191,102,976}$	8	$\frac{30,074,141,972,875}{6,115,295,232}$
9	$\frac{27,096,801,917,560,375}{660,451,885,056}$	10	$\frac{6,075,102,989,917,036,075}{15,850,845,241,344}$

(a) A 'balanced' quotient when $p = q = 1$. The situation involving two gamma functions with $\alpha_1 = \beta_1 = 1$ corresponds to a 'balanced' quotient (since $\kappa = 0$) and is more straightforward than that in (2.2.27). In this case, the asymptotic expansion is well known and explicit values of the coefficients are given by [Copson (1965, p. 61); Olver (1974, p. 119); Temme (1996, p. 67)]

$$\frac{\Gamma(s+a)}{\Gamma(s+b)} \sim s^{a-b} \sum_{j=0}^{\infty} C_j(a,b) s^{-j} \qquad (2.2.30)$$

as $|s| \to \infty$ in $|\arg s| < \pi$, where the coefficients $C_j(a, b)$ are expressed in terms of the generalised Bernoulli polynomial $B_j^{(m)}(a)$ by

$$C_j(a,b) = \frac{(-)^j}{j!} (b-a)_j B_j^{(a-b+1)}(a).$$

The explicit representation of the first few coefficients is

$$C_0(a,b) = 1, \qquad C_1(a,b) = \tfrac{1}{2}(a-b)(a+b-1),$$

$$C_2(a,b) = \tfrac{1}{24}(b-a)_2[3(a+b)^2 - 7a - 5b + 2],$$

$$C_3(a,b) = \tfrac{1}{48}(b-a)_3(1-a-b)[(a+b)^2 - 3a - b],$$

$$C_4(a,b) = \tfrac{1}{5760}(b-a)_4[15(a^2+b^2)^2 - 8 - 18a + 125a^2 - 90a^3$$
$$+ 18b + 110ab - 210a^2b + 60a^3b + 5b^2 - 150ab^2$$
$$+ 60a^2b^2 - 30b^3 + 60ab^3], \dots,$$

where $(\alpha)_n = \alpha(\alpha + 1) \cdots (\alpha + n - 1)$. An alternative, more efficient expansion, in inverse even powers of $s + \tfrac{1}{2}(a + b - 1)$ (when $\mathrm{Re}(b - a) > 0$) has been given by Fields (1966) and discussed for complex a, b by Frenzen (1992); see Temme (1996, p. 68).

The special cases when $a = 0$, $b = \tfrac{1}{2}$ and $a = \tfrac{1}{2}$, $b = 0$ are worth mentioning. We have the expansions

$$\frac{\Gamma(s)}{\Gamma(s + \tfrac{1}{2})} \sim s^{-\frac{1}{2}} \sum_{j=0}^{\infty} C_j(0, \tfrac{1}{2}) s^{-j}, \qquad (2.2.31)$$

$$\frac{\Gamma(s + \tfrac{1}{2})}{\Gamma(s)} \sim s^{\frac{1}{2}} \sum_{j=0}^{\infty} (-)^j C_j(0, \tfrac{1}{2}) s^{-j} \qquad (2.2.32)$$

for $|s| \to \infty$ in $|\arg s| < \pi$, since it is readily established that $C_j(\tfrac{1}{2}, 0) = (-)^j C_j(0, \tfrac{1}{2})$. The values of the coefficients $C_j(0, \tfrac{1}{2})$ for $1 \le j \le 10$ are listed in Table 2.3, where $C_0(0, \tfrac{1}{2}) = 1$; a more detailed discussion of these coefficients can be found in McCabe (1983).

Table 2.3. *The coefficients* $C_j(0, \frac{1}{2})$
for $1 \le j \le 10$

j	$C_j(0, \frac{1}{2})$	j	$C_j(0, \frac{1}{2})$
1	$\frac{1}{8}$	2	$\frac{1}{128}$
3	$-\frac{5}{1024}$	4	$-\frac{21}{32768}$
5	$\frac{399}{262,144}$	6	$\frac{869}{4,194,304}$
7	$-\frac{39,325}{33,554,432}$	8	$-\frac{334,477}{2,147,483,648}$
9	$\frac{28,717,403}{17,179,869,184}$	10	$\frac{59,697,183}{274,877,906,944}$

(*b*) *Products of two gamma functions.* The product of two gamma functions

$$\Gamma(s+a)\Gamma(s+b)$$

and the associated 'reciprocal' product $\{\Gamma(1-a+s)\Gamma(1-b+s)\}^{-1}$ can be identified with $Q(s)$ in (2.2.7) and $P(s)$ in (2.2.1), respectively. The parameters are accordingly given by $\kappa = 2$, $h = 1$ and $\vartheta = a+b-\frac{1}{2}$. Then, from Lemmas 2.1 and 2.2, we obtain the inverse factorial expansions

$$\Gamma(s+a)\Gamma(s+b) = 2^{1-2s}\pi \left\{ \sum_{j=0}^{M-1}(-)^j A_j \Gamma(2s + \vartheta - j) \right.$$

$$\left. +\rho_M(s)\Gamma(2s + \vartheta - M) \right\} \qquad (2.2.33)$$

and

$$\frac{1}{\Gamma(1-a+s)\Gamma(1-b+s)} = 2^{2s} \left\{ \sum_{j=0}^{M-1} \frac{A_j}{\Gamma(2s + \vartheta' + j)} + \frac{\sigma_M(s)}{\Gamma(2s + \vartheta' + M)} \right\},$$

$$(2.2.34)$$

where $\sigma_M(s)$ and $\rho_M(s)$ are $O(1)$ as $|s| \to \infty$ in $|\arg s| \le \pi - \epsilon$. The coefficients A_j are the same in both expansions (2.2.33) and (2.2.34) and, with the normalised coefficients $c_j = A_j/A_0$, where $A_0 = (2\pi)^{-\frac{1}{2}}2^{1-a-b}$, we have

$$c_j = \frac{(-2)^{-j}}{j!} \prod_{r=1}^{j} \{(a-b)^2 - (r-\tfrac{1}{2})^2\} \quad (j \ge 1). \qquad (2.2.35)$$

To establish the above form of the coefficients we use the fact that

$$\frac{\Gamma(2s + \vartheta - j)}{\Gamma(2s + \vartheta)} = \frac{(-)^j}{(\vartheta' - 2s)_j},$$

where we recall that $\vartheta' = 1 - \vartheta$, together with the duplication formula for the gamma function (2.2.23) applied to $\Gamma(2s + \vartheta)$. This yields the expansion in the form

$$
\Upsilon(s) = \frac{\Gamma(s+a)\Gamma(s+b)}{\Gamma\left(s+\frac{1}{2}\vartheta\right)\Gamma\left(s+\frac{1}{2}\vartheta+\frac{1}{2}\right)}
$$

$$
= \sum_{j=0}^{M-1} \frac{c_j}{(\vartheta'-2s)_j} + \frac{(-)^M \rho_M(s)}{(\vartheta'-2s)_M}. \tag{2.2.36}
$$

Rather than expand $\Upsilon(s)$ by Stirling's formula, as in the preceding example, we proceed to express it in terms of a Gauss hypergeometric function by means of the standard summation formula [Abramowitz & Stegun (1965, p. 557)]

$$
{}_2F_1\left(\alpha, 1-\alpha; \beta; \tfrac{1}{2}\right) = \frac{2^{1-\beta}\pi^{\frac{1}{2}}\Gamma(\beta)}{\Gamma\left(\frac{1}{2}\alpha+\frac{1}{2}\beta\right)\Gamma\left(\frac{1}{2}-\frac{1}{2}\alpha+\frac{1}{2}\beta\right)} \tag{2.2.37}
$$

$$
= \frac{\Gamma\left(\frac{1}{2}\beta\right)\Gamma\left(\frac{1}{2}\beta+\frac{1}{2}\right)}{\Gamma\left(\frac{1}{2}\alpha+\frac{1}{2}\beta\right)\Gamma\left(\frac{1}{2}-\frac{1}{2}\alpha+\frac{1}{2}\beta\right)}
$$

provided $\beta \neq 0, -1, -2, \ldots$. Then, setting $\alpha = \frac{1}{2} + a - b$ and $\beta = \vartheta' - 2s$ (where $\vartheta' = \frac{3}{2} - a - b$), we find upon use of the reflection formula for the gamma function

$$
\Upsilon(s) = \frac{\Gamma\left(\frac{1}{2}\vartheta'-s\right)\Gamma\left(\frac{1}{2}+\frac{1}{2}\vartheta'-s\right)}{\Gamma(1-a-s)\Gamma(1-b-s)} H(s)
$$

$$
= {}_2F_1\left(\tfrac{1}{2}+a-b, \tfrac{1}{2}-a+b; \vartheta'-2s; \tfrac{1}{2}\right) H(s),
$$

where

$$
H(s) = \frac{-\cos\pi(2s+a+b)}{2\sin\pi(s+a)\sin\pi(s+b)} = 1 - \frac{\cos\pi(a-b)}{2\sin\pi(s+a)\sin\pi(s+b)}.
$$

Since $H(s) = 1 + O(e^{-2\pi|t|})$ as $t = \mathrm{Im}(s) \to \pm\infty$, it is easily seen from the series expansion of the hypergeometric function that when $\mathrm{Im}(s) \to \pm\infty$

$$
\Upsilon(s) = \sum_{j=0}^{M-1} \frac{(-2)^{-j}}{j!\,(\vartheta'-2s)_j} \prod_{r=1}^{j}\left\{(a-b)^2 - \left(r-\tfrac{1}{2}\right)^2\right\} + \frac{O(1)}{(\vartheta'-2s)_M}.
$$

Comparison of this form with (2.2.36), where $\rho_M(s) = O(1)$ as $\mathrm{Im}(s) \to \pm\infty$, then shows that the coefficients c_j ($1 \leq j \leq M-1$) are given by (2.2.35).

We observe that when $b = a + \frac{1}{2}$, the product

$$
\Gamma(s+a)\Gamma\left(s+a+\tfrac{1}{2}\right) = 2^{1-2s-2a}\pi^{\frac{1}{2}}\Gamma(2s+2a)
$$

by the duplication formula in (2.2.23). In this case, the coefficients c_j ($j \geq 1$) in (2.2.35) correctly reduce to zero; a similar result applies when $a = b + \frac{1}{2}$.

(*c*) *Quotients of three gamma functions.* The quotient of three gamma functions

$$\frac{\Gamma(s+a)\Gamma(s+b)}{\Gamma(s+c)}$$

is of importance in the construction of exponentially-improved asymptotic solutions of linear differential equations, particularly the confluent hypergeometric function discussed in Chapter 6. The quotient is identified with $Q(s)$ in (2.2.7) and we have $p = 1$, $q = 2$, with $\kappa = h = 1$ and $\vartheta = a + b - c$. Then, from Lemma 2.2, we find the expansion† in the form

$$\frac{\Gamma(s+a)\Gamma(s+b)}{\Gamma(s+c)} = \sum_{j=0}^{M-1}(-)^j A_j \Gamma(s+\vartheta-j) + \rho_M(s)\Gamma(s+\vartheta-M),$$

$$(2.2.38)$$

together with the expansion of the associated 'reciprocal' quotient from Lemma 2.1

$$\frac{\Gamma(1-c+s)}{\Gamma(1-a+s)\Gamma(1-b+s)} = \sum_{j=0}^{M-1}\frac{A_j}{\Gamma(s+\vartheta'+j)} + \frac{\sigma_M(s)}{\Gamma(s+\vartheta'+M)},$$

$$(2.2.39)$$

where $\rho_M(s)$ and $\sigma_M(s)$ are $O(1)$ as $|s| \to \infty$ in $|\arg s| \leq \pi - \epsilon$. The coefficients A_j in *both* expansions are given by

$$A_j = \frac{(c-a)_j(c-b)_j}{j!}.\qquad(2.2.40)$$

To establish the form of the coefficients in (2.2.40) we follow the method used by Olver (1995) in the case $c = 1$; an advantage of this approach is that it also furnishes an explicit integral representation for the remainder term in the expansion (2.2.38).

A straightforward application of Barnes' lemma [see Lemma 3.5 in §3.3.6] shows that

$$\frac{\Gamma(s+a)\Gamma(s+b)}{\Gamma(s+c)} = \frac{1}{2\pi i}\int_{-\infty i}^{\infty i}\phi(\tau)\,d\tau,$$

where

$$\phi(\tau) = \frac{\Gamma(\tau+c-a)\Gamma(\tau+c-b)\Gamma(s+\vartheta-\tau)\Gamma(-\tau)}{\Gamma(c-a)\Gamma(c-b)}\qquad(2.2.41)$$

and the path of integration is suitably indented to separate the sequences of poles at $\tau = a - c - k$ and $\tau = b - c - k$ from those at $\tau = k$ and $\tau = s + \vartheta + k$

---

† The expansion (2.2.38) is quoted in Dingle (1973, p. 15) and proved in Paris (1992a) using an approach similar to that in §2.2.5(b), and Olver (1995) only in the special case $c = 1$. A different proof of this expansion when $c \neq 1$ has been given by Bühring (2000).

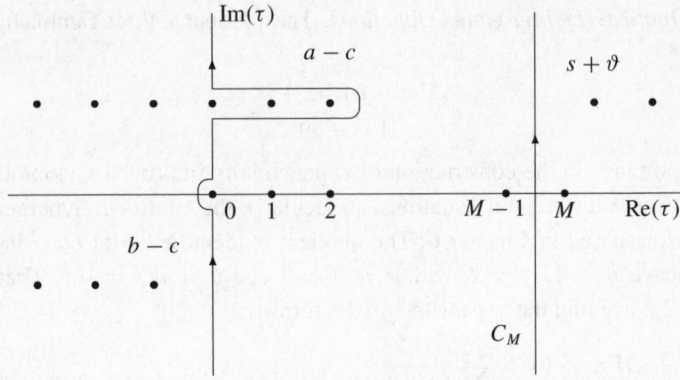

Fig. 2.1. The path of integration in the τ plane and the path C_M given by $\mathrm{Re}(\tau) = M - \delta$, $0 < \delta < 1$. The heavy points denote poles of $\phi(\tau)$.

($k = 0, 1, 2, \ldots$); see Fig. 2.1. This separation requires the temporary restriction that neither $a - c$ nor $b - c$ equals a nonnegative integer. In addition, we further suppose that M denotes a positive integer chosen such that $M > \mathrm{Re}(a-c)+\delta$ and $M > \mathrm{Re}(b-c)+\delta$, where $0 < \delta < 1$, but subject to the restriction $\mathrm{Re}(s+\vartheta) > M$ as $|s| \to \infty$ in $|\arg s| < \frac{1}{2}\pi$.

Then the only poles of $\phi(\tau)$ located between the path of integration and the path C_M defined by the vertical line $\mathrm{Re}(\tau) = M - \delta$ are those situated at $\tau = j$ ($j = 0, 1, \ldots, M - 1$) with residue

$$(-)^{j-1}(c - a)_j(c - b)_j\Gamma(s + \vartheta - j)/j!.$$

Since the modulus of the integrand is $O(|y|^{\mathrm{Re}\,(c+s)-2}e^{-2\pi|y|})$ as $\mathrm{Im}(\tau) = y \to \infty$ by Stirling's formula (2.1.8), the path can be displaced over the first M poles of $\phi(\tau)$ to produce the expansion in $|\arg s| < \frac{1}{2}\pi$

$$\frac{\Gamma(s + a)\Gamma(s + b)}{\Gamma(s + c)} = \sum_{j=0}^{M-1} \frac{(-)^j}{j!}(c - a)_j(c - b)_j\Gamma(s + \vartheta - j) + r_1(s),$$

$$(2.2.42)$$

where

$$r_1(s) = \frac{1}{2\pi i}\int_{C_M} \phi(\tau)\,d\tau. \qquad (2.2.43)$$

The remainder integral $r_1(s)$ can be shown to be $O(\Gamma(s + \vartheta - M))$ as $|s| \to \infty$ in accordance with (2.2.38). Analytic continuation with respect to $a - c$ and $b - c$ removes the restriction that neither $a - c$ nor $b - c$ equals a nonnegative integer. This is the desired expansion which shows that the coefficients A_j are given by (2.2.40); extension to the wider sector $|\arg s| \le \pi - \epsilon$ follows from Lemma 2.2.

2.3 The Asymptotic Expansion of Integral Functions

The determination of the asymptotic expansion for large $|z|$ of the integral function $f(z)$ defined by the Maclaurin series

$$f(z) = \sum_{n=0}^{\infty} g(n) z^n \qquad (|z| < \infty),$$

for suitably defined coefficients $g(n)$, is a problem of considerable interest which has been discussed by various authors; see the references given in Paris & Wood (1986, §2.3.1). The important special class of such functions, known as the generalised hypergeometric function has the Maclaurin coefficients given by

$$g(n) = \frac{1}{n!} \frac{\prod_{r=1}^{p} \Gamma(\alpha_r n + a_r)}{\prod_{r=1}^{q} \Gamma(\beta_r n + b_r)}, \tag{2.3.1}$$

where p and q are nonnegative integers; compare (2.2.1). The parameters α_r $(1 \leq r \leq p)$ and β_r $(1 \leq r \leq q)$ are supposed to be real and positive†, while a_r and b_r are arbitrary complex numbers. It will be supposed that the α_r and a_r are subject to the restriction

$$\alpha_r n + a_r \neq 0, -1, -2, \ldots \qquad (n = 0, 1, 2, \ldots \; ; \; 1 \leq r \leq p) \tag{2.3.2}$$

so that no gamma function in the numerator is singular.

From (2.2.2), we have the parameters associated with $g(n)$ given by

$$h = \prod_{r=1}^{p} \alpha_r^{\alpha_r} \prod_{r=1}^{q} \beta_r^{-\beta_r},$$

$$\vartheta = \sum_{r=1}^{p} a_r - \sum_{r=1}^{q} b_r + \tfrac{1}{2}(q - p), \qquad \vartheta' = 1 - \vartheta,$$

$$\kappa = 1 + \sum_{r=1}^{q} \beta_r - \sum_{r=1}^{p} \alpha_r.$$

If it is supposed that α_r and β_r are such that $\kappa > 0$ then, subject to (2.3.2), the generalised hypergeometric, or Wright, function defined by

$$_p f_q(z) = \sum_{n=0}^{\infty} \frac{\prod_{r=1}^{p} \Gamma(\alpha_r n + a_r)}{\prod_{r=1}^{q} \Gamma(\beta_r n + b_r)} \frac{z^n}{n!} \qquad (\kappa > 0; \; |z| < \infty) \tag{2.3.3}$$

is uniformly and absolutely convergent for all finite z. If $\kappa = 0$, the sum on the right-hand side of (2.3.3) has a finite radius of convergence equal to h^{-1}, while for $\kappa < 0$ the sum is divergent for all nonzero values of z. The parameter κ will be seen to play a critical role in the asymptotic theory of $_p f_q(z)$ by determining the sectors in the z plane in which the behaviour of $_p f_q(z)$ is either exponentially

---

† The case of complex α_r and β_r has been considered in Wright (1940).

large or algebraic in character as $|z| \to \infty$. We note that the familiar generalised hypergeometric function, denoted by ${}_pF_q(z)$ [Slater (1966, p. 40)], corresponds to (2.3.3) with $\alpha_r = \beta_r = 1$, namely

$$
{}_pF_q \begin{pmatrix} a_1, \ldots, a_p; \\ b_1, \ldots, b_q; \end{pmatrix} z \end{pmatrix} = C \sum_{n=0}^{\infty} \frac{\prod_{r=1}^{p} \Gamma(n + a_r)}{\prod_{r=1}^{q} \Gamma(n + b_r)} \frac{z^n}{n!},
$$

where $C = \prod_{r=1}^{q} \Gamma(b_r) / \prod_{r=1}^{p} \Gamma(a_r)$.

An account of the derivation of the asymptotic expansion of ${}_pf_q(z)$ for large $|z|$ is given in Paris & Wood (1986, §2.3); see also Braaksma (1963, §12) where the function ${}_pf_q(z)$ is denoted by ${}_p\psi_q(z)$. A detailed survey of the particular function corresponding to $p = 0, q = 1$ has been given in Gerenflo *et al.* (1999). Here we content ourselves with a brief statement of the results. The algebraic expansion of ${}_pf_q(z)$ follows from the Mellin-Barnes integral representation (when $\kappa > 0$)

$$
{}_pf_q(z) = \frac{1}{2\pi i} \int_{-\infty i}^{\infty i} \Gamma(-s)\Gamma(s + 1)g(s)(ze^{\mp \pi i})^s ds, \qquad (2.3.4)
$$

where the path of integration is indented near $s = 0$ to separate† the poles at $s = k$ from those of $\Gamma(1 + s)g(s)$ situated at

$$
s = -(a_r + k)/\alpha_r, \qquad k = 0, 1, 2, \ldots \ (1 \le r \le p). \qquad (2.3.5)
$$

In general there will be p such sequences of simple poles though, depending on the values of α_r and a_r, some of these poles could be multiple poles or even ordinary points if any of the $\Gamma(\beta_r s + b_r)$ are singular there. Application of the convergence Rule 1 in §2.4 shows that the above integral defines ${}_pf_q(z)$ only in the sector $|\arg(-z)| < \frac{1}{2}\pi(2 - \kappa)$. Displacement of the contour to the left over the poles of $\Gamma(1 + s)g(s)$ then yields the algebraic expansion of ${}_pf_q(z)$ valid in this sector. If it is assumed that the parameters are such that the poles in (2.3.5) are all simple we obtain the algebraic expansion given by $H(ze^{\mp \pi i})$, where

$$
H(z) = \sum_{m=1}^{p} \alpha_m^{-1} z^{-a_m/\alpha_m} S_{p,q}(z; m) \qquad (2.3.6)
$$

and $S_{p,q}(z; m)$ denotes the formal asymptotic sum

$$
S_{p,q}(z; m) = \sum_{k=0}^{\infty} \frac{(-)^k}{k!} \Gamma\left(\frac{k + a_m}{\alpha_m}\right) \frac{\prod_{r=1}^{p} \Gamma(a_r - \alpha_r(k + a_m)/\alpha_m)}{\prod_{r=1}^{q} \Gamma(b_r - \beta_r(k + a_m)/\alpha_m)} z^{-k/\alpha_m},
$$

$$(2.3.7)$$

with the prime indicating the omission of the term corresponding to $r = m$ in the product. This expression consists of p expansions each with the leading behaviour z^{-a_m/α_m} $(1 \le m \le p)$ and yields the algebraic expansion of ${}_pf_q(z)$ valid as

---

† This is always possible when the condition (2.3.2) is satisfied.

$|z| \to \infty$ in the sector $|\arg(-z)| < \frac{1}{2}\pi(2-\kappa)$. The upper or lower sign in $H(ze^{\mp\pi i})$ is chosen according as $\arg z > 0$ or $\arg z < 0$, respectively. When the parameters α_r and a_r are such that some of the poles are of higher order, the expansion (2.3.6) becomes nugatory and the residues must then be evaluated according to the multiplicity of the poles concerned.

The exponential expansion of $_p f_q(z)$ is represented in terms of the formal asymptotic sum

$$E(z) = X^\vartheta e^X \sum_{k=0}^{\infty} A_k X^{-k}, \qquad X = \kappa(hz)^{1/\kappa}. \tag{2.3.8}$$

The coefficients A_k are those appearing in the inverse factorial expansion of $g(s)$ given by† (see Lemma 2.1)

$$g(s) = \kappa(h\kappa^\kappa)^s \left\{ \sum_{j=0}^{M-1} \frac{A_j}{\Gamma(\kappa s + \vartheta' + j)} + \frac{O(1)}{\Gamma(\kappa s + \vartheta' + M)} \right\}$$

for $|s| \to \infty$ uniformly in $|\arg s| \leq \pi - \epsilon$, where

$$A_0 = (2\pi)^{\frac{1}{2}(p-q)} \kappa^{-\frac{1}{2}-\vartheta} \prod_{r=1}^{p} \alpha_r^{a_r - \frac{1}{2}} \prod_{r=1}^{q} \beta_r^{\frac{1}{2}-b_r}. \tag{2.3.9}$$

These coefficients are independent of s and depend only on the parameters p, q, α_r, β_r, a_r and b_r. As pointed out in §2.2.1, their actual evaluation turns out to be the most difficult part of the theory.

Since $_p f_q(z)$ in (2.3.3) is an integral function of z when $\kappa > 0$, we need only consider the asymptotic expansion in $|\arg z| \leq \pi$. The asymptotic character of $_p f_q(z)$ depends critically upon the value of the parameter κ: three different cases arise according as (i) $0 < \kappa < 2$, (ii) $\kappa = 2$ and (iii) $\kappa > 2$.

Case (i): $0 < \kappa < 2$. The z plane is divided into two sectors, with a common vertex at $z = 0$, by the rays (the anti-Stokes lines) $\arg z = \pm\frac{1}{2}\pi\kappa$. In the sector $|\arg z| < \frac{1}{2}\pi\kappa$, the asymptotic character of $_p f_q(z)$ is exponentially large while in the complementary sector $|\arg(-z)| < \frac{1}{2}\pi(2-\kappa)$, $_p f_q(z)$ is algebraic in character. As $|z| \to \infty$ we obtain the expansion (in the Poincaré sense)

$$_p f_q(z) \sim \begin{cases} E(z) + H(ze^{\mp\pi i}) & \text{in} \quad |\arg z| \leq \frac{1}{2}\pi\kappa \\ H(ze^{\mp\pi i}) & \text{in} \quad |\arg(-z)| < \frac{1}{2}\pi(2-\kappa), \end{cases} \tag{2.3.10}$$

where $E(z)$ is defined in (2.3.8) and, in the case when all the poles of $g(s)$ at the points (2.3.5) are simple, $H(z)$ is given by the p asymptotic sums in (2.3.6). The upper or lower sign in $H(ze^{\mp\pi i})$ is chosen according as z lies in the upper or lower half-plane, respectively. It is seen that $\arg z = 0$ is a Stokes line, where the

---

† For convenience, a factor κ has been extracted from the coefficients A_k in Lemma 2.1.

algebraic expansion is of maximum degree of subdominancy. In the neighbourhood of this ray, the leading terms of $H(ze^{\mp\pi i})$ change by the factor $\exp(2\pi i a_m/\alpha_m)$ $(1 \leq m \leq p)$. This is the Stokes phenomenon where the coefficients of the subdominant algebraic terms change in the neighbourhood of arg $z = 0$ in order to preserve the single-valuedness of $_p f_q(z)$ as z describes a circuit about the origin.

If $p = 0$ in (2.3.1) (so that $\kappa > 1$), then $g(s)$ has no poles in the finite s plane and $H(z)$ is accordingly identically zero. In this case, the asymptotic expansion of $_p f_q(z)$ for $1 < \kappa < 2$ is described by $E(z)$ as $|z| \to \infty$ in $|\arg z| \leq \pi$, and is therefore exponentially small in the sector $|\arg(-z)| < \frac{1}{2}\pi(2 - \kappa)$.

Case (ii): $\kappa = 2$. The rays arg $z = \pm\frac{1}{2}\pi\kappa$ now coincide with the negative real axis. It follows that $_p f_q(z)$ is exponentially large in character as $|z| \to \infty$ except in the neighbourhood of arg $z = \pm\pi$, where it is of the mixed type with the algebraic expansion becoming asymptotically significant. In this case we have

$$_p f_q(z) \sim E(z) + E(ze^{\mp 2\pi i}) + H(ze^{\mp\pi i}) \qquad (|\arg z| \leq \pi), \qquad (2.3.11)$$

where again the upper or lower sign is chosen according as arg $z > 0$ or arg $z < 0$, respectively. We remark that the expansions $E(ze^{\mp 2\pi i})$ only become significant (in the Poincaré sense) in the neighbourhood of arg $z = \pm\pi$.

Case (iii): $\kappa > 2$. In this case the asymptotic behaviour of $_p f_q(z)$ is exponentially large for all values of arg z. For $|z| \to \infty$ we have

$$_p f_q(z) \sim \sum_{r=-P}^{P} E(ze^{2\pi i r}) \qquad (|\arg z| \leq \pi), \qquad (2.3.12)$$

where P is chosen such that $2P + 1$ is the smallest odd integer satisfying $2P + 1 > \frac{1}{2}\kappa$. The exponential sums $E(ze^{2\pi i r})$ are exponentially large as $|z| \to \infty$ for values of arg z satisfying $|\arg z + 2\pi r| < \frac{1}{2}\pi\kappa$ and $|\arg z| \leq \pi$.

2.3.1 An Example

As an application of the above theory let us consider the asymptotic expansion of the double Laplace-type integral

$$I(\lambda) = \lambda^{1/\mu+1/\nu} \int_0^\infty \int_0^\infty x^\alpha y^\beta e^{-\lambda f(x,y)} \, dx dy$$

for $\lambda \to +\infty$, where the phase function $f(x, y)$ is given by

$$f(x, y) = x^\mu - cx^m y^n + y^\nu.$$

The parameters $\mu, \nu, m, n, 1+\alpha$ and $1+\beta$ are assumed positive with c an arbitrary complex constant, so that for convergence we require

$$1 - \frac{m}{\mu} - \frac{n}{\nu} > 0. \qquad (2.3.13)$$

Integrals of this type (with $c < 0$) are discussed at length in Chapter 7, where it is shown that the above condition corresponds to the single internal point (m, n) lying in front of the back face of the associated Newton diagram; see §7.3.2.

In §7.7 it is shown that $I(\lambda)$ can be expressed as a Mellin-Barnes integral [cf. (7.7.2)], which in turn can be evaluated as the absolutely convergent series

$$I(\lambda) = \frac{\lambda^{-\alpha/\mu - \beta/\nu}}{\mu\nu} \sum_{k=0}^{\infty} \Gamma\left(\frac{1 + \alpha + mk}{\mu}\right) \Gamma\left(\frac{1 + \beta + nk}{\nu}\right) \frac{(c\lambda^\kappa)^k}{k!}. \quad (2.3.14)$$

This series is an integral function of the type in (2.3.3) and is associated with the parameters

$$\kappa = 1 - \frac{m}{\mu} - \frac{n}{\nu}, \quad h = (m/\mu)^{m/\mu}(n/\nu)^{n/\nu}, \quad X = \kappa\lambda(hc)^{1/\kappa}$$

and, when $\alpha = \beta = 0$,

$$\vartheta = \frac{1}{\mu} + \frac{1}{\nu} - 1, \quad A_0 = 2\pi\kappa^{-\frac{1}{2} - \vartheta}(m/\mu)^{\frac{1}{\mu} - \frac{1}{2}}(n/\nu)^{\frac{1}{\nu} - \frac{1}{2}}.$$

Since $0 < \kappa < 1$, it then follows from (2.3.10) that $I(\lambda)$ possesses the asymptotic expansion (in the Poincaré sense) as $\lambda \to +\infty$ given by†

$$I(\lambda) \sim \begin{cases} \hat{H}(\lambda) & \text{in } |\arg(-c)| < \frac{1}{2}\pi(2 - \kappa) \\ \hat{E}(\lambda) + \hat{H}(\lambda) & \text{in } |\arg c| \leq \frac{1}{2}\pi\kappa. \end{cases} \quad (2.3.15)$$

From (2.3.6) (when it is assumed the parameters are such that all poles in (2.3.5) are simple) the algebraic expansion is $\hat{H}(\lambda) = H_{\alpha,\beta}(\lambda)$, where

$$H_{\alpha,\beta}(\lambda) = \lambda^{-\alpha/\mu - \beta/\nu} \left\{ \frac{1}{m\nu} \sum_{k=0}^{\infty} \frac{(-)^k}{k!} \Gamma\left(\frac{1 + \alpha + \mu k}{m}\right) \right.$$

$$\times \Gamma\left(\frac{m(1 + \beta) - n(1 + \alpha + \mu k)}{m\nu}\right)(-c\lambda^\kappa)^{-(1+\alpha+\mu k)/m}$$

$$+ \frac{1}{\mu n} \sum_{k=0}^{\infty} \frac{(-)^k}{k!} \Gamma\left(\frac{1 + \beta + \nu k}{n}\right)$$

$$\left. \times \Gamma\left(\frac{n(1 + \alpha) - m(1 + \beta + \nu k)}{\mu n}\right)(-c\lambda^\kappa)^{-(1+\beta+\nu k)/n} \right\}. \quad (2.3.16)$$

The minus sign in the arguments $-c\lambda^\kappa$ of these expansions is to be interpreted as $e^{\mp\pi i}$ according as $\arg(c\lambda^\kappa) > 0$ or < 0, respectively. When some of the poles on the left of the contour in (2.3.4) are higher-order poles, then it is necessary to

---

† The expansion (2.3.15) also holds when λ is complex; see Paris & Liakhovetski (2000a) for details.

evaluate the residues of the associated integrand accordingly; this will lead to the generation of terms involving log λ in the algebraic expansion.

The exponential expansion $\hat{E}(\lambda)$ can be obtained from (2.3.8) and (2.3.9) in the form

$$\hat{E}(\lambda) = x_0^\alpha y_0^\beta \frac{A_0}{\mu\nu} X^\vartheta e^X \sum_{j=0}^{\infty} c_j X^{-j}, \qquad (2.3.17)$$

where† x_0, y_0 are the coordinates of the saddle point defined in (2.3.18) below and $c_0 = 1$. The coefficients c_j are defined as the coefficients in the inverse factorial expansion of the quotient of gamma functions in the sum on the right-hand side of (2.3.14) and the procedure for their evaluation has been discussed in §2.2.4. In the particular case $\mu = 3, \nu = 6, m = 1, n = 2$ and $\alpha = \beta = 0$, for example, we have the exponential expansion

$$\hat{E}(\lambda) = \frac{\pi c^{-\frac{3}{2}}}{\lambda^{\frac{1}{2}}} \exp(\lambda c^3/27) \sum_{j=0}^{\infty} c_j (\lambda c^3/27)^{-j},$$

where the corresponding coefficients c_j for $0 \le j \le 10$ are listed in Table 2.2.

We note that, when $c > 0$, the phase function $f(x, y)$ has a saddle point, corresponding to $\partial f/\partial x = \partial f/\partial y = 0$, in the positive quadrant at the point‡

$$x_0 = (m/\mu)^{1/\mu}(hc)^{1/(\mu\kappa)}, \quad y_0 = (n/\nu)^{1/\nu}(hc)^{1/(\nu\kappa)}, \qquad (2.3.18)$$

and that $f(x_0, y_0) = -\kappa(hc)^{1/\kappa}$. The leading term of the exponential expansion in (2.3.17) is then seen to agree with the standard saddle-point approximation for double integrals given by [see, for example, Wong (1989, p. 461)]

$$\lambda^{-1/\mu-1/\nu} I(\lambda) \sim \frac{2\pi}{\lambda} [\det f''(x_0, y_0)]^{-\frac{1}{2}} x_0^\alpha y_0^\beta e^{-\lambda f(x_0, y_0)},$$

where the Hessian $\det f'' = f_{xx} f_{yy} - f_{xy}^2$ is evaluated at the saddle point. We remark, however, that the exponential expansion in (2.3.17) is valid in the sector $|\arg c| \le \frac{1}{2}\pi\kappa$, where the saddle point (x_0, y_0) moves into the *complex* x, y planes when c becomes complex. The algebraic expansion $\hat{H}(\lambda)$ originates from the neighbourhood of the origin.

Similar arguments have been given in Paris & Liakhovetski (2000a) to derive the expansion of the n-dimensional integral for large complex λ

$$I(\lambda) = \lambda^d \int_0^\infty \cdots \int_0^\infty x_1^{\alpha_1} x_2^{\alpha_2} \cdots x_n^{\alpha_n} e^{-\lambda f(x_1, \ldots, x_n)} dx_1 \ldots dx_n, \qquad (2.3.19)$$

---

† We remark that to make apparent the factor $x_0^\alpha y_0^\beta$ we have employed the values of ϑ and A_0 corresponding to $\alpha = \beta = 0$.

‡ There are other saddle points in the complex x, y planes.

where $d = \sum_{j=1}^{n}(1/\mu_j)$ and the phase function

$$f(x_1, \ldots, x_n) = \sum_{j=1}^{n} x_j^{\mu_j} - cx_1^{m_1} x_2^{m_2} \cdots x_n^{m_n}$$

corresponds to a single internal point in the associated Newton diagram; see §7.3.2. We suppose that the exponents (not necessarily integer values) satisfy $\mu_j > 0$, $m_j > 0$ and $\text{Re}(\alpha_j) > -1$ $(1 \leq j \leq n)$. To secure convergence when $\text{Re}(c) > 0$ we require that

$$\kappa = 1 - \sum_{j=1}^{n} \frac{m_j}{\mu_j} > 0,$$

so that the internal point $(m_1, \ldots, m_n)$ in the Newton diagram is situated in front of the back face (which is a hyperplane in n dimensions). When $n = 1$, we have

$$I(\lambda) = \lambda^{1/\mu_1} \int_0^\infty x^{\alpha_1} \exp\{-\lambda(x^{\mu_1} - cx^{m_1})\}dx$$

$$= \frac{\lambda^{-\alpha_1/\mu_1}}{\mu_1} \text{Fi}(m_1/\mu_1, (1+\alpha_1)/\mu_1; c\lambda^\kappa),$$

where $\kappa = 1 - m_1/\mu_1$ and $\text{Fi}(a, b; z)$ denotes Faxén's integral; see Olver (1974, p. 332) and §5.5. Accordingly, the integral $I(\lambda)$ in (2.3.19) can be regarded as an n-dimensional extension of Faxén's integral. Special cases of this integral when $n = 1$ and $\alpha_1 = 0$ were first studied asymptotically in Brillouin (1916), Burwell (1924), and more generally in Bakhoom (1933), by the method of steepest descent. A study of a generalisation of the one-dimensional Faxén integral with more than one internal point in the phase function f has recently been given in Kaminski & Paris (1997). In the n-dimensional case, it is found that $I(\lambda)$ possesses the asymptotic expansion given in (2.3.15) as $\lambda \to +\infty$, where κ is as defined above and $\hat{E}(\lambda)$ and $\hat{H}(\lambda)$ are the n-dimensional analogues of the exponential and algebraic expansions in (2.3.17) and (2.3.16).

This approach can also be extended to deal with the more general integral

$$J(\lambda) = \lambda^{1/\mu+1/\nu} \int_0^\infty \int_0^\infty g(x, y)e^{-\lambda f(x,y)}\,dx\,dy,$$

where $g(x, y)$ is a smoothly varying amplitude function chosen such that the integral converges absolutely. It is found that the expansion of $J(\lambda)$ for $\lambda \to +\infty$ is again given by (2.3.15), with the algebraic expansion now taking the form

$$\hat{H}(\lambda) = \sum_{p,q} g_{pq}^{(0)} H_{p,q}(\lambda), \tag{2.3.20}$$

where $g(x, y) = \sum_{p,q} g_{pq}^{(0)} x^p y^q$ in some disc centred at the origin, p and q are nonnegative integers and $H_{p,q}(\lambda)$ is defined in (2.3.16). The exponential expansion is

$$\hat{E}(\lambda) = \frac{A_0}{\mu\nu} X^\vartheta e^X \sum_{j=0}^{\infty} B_j X^{-j}, \tag{2.3.21}$$

where

$$B_j = \sum_{r=0}^{2j} \sum_{k=0}^{2j-r} (-)^{r+k} g_{rk}^{(s)} x_0^r y_0^k \, C_j^{(r,k)}$$

$$C_j^{(r,k)} = \sum_{p=0}^{r} \sum_{q=0}^{k} (-)^{p+q} \binom{r}{p} \binom{k}{q} c_j^{(p,q)}$$

and we have put $g(x, y) = \sum_{p,q} g_{pq}^{(s)} (x - x_0)^p (y - y_0)^q$ in the neighbourhood of the saddle point (x_0, y_0). To denote the dependence on the integers p and q, we have written the coefficients in the inverse factorial expansion of the quotient of gamma functions

$$\frac{1}{s!} \Gamma\left(\frac{1 + p + ms}{\mu}\right) \Gamma\left(\frac{1 + q + ns}{\nu}\right) \tag{2.3.22}$$

as $c_j \equiv c_j^{(p,q)}$, with $c_0^{(p,q)} = 1$. The details of these calculations can be found in Paris & Liakhovetski (2000b).

Higher dimensional Laplace integrals with the phase function f containing only one internal point in the Newton diagram can also be dealt with in the same manner by employing the asymptotics of the integral $I(\lambda)$ in (2.3.19) given in Paris & Liakhovetski (2000a), but the computational effort involved in the calculation of the coefficients in the exponential expansion rapidly increases with the dimension of the integral. The same procedure, of course, can also be brought to bear on one-dimensional integrals of the form

$$J(\lambda) = \lambda^{1/\mu} \int_0^\infty g(x) e^{-\lambda(x^\mu - cx^m)} dx,$$

where $\mu > m > 0$ so that the parameter $\kappa = 1 - m/\mu$ satisfies $0 < \kappa < 1$. The analysis of this integral is carried out in terms of the one-dimensional analogue of $I(\lambda)$ (which corresponds to Faxén's integral). It is found that the expansion of $J(\lambda)$ for $\lambda \to +\infty$ is again described by (2.3.15), where the algebraic expansion is now given by the one-dimensional analogue of that in (2.3.20), namely $\hat{H}(\lambda) = \sum_p g_p^{(0)} H_p(\lambda)$, with

$$H_\alpha(\lambda) = \frac{\lambda^{-\alpha/\mu}}{m} \sum_{k=0}^{\infty} \frac{(-)^k}{k!} \Gamma\left(\frac{1 + \alpha + \mu k}{m}\right) (-c\lambda^\kappa)^{-(1+\alpha+\mu k)/m}.$$

The exponential expansion has the same form as (2.3.21), but with the factor $A_0/\mu\nu$ replaced by A_0/μ and the parameters now defined by $\kappa = 1 - m/\mu$,

$\vartheta = 1/\mu - \frac{1}{2}$, $h = (m/\mu)^{m/\mu}$ and $A_0 = (2\pi)^{\frac{1}{2}}\kappa^{-\frac{1}{2}-\vartheta}(m/\mu)^{\vartheta}$. The coefficients B_j in the one-dimensional case are given by

$$B_j = \sum_{r=0}^{2j}(-)^r g_r^{(s)} x_0^r C_j^{(r)}, \qquad C_j^{(r)} = \sum_{p=0}^{r}(-)^p \binom{r}{p} c_j^{(p)},$$

where $x_0 = (cm/\mu)^{1/(\mu\kappa)}$ denotes the saddle point of the phase function that is situated in $x > 0$ when $c > 0$ and the coefficients $c_j^{(p)}$ appear in the inverse factorial expansion of the one-dimensional analogue of (2.3.22), namely the ratio $\Gamma((1 + p + ms)/\mu)/s!$ for large s.

2.4 Convergence of Mellin-Barnes Integrals

In this section we develop convenient rules for determining, at a glance, the convergence of a class of frequently occurring Mellin-Barnes integrals. These integrals typically have the form

$$\frac{1}{2\pi i}\int_C g(s)z^s ds, \tag{2.4.1}$$

where the contour C is usually either a loop in the complex s plane, a vertical line indented to avoid certain poles of the integrand, or a curve midway between these two, in the sense of avoiding certain poles of the integrand and tending to infinity in certain fixed directions.† For ease of reference, let us label a contour C that is a loop beginning at $+\infty$, travelling in a clockwise sense and returning to $+\infty$ (with suitable indentations to avoid poles) as C_1, and use the label C_2 to indicate a (possibly) indented vertical line; see Fig. 2.2.

The integrand $g(s)$ in (2.4.1) is assumed to have the form

$$g(s) = \frac{\prod_{r=1}^{m}\Gamma(b_r - \beta_r s)\prod_{r=1}^{n}\Gamma(1 + \alpha_r s - a_r)}{\prod_{r=m+1}^{q}\Gamma(1 + \beta_r s - b_r)\prod_{r=n+1}^{p}\Gamma(a_r - \alpha_r s)} \tag{2.4.2}$$

$$(0 \le m \le q, \ 0 \le n \le p).$$

The parameters α_r, β_r are restricted to be positive numbers and it is assumed that the parameters are such that the path C may be chosen to separate the sequences of poles resulting from $\Gamma(b_r - \beta_r s)$ $(1 \le r \le m)$ from those of $\Gamma(1 + \alpha_r s - a_r)$ $(1 \le r \le n)$. The coefficients α_r, β_r of the integration variable in the gamma functions in (2.4.2) are termed the *multiplicities* of the associated gamma functions. For later convenience, we also set

$$h = \prod_{r=1}^{p}\alpha_r^{\alpha_r}\prod_{r=1}^{q}\beta_r^{-\beta_r}; \tag{2.4.3}$$

---

† Integrals of this type are known as the Fox H-function; see also Braaksma (1963, p. 239). When $\alpha_r = \beta_r = 1$, the integral reduces to the Meijer G-function.

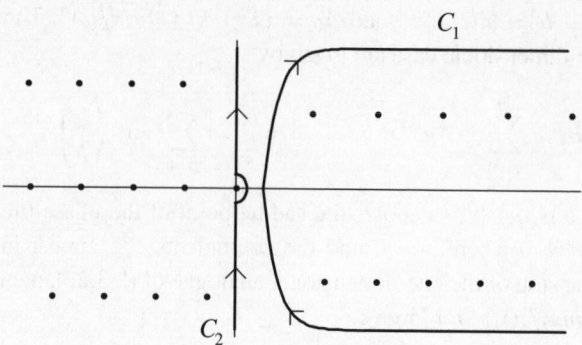

Fig. 2.2. The contours of integration C_1 and C_2 in the complex s plane. The heavy points represent poles.

recall (2.2.2). In order to determine convergence of the Mellin-Barnes integral (2.4.1), we shall need to examine the behaviour of the integrand $g(s)z^s$ for $|s| \to \infty$ along the contours C_k, $k = 1, 2$. In the process, we will obtain two 'rules of thumb' for determining the domains of convergence by simple inspection.

Let us begin by recalling (2.1.1),

$$\log \Gamma(z) = \left(z - \tfrac{1}{2}\right) \log z - z + \tfrac{1}{2} \log 2\pi + O(z^{-1})$$

for $|z| \to \infty$ in $|\arg z| < \pi$, where the O-estimate is immediate from (2.1.5). With an application of the reflection formula (2.1.20), this permits us to write, for $\beta > 0$ and arbitrary complex α,

$$\log \Gamma(\alpha + \beta s) = \left(\alpha + \beta s - \tfrac{1}{2}\right) \log \beta s - \beta s + O(1),$$
$$\log \Gamma(\alpha - \beta s) = -\log \sin \pi(\alpha - \beta s) + \left(\alpha - \beta s - \tfrac{1}{2}\right) \log \beta s + \beta s + O(1)$$

for $|s| \to \infty$ in $|\arg(\alpha \pm \beta s)| < \pi$. With $s = Re^{i\theta}$, the real part of $\log \Gamma(\alpha \pm \beta s)$ as $R \to \infty$ is therefore given by

$$\left. \begin{aligned} \log |\Gamma(\alpha + \beta s)| &\sim \beta R \cos \theta \log \beta R - \beta R(\theta \sin \theta + \cos \theta) \\ &\quad + \left(\mathrm{Re}(\alpha) - \tfrac{1}{2}\right) \log \beta R, \\ \log |\Gamma(\alpha - \beta s)| &\sim -\beta R \cos \theta \log \beta R + \beta R(\theta \sin \theta + \cos \theta) \\ &\quad + \left(\mathrm{Re}(\alpha) - \tfrac{1}{2}\right) \log \beta R - \log |\sin \pi(\alpha - \beta s)|. \end{aligned} \right\} \quad (2.4.4)$$

Note that for $s = Re^{i\theta}$, the term $\log |\sin \pi(\alpha - \beta s)| \sim \pi \beta R |\sin \theta|$ for $\sin \theta \neq 0$, and is bounded otherwise.

Along the loop C_1, we have $\theta \to 0$ for large R and the estimates (2.4.4) accordingly simplify considerably to

$$\log |\Gamma(\alpha \pm \beta s)| = \pm \beta R \log R + O(R).$$

Thus, the term $g(s)$ in (2.4.1) has the leading behaviour as $R \to \infty$ given by

$$\log |g(s)| = \left\{ \sum_{r=1}^{p} \alpha_r - \sum_{r=1}^{q} \beta_r \right\} R \log R + O(R),$$

so that, for all finite $z \neq 0$, the integrand is seen to decay exponentially provided

$$\kappa = \sum_{r=1}^{q} \beta_r - \sum_{r=1}^{p} \alpha_r > 0. \tag{2.4.5}$$

It therefore follows that the integral (2.4.1) with $C = C_1$ converges absolutely when this condition is satisfied.

For the situation where $\kappa = 0$, finer estimates must be made. For this case, we find that $\log |g(s)|$ has the large-R behaviour along C_1

$$\left\{ \sum_{r=1}^{p} \alpha_r \log \alpha_r - \sum_{r=1}^{q} \beta_r \log \beta_r \right\} R + O(\log R) = R \log h + O(\log R)$$

which, when coupled with the large-R behaviour of z^s, yields an integrand for (2.4.1) with the asymptotic behaviour

$$\log |g(s)z^s| = R \log(h|z|) + O(\log R) \quad (R \to \infty).$$

In order to ensure an exponentially decaying integrand, we see that z must be restricted to the domain $|z| < h^{-1}$. Should it happen that $h = 1$, then convergence will be possible only inside the unit disc $|z| < 1$.

A remaining possibility for the case $\kappa = 0$ is that $\log(h|z|)$ vanishes. In this event, it is the terms of order $\log R$ that control convergence of the integral. For these circumstances, $\log |g(s)z^s|$ has the large-R behaviour along C_1

$$\text{Re} \left\{ \sum_{r=1}^{m} (b_r - \tfrac{1}{2}) + \sum_{r=1}^{n} (\tfrac{1}{2} - a_r) - \sum_{r=m+1}^{q} (\tfrac{1}{2} - b_r) - \sum_{r=n+1}^{p} (a_r - \tfrac{1}{2}) \right\} \log R$$

$$= (\tfrac{1}{2} - \text{Re}(\vartheta)) \log R,$$

where

$$\vartheta = \left\{ \sum_{r=1}^{p} a_r - \sum_{r=1}^{q} b_r \right\} + \tfrac{1}{2}(q - p + 1).$$

Convergence then depends on the algebraic decay of the integrand, which is controlled by the behaviour $R^{\frac{1}{2} - \vartheta}$.

Along the path C_2, we have $\theta = \pm \frac{1}{2}\pi$ as $s \to \infty$. From (2.4.4), we then obtain the dominant behaviour of $g(s)$ given by

$$\log |g(s)| = -\tfrac{1}{2}\pi R \left\{ \sum_{r=1}^{n} \alpha_r + \sum_{r=1}^{m} \beta_r - \sum_{r=n+1}^{p} \alpha_r - \sum_{r=m+1}^{q} \beta_r \right\} + O(\log R).$$

When coupled with the large-R behaviour of z^s, we find the integrand of (2.4.1) taken along C_2 has the asymptotic behaviour

$$\log|g(s)z^s| = -\tfrac{1}{2}\pi R \left\{ \sum_{r=1}^{n} \alpha_r + \sum_{r=1}^{m} \beta_r - \sum_{r=n+1}^{p} \alpha_r - \sum_{r=m+1}^{q} \beta_r \right\}$$

$$\mp R \arg z + O(\log R),$$

where the upper choice of sign is taken when $\theta = \tfrac{1}{2}\pi$, and the lower choice of sign taken when $\theta = -\tfrac{1}{2}\pi$. For the integral to converge absolutely over the whole path C_2 therefore requires that

$$|\arg z| < \tfrac{1}{2}\pi \left\{ \sum_{r=1}^{n} \alpha_r + \sum_{r=1}^{m} \beta_r - \sum_{r=n+1}^{p} \alpha_r - \sum_{r=m+1}^{q} \beta_r \right\}. \qquad (2.4.6)$$

These convergence results can be encapsulated into two easily applied rules that, at a glance, can determine the convergence of (2.4.1) taken over C_1 or C_2; see Paris & Wood (1986, §2.1.3). The first rule is a reformulation of (2.4.6), whilst the second is a reformulation of (2.4.5) and special cases following that criterion.

Lemma 2.4. [Rule 1]. *The Mellin-Barnes integral (2.4.1) taken along C_2, the vertical line* $\mathrm{Re}(s) = 0$, *suitably indented (if necessary) to avoid poles of the integrand (2.4.2), converges in the sector defined by*

$$|\arg z| < \tfrac{1}{2}\pi \left\{ \begin{array}{cc} \textit{number of} & \textit{number of} \\ \textit{gamma functions in} & - \textit{ gamma functions in} \\ \textit{the numerator} & \textit{the denominator} \end{array} \right\},$$

where each gamma function in the numerator and denominator of (2.4.2) is counted according to its multiplicity.

Lemma 2.5. [Rule 2]. *The Mellin-Barnes integral (2.4.1) taken along the loop C_1, beginning at $+\infty$ and encircling in the negative sense only the poles of the integrand located at $\beta_r s = b_r + k$, $(1 \le r \le m)$ and k a nonnegative integer, converges for all finite, nonzero values of z provided*

$$\left\{ \begin{array}{cc} \textit{number of gamma functions} & \textit{number of gamma functions} \\ \textit{with negative multiplicity in} & \textit{with positive multiplicity in} \\ \textit{the numerator and with} & - \quad \textit{the numerator and with} \\ \textit{positive multiplicity in} & \textit{negative multiplicity in} \\ \textit{the denominator} & \textit{the denominator} \end{array} \right\} > 0,$$

each gamma function being counted according to its multiplicity. When this quantity vanishes, the integral converges in $|z| < h^{-1}$.

A simple mnemonic summarises the content of these two rules. Consider the following diagram:

$$
\frac{\displaystyle\prod_{r=1}^{m}\Gamma(b_r - \beta_r s)\;\left|\;\displaystyle\prod_{r=1}^{n}\Gamma(1+\alpha_r s - a_r)\right.}{\displaystyle\prod_{r=m+1}^{q}\Gamma(1+\beta_r s - b_r)\;\left|\;\displaystyle\prod_{r=n+1}^{p}\Gamma(a_r - \alpha_r s)\right.}\;\cdot\; C_2
$$
$$C_1$$

For the integral taken over the loop C_1, we use the vertical line to separate components of the integrand: gamma functions to the left of the line are balanced against those appearing on the right of the line, counted in each case according to their multiplicities. For the integral taken over the vertical contour C_2, we use the horizontal line in our deliberations, balancing the gamma functions above the line against those below it, again counted according to their multiplicities.

We observe that similar convergence rules can be constructed for other choices of integration contours C. By way of illustration, if loops that began and ended at $-\infty$ were of interest, then a slight modification of the argument leading to (2.4.5) would yield a rule of the form: the integral

$$
\frac{1}{2\pi i}\int_{-\infty}^{(0+)} g(s)z^s\,ds
$$

converges absolutely for all finite, nonzero z when

$$
\kappa = \sum_{r=1}^{q}\beta_r - \sum_{r=1}^{p}\alpha_r < 0.
$$

A discussion of the convergence of Mellin-Barnes integrals of hypergeometric type can also be found in Marichev (1982, §4).

Example 1. The exponential function has the Mellin-Barnes integral representation [see (3.3.2) *et seq.*]

$$
e^{-z} = \frac{1}{2\pi i}\int_{c-\infty i}^{c+\infty i}\Gamma(s)z^{-s}\,ds \qquad (c > 0). \tag{2.4.7}
$$

Upon putting $s = \sigma + it$ (with $\sigma = c > 0$) and $\theta = \arg z$, we see that the integrand has the controlling behaviour (we omit unimportant factors)

$$
|z|^{-s}e^{\theta t}e^{-s}s^{s-\frac{1}{2}} \sim (\pm i)^{\sigma-\frac{1}{2}}(e|z|)^{-\sigma-it}|t|^{\sigma-\frac{1}{2}\pm i|t|}e^{\theta t-\frac{1}{2}\pi|t|}
$$

along the path of integration for $t \to \pm\infty$.

The integral (2.4.7) converges for $|\theta| < \frac{1}{2}\pi$; compare Rule 1. When $\theta = \frac{1}{2}\pi$, the integrand has the large-t controlling behaviour $t^{\sigma-\frac{1}{2}}e^{i(t\log(t/|z|)-t)}$ as $t \to +\infty$, the behaviour at the lower limit $t \to -\infty$ being dominated by the factor $e^{-\pi|t|}$. Convergence is consequently assured if $\sigma < \frac{1}{2}$; a similar argument applies when $\theta = -\frac{1}{2}\pi$. For the special case $\sigma = \frac{1}{2}$, however, convergence is still assured, since the convergence is now determined by that of

$$\int_1^\infty e^{i|z|(\tau \log \tau - \tau)} \, d\tau = \int_{-1}^\infty e^{i|z|\psi} \frac{d\psi}{\log \tau},$$

where $\psi = \tau \log \tau - \tau, \tau = t/|z|$. This latter integral can be seen to be convergent by virtue of Dirichlet's test† [Bromwich (1926), p. 477].

We conclude, therefore, that (2.4.7) converges for $|\arg z| < \frac{1}{2}\pi$, and when $|\arg z| = \frac{1}{2}\pi$, we further require $c \le \frac{1}{2}$ to ensure convergence. The representation (2.4.7) occupies a central role in several arguments later in the book.

Example 2. The Gauss hypergeometric function has the Mellin-Barnes integral representation

$$\frac{\Gamma(a)\Gamma(b)}{\Gamma(c)} {}_2F_1(a, b; c; z) = \frac{1}{2\pi i} \int_{-\infty i}^{\infty i} \Gamma(-s) \frac{\Gamma(s+a)\Gamma(s+b)}{\Gamma(s+c)} (-z)^s \, ds,$$

where the contour of integration separates the poles of $\Gamma(-s)$ from those of $\Gamma(s+a)\Gamma(s+b)$, for a and $b \ne 0, -1, -2, \ldots$; see §3.4.2. Rule 1 applied to this integral shows that it is convergent for

$$|\arg(-z)| < \tfrac{1}{2}\pi(3 - 1) = \pi.$$

If instead of the vertical integration contour we use a loop C_1, then Rule 2 would reveal convergence only in the open unit disc.

Example 3. Consider the integral

$$\frac{1}{2\pi i} \int_C \frac{\Gamma\left(a - \frac{1}{2}s\right)\Gamma(\lambda s + b)\Gamma\left(\frac{1}{4}s + c\right)}{\Gamma\left(\frac{1}{4}s + d\right)\Gamma(s + 1)} z^s \, ds \qquad (\lambda > 0),$$

where the parameters a, b, c and λ are such that the poles of $\Gamma(a - \frac{1}{2}s)$ can be separated from those of $\Gamma(\lambda s + b)\Gamma(\frac{1}{4}s + c)$. When the contour C is the path C_2 parallel to the imaginary axis, this integral defines a holomorphic function of z only when $\lambda > \frac{1}{2}$ since, by Rule 1, the integral converges in the sector

$$|\arg z| < \tfrac{1}{2}\pi\left(\tfrac{1}{2} + \lambda + \tfrac{1}{4} - \tfrac{1}{4} - 1\right) = \tfrac{1}{2}\pi\left(\lambda - \tfrac{1}{2}\right).$$

When the path is taken to be the loop C_1, the integral converges for *all values* of z when $\frac{1}{2} + \frac{1}{4} + 1 - \lambda - \frac{1}{4} > 0$ by Rule 2; that is, for $\lambda < \frac{3}{2}$. When $\lambda = \frac{3}{2}$, the

---

† An infinite integral that oscillates finitely (i.e., undergoes bounded oscillation) becomes convergent after the insertion of a monotonic factor which tends to zero in the limit.

integral over the loop defines an analytic function in $|z| < h^{-1}$, where, from (2.4.3),

$$h = \sqrt{2}\lambda^{\lambda} = \frac{3\sqrt{3}}{2}.$$

2.5 Order Estimates for Remainder Integrals

This section is concerned with the construction of bounds on various types of remainder integrals that appear in the development of asymptotic expansions by Mellin-Barnes integrals in subsequent sections of this book. Estimates for these remainder integrals are essential in the development of the theory as they establish the asymptotic nature of the various expansions so obtained. In the case when the only large variable is z, estimates for these remainder integrals are relatively straightforward; see the examples in §5.1. More difficult cases arise when combining the Mellin-Barnes approach with inverse factorial expansions for products or quotients of gamma functions. Another generalisation occurs in the discussion of the Stokes phenomenon in Chapter 6 where remainder integrals involving both a large parameter and the variable z are encountered.

2.5.1 An Example

Before considering these remainder integrals, we discuss a simple example which contains some of the main features of these more involved cases and which also helps to establish the subsequent analysis. Let us consider a bound for the modulus of the integral

$$J(z) = \frac{1}{2\pi i} \int_{c-\infty i}^{c+\infty i} \Gamma(s)z^{-s}\, ds \qquad (c > 0), \qquad (2.5.1)$$

as $|z| \to \infty$ in $|\arg z| < \frac{1}{2}\pi$. In this case, of course, the answer is simple, since we know from (2.4.7) that $J(z) = e^{-z}$ and hence that (for all $|z|$)

$$|J(z)| = e^{-|z|\cos\theta}, \qquad (2.5.2)$$

where $\theta = \arg z$. This simple result will provide a basis for comparison with the method given below that exploits properties of the gamma function. With $s = \sigma + it$ (where $\sigma = c$) we then find

$$|J(z)| \leq \frac{|z|^{-\sigma}}{2\pi} \int_{-\infty}^{\infty} |\Gamma(\sigma + it)|e^{\theta t}\, dt.$$

It is clear that the exponential decay of $|\Gamma(\sigma + it)|$ as $t \to \pm\infty$ has to be taken into account† in order to secure a bound valid in $|\arg z| < \frac{1}{2}\pi$. We note at this point that the vertical line $\mathrm{Re}(s) = \sigma$ can be displaced as far to the right as we please,

---

† The simple bound $|\Gamma(\sigma + it)| \leq \Gamma(\sigma)$ ($\sigma > 0$) is too crude here as the resulting integral would diverge for all $\arg z$.

since the integrand is holomorphic in $\text{Re}(s) > 0$: eventually σ will be chosen to scale like $|z|$.

From Stirling's formula (2.1.8), we deduce that there exist positive constants K_1 and K_2 independent of s such that if $|s| \geq K_1$ and $|\arg s| \leq \frac{1}{2}\pi$ then

$$|\Gamma(\sigma + it)| \leq K_2 e^{-\sigma} |(\sigma + it)^{\sigma + it - \frac{1}{2}}|$$

$$\leq K_2 e^{-\sigma} (\sigma^2 + t^2)^{\frac{1}{2}(\sigma - \frac{1}{2})} e^{-t\psi(s)}$$

$$= K_2 e^{-\sigma} \sigma^{\sigma - \frac{1}{2}} (1 + \tau^2)^{\frac{1}{2}(\sigma - \frac{1}{2})} \exp\{-\sigma\tau \arctan \tau\}, \qquad (2.5.3)$$

where $\psi(s) = \arg s = \arctan(t/\sigma)$ (when $\sigma \geq 0$) and we have put $\tau = t/\sigma$. Then we find

$$|J(z)| \leq \frac{K_2}{2\pi} e^{-\sigma} \sigma^{\sigma + \frac{1}{2}} |z|^{-\sigma} \int_{-\infty}^{\infty} e^{\sigma f(\tau)} \frac{d\tau}{(1 + \tau^2)^{\frac{1}{4}}},$$

where

$$f(\tau) = \frac{1}{2} \log(1 + \tau^2) - \tau \arctan \tau + \theta\tau. \qquad (2.5.4)$$

The function $f(\tau)$ has a saddle point where $f'(\tau) = \theta - \arctan \tau = 0$, that is at the point $\tau_0 = \tan \theta$. With the values $f(\tau_0) = \log \sec \theta$ and $f''(\tau_0) = -\cos^2 \theta$, we can apply the saddle point method for $\sigma \gg 1$ to find the estimate

$$|J(z)| = \frac{K_2}{2\pi} e^{-\sigma} \sigma^{\sigma + \frac{1}{2}} |z|^{-\sigma} O\left((2\pi/\sigma)^{\frac{1}{2}} (\sec \theta)^{\sigma + \frac{1}{2}}\right)$$

$$= O\left(B(\sigma) \sec^{\frac{1}{2}} \theta\right), \qquad (2.5.5)$$

where

$$B(\sigma) = e^{-\sigma} |z|^{-\sigma} (\sigma \sec \theta)^{\sigma}.$$

Thus we have the following lemma (obtained by putting $|z| = 1$):

Lemma 2.6. *Let $|\theta| < \frac{1}{2}\pi$ and the integral I_1 be defined by*

$$I_1 = \int_{-\infty}^{\infty} |\Gamma(\sigma + it)| e^{\theta t} \, dt,$$

where $\sigma > 0$. Then, for $\sigma \to \infty$, we have the estimate

$$I_1 = O(e^{-\sigma} \sigma^{\sigma} (\sec \theta)^{\sigma + \frac{1}{2}}).$$

Also, if I_2 denotes the integral

$$I_2 = \frac{1}{2\pi i} \int_{-c+n-\infty i}^{-c+n+\infty i} \Gamma(s + \alpha) \, ds,$$

where α and c are fixed constants and n is supposed sufficiently large for the integration path to lie to the right of the poles of $\Gamma(s + \alpha)$, then, as $n \to \infty$, we have

$$|I_2| = O(e^{-n}n^{n+\alpha-c}).$$

The second part of this lemma follows immediately from the first part by putting $\theta = 0$ and $\sigma = \text{Re}(s + \alpha) = n - c + \alpha$.

The result in (2.5.5) depends on $\sigma = \text{Re}(s)$, that is, on the position of the path of integration in (2.5.1). The value of this parameter is free to be chosen: for example, if we let $\sigma = |z|$ then we obtain $|J(z)| = O(e^{-|z|}(\sec\theta)^{|z|+\frac{1}{2}})$. When $\theta \neq 0$ this bound is clearly not very good as $|z| \to \infty$. The optimal choice for σ corresponds to the minimum of $B(\sigma)$, given by $e^{-|z|\cos\theta}$ when $\sigma = |z|\cos\theta$. Then we obtain the estimate

$$|J(z)| = O(\sec^{\frac{1}{2}}\theta\, e^{-|z|\cos\theta}) \tag{2.5.6}$$

as $|z| \to \infty$ in $|\arg z| < \frac{1}{2}\pi$, which compares well with the exact result in (2.5.2).

We shall find that this process of path displacement to an optimal position will be employed in the first two of the remainder integrals to be discussed in the next section.

2.5.2 Lemmas

In this section we assemble a collection of lemmas concerning order estimates for certain types of remainder integrals. Throughout we shall write $s = \sigma + it$, where σ and t are real, and let $\theta = \arg z$. In the first three lemmas, the remainder integrals contain the function $\rho_M(s)$, for $M = 1, 2, \ldots$, which appears as a remainder in the inverse factorial expansion of Lemma 2.2. An analogous procedure can be employed to deal with remainder integrals involving the function $\sigma_M(s)$ in Lemma 2.1.

Lemma 2.7. *Let M denote a positive integer and α an arbitrary complex constant. Let the remainder integral $R_M^{(1)}(z)$ be defined by*

$$R_M^{(1)}(z) = \frac{1}{2\pi i} \int_{c-\infty i}^{c+\infty i} \rho_M(s)\Gamma(s+\alpha)z^{-s}\, ds,$$

where $c > 0$ is supposed sufficiently large so that the integration path lies to the right of all the poles of the integrand. The function $\rho_M(s)$ is a remainder term appearing in the inverse factorial expansion in Lemma 2.2 and is such that $\rho_M(s) = O(1)$ as $|s| \to \infty$ uniformly in $|\arg s| \leq \pi - \epsilon$, $\epsilon > 0$. Then

$$|R_M^{(1)}(z)| = O(z^\alpha e^{-z})$$

as $|z| \to \infty$ in $|\arg z| < \frac{1}{2}\pi$.

Proof. The integral for $R_M^{(1)}(z)$ converges in the sector $|\arg z| < \frac{1}{2}\pi$, since $\rho_M(s) = O(1)$ and the gamma function decays like $e^{-\frac{1}{2}\pi|t|}$ as $t \to \pm\infty$ on the integration path. By hypothesis the path of integration lies to the right of all poles of the integrand so that, because of the exponential decay of the integrand as $t \to \pm\infty$, the parameter c can be chosen as large as we please. From Lemma 2.2 and Stirling's formula (2.1.8), we deduce that there exist positive constants K_1 and K_2 such that if $|s| \geq K_1$ and $|\arg s| \leq \frac{1}{2}\pi$ then

$$\begin{cases} |\rho_M(s)| \leq K_2 \\ |\Gamma(s+\alpha)| \leq K_2|s|^{\sigma+\mathrm{Re}(\alpha)-\frac{1}{2}}e^{-\sigma-t\psi(s)} \end{cases} \tag{2.5.7}$$

$$\leq K_2|s|^{\mathrm{Re}(\alpha)-\frac{1}{2}}e^{-\sigma}\sigma^\sigma \exp\left\{\sigma\left(\tfrac{1}{2}\log(1+\tau^2) - \tau\arctan\tau\right)\right\},$$

where $\psi(s) = \arctan\tau$ with $\tau = t/\sigma$; compare (2.5.3).

Then we find

$$|R_M^{(1)}(z)| \leq \frac{K_2^2}{2\pi}|z|^{-\sigma}e^{-\sigma}\sigma^{\sigma+\gamma+1}\int_{-\infty}^{\infty}(1+\tau^2)^{\frac{1}{2}\gamma}e^{\sigma f(\tau)}\,d\tau,$$

where $f(\tau)$ is defined in (2.5.4) and we have put $\gamma = \mathrm{Re}(\alpha) - \frac{1}{2}$. The exponential factor in the integrand has a saddle point at $\tau_0 = \tan\theta$ so that, as in (2.5.5), we find

$$|R_M^{(1)}(z)| = O\left(B(\sigma)(\sigma\sec\theta)^{\mathrm{Re}(\alpha)}\sec^{\frac{1}{2}}\theta\right),$$

where $B(\sigma) = e^{-\sigma}|z|^{-\sigma}(\sigma\sec\theta)^\sigma$. The minimum value of $B(\sigma)$ occurs when $\sigma = |z|\cos\theta$ and we therefore obtain

$$|R_M^{(1)}(z)| = O(|z|^{\mathrm{Re}(\alpha)}e^{-|z|\cos\theta})$$

as $|z| \to \infty$ in the sector $|\arg z| < \frac{1}{2}\pi$, which establishes the lemma. $\qquad\square$

As a corollary, we observe that if

$$R_M^{(1)}(z) = \frac{1}{2\pi i}\int_{c-\infty i}^{c+\infty i}\rho_M(s)\Gamma(\kappa s + \alpha)z^{-s}\,ds,$$

where $\kappa > 0$, then by the obvious change of variable $s \to s/\kappa$ we find that

$$|R_M^{(1)}(z)| = O(z^{\alpha/\kappa}e^{-z^{1/\kappa}}) \tag{2.5.8}$$

as $|z| \to \infty$ in the sector $|\arg z| < \frac{1}{2}\pi\kappa$. This lemma and its method of proof were given in Braaksma (1963, §10.1).

Lemma 2.8. *Let M denote a positive integer and α an arbitrary complex constant. Let the remainder integral $R_M^{(2)}(z)$ be defined by*

$$R_M^{(2)}(z) = \frac{1}{2\pi i}\int_C \rho_M(s)\Gamma(s+\alpha)z^{-s}\,ds,$$

where C denotes a loop that embraces all the poles of the integrand and has end-points at infinity in $\text{Re}(s) < 0$. *The function* $\rho_M(s)$ *is a remainder term appearing in the inverse factorial expansion in Lemma 2.2 and is such that* $\rho_M(s) = O(1)$ *as* $|s| \to \infty$ *uniformly in* $|\arg s| \le \pi - \epsilon, \epsilon > 0$. *Then*

$$|R_M^{(2)}(z)| = O(z^\alpha e^{-z})$$

as $|z| \to \infty$ *in* $|\arg z| < \pi$.

Proof. Since the loop C passes to infinity in the directions $\pm\theta_0$, where $\frac{1}{2}\pi < |\theta_0| \le \pi$, the convergence of the integral as $|s| \to \infty$ is controlled by the term $\exp\{|s|\cos\theta_0 \log|s|\}$ in the behaviour of the gamma function for large $|s|$. This results in the above integral for $R_M^{(2)}(z)$ being convergent *without restriction* on $\arg z$. As in (2.5.7), there exist positive constants K_1 and K_2 such that when $|s| \ge K_1$ and $|\arg s| < \pi$ then

$$|\rho_M(s)| \le K_2, \qquad |\Gamma(s+\alpha)| \le K_2|s|^{\sigma+\gamma} e^{-\sigma - t\psi(s)},$$

where

$$\psi(s) = \arg s = \begin{cases} \arctan(t/\sigma) & (\sigma \ge 0) \\ \pm\pi + \arctan(t/\sigma) & (\sigma < 0, \ t \gtrless 0) \end{cases}$$

and $\gamma = \text{Re}(\alpha) - \frac{1}{2}$, with the upper or lower sign being chosen according as $t > 0$ or $t < 0$, respectively. The modulus of the integrand on C is consequently

$$\le K_2^2|s|^\gamma \exp\left\{-\sigma + \frac{1}{2}\sigma \log(\sigma^2 + t^2) - t\psi(s) - \sigma \log|z| + \theta t\right\}$$

$$\le K_2^2|z|^\gamma (x^2+y^2)^{\frac{1}{2}\gamma} e^{|z|F(x,y)},$$

where we have put $\sigma = x|z|, t = y|z|$ and

$$F(x, y) = -x + \frac{1}{2}x \log(x^2 + y^2) - y\psi(s) + \theta y.$$

As a function of the real variables x and y, $F(x, y)$ has a stationary point when $\partial F/\partial x = \partial F/\partial y = 0$, where

$$\frac{\partial F}{\partial x} = -1 + \frac{1}{2}\log(x^2 + y^2) + \frac{x^2}{x^2 + y^2} - y\frac{\partial \psi}{\partial x} = \frac{1}{2}\log(x^2 + y^2),$$

$$\frac{\partial F}{\partial y} = \frac{xy}{x^2 + y^2} - \psi(s) - y\frac{\partial \psi}{\partial y} + \theta = \theta - \psi(s);$$

that is, when $x^2 + y^2 = 1$ and $y/x = \tan\theta$ $(x \ge 0)$ or $y/x = \tan(\theta \mp \pi)$ $(x < 0, \ y \gtrless 0)$. Thus the stationary point is given by $(x_0, y_0) = (\cos\theta, \sin\theta)$ and corresponds to a saddle, since† $F_{xx}F_{yy} - F_{xy}^2 = -1$. In the neighbourhood of the saddle, we let $x - x_0 = u = w\cos\Psi$ and $y - y_0 = v = w\sin\Psi$, where Ψ denotes

---

† The second derivatives are given by $F_{xx} = -F_{yy} = x/(x^2 + y^2)$ and $F_{xy} = y/(x^2 + y^2)$.

the direction of steepest descent through the saddle and w measures distance from the saddle. Then we have

$$
\begin{aligned}
F(x, y) &= -\cos\theta + \tfrac{1}{2}(u^2 F_{xx} + 2uv F_{xy} + v^2 F_{yy}) + \cdots \\
&= -\cos\theta + \tfrac{1}{2}[(u^2 - v^2)\cos\theta + 2uv\sin\theta] + \cdots \\
&= -\cos\theta + \tfrac{1}{2}w^2 \cos(2\Psi - \theta) + \cdots ,
\end{aligned}
$$

so that the steepest descent directions at the saddle are therefore given by $\cos(2\Psi - \theta) = -1$, or $\Psi = \tfrac{1}{2}\theta \pm \tfrac{1}{2}\pi$.

Provided $|\theta| < \pi$, the path of steepest descent† from the saddle point (x_0, y_0) can be shown to be topologically similar to the contour C and to pass to infinity in $\mathrm{Re}(s) < 0$ in the directions parallel to the negative real axis (we omit these details). On this path, the function $F(x, y)$ decreases monotonically to $-\infty$ in both directions from the saddle. With $s' = x + iy$ and C' denoting the path of steepest descent in the s' plane, we can deform the map of the contour C in the s' plane into the path C' to find

$$
|R_M^{(2)}(z)| \leq \frac{K_2^2}{2\pi}|z|^{\gamma+1} \int_{C'} (x^2 + y^2)^{\frac{1}{2}\gamma} e^{|z|F(x,y)} \, |ds'|.
$$

Laplace's method then shows that, for large $|z|$ in $|\arg z| < \pi$, this last integral is approximated by

$$
e^{-|z|\cos\theta} \int_{-\infty}^{\infty} (x^2 + y^2)^{\frac{1}{2}\gamma} e^{-\frac{1}{2}|z|w^2} \, dw \simeq (x_0^2 + y_0^2)^{\frac{1}{2}\gamma} (2\pi/|z|)^{\frac{1}{2}} e^{-|z|\cos\theta}
$$

$$
= (2\pi/|z|)^{\frac{1}{2}} e^{-|z|\cos\theta}.
$$

Hence we find

$$
|R_M^{(2)}(z)| = O(|z|^{\mathrm{Re}(\alpha)} e^{-|z|\cos\theta})
$$

as $|z| \to \infty$ in the sector $|\arg z| < \pi$, which establishes the lemma.   □

The extension of this lemma to the case when

$$
R_M^{(2)}(z) = \frac{1}{2\pi i} \int_C \rho_M(s) \Gamma(\kappa s + \alpha) z^{-s} \, ds
$$

with $\kappa > 0$, then immediately follows by the change of variable $s \to s/\kappa$ and we consequently find the order estimate

$$
|R_M^{(2)}(z)| = O(z^{\alpha/\kappa} e^{-z^{1/\kappa}}) \tag{2.5.9}
$$

as $|z| \to \infty$ in the sector $|\arg z| < \pi\kappa$.

---

† If we set $f(s') = s'\log s' - (1 + i\theta)s'$ $(s' = x + iy)$, so that $F(x, y) = \mathrm{Re}(f(s'))$, then the path of steepest descent through the saddle $s' = e^{i\theta}$ is determined by the condition $\mathrm{Im}(f(s')) = -\sin\theta$.

Lemma 2.9. *Let M and n denote positive integers and α be an arbitrary complex constant. Let the remainder integral $R_M^{(3)}(z)$ be defined by*

$$R_M^{(3)}(z) = \frac{1}{2\pi i} \int_{-c+n-\infty i}^{-c+n+\infty i} \rho_M(s)\Gamma(s+\alpha)\frac{z^{-s}}{\sin \pi s}\, ds \quad (0 < c < 1),$$

where n is supposed sufficiently large for the integration path to lie to the right of the poles of $\rho_M(s)\Gamma(s+\alpha)$ and be a vertical line without any indentations. The function $\rho_M(s)$ is a remainder term appearing in the inverse factorial expansion in Lemma 2.2 and is such that $\rho_M(s) = O(1)$ as $|s| \to \infty$ uniformly in $|\arg s| \le \pi - \epsilon$, $\epsilon > 0$. Then, as $n \to \infty$, we have†

$$|R_M^{(3)}(z)| = \begin{cases} |z|^{-n+c}O(e^{-n}n^{n+\alpha-c-\frac{1}{2}}) & (|\arg z| < \pi) \\ |z|^{-n+c}O(e^{-n}n^{n+\alpha-c}) & (|\arg z| \le \pi). \end{cases} \tag{2.5.10}$$

If, when $|z|$ is large, n is chosen to be the integer part of $|z|$, i.e., if $n = |z| + \beta$, where $|\beta| < 1$, then we have as $|z| \to \infty$

$$|R_M^{(3)}(z)| = \begin{cases} O(z^{\alpha-\frac{1}{2}}e^{-|z|}) & (|\arg z| < \pi) \\ O(z^\alpha e^{-|z|}) & (|\arg z| \le \pi). \end{cases} \tag{2.5.11}$$

Proof. Although the integral defining $R_M^{(3)}(z)$ converges in $|\arg z| < \frac{3}{2}\pi$, we consider the bound only in the sector $|\arg z| \le \pi$. We let $s = n - c + it$, where $0 < c < 1$, and deal first with the more straightforward case when $|\arg z| < \pi$. We shall use the inequalities $|\Gamma(\sigma + it)| \le \Gamma(\sigma)$ (for $\sigma > 0$) and, for $t \in (-\infty, \infty)$,

$$\frac{e^{\theta t}}{(\cosh^2 \pi t - \cos^2 \pi c)^{\frac{1}{2}}} \le \frac{e^{\theta t}}{\cosh \pi t}\, \text{cosec}\, \pi c \le 2\,\text{cosec}\,\pi c, \tag{2.5.12}$$

when $0 < c < 1$ and $|\theta| \le \pi$, together with the integral

$$\int_0^\infty \frac{\cosh \theta t}{\cosh \pi t}\, dt = \frac{1}{2}\sec\frac{1}{2}\theta \quad (|\theta| < \pi). \tag{2.5.13}$$

As in (2.5.7), there exist positive constants K_1 and K_2 such that $|\rho_M(s)| \le K_2$ for $|s| \ge K_1$, and hence

$$|R_M^{(3)}(z)| \le \frac{K_2}{2\pi}|z|^{-n+c}\int_{-\infty}^\infty |\Gamma(n-c+\alpha+it)|\frac{e^{\theta t}}{(\cosh^2 \pi t - \cos^2 \pi c)^{\frac{1}{2}}}\, dt$$

$$\le \frac{K_2}{\pi \sin \pi c}|z|^{-n+c}\Gamma(n-c+\text{Re}(\alpha))\int_0^\infty \frac{\cosh \theta t}{\cosh \pi t}\, dt$$

$$= \frac{K_2 \sec\frac{1}{2}\theta}{2\pi \sin \pi c}|z|^{-n+c}\Gamma(n-c+\text{Re}(\alpha)).$$

† Note that in the estimates in (2.5.10) $|z|$ is arbitrary.

Use of Stirling's formula (2.1.8) then enables us to conclude that

$$|R_M^{(3)}(z)| = |z|^{-n+c} O(e^{-n} n^{n+\mathrm{Re}(\alpha)-c-\frac{1}{2}}) \tag{2.5.14}$$

as $n \to \infty$ when $|\arg z| < \pi$.

To extend this estimate to the sector $|\arg z| \leq \pi$ we must retain the modulus of the gamma function in the integrand. Thus, using the second inequality in (2.5.12), we find, when $|\arg z| \leq \pi$,

$$|R_M^{(3)}(z)| \leq \frac{K_2}{\pi \, \sin \pi c} |z|^{-n+c} \int_{-\infty}^{\infty} |\Gamma(n-c+\alpha+it)| \, dt. \tag{2.5.15}$$

By Lemma 2.6 with $\theta = 0$ and $\sigma = n - c + \mathrm{Re}(\alpha)$, the last integral is $O(e^{-n} n^{n+\mathrm{Re}(\alpha)-c})$ for $n \to \infty$. Hence we find the result

$$|R_M^{(3)}(z)| = |z|^{-n+c} O(e^{-n} n^{n+\mathrm{Re}(\alpha)-c}) \tag{2.5.16}$$

as $n \to \infty$ when $|\arg z| \leq \pi$, which establishes the first part of the lemma.

If we now suppose that $|z| \to \infty$ and choose n to be the integer part of $|z|$, i.e., if $n = |z| + \beta$, where $|\beta| < 1$, we immediately obtain from (2.5.14) and (2.5.16) the bounds†

$$|R_M^{(3)}(z)| = \begin{cases} O(z^{\alpha-\frac{1}{2}} e^{-|z|}) & (|\arg z| < \pi) \\[2mm] O(z^{\alpha} e^{-|z|}) & (|\arg z| \leq \pi) \end{cases}$$

as $|z| \to \infty$.   □

As a corollary, it is evident that the estimates in (2.5.10) and (2.5.11) also apply to the remainder integral of the type

$$R^{(4)}(z) = \frac{1}{2\pi i} \int_{-c+n-\infty i}^{-c+n+\infty i} \Gamma(s+\alpha) \frac{z^{-s}}{\sin \pi s} \, ds \qquad (0 < c < 1).$$

This follows at once from the above by putting $\rho_M(s) \equiv 1$.

Lemma 2.10. *Let the remainder integral $J(z)$ be defined by*

$$J(z) = \frac{1}{2\pi i} \int_{-c+n-\infty i}^{-c+n+\infty i} \frac{\Gamma(s+\alpha)\Gamma(s+\beta)}{\Gamma(s+1) \sin \pi s} z^{-s} ds \qquad (0 < c < 1),$$

where n is a positive integer supposed sufficiently large for the integration path to lie to the right of the poles of $\Gamma(s+\alpha)\Gamma(s+\beta)$. Then, as $n \to \infty$, we have for $|\arg z| \leq \pi$ and $\omega = \alpha + \beta - 1$

$$|J(z)| = |z|^{-n+c} O(e^{-n} n^{n+\omega-c}).$$

---

† We observe that the bound in the sector $|\arg z| < \pi$ is sharper than that in the sector $|\arg z| \leq \pi$ by the factor $1/\sqrt{|z|}$. This fact will prove to be of significance in the discussion of the Stokes phenomenon in Chapter 6.

If $|\arg z| < \frac{1}{2}\pi$, this order estimate can be strengthened to

$$|J(z)| = |z|^{-n+c} O(e^{-n} n^{n+\omega-c-\frac{1}{2}})$$

as $n \to \infty$.

Proof. With $s = n - c + it$ and $\theta = \arg z$ we have

$$|J(z)| \leq \frac{|z|^{-n+c}}{2\pi} \int_{-\infty}^{\infty} \left| \frac{\Gamma(n-c+\alpha+it)\Gamma(n-c+\beta+it)}{\Gamma(n+1-c+it)} \right|$$

$$\times \frac{e^{\theta t}\, dt}{(\cosh^2 \pi t - \cos^2 \pi c)^{\frac{1}{2}}} \qquad (2.5.17)$$

$$\leq \frac{|z|^{-n+c}}{\pi \sin \pi c} \int_{-\infty}^{\infty} \left| \frac{\Gamma(n-c+\alpha+it)\Gamma(n-c+\beta+it)}{\Gamma(n+1-c+it)} \right| dt$$

by (2.5.12) when $|\arg z| \leq \pi$. For $n \to \infty$, the gamma function ratio is from (2.5.7)

$$O(e^{-n} n^{n+\gamma'} (1+\tau^2)^{\frac{1}{2}\gamma'} e^{nf(\tau)}), \qquad \gamma' = \mathrm{Re}(\omega) - c - \frac{1}{2},$$

where $f(\tau) = \frac{1}{2}\log(1+\tau^2) - \tau \arctan \tau$, with $\tau = t/n$, so that

$$|J(z)| = |z|^{-n+c} O\left(e^{-n} n^{n+\gamma'+1} \int_{-\infty}^{\infty} (1+\tau^2)^{\frac{1}{2}\gamma'} e^{nf(\tau)}\, d\tau \right).$$

The exponential factor in the integrand has a saddle point at $\tau = 0$ and the integral is therefore $O((2\pi/n)^{\frac{1}{2}})$ for $n \to \infty$. Hence we find the result

$$|J(z)| = |z|^{-n+c} O(e^{-n} n^{n+\omega-c})$$

as $n \to \infty$ when $|\arg z| \leq \pi$, which establishes the first part of the lemma.

When $|\arg z| < \frac{1}{2}\pi$ we can obtain a sharper estimate by retaining the term $\mathrm{sech}\, \pi t$ in the integrand. From (2.5.17) and the first inequality in (2.5.12) we have

$$|J(z)| \leq |z|^{-n+c} \frac{\Gamma(n-c+\mathrm{Re}(\alpha))\Gamma(n-c+\mathrm{Re}(\beta))}{2\pi \sin \pi c\, \Gamma(n+1-c)}$$

$$\times \int_{-\infty}^{\infty} \frac{\Gamma(1-c)}{|\Gamma(1-c+it)|} \frac{e^{\theta t}\, dt}{\cosh \pi t},$$

where repeated use has been made of $\Gamma(z+1) = z\Gamma(z)$ to show that, for $t \in (-\infty, \infty)$,

$$\frac{1}{|\Gamma(n+1-c+it)|} = \frac{\prod_{j=0}^{n-1} |(n-c-j+it)|^{-1}}{|\Gamma(1-c+it)|}$$

$$\leq \frac{\Gamma(1-c)}{\Gamma(n+1-c)} \frac{1}{|\Gamma(1-c+it)|}.$$

Since the last integrand has the behaviour $|t|^{c-\frac{1}{2}} e^{\theta t - \frac{1}{2}\pi|t|}$ as $t \to \pm\infty$, we see that, when $0 < c < 1$, the integral is bounded when $|\arg z| < \frac{1}{2}\pi$. It follows that

$$|J(z)| = |z|^{-n+c} O \left(\frac{\Gamma(n-c+\mathrm{Re}(\alpha))\Gamma(n-c+\mathrm{Re}(\beta))}{\Gamma(n+1-c)} \right)$$

$$= |z|^{-n+c} O(e^{-n} n^{n+\omega-c-\frac{1}{2}}) \qquad (|\arg z| < \tfrac{1}{2}\pi)$$

as $n \to \infty$, by a straightforward application of Stirling's formula (2.1.8). $\qquad\square$

3

Properties of Mellin Transforms

3.1 Basic Properties

In this introductory section we give the definition of the Mellin transform and some of its elementary translational and differential properties. We also include a statement of the important Mellin inversion theorem and the Parseval formula together with some of its variants.

3.1.1 Definition

The Mellin transform of a locally integrable function† $f(x)$ on $(0, \infty)$ is defined by

$$M[f; s] = F(s) = \int_0^\infty x^{s-1} f(x) \, dx \qquad (3.1.1)$$

when the integral converges. The basic properties of the Mellin transform follow immediately from those of the Laplace transform since these transforms are intimately connected. This can be seen by making a suitable change of variable in the two-sided (bilateral) Laplace transform given by

$$\mathcal{L}[g; s] = \int_{-\infty}^\infty e^{-s\tau} g(\tau) \, d\tau$$

which converges absolutely and is holomorphic in the strip $a < \text{Re}(s) < b$, where a and b are real constants (with $a < b$) such that

$$g(\tau) = \begin{cases} O(e^{(a+\epsilon)\tau}) & \text{as} \quad \tau \to +\infty \\ O(e^{(b-\epsilon)\tau}) & \text{as} \quad \tau \to -\infty \end{cases}$$

---

† A locally integrable function on $(0, \infty)$ is one that is absolutely integrable on all closed subintervals of $(0, \infty)$.

for every (small) positive ϵ. Then, with the new variable $\tau = -\log x$ and $f(x) \equiv g(-\log x)$, we find

$$\mathcal{L}[g; s] = \int_0^\infty x^{s-1} g(-\log x)\, dx = M[f; s], \qquad (3.1.2)$$

thereby establishing the connection between these two types of transform.

The integral (3.1.1) defines the Mellin transform in a vertical strip in the s plane whose boundaries are determined by the analytic structure of $f(x)$ as $x \to 0+$ and $x \to +\infty$. If we suppose that

$$f(x) = \begin{cases} O(x^{-a-\epsilon}) & \text{as } x \to 0+ \\ O(x^{-b+\epsilon}) & \text{as } x \to +\infty, \end{cases} \qquad (3.1.3)$$

where $\epsilon > 0$ and $a < b$, then the integral (3.1.1) converges absolutely and defines an analytic function in the strip

$$a < \text{Re}(s) < b. \qquad (3.1.4)$$

This strip is known as the *strip of analyticity* of $M[f; s]$.

The inversion formula for (3.1.1) follows directly from the corresponding inversion formula for the bilateral Laplace transform. Thus, with $\mathcal{L}[g; s]$ defined above, we have for continuous $g(\tau)$ the inversion

$$g(\tau) = \frac{1}{2\pi i} \int_{c-\infty i}^{c+\infty i} e^{s\tau} \mathcal{L}[g; s]\, ds,$$

where $a < c < b$. With the change of variables in (3.1.2) we therefore find the result

$$f(x) = \frac{1}{2\pi i} \int_{c-\infty i}^{c+\infty i} x^{-s} M[f; s]\, ds \quad (a < c < b). \qquad (3.1.5)$$

This is the inversion formula for the Mellin transform which is valid at all points $x \geq 0$ where $f(x)$ is continuous. A proof of this theorem can be found in Sneddon (1972, pp. 273–275) and McLachlan (1963, pp. 341–343). Although this result[†] was used by Riemann (1859) in his famous paper on prime numbers, and even as early as 1815 by Poisson (for Fourier integrals), the first rigorous justification of this formula was supplied by Mellin (1896, 1902).

When the function $f(x)$ has a discontinuity the inversion theorem takes on the following form [Titchmarsh (1975, p. 46); Wong (1989, p. 151)]. Suppose (3.1.1) converges absolutely on the line $\text{Re}(s) = c$ and let $f(y)$ be of bounded variation in the neighbourhood of the point $y = x$. Then

$$\tfrac{1}{2}\{f(x+0) + f(x-0)\} = \frac{1}{2\pi i} \lim_{T \to \infty} \int_{c-iT}^{c+iT} x^{-s} M[f; s]\, ds. \qquad (3.1.6)$$

---

† For additional historical information, see Bateman (1942).

3.1.2 Translational and Differential Properties

In common with other integral transforms, the Mellin transform possesses a series of simple translational properties which greatly facilitate the evaluation of transforms of more involved functions. All these results can be obtained by straight-forward manipulation of the definition (3.1.1); proofs can be found in Sneddon (1972, §4.1). We let

$$M[f(x); s] = F(s) = \int_0^\infty x^{s-1} f(x) \, dx,$$

where, for convenience, we have explicitly shown the argument of the function f appearing in M. Then we have

$$
\left.
\begin{array}{ll}
(a) & M[f(ax); s] = a^{-s} F(s) \quad (a > 0) \\
(b) & M[x^a f(x); s] = F(s + a) \\
(c) & M[f(x^a); s] = a^{-1} F(s/a) \quad (a > 0) \\
(d) & M[f(x^{-a}); s] = a^{-1} F(-s/a) \quad (a > 0) \\
(e) & M[x^\alpha f(x^\mu); s] = \mu^{-1} F((s + \alpha)/\mu) \quad (\mu > 0) \\
(f) & M[x^\alpha f(x^{-\mu}); s] = \mu^{-1} F(-(s + \alpha)/\mu) \quad (\mu > 0) \\
(g) & M[(\log x)^n f(x); s] = F^{(n)}(s) \quad (n = 1, 2, \ldots).
\end{array}
\right\} \quad (3.1.7)
$$

Similarly there exists a corpus of results for the Mellin transform of derivatives and integrals of $f(x)$. The Mellin transform of $f'(x)$ can be found by integration by parts to yield

$$
\begin{aligned}
M[f'(x); s] &= \int_0^\infty x^{s-1} f'(x) \, dx \\
&= \left[x^{s-1} f(x) \right]_0^\infty - (s - 1) \int_0^\infty x^{s-2} f(x) \, dx.
\end{aligned}
$$

If $f(x)$ satisfies (3.1.3), we have

$$\lim_{x \to 0} x^{s-1} f(x) = 0 \quad \text{for} \quad \text{Re}(s) > a + 1,$$

$$\lim_{x \to \infty} x^{s-1} f(x) = 0 \quad \text{for} \quad \text{Re}(s) < b + 1,$$

and hence

$$M[f'(x); s] = -(s - 1) F(s - 1) \quad (a < \text{Re}(s - 1) < b). \quad (3.1.8)$$

In a similar manner we obtain by induction

$$M[f^{(n)}(x); s] = (-)^n \frac{\Gamma(s)}{\Gamma(s - n)} F(s - n) \quad (a < \text{Re}(s - n) < b) \quad (3.1.9)$$

for positive integer n, provided

$$\lim_{x \to 0, \infty} x^{s-r-1} f^{(n-r-1)}(x) = 0 \quad (r = 0, 1, \ldots, n - 1).$$

From (3.1.7b) and (3.1.8) we find $M[xf'(x); s] = -sF(s)$, and hence that

$$M\left[\left(x\frac{d}{dx}\right)^n f(x); s\right] = (-s)^n F(s) \qquad (n = 1, 2, \ldots). \tag{3.1.10}$$

Further results on the transformation of derivatives and integrals involving $f(x)$ are detailed in Sneddon (1972, pp. 269–271).

The results in this section have tacitly assumed that the strip of analyticity exists, i.e., that $a < b$ in (3.1.3). In the examples considered in this book this will always be found to be the situation. In cases where $a \geq b$, the Mellin transform as defined in (3.1.1) does not exist. For example, if we take the simple function $f(x) = (1 + x)^v$ ($v > 0$) then from (3.1.3) we have $a = 0$ and $b = -v$, so that the strip (3.1.4) is not defined. An extension of the Mellin transform to cover such situations has been discussed in Bleistein & Handelsman (1975, p. 115); see also Wong (1989, §3.4). The procedure is to decompose the function $f(x)$ into two functions $f_1(x)$ and $f_2(x)$ defined on disjoint intervals $[0, 1)$ and $[1, \infty)$, say. Then, with

$$f_1(x) = \begin{cases} f(x) & x \in [0, 1) \\ 0 & x \in [1, \infty) \end{cases}, \qquad f_2(x) = \begin{cases} 0 & x \in [0, 1) \\ f(x) & x \in [1, \infty), \end{cases}$$

we have $f(x) = f_1(x) + f_2(x)$. It then follows that

$$M[f; s] = M[f_1; s] + M[f_2; s],$$

where $M[f_1; s]$ is holomorphic in $\mathrm{Re}(s) > a$ and $M[f_2; s]$ is holomorphic in $\mathrm{Re}(s) < b$. If $a < b$ then, of course, $M[f; s]$ is defined and holomorphic in the strip (3.1.4). In this manner, by appropriate analytic continuation, it is possible to extend the definition of the Mellin transform to produce a transform that exists in the entire s plane. We do not discuss this generalised transform any further.

3.1.3 The Parseval Formula

A fundamental result in Mellin transform theory is an identity known as the *Parseval formula*. Suppose the functions $f(x)$ and $g(x)$ are such that the integral

$$I = \int_0^\infty f(x)g(x)\,dx$$

exists. We assume that the Mellin transforms $M[f; 1 - s]$ and $M[g; s]$ have a common strip of analyticity (which will be the case when I is absolutely convergent) and we take the vertical line $\mathrm{Re}(s) = c$ to lie in this common strip.

Then, proceeding formally, we have

$$J = \frac{1}{2\pi i} \int_{c-\infty i}^{c+\infty i} M[g; s] M[f; 1-s] \, ds$$

$$= \frac{1}{2\pi i} \int_{c-\infty i}^{c+\infty i} M[f; 1-s] \left\{ \int_0^\infty x^{s-1} g(x) \, dx \right\} ds$$

$$= \int_0^\infty g(x) \left\{ \frac{1}{2\pi i} \int_{c-\infty i}^{c+\infty i} x^{s-1} M[f; 1-s] \, ds \right\} dx$$

upon interchanging the order of integration. By the inversion formula (3.1.5) we consequently find that $J = I$, which is the desired Parseval formula:

$$\int_0^\infty f(x) g(x) \, dx = \frac{1}{2\pi i} \int_{c-\infty i}^{c+\infty i} M[f; 1-s] M[g; s] \, ds$$

$$= \frac{1}{2\pi i} \int_{c-\infty i}^{c+\infty i} F(1-s) G(s) \, ds, \qquad (3.1.11)$$

where $F(s)$ and $G(s)$ denote the Mellin transforms of $f(x)$ and $g(x)$. This result is valid provided the above interchange in the order of integration can be justified. There are several sets of conditions that are sufficient for this purpose, the simplest being that if $M[f; 1-c-it] \in L(-\infty, \infty)$ and $x^{c-1} g(x) \in L[0, \infty)$ then the interchange is justified by absolute convergence.

We mention some convolution integrals which are essentially variants of Parseval's formula. From (3.1.7a) we find, for $\alpha > 0$, $\beta > 0$,

$$\int_0^\infty f(\alpha x) g(\beta x) \, dx = \frac{1}{2\pi i} \int_{c-\infty i}^{c+\infty i} F(s) G(1-s) \alpha^{-s} \beta^{s-1} ds \qquad (3.1.12)$$

and, from (3.1.7a) and (3.1.7f),

$$\int_0^\infty f(u/x) g(x) \frac{dx}{x} = \frac{1}{2\pi i} \int_{c-\infty i}^{c+\infty i} M[x^{-1} f(x^{-1}); 1-s] G(s) u^{-s} ds$$

$$= \frac{1}{2\pi i} \int_{c-\infty i}^{c+\infty i} F(s) G(s) u^{-s} ds. \qquad (3.1.13)$$

Application of (3.1.7b) in Parseval's formula shows that

$$\int_0^\infty f(x) g(x) x^{z-1} dx = \frac{1}{2\pi i} \int_{c-\infty i}^{c+\infty i} M[f(x); s] M[x^{z-1} g(x); 1-s] \, dx$$

$$= \frac{1}{2\pi i} \int_{c-\infty i}^{c+\infty i} F(s) G(z-s) \, ds. \qquad (3.1.14)$$

This last result represents $M[f(x) g(x); z]$ and reduces to Parseval's formula when $z = 1$. The identities in (3.1.12)–(3.1.14) are established from first principles in Titchmarsh (1975, pp. 51–54). An application of the Parseval formula for Mellin transforms in the evaluation of various integrals can be found in Marichev (1982).

Finally, if we let $f(x) = g(x)$ be a real function and put $z = 2\sigma, c = \sigma$ in (3.1.14), we find

$$\int_0^\infty f^2(x) x^{2\sigma-1} dx = \frac{1}{2\pi} \int_{-\infty}^\infty |F(\sigma + it)|^2 dt. \qquad (3.1.15)$$

To conclude this section, we note that a result closely related to the Mellin inversion theorem is Ramanujan's formula†

$$\int_0^\infty x^{s-1}\{\phi(0) - x\phi(1) + x^2\phi(2) - \cdots\} dx = \frac{\pi}{\sin \pi s}\phi(-s); \qquad (3.1.16)$$

for a discussion of this result see Edwards (1974, pp. 218–225). If we let $f(x) = \sum_{n=0}^\infty (-)^n \phi(n) x^n$ and suppose that $f(x) = O(x^{-a})$, where $a > 0$, as $x \to +\infty$ then the strip of analyticity for the Mellin transform of $f(x)$ is $0 < \text{Re}(s) < a$. An obvious set of conditions for the validity of (3.1.16), established using the Mellin inversion theorem in Hardy (1920), requires that $\phi(s)$ be a holomorphic function of the complex variable s in the sector $-\alpha \le \arg s \le \alpha$, where $\frac{1}{2}\pi \le \alpha < \pi$, and satisfy the bound $|\phi(s)| < Ae^{k|s|}$, where A is a constant and $k < \pi$ throughout this sector.

A simple example verifying (3.1.16) is given by taking $\phi(s) = 1/\Gamma(s + 1)$ which satisfies the above conditions. Then $f(x) = e^{-x}$ and

$$\int_0^\infty x^{s-1} e^{-x} dx = \frac{\pi}{\sin \pi s \, \Gamma(1 - s)} = \Gamma(s).$$

Note that the integral is convergent only for $\text{Re}(s) > 0$ and that the right-hand side of (3.1.16) gives the analytic continuation of the integral into $\text{Re}(s) \le 0$. Another example is given by taking $\phi(s) = \{\Gamma(s + 1)\Gamma(s + \nu + 1)\}^{-1}$, so that $f(x) = x^{-\nu/2} J_\nu(2x^{1/2})$, where J_ν denotes the Bessel function of order ν. Ramanujan's formula then gives

$$\int_0^\infty x^{s-1-\frac{1}{2}\nu} J_\nu\left(2x^{\frac{1}{2}}\right) dx = \frac{\pi}{\sin \pi s} \frac{1}{\Gamma(1 - s)\Gamma(1 + \nu - s)} = \frac{\Gamma(s)}{\Gamma(1 + \nu - s)}.$$

Upon replacement of x by $\frac{1}{4}x^2$ and s by $\frac{1}{2}s$, this finally yields the result

$$M[x^{-\nu} J_\nu(x); s] = 2^{s-\nu-1} \frac{\Gamma\left(\frac{1}{2}s\right)}{\Gamma\left(1 + \nu - \frac{1}{2}s\right)},$$

for $0 < \text{Re}(s) < \text{Re}(\nu) + \frac{3}{2}$.

---

† An extension of the result (3.1.16) in the form of a Fourier integral is

$$\int_0^\infty \left\{\phi(0) - \frac{\phi(1)}{1!}x + \frac{\phi(2)}{2!}x^2 - \cdots\right\} e^{-ixt} dx = \sum_{n=1}^\infty \phi(-n)(-it)^{n-1}.$$

The conditions of validity of this formula have been discussed by Hardy (1937).

It is easily seen, however, that Ramanujan's formula is not valid when $\phi(s) = \sin \pi s$, since the integral is identically zero while the right-hand side equals $-\pi$. This case is seen to violate the condition on the growth of $\phi(s)$ as $|s| \to \infty$.

3.2 Analytic Properties

We now discuss the analytic properties of the Mellin transform

$$M[f; s] = F(s) = \int_0^\infty x^{s-1} f(x) \, dx \qquad (3.2.1)$$

as a function of the complex variable s, where, as usual, we let $s = \sigma + it$ with σ and t being real. If the behaviour of $f(x)$ in the limits $x \to 0+$ and $x \to +\infty$ is specified by (3.1.3), the above integral is absolutely convergent and defines $M[f; s]$ as an analytic function of s in the strip $a < \mathrm{Re}(s) < b$. The behaviour of $M[f; s]$ as $t \to \pm\infty$ in the strip of analyticity is given by the following lemma:

Lemma 3.1. *For $a < \mathrm{Re}(s) < b$, we have*

$$M[f; s] \to 0 \quad as \quad t \to \pm\infty. \qquad (3.2.2)$$

This result is readily established by application of the Riemann-Lebesgue lemma.

More precise statements about the rate of decay of $M[f; s]$ in the strip can be made with additional information on the behaviour of $f(x)$. Such a result is given in Titchmarsh (1975, p. 47) [see also Bleistein & Handelsman (1975, p. 138 *et seq.*)] as follows:

Lemma 3.2. *Let $f(x)$ be a holomorphic function of the complex variable x in the sector $-\alpha < \arg x < \beta$, where $0 < \alpha, \beta \le \pi$. Let $f(x)$ satisfy (3.1.3) with $a < b$, uniformly in any sector interior to the above sector.*

Then $M[f; s]$, defined in (3.2.1), is a holomorphic function of s in the strip $a < \mathrm{Re}(s) < b$ and

$$M[f; s] = \begin{cases} O\left(e^{-(\beta-\epsilon)t}\right) & as \quad t \to \infty \\ O\left(e^{(\alpha-\epsilon)t}\right) & as \quad t \to -\infty \end{cases}$$

for any (small) positive ϵ, uniformly in any strip interior to $a < \mathrm{Re}(s) < b$.

This result enables us to establish the nature of the decay of the Mellin transform as $t \to \pm\infty$ in the strip of analyticity. To illustrate the application of Lemma 3.2 we consider the functions e^{-x} and $(1 + x)^{-1}$, which correspond to simple examples of functions exhibiting exponential and algebraic decay as $x \to \infty$. These functions have the Mellin transforms $M[e^{-x}; s] = \Gamma(s)$ and $M[(1 + x)^{-1}; s] = \pi/\sin \pi s$, together with the associated strips of analyticity given by the right half-plane $\mathrm{Re}(s) > 0$ and $0 < \mathrm{Re}(s) < 1$, respectively [see (3.2.6)]. Then, for $f(x) = e^{-x}$, we have $\alpha = \beta = \frac{1}{2}\pi$, while for $f(x) = (1+x)^{-1}$

(which has a pole at $x = -1$) $\alpha = \beta = \pi$. By Lemma 3.2, the decay of the Mellin transform of these functions in their respective strips of analyticity is then controlled by $O(e^{-\frac{1}{2}\pi|t|})$ and $O(e^{-\pi|t|})$, respectively. The predicted exponential decay in these examples can be verified by examination of the behaviour of the Mellin transforms as $t \to \pm\infty$, through use of Stirling's formula given in (2.1.8).

We remark that, for functions with algebraic behaviour as $x \to \infty$, the rate of decay of the Mellin transform in the strip of analyticity depends on the location of the singularities of $f(x)$. In the above example with $f(x) = (1 + x)^{-1}$, there is a simple pole at $x = -1$, so that $\alpha = \beta = \pi$, which gives a maximal rate of decay. To demonstrate the dependence of the decay rate on the singularity structure of $f(x)$, let us consider

$$f(x) = (1 + 2x \cos \psi + x^2)^{-1}, \qquad 0 \le \psi \le \pi,$$

which has two simple poles at $x = -e^{\pm i\psi}$. In this case $\alpha = \beta = \pi - \psi$, so that Lemma 3.2 yields the controlling behaviour of the Mellin transform given by $O(e^{-(\pi-\psi)|t|})$ as $t \to \pm\infty$ in the strip of analyticity $0 < \mathrm{Re}(s) < 2$. Thus, when $\psi = 0$ (so that $f(x) = (1 + x)^{-2}$ with a double pole at $x = -1$) the decay is controlled by $O(e^{-\pi|t|})$; as ψ increases towards the value π, the location of the two poles moves round the unit circle in the x plane towards the positive axis and the decay of the Mellin transform accordingly progressively decreases. From the Appendix, we have

$$M[(1 + 2x \cos \psi + x^2)^{-1}; s] = \frac{\pi}{\sin \psi} \frac{\sin[(1 - s)\psi]}{\sin \pi s},$$

when $0 < \mathrm{Re}(s) < 2$, from which these conclusions may be readily verified.

The results so far presented only apply to the behaviour of the Mellin transform $M[f; s]$ inside the strip of analyticity $a < \mathrm{Re}(s) < b$. We now discuss the analytic continuation of $M[f; s]$ outside this strip which will enable us to determine its analytical structure in the complex s plane. A thorough discussion of this aspect of the theory has been given in Bleistein & Handelsman (1975, pp. 109–114) which we summarise below. The analytic continuation of $M[f; s]$ into the half-planes $\mathrm{Re}(s) \le a$ and $\mathrm{Re}(s) \ge b$ is found to depend critically on the structure of $f(x)$ in the limits $x \to 0+$ and $x \to \infty$, respectively. We shall assume that $f(x)$ possesses the asymptotic forms

$$f(x) \sim \begin{cases} \exp(-d_1 x^{-\mu_1}) \displaystyle\sum_{k=0}^{\infty} A_k x^{a_k} & (x \to 0+) \\[2em] \exp(-d_2 x^{\mu_2}) \displaystyle\sum_{k=0}^{\infty} B_k x^{-b_k} & (x \to \infty) \end{cases} \tag{3.2.3}$$

in these limits, where we suppose that† $\text{Re}(d_{1,2}) \geq 0$, $\mu_{1,2} > 0$ and that $\text{Re}(a_k, b_k)$ increase monotonically to ∞ as $k \to \infty$. There are three cases according as $\text{Re}(d_{1,2}) > 0$, $\text{Re}(d_{1,2}) = 0$ and $d_{1,2} = 0$. Then the analytic continuation into the half-planes $\text{Re}(s) \leq a$ and $\text{Re}(s) \geq b$ is given as follows:

Lemma 3.3. *If $\text{Re}(d_1) > 0$ then $a = -\infty$ in (3.1.3). The strip of analyticity in this case is the left half-plane $\text{Re}(s) < b$ and no analytic continuation to the left is required. If $\text{Re}(d_1) = 0$ with $\text{Im}(d_1) \neq 0$, then $f(x)$ is oscillatory as $x \to 0+$ and $M[f; s]$ can be analytically continued into $\text{Re}(s) \leq a = -\text{Re}(a_0)$ as a holomorphic function. The continued Mellin transform possesses the growth*

$$M[f; s] = O(|t|^{m_1(\sigma)}) \quad (t \to \pm\infty, \quad \text{Re}(s) < a),$$

where $m_1(\sigma) = -(\sigma - a)/\mu_1 - \frac{1}{2}$.

Finally, if $d_1 = 0$ so that $f(x)$ is algebraic as $x \to 0+$, then $M[f; s]$ can be analytically continued into the left half-plane $\text{Re}(s) \leq a$ as a meromorphic function with simple poles at the points $s = -a_k$, $k = 0, 1, 2, \ldots$. For $\text{Re}(s) < b$ (3.2.2) holds.

Lemma 3.4. *If $\text{Re}(d_2) > 0$ then $b = \infty$ in (3.1.3). The strip of analyticity in this case is the right half-plane $\text{Re}(s) > a$ and no analytic continuation to the right is required. If $\text{Re}(d_2) = 0$ with $\text{Im}(d_2) \neq 0$, then $f(x)$ is oscillatory at infinity and $M[f; s]$ can be analytically continued into $\text{Re}(s) \geq b = \text{Re}(b_0)$ as a holomorphic function. The continued Mellin transform possesses the growth*

$$M[f; s] = O(|t|^{m_2(\sigma)}) \quad (t \to \pm\infty, \quad \text{Re}(s) > b),$$

where $m_2(\sigma) = (\sigma - b)/\mu_2 - \frac{1}{2}$.

Finally, if $d_2 = 0$ so that $f(x)$ is algebraic at infinity, then $M[f; s]$ can be analytically continued into the right half-plane $\text{Re}(s) \geq b$ as a meromorphic function with simple poles at the points $s = b_k$, $k = 0, 1, 2, \ldots$. For $\text{Re}(s) > a$ (3.2.2) holds.

The continuation and the growth of $M[f; s]$ in the case $\text{Re}(d_{1,2}) = 0$ are established in Bleistein & Handelsman (1975, pp. 111–112) and these arguments are not reproduced here. The meromorphic nature of the continuation when $d_1 = 0$ is established by putting

$$f_n(x) = f(x) - \sum_{k=0}^{n-1} A_k x^{a_k}.$$

---

† The Mellin transform is not defined if $\text{Re}(d_{1,2}) < 0$.

Then, for $a < \mathrm{Re}(s) < b$ we have, upon decomposing the integration path in (3.2.1) into $[0, 1]$ and $[1, \infty)$,

$$M[f; s] = \int_0^1 x^{s-1} f_n(x)\,dx + \sum_{k=0}^{n-1} \frac{A_k}{s + a_k} + \int_1^\infty x^{s-1} f(x)\,dx. \qquad (3.2.4)$$

The first integral on the right is analytic in the strip $\mathrm{Re}(s) > -\mathrm{Re}(a_n)$, whereas the second integral is analytic for $\mathrm{Re}(s) < b$. Then the right-hand side of (3.2.4) represents the continuation of $M[f; s]$ to the strip $-\mathrm{Re}(a_n) < \mathrm{Re}(s) < b$, with simple poles at $s = -a_k$ of residue A_k, $k = 0, 1, 2, \ldots, n - 1$. Since $\mathrm{Re}(a_n) \to +\infty$ as $n \to \infty$, this establishes the result. An analogous argument applies when $d_2 = 0$.

Lemmas 3.3 and 3.4, with the asymptotic forms in (3.2.3), enable the analytic continuation of the Mellin transform $M[f; s]$ to be carried out over the whole complex s plane as a meromorphic function at worst. The poles that arise in the continuation process are related to the exponents a_k and b_k in the expansions (3.2.3). The proof of these results in a more general setting, where $f(x)$ is allowed to have logarithmic growth as $x \to 0+$ and $x \to \infty$, is given in Bleistein & Handelsman (1975, pp. 111–114). There, the more general forms of expansion

$$f(x) \sim \begin{cases} \exp(-d_1 x^{-\mu_1}) \displaystyle\sum_{k=0}^{\infty} \sum_{n=0}^{N_1(k)} A_{nk} (\log x)^n x^{a_k} & (x \to 0+) \\[2em] \exp(-d_2 x^{\mu_2}) \displaystyle\sum_{k=0}^{\infty} \sum_{n=0}^{N_2(k)} B_{nk} (\log x)^n x^{-b_k} & (x \to \infty) \end{cases} \qquad (3.2.5)$$

are assumed, where the integers $N_{1,2}(k), k = 0, 1, 2, \ldots$ are all positive and finite. The above form is very general and allows $f(x)$ to possess almost any behaviour in these two limits that might reasonably be expected in applications. When logarithmic terms are present, the order of any given pole arising in the meromorphic extensions is determined solely by the highest power of $\log x$ appearing in (3.2.5).

The above basic features of the analytic continuation of $M[f; s]$ can be verified by considering archetypal examples of the three basic types of limiting behaviour of $f(x)$, namely (i) exponential decay, (ii) exponential oscillation and (iii) algebraic decay. Simple examples of such behaviour as $x \to \infty$ are supplied by (i) e^{-x}, (ii) e^{ix} and (iii) $(1+x)^{-1}$ for which, from the Appendix, we have the Mellin transforms

$$\left. \begin{aligned} M[e^x; s] &= \Gamma(s) & (\mathrm{Re}(s) > 0) \\[1em] M[e^{ix}; s] &= e^{\frac{1}{2}\pi i s} \Gamma(s) & (0 < \mathrm{Re}(s) < 1) \\[1em] M[(1 + x)^{-1}; s] &= \frac{\pi}{\sin \pi s} & (0 < \mathrm{Re}(s) < 1). \end{aligned} \right\} \qquad (3.2.6)$$

We note that, although holomorphic in the strip $0 < \mathrm{Re}(s) < 1$, $M[e^{ix}; s]$ exists only as a conditionally convergent integral. The analytic continuation of these three Mellin transforms is, of course, immediate and supplied by the right-hand sides of (3.2.6). It will be seen that both $M[e^{-x}; s]$ and $M[e^{ix}; s]$ are holomorphic in $\mathrm{Re}(s) \geq 1$, whereas $M[(1+x)^{-1}; s]$ is meromorphic there possessing simple poles at $s = 1, 2, \ldots$. The analytic continuations in $\mathrm{Re}(s) \leq 0$, however, are all meromorphic, possessing simple poles at $s = 0, -1, -2, \ldots$. This follows because all three functions possess algebraic behaviour as $x \to 0+$.

In conclusion, we remark that if $f(x)$ has algebraic behaviour in both limits $x \to 0+$ and $x \to \infty$, then poles will arise in the analytic continuation of $M[f; s]$ on both sides of the strip of analyticity; compare (3.2.6c). However, if $f(x)$ has exponential behaviour (whether exponential decay or oscillation) in *both* limits, the analytic continuation of $M[f; s]$ will have *no singularities in the finite plane*.

3.3 Inverse Mellin Transforms

The inversion formula for the Mellin transform is given in (3.1.5). This states that if a function $f(x)$ is locally integrable on $(0, \infty)$ and has the Mellin transform $F(s) = M[f; s]$ in the strip of analyticity $a < \mathrm{Re}(s) < b$, then the inverse Mellin transform is given by

$$f(x) = M^{-1}[F(s); x] = \frac{1}{2\pi i} \int_{c-\infty i}^{c+\infty i} x^{-s} F(s)\, ds \quad (a < c < b). \qquad (3.3.1)$$

In this section we discuss some important inverse Mellin transforms, which are employed extensively in later chapters, together with a collection of useful results. A short table of Mellin transforms is given in the Appendix; a more complete set of tables can be found in Erdélyi (1954), Oberhettinger (1974) and Marichev (1982).

3.3.1 Integrals Connected with e^{-z}

The most important result in the development of asymptotic expansions by means of Mellin-Barnes integrals is the integral

$$e^{-z} = \frac{1}{2\pi i} \int_{c-\infty i}^{c+\infty i} \Gamma(s) z^{-s} ds \quad \left(|\arg z| < \tfrac{1}{2}\pi; \; z \neq 0 \right), \qquad (3.3.2)$$

where the path of integration is the vertical line $\mathrm{Re}(s) = c$, with $c > 0$, lying to the right of all the poles of $\Gamma(s)$. This integral is known as the Cahen-Mellin integral. A discussion of the convergence of this integral has been given in §2.4, where it was shown that the modulus of the integrand has the control-ling behaviour $|z|^{-\sigma} O(|t|^{\sigma-\frac{1}{2}} e^{t \arg z - \frac{1}{2}\pi|t|})$ as $t \to \pm\infty$, where $s = \sigma + it$. Because of this exponential decay, we are free to displace the contour of inte-gration (when $|\arg z| < \frac{1}{2}\pi$) over the poles at $s = -k$ of $\Gamma(s)$, with residue $(-)^k/k!$ ($k = 0, 1, 2, \ldots$), to produce the exponential series. An alternative proof is to use the Mellin inversion formula applied to the integral representation for the

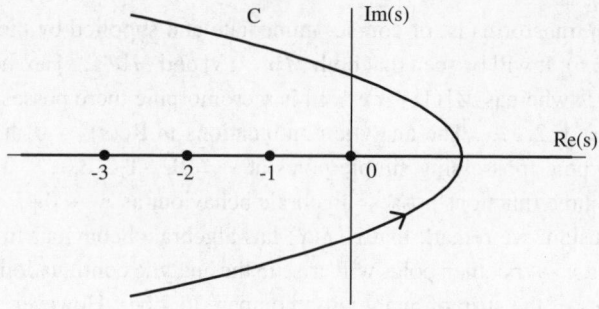

Fig. 3.1. The loop C in the complex s plane.

gamma function

$$\Gamma(x) = \int_0^\infty t^{x-1}e^{-t}dt \quad (\mathrm{Re}(x) > 0).$$

The above representation has the inconvenience of defining e^{-z} only in the sector† $|\arg z| < \frac{1}{2}\pi$. The analytic continuation of (3.3.2) is given by

$$e^{-z} = \frac{1}{2\pi i}\int_C \Gamma(s)z^{-s}ds \quad (z \neq 0), \tag{3.3.3}$$

where C denotes a loop in the complex s plane that encircles the poles of $\Gamma(s)$ (in the positive sense) with endpoints at infinity in $\mathrm{Re}(s) < 0$; see Fig. 3.1. If C approaches ∞ in the directions $\pm\theta_0$ say, where $\frac{1}{2}\pi < \theta_0 \leq \pi$, then from (2.4.4) the logarithm of the modulus of the integrand on C (where $s = Re^{i\theta}$) behaves like $R\cos\theta_0 \log R + O(R) \to -\infty$ as $R \to \infty$. Consequently (3.3.3) defines e^{-z} *without restriction* on $\arg z$ and thereby represents the analytic continuation of (3.3.2).

Straightforward extensions of (3.3.3) are

$$z^{-\alpha}e^{-z} = \frac{1}{2\pi i}\int_C \Gamma(s - \alpha)z^{-s}ds \tag{3.3.4}$$

and

$$z^{-\alpha}\begin{pmatrix}\sin\\\cos\end{pmatrix}(z + \tfrac{1}{2}\pi\alpha) = \frac{1}{2\pi i}\int_C \Gamma(s - \alpha)z^{-s}\begin{pmatrix}\sin\\\cos\end{pmatrix}(\tfrac{1}{2}\pi s)\,ds, \tag{3.3.5}$$

which hold for all $\arg z$ and arbitrary α provided the loop C embraces all the poles of $\Gamma(s - \alpha)$. We note that if C is bent back into the vertical path $\mathrm{Re}(s) = c$,

---

† The integral (3.3.2) also holds when $|\arg z| = \frac{1}{2}\pi$ provided $0 < c \leq \frac{1}{2}$; see §2.4.

with $c > \mathrm{Re}(\alpha)$, (3.3.4) is valid for $|\arg z| < \frac{1}{2}\pi$, while from (3.3.5) we have

$$x^{-\alpha} \begin{pmatrix} \sin \\ \cos \end{pmatrix} (x + \tfrac{1}{2}\pi\alpha) = \frac{1}{2\pi i} \int_{c-\infty i}^{c+\infty i} \Gamma(s - \alpha) x^{-s} \begin{pmatrix} \sin \\ \cos \end{pmatrix} \left(\tfrac{1}{2}\pi s\right) ds. \quad (3.3.6)$$

In this case, the exponential decay contained in the gamma function, given by $O(t^{c-\mathrm{Re}(\alpha)-\frac{1}{2}} e^{-\frac{1}{2}\pi|t|})$ as $t \to \pm\infty$, is offset by the $e^{\frac{1}{2}\pi|t|}$ growth in the trigonometric function in the integrand, with the result that the integrals converge only when $x > 0$ and $\mathrm{Re}(\alpha) < c \le \mathrm{Re}(\alpha) + \frac{1}{2}$.

Another straightforward extension of (3.3.3) is given by the integrals

$$e^{z\cos\pi\beta} \begin{pmatrix} \sin \\ \cos \end{pmatrix} (z\sin\pi\beta) = \frac{1}{2\pi i} \int_C \Gamma(s) z^{-s} \begin{pmatrix} \sin \\ \cos \end{pmatrix} (\pi(1-\beta)s) \, ds, \quad (3.3.7)$$

which are also valid for all $\arg z$ and arbitrary β, provided the loop C embraces all the poles of $\Gamma(s)$.

3.3.2 Some Standard Integrals

Numerous special evaluations of inverse Mellin integrals (or Mellin-Barnes integrals) can be generated by means of the Mellin inversion theorem in (3.1.5) and (3.1.6). We collect together here some useful examples for reference.

From the standard formula for the beta function $B(x, y)$ [Abramowitz & Stegun (1965, p. 258)]

$$B(x, y) = \int_0^1 \tau^{x-1}(1-\tau)^{y-1} d\tau = \int_0^\infty \frac{\tau^{x-1}}{(1+\tau)^{x+y}} d\tau$$

$$= \frac{\Gamma(x)\Gamma(y)}{\Gamma(x+y)} \quad (\mathrm{Re}(x, y) > 0), \quad (3.3.8)$$

we have

$$M[(1 + x)^{-a}; s] = \frac{\Gamma(s)\Gamma(a - s)}{\Gamma(a)}$$

when $0 < \mathrm{Re}(s) < \mathrm{Re}(a)$. It then follows from (3.1.5) that

$$\frac{1}{2\pi i} \int_{c-\infty i}^{c+\infty i} \Gamma(s)\Gamma(a - s) x^{-s} ds = \frac{\Gamma(a)}{(1 + x)^a} \quad (3.3.9)$$

when $0 < c < \mathrm{Re}(a)$ and, from Rule 1 in §2.4, in the sector $|\arg x| < \pi$.

When $a = 1$ in (3.3.9) we obtain the formula

$$\frac{1}{2\pi i} \int_{c-\infty i}^{c+\infty i} \frac{\pi}{\sin\pi s} x^{-s} ds = \frac{1}{1 + x} \quad (0 < c < 1; \, |\arg x| < \pi). \quad (3.3.10)$$

Generalisations of this last result are given by

$$\frac{1}{2\pi i}\int_{c-\infty i}^{c+\infty i}\frac{\cos s\phi}{\sin \pi s}\pi x^{-s}\,ds = \frac{1+x\cos\phi}{1+2x\cos\phi+x^2}, \qquad (3.3.11)$$

$$\frac{1}{2\pi i}\int_{c-\infty i}^{c+\infty i}\frac{\sin s\phi}{\sin \pi s}\pi x^{-s}\,ds = \frac{x\sin\phi}{1+2x\cos\phi+x^2}, \qquad (3.3.12)$$

where $0 < c < 1$ and $-\pi < \phi < \pi$. These results are easily deduced by writing $\cos s\phi$ and $\sin s\phi$ in terms of exponentials and employing (3.3.10).

From the result† $M[(1-x)^{-1}; s] = \pi \cot \pi s$ for $0 < \operatorname{Re}(s) < 1$ [see the Appendix], we find

$$\frac{1}{2\pi i}\int_{c-\infty i}^{c+\infty i}\frac{\pi}{\tan \pi s}x^{-s}\,ds = \frac{1}{1-x} \quad (0 < c < 1;\ x > 0). \qquad (3.3.13)$$

Similarly, since

$$M[\log(1+x); s] = \frac{\pi}{s\sin \pi s}, \quad M[\log|1-x|; s] = \frac{\pi}{s\tan \pi s}$$

for $-1 < \operatorname{Re}(s) < 0$, we consequently find from (3.1.5) (upon replacing the variable s by $-s$)

$$\frac{1}{2\pi i}\int_{c-\infty i}^{c+\infty i}\frac{\pi}{\sin \pi s}x^s\frac{ds}{s} = \log(1+x) \quad (|\arg x| < \pi), \qquad (3.3.14)$$

$$\frac{1}{2\pi i}\int_{c-\infty i}^{c+\infty i}\frac{\pi}{\tan \pi s}x^s\frac{ds}{s} = \log|1-x| \quad (x > 0), \qquad (3.3.15)$$

where $0 < c < 1$. From (3.3.14) and the fact that $M[\operatorname{arccot} x; s] = (\pi/2s)\sec\frac{1}{2}\pi s$ when $0 < \operatorname{Re}(s) < 1$, we also have

$$\frac{1}{2\pi i}\int_{c-\infty i}^{c+\infty i}\frac{\pi}{\sin \frac{1}{2}\pi s}x^{-s}\frac{ds}{s} = \log(1+x^{-2}) \quad (0 < c < 2), \qquad (3.3.16)$$

$$\frac{1}{2\pi i}\int_{c-\infty i}^{c+\infty i}\frac{\pi}{\cos \frac{1}{2}\pi s}x^{-s}\frac{ds}{s} = 2\operatorname{arccot} x \quad (0 < c < 1) \qquad (3.3.17)$$

when $|\arg x| < \frac{1}{2}\pi$.

Further formulas of this type can be obtained from the results [Erdélyi (1954, p. 314)]

$$M[-(1+x)^{-1}\log x; s] = \left(\frac{\pi}{\sin \pi s}\right)^2 \cos \pi s,$$

$$M[(x-1)^{-1}\log x; s] = \left(\frac{\pi}{\sin \pi s}\right)^2$$

† This integral is a Cauchy principal value.

when $0 < \mathrm{Re}(s) < 1$, so that

$$\frac{1}{2\pi i} \int_{c-\infty i}^{c+\infty i} \left(\frac{\pi}{\sin \pi s}\right)^2 x^{-s} ds = \frac{\log x}{x-1}, \tag{3.3.18}$$

$$\frac{1}{2\pi i} \int_{c-\infty i}^{c+\infty i} \left(\frac{\pi}{\sin \pi s}\right)^2 \cos \pi s \, x^{-s} ds = -\frac{\log x}{1+x}, \tag{3.3.19}$$

which are both valid when $0 < c < 1$ and $|\arg x| < \pi$ [Nielsen (1906, p. 232)].

3.3.3 Discontinuous Integrals

As an example of a discontinuous integral we let $f(x)$ denote the simple step function given by $f(x) = 1$ for $0 < x < 1$ and $f(x) = 0$ for $x > 1$. The Mellin transform of this function is easily shown to be $M[f; s] = s^{-1}$ when $\mathrm{Re}(s) > 0$. Then from (3.1.6) we obtain the result

$$\frac{1}{2\pi i} \int_{c-\infty i}^{c+\infty i} x^s \frac{ds}{s} = \begin{cases} 0 & (0 < x < 1) \\ \frac{1}{2} & (x = 1) \\ 1 & (x > 1) \end{cases} \tag{3.3.20}$$

provided $c > 0$. An alternative representation of this well-known result is found by replacing x by e^τ to yield, when $c > 0$,

$$\frac{1}{2\pi i} \int_{c-\infty i}^{c+\infty i} e^{s\tau} \frac{ds}{s} = \begin{cases} 0 & (\tau < 0) \\ \frac{1}{2} & (\tau = 0) \\ 1 & (\tau > 0). \end{cases} \tag{3.3.21}$$

An obvious extension of (3.3.20) is given by Titchmarsh (1975, p. 202) in the form

$$\frac{1}{2\pi i} \int_{c-\infty i}^{c+\infty i} \frac{x^{-s}}{s+a} ds = \begin{cases} x^a & (0 < x < 1) \\ \frac{1}{2} & (x = 1) \\ 0 & (x > 1), \end{cases} \tag{3.3.22}$$

where $c > -a$. This last integral is related to the discontinuous integrals involving products of Bessel functions discussed in Watson (1966, p. 406).

We now examine in more detail the derivation of the discontinuous integral in (3.3.20). This can be seen directly by application of Cauchy's theorem since the integration contour can be completed by a large semi-circle on the right or left according as $0 < x < 1$ or $x > 1$, respectively. In the special case $x = 1$ the integral in (3.3.20) becomes

$$\lim_{T \to \infty} \frac{1}{2\pi} \int_{-T}^{T} \frac{dt}{c+it} = \lim_{T \to \infty} \frac{1}{2\pi} \int_{-T}^{T} \frac{c-it}{c^2+t^2} dt$$

$$= \frac{1}{2\pi} \int_{-\infty}^{\infty} \frac{c \, dt}{c^2+t^2} = \frac{1}{2}. \tag{3.3.23}$$

An equivalent way of establishing (3.3.20), and which yields as a by-product an estimate for the integral

$$\frac{1}{2\pi i} \int_{c-iT}^{c+iT} x^s \frac{ds}{s} \qquad (c > 0),\tag{3.3.24}$$

is described by Edwards (1974, p. 54). Consider a rectangle in the s plane with vertices at the points $c \pm iT$ and $M \pm iT$, where $M > c > 0$ and $T > 0$. By Cauchy's theorem the integral of x^s/s round this rectangle is zero and hence

$$\frac{1}{2\pi i} \int_{c-iT}^{c+iT} x^s \frac{ds}{s} = \frac{1}{2\pi i} \left\{ -\int_{c+iT}^{M+iT} + \int_{c-iT}^{M-iT} + \int_{M-iT}^{M+iT} \right\} x^s \frac{ds}{s}.$$

The last integral on the right-hand side is bounded by $(2\pi)^{-1}(x^M/M)2T$ while the modulus of the first two integrals is bounded by

$$\frac{1}{2\pi} \int_c^M \frac{x^\sigma}{\sqrt{\sigma^2 + T^2}} \, d\sigma < \frac{1}{2\pi T} \int_c^M x^\sigma \, d\sigma = \frac{1}{2\pi T} \frac{x^M - x^c}{\log x}.$$

This then yields the result

$$\left| \frac{1}{2\pi i} \int_{c-iT}^{c+iT} x^s \frac{ds}{s} \right| < \frac{|x^M - x^c|}{\pi T |\log x|} + \frac{x^M T}{\pi M}.$$

If we now allow $M \to \infty$ when $0 < x < 1$, we find the bound

$$\left| \frac{1}{2\pi i} \int_{c-iT}^{c+iT} x^s \frac{ds}{s} \right| < \frac{x^c}{\pi T |\log x|} \qquad (c > 0;\ 0 < x < 1).\tag{3.3.25}$$

A similar procedure for $x > 1$ consists of considering the integral of x^s/s taken round a rectangle with vertices at $c \pm iT$ and $-M \pm iT$, where again $M > 0$. In this case, the integral round the rectangle equals 1 since the simple pole at $s = 0$ lies inside the contour. Similar reasoning as in the case $0 < x < 1$ then shows that

$$\left| \frac{1}{2\pi i} \int_{c-iT}^{c+iT} x^s \frac{ds}{s} - 1 \right| = \left| \frac{1}{2\pi i} \left\{ \int_{-M+iT}^{c+iT} - \int_{-M-iT}^{c-iT} + \int_{-M-iT}^{-M+iT} \right\} x^s \frac{ds}{s} \right|$$

$$< \frac{x^c - x^{-M}}{\pi T \log x} + \frac{x^{-M} T}{\pi M}.$$

As $M \to \infty$, we therefore find that

$$\left| \frac{1}{2\pi i} \int_{c-iT}^{c+iT} x^s \frac{ds}{s} - 1 \right| < \frac{x^c}{\pi T \log x} \qquad (c > 0;\ x > 1).\tag{3.3.26}$$

If we now let $T \to \infty$ in (3.3.25) and (3.3.26) we obtain the limits 0 and 1 for the integral (3.3.24) when $0 < x < 1$ and $x > 1$, which, together with (3.3.23), establishes (3.3.20).

Other discontinuous integrals can be generated by taking, for example,

$$f(x) = \log^n x \quad (0 < x < 1), \qquad f(x) = 0 \quad (x \geq 1)$$

and

$$f(x) = (1 - x)^a \quad (0 < x < 1), \quad f(x) = 0 \quad (x \geq 1),$$

for positive integer n and $a > 0$. The Mellin transforms of these functions are $M[f; s] = (-)^n s^{-n-1} n!$ and $M[f; s] = \Gamma(a+1)\Gamma(s)/\Gamma(s+a+1)$ for $\mathrm{Re}(s) > 0$; see the Appendix. The inversion formula (3.1.6) therefore yields the results, when $c > 0$,

$$\frac{1}{2\pi i} \int_{c-\infty i}^{c+\infty i} x^{-s} \frac{ds}{s^{n+1}} = \begin{cases} \dfrac{(-)^n}{n!} \log^n x & (0 < x \leq 1) \\ 0 & (x > 1) \end{cases} \tag{3.3.27}$$

and

$$\frac{1}{2\pi i} \int_{c-\infty i}^{c+\infty i} x^{-s} \frac{\Gamma(s)}{\Gamma(s + a + 1)} ds = \begin{cases} \dfrac{(1 - x)^a}{\Gamma(a + 1)} & (0 < x \leq 1) \\ 0 & (x > 1). \end{cases} \tag{3.3.28}$$

A useful Mellin-Barnes integral representation for a sum of a finite number of terms is given by *Perron's formula* [Titchmarsh (1939, p. 300); Ivić (1985, p. 485)], which can be established by means of the result (3.3.20). Given a Dirichlet series of the form

$$D(s) = \sum_{n=1}^{\infty} a_n n^{-s},$$

with a finite abscissa of absolute convergence $\mathrm{Re}(s) = \sigma_0$, say, then we have the result

$$\sum_{n \leq x}' a_n = \frac{1}{2\pi i} \int_{c-\infty i}^{c+\infty i} D(s) x^s \frac{ds}{s}, \tag{3.3.29}$$

where $c > \sigma_0$ so that $D(s)$ is absolutely convergent on the integration path $\mathrm{Re}(s) = c$. Here the prime on the summation sign means that when x is an integer the last term in the sum is replaced by $\frac{1}{2}a_x$. Perron's formula (3.3.29) can be derived by using the fact that termwise integration of $D(s)$ on $\mathrm{Re}(s) = c$ is permissible by absolute convergence, to obtain

$$\frac{1}{2\pi i} \int_{c-\infty i}^{c+\infty i} D(s) x^s \frac{ds}{s} = \sum_{n=1}^{\infty} a_n \frac{1}{2\pi i} \int_{c-\infty i}^{c+\infty i} (x/n)^s \frac{ds}{s} = \sum_{n \leq x}' a_n$$

by virtue of (3.3.20).

Another formula of this type (also known by the generic name of Perron's formula) is given by

$$\sum_{n \leq x}' a_n n^{-w} = \frac{1}{2\pi i} \int_{c-\infty i}^{c+\infty i} D(s + w) x^s \frac{ds}{s}, \tag{3.3.30}$$

where $c + \text{Re}(w)$ exceeds the abscissa of absolute convergence of $D(s)$. This can be established in a similar fashion to (3.3.29); for details, and for other formulas of this type, see Titchmarsh (1939, p. 300) and Ivić (1987, pp. 486–487).

3.3.4 Gamma-Function Integrals

Some useful evaluations of Mellin-Barnes integrals involving gamma functions can also be carried out with the aid of Parseval's formula in (3.1.11) and its generalisation in (3.1.14).

If we let $f(x) = x^a e^{-x}$ and $g(x) = z^{-b} x^{b-1} e^{-x/z}$, where $z > 0$, we have the Mellin transforms $M[f; s] = \Gamma(s + a)$ when $\text{Re}(s + a) > 0$ and

$$M[g; s] = z^{-b} \int_0^\infty x^{s+b-2} e^{-x/z} \, dx = z^{s-1} \Gamma(s + b - 1)$$

when $\text{Re}(1 - s) < \text{Re}(b)$. From (3.1.11), we then find that

$$\frac{1}{2\pi i} \int_{c-\infty i}^{c+\infty i} \Gamma(s + a) \Gamma(b - s) z^{-s} \, ds = z^{-b} \int_0^\infty x^{a+b-1} \exp\{-x(1 + z^{-1})\} \, dx$$

$$= z^a (1 + z)^{-a-b} \Gamma(a + b), \qquad (3.3.31)$$

provided $\text{Re}(a + b) > 0$ and $-\text{Re}(a) < c < \text{Re}(b)$. This result can be extended to complex values of z satisfying $|\arg z| < \pi$ by analytic continuation; see also Rule 1 in §2.4. Special cases of (3.3.31) are found by putting $a = 0$, which gives (3.3.9), and $z = 1$ to yield

$$\frac{1}{2\pi i} \int_{c-\infty i}^{c+\infty i} \Gamma(s + a) \Gamma(b - s) \, ds = 2^{-a-b} \Gamma(a + b) \qquad (3.3.32)$$

and, with $a = b > 0$, to yield

$$\int_0^\infty |\Gamma(a + it)|^2 \, dt = 2^{-2a} \pi \Gamma(2a). \qquad (3.3.33)$$

If we take $f(x)$ to be defined as in the example above but now let $g(x) = z^b x^{-1-b} e^{-z/x}$, where $z > 0$, so that

$$M[g; s] = z^b \int_0^\infty x^{s-b-2} e^{-z/x} \, dx = z^{s-1} \Gamma(1 + b - s)$$

when† $\text{Re}(1 - s) > -\text{Re}(b)$, we have

$$\frac{1}{2\pi i} \int_{c-\infty i}^{c+\infty i} \Gamma(s + a) \Gamma(s + b) z^{-s} \, ds = z^b \int_0^\infty x^{a-b-1} \exp\left\{-x - \frac{z}{x}\right\} \, dx$$

$$= 2 z^{\frac{1}{2}a + \frac{1}{2}b} K_{a-b}(2z^{\frac{1}{2}}), \qquad (3.3.34)$$

---

† This convergence condition originates from the behaviour of the integrand at the upper limit of integration.

provided $c > \max\{-\mathrm{Re}(a), -\mathrm{Re}(b)\}$. The above integral has been evaluated in terms of the modified Bessel function of the second kind by means of the representation in Watson (1966, p. 183)

$$K_\nu(x) = \tfrac{1}{2}\left(\tfrac{1}{2}x\right)^\nu \int_0^\infty \exp\left\{-\tau - \frac{x^2}{4\tau}\right\} \tau^{-\nu-1}d\tau \quad (\mathrm{Re}(x^2) > 0)$$

and use of the fact that $K_{-\nu}(x) = K_\nu(x)$. When $a = b$ we find the formula

$$\frac{1}{2\pi i}\int_{c-\infty i}^{c+\infty i} \Gamma^2(s + a)z^{-s}ds = 2z^a K_0\big(2z^{\frac{1}{2}}\big), \tag{3.3.35}$$

provided $c > -\mathrm{Re}(a)$. Analytic continuation enables both results in (3.3.34) and (3.3.35) to be extended to $|\arg z| < \pi$.

A more involved evaluation concerns the integral

$$I_1 = \frac{1}{2\pi i}\int_{k-\infty i}^{k+\infty i} \frac{\Gamma(s + a)\Gamma(c - s)}{\Gamma(s + b)\Gamma(d - s)}\,ds. \tag{3.3.36}$$

In this case the integrand decays algebraically like $O\big(|t|^{\mathrm{Re}(a-b+c-d)}\big)$ as $t \to \pm\infty$ on the path of integration. Thus I_1 converges absolutely when

$$\mathrm{Re}(a - b + c - d) < -1. \tag{3.3.37}$$

To evaluate the integral using Parseval's formula we let

$$f(x) = x^a(1 - x)^{b-a-1} \quad (0 < x < 1), \quad f(x) = 0 \quad (x > 1),$$

with $g(x)$ similarly defined with the parameters a and b replaced by $c - 1$ and $d - 1$, respectively. Then, by the first integral in (3.3.8), we have

$$M[f; s] = \frac{\Gamma(s + a)\Gamma(b - a)}{\Gamma(s + b)}, \quad M[g; s] = \frac{\Gamma(s + c - 1)\Gamma(d - c)}{\Gamma(s + d - 1)}$$

provided $\mathrm{Re}(s + a) > 0$, $\mathrm{Re}(b - a) > 0$ for $M[f; s]$ and $\mathrm{Re}(1 - s) < \mathrm{Re}(c)$, $\mathrm{Re}(d - c) > 0$ for $M[g; s]$. Substitution of these expressions in Parseval's formula (3.1.11) then produces

$$I_1 = \frac{1}{\Gamma(b - a)\Gamma(d - c)}\int_0^1 x^{a+c-1}(1 - x)^{b+d-a-c-2}\,dx$$

$$= \frac{\Gamma(a + c)\Gamma(b + d - a - c - 1)}{\Gamma(b - a)\Gamma(d - c)\Gamma(b + d - 1)}, \tag{3.3.38}$$

provided (3.3.37) holds and, in addition, $\mathrm{Re}(a + c) > 0$. This last condition, combined with the two conditions above on the parameters, yields $\mathrm{Re}(d) > \mathrm{Re}(c) > -\mathrm{Re}(a) > -\mathrm{Re}(b)$. The path of integration $\mathrm{Re}(s) = k$ in I_1 then lies in the common strip of analyticity given by $-\mathrm{Re}(a) < \mathrm{Re}(s) < \mathrm{Re}(c)$.

We note that these conditions on the validity of (3.3.38) are unnecessarily restrictive and can be removed by analytic continuation in the parameters. The integral I_1 can, of course, also be evaluated by completing the integration contour by a

circular arc on the right and straightforward evaluation of the residues at the poles
of $\Gamma(c - s)$ at $s = c + n$, $n = 0, 1, 2, \ldots$. Provided the arc is chosen not to pass
through a neighbourhood of a pole, the contribution round the arc will tend to zero
as $|s| \to \infty$ when (3.3.37) holds, so that

$$
\begin{aligned}
I_1 &= \sum_{n=0}^{\infty} \frac{(-)^n \Gamma(a + c + n)}{n!\, \Gamma(b + c + n)\Gamma(d - c - n)} \\
&= \frac{\sin \pi(d - c)}{\pi} \sum_{n=0}^{\infty} \frac{\Gamma(a + c + n)\Gamma(1 + c - d + n)}{n!\, \Gamma(b + c + n)} \\
&= \frac{\Gamma(a + c)}{\Gamma(b + c)\Gamma(d - c)}\, {}_2F_1(a + c, 1 + c - d; b + c; 1).
\end{aligned}
$$

This yields (3.3.38) upon application of Gauss' summation formula for the
hypergeometric function of unit argument given by

$$
{}_2F_1(\alpha, \beta; \gamma; 1) = \frac{\Gamma(\gamma)\Gamma(\gamma - \alpha - \beta)}{\Gamma(\gamma - \alpha)\Gamma(\gamma - \beta)}, \tag{3.3.39}
$$

when $\gamma \neq 0, -1, -2, \ldots$ and $\mathrm{Re}(\gamma - \alpha - \beta) > 0$; see Abramowitz & Stegun
(1965, p. 556). The result (3.3.38) will therefore be valid when (3.3.37) holds provided the path of integration in (3.3.36) (suitably indented if necessary) separates[†]
the poles of $\Gamma(s + a)$ from those of $\Gamma(c - s)$.

Another result of this type can be obtained by letting

$$
f(x) = x^a (1 - x)^{-b-c} \quad (0 < x < 1), \quad f(x) = 0 \quad (x > 1)
$$

and $g(x) = x^{c-1}(1 + x)^{-b-c}$ $(x \geq 0)$. Then we have the Mellin transforms given
by

$$
M[f; s] = \frac{\Gamma(s + a)\Gamma(1 - b - c)}{\Gamma(s + a - b - c + 1)}, \quad M[g; s] = \frac{\Gamma(s + c - 1)\Gamma(b + 1 - s)}{\Gamma(b + c)}
$$

provided $\mathrm{Re}(s + a) > 0$, $\mathrm{Re}(b + c) < 1$ for $M[f; s]$ and $\mathrm{Re}(c) > \mathrm{Re}(1 - s)$,
$\mathrm{Re}(1 - s) > -\mathrm{Re}(b)$ for $M[g; s]$. We therefore find from Parseval's formula

$$
\begin{aligned}
I_2 &= \frac{1}{2\pi i} \int_{k-\infty i}^{k+\infty i} \frac{\Gamma(s + a)\Gamma(s + b)\Gamma(c - s)}{\Gamma(s + a - b - c + 1)}\, ds \\
&= \frac{\Gamma(b + c)}{\Gamma(1 - b - c)} \int_0^1 x^{a+c-1}(1 - x^2)^{-b-c}\, dx \\
&= \frac{\Gamma(b + c)}{2\Gamma(1 - b - c)} \int_0^1 u^{\frac{1}{2}a + \frac{1}{2}c - 1}(1 - u)^{-b-c}\, du \\
&= \frac{\Gamma(b + c)\Gamma\left(\frac{1}{2}a + \frac{1}{2}c\right)}{2\,\Gamma\left(1 - b + \frac{1}{2}a - \frac{1}{2}c\right)}, \tag{3.3.40}
\end{aligned}
$$

† This will always be possible when $a + c \neq 0, -1, -2, \ldots$.

where $\mathrm{Re}(a + c) > 0$ and $0 < \mathrm{Re}(b + c) < 1$. The integration path $\mathrm{Re}(s) = k$ in
(3.3.40) lies in the common strip of analyticity

$$\max\{-\mathrm{Re}(a), -\mathrm{Re}(b)\} < \mathrm{Re}(s) < \mathrm{Re}(c),$$

where the conditions $\mathrm{Re}(a + c) > 0$ and $0 < \mathrm{Re}(b + c) < 1$ on the parameters
ensure that this strip is of finite width.

As in the preceding example, these restrictions are unnecessarily restrictive and
can be considerably relaxed. On a circular arc on the right of the path the modulus
of the integrand has the behaviour $O(s^{2b+2c-2}e^{-\pi|\mathrm{Im}(s)|})$ so that, provided the arc
does not pass through a neighbourhood of a pole, the contribution round this arc
tends to zero as $|s| \to \infty$ when $\mathrm{Re}(b + c) < \frac{1}{2}$. Evaluation of the residues at the
poles of $\Gamma(c - s)$ then yields

$$I_2 = \sum_{n=0}^{\infty} \frac{(-)^n \Gamma(a + c + n)\Gamma(b + c + n)}{n!\,\Gamma(a - b + 1 + n)}$$

$$= \frac{\Gamma(a + c)\Gamma(b + c)}{\Gamma(a - b + 1)}\,{}_2F_1(a + c; b + c; a - b + 1; -1)$$

which reduces to the result in (3.3.40) upon use of the summation formula
[Abramowitz & Stegun (1965, p. 557)]

$$_2F_1(\alpha, \beta; \alpha - \beta + 1; -1) = \frac{2^{-\alpha}\pi^{\frac{1}{2}}\Gamma(\alpha - \beta + 1)}{\Gamma(1 + \frac{1}{2}\alpha - \beta)\Gamma(\frac{1}{2} + \frac{1}{2}\alpha)},$$

provided $\alpha - \beta + 1 \neq 0, -1, -2, \ldots$, and the duplication formula for the gamma
function in (2.2.23). The above result has been established when $\mathrm{Re}(b + c) < \frac{1}{2}$;
but, by the theory of analytic continuation, it is true throughout the domain in which
both sides of the equation are analytic functions of the parameters. Hence, (3.3.40)
holds for all values of a, b and c for which none of the poles† of $\Gamma(s + a)\Gamma(s + b)$
coincides with any of the poles of $\Gamma(c - s)$.

3.3.5 Ramanujan-Type Integrals

We briefly mention some other integrals involving gamma functions where the
path of integration is no longer parallel to the imaginary axis but along the real
axis. Although these integrals are not of Mellin-Barnes type, they are included here
as they bear some resemblance to those discussed in the preceding section and can
be evaluated in an analogous fashion using the theory of Fourier transforms. The
integrals presented here have been discussed at length in Ramanujan (1920); see
also Titchmarsh (1975, §7.6).

---

† This will be the case when $a + c \neq 0, -1, -2, \ldots$ and $b + c \neq 0, -1, -2, \ldots$.

From the well-known formula

$$\int_{-\pi/2}^{\pi/2} (\cos t)^{a+b-2}\, e^{i(a-b+2x)t}\, dt = \frac{\pi \Gamma(a+b-1)}{2^{a+b-2}\Gamma(a+x)\Gamma(b-x)} \tag{3.3.41}$$

valid when $\mathrm{Re}(a+b) > 1$, we have by the Fourier inversion theorem (for real t)

$$\int_{-\infty}^{\infty} \frac{e^{-ixt}}{\Gamma(a+x)\Gamma(b-x)}\, dx = \begin{cases} \dfrac{\left(2\cos\frac{1}{2}t\right)^{a+b-2}}{\Gamma(a+b-1)} e^{\frac{1}{2}i(a-b)t} & (|t| < \pi) \\ 0 & (|t| \geq \pi). \end{cases} \tag{3.3.42}$$

Stirling's formula in (2.1.8) shows that the behaviour of the integrand in (3.3.42) is controlled by $O(x^{1-a-b}\sin\pi(b-x))$ as $x \to \infty$ and $O(|x|^{1-a-b}\sin\pi(a+x))$ as $x \to -\infty$. Consequently, the integral converges when $\mathrm{Re}(a+b) > 1$ for all real t except when $t = \pm\pi$, when we require $\mathrm{Re}(a+b) > 2$ for convergence[†] (the convergence in this case being absolute). The particular case $t = 0$ yields

$$\int_{-\infty}^{\infty} \frac{dx}{\Gamma(a+x)\Gamma(b-x)} = \frac{2^{a+b-2}}{\Gamma(a+b-1)} \tag{3.3.43}$$

when $\mathrm{Re}(a+b) > 1$.

Application of Parseval's formula for Fourier integrals [Titchmarsh (1975, p. 50)]

$$\int_{-\infty}^{\infty} F(x)G(x)\, dx = \int_{-\infty}^{\infty} f(u)g(-u)\, du,$$

where F and G here denote the Fourier transforms of $f(x)$ and $g(x)$, yields the evaluation

$$\int_{-\infty}^{\infty} \frac{dx}{\Gamma(a+x)\Gamma(b-x)\Gamma(c+x)\Gamma(d-x)}$$

$$= \frac{1}{\pi\Gamma(a+b-1)\Gamma(c+d-1)} \int_{-\frac{1}{2}\pi}^{\frac{1}{2}\pi} (2\cos u)^{a+b+c+d-4} e^{i(a-b-c+d)u}\, du$$

$$= \frac{\Gamma(a+b+c+d-3)}{\Gamma(a+b-1)\Gamma(b+c-1)\Gamma(c+d-1)\Gamma(d+a-1)} \tag{3.3.44}$$

provided *either* (i) $\mathrm{Re}(a+b+c+d) > 3$ *or* (ii) $2(a-c)$ and $2(b-d)$ are odd integers and $\mathrm{Re}(a+b+c+d) > 2$. The second of these conditions[‡] results from consideration of the large-$|x|$ behaviour of the integrand and noting that

---

[†] This follows from the fact that, when $t = \pm\pi$, products like $\sin\pi x \sin\pi(a+x)$ contain a non-oscillatory term.

[‡] The formula fails when $a+b+c+d = 3$ and $2(a-c)$, $2(b-d)$ are odd integers. In this case, one of the gamma functions in the denominator is singular. The limiting value of the expression on the right of (3.3.44) then has the value $\pm 1/2\pi$ according as $2(b-d) \equiv \mp 1 \pmod 4$, when one of the first pair of gamma functions is singular, and $2(b-d) \equiv \pm 1 \pmod 4$, when one of the second pair of gamma functions is singular.

products like $\sin \pi (a+x) \sin \pi (c+x)$ and $\sin \pi (b-x) \sin \pi (d-x)$ will contain non-oscillatory terms unless $\cos \pi (a - c)$ and $\cos \pi (b - d)$ vanish.

In a similar manner we find

$$\int_{-\infty}^{\infty} \frac{e^{\pm ixt}}{\Gamma(a+x)\Gamma(b-x)\Gamma(c+x)\Gamma(d-x)} \, dx = 0 \qquad (3.3.45)$$

for $\operatorname{Re}(a + b + c + d) > 2$ when $|t| > 2\pi$ and for $\operatorname{Re}(a + b + c + d) > 3$ when $|t| = 2\pi$, and

$$\int_{-\infty}^{\infty} \frac{e^{\pm \pi ix}}{\Gamma(a+x)\Gamma(b-x)\Gamma(c+x)\Gamma(d-x)} \, dx$$

$$= \frac{e^{\mp \frac{1}{2}\pi i(a-b)}}{2\,\Gamma\!\left(\frac{1}{2}a + \frac{1}{2}b\right)\Gamma\!\left(\frac{1}{2}c + \frac{1}{2}d\right)\Gamma(a+d-1)}, \qquad (3.3.46)$$

if $a + d = b + c$ and $\operatorname{Re}(a + b + c + d) > 2$. Particular cases of these integrals are given by

$$\int_{-\infty}^{\infty} \frac{dx}{\{\Gamma(a+x)\Gamma(b-x)\}^2} = \frac{\Gamma(2a + 2b - 3)}{\Gamma^4(a+b-1)} \qquad (3.3.47)$$

when $\operatorname{Re}(a + b) > \frac{3}{2}$,

$$\int_{0}^{\infty} \frac{dx}{\Gamma(a+x)\Gamma(a-x)\Gamma(b+x)\Gamma(b-x)}$$

$$= \frac{\Gamma(2a + 2b - 3)}{2\,\Gamma(2a - 1)\Gamma(2b - 1)\Gamma^2(a+b-1)}, \qquad (3.3.48)$$

when either (i) $\operatorname{Re}(a+b) > \frac{3}{2}$ or (ii) $2(a-b)$ is an odd integer and $\operatorname{Re}(a+b) > 1$, and

$$\int_{-\infty}^{\infty} \frac{\binom{\sin}{\cos} \pi x}{\{\Gamma(a-x)\Gamma(b+x)\}^2} \, dx = \frac{\binom{\sin}{\cos} \frac{1}{2}\pi(a-b)}{2\,\Gamma(a+b-1)\Gamma^2\!\left(\frac{1}{2}a + \frac{1}{2}b\right)} \qquad (3.3.49)$$

when $\operatorname{Re}(a + b) > 1$. Other integrals of this type involving $\Gamma(c + nx)\Gamma(d - nx)$ $(n = 1, 2, \dots)$ in the denominator are detailed in Ramanujan (1920).

Integrals involving gamma functions in the numerator can also be evaluated by means of (3.3.42). Consider the integral

$$I = \int_{-\infty}^{\infty} \frac{\Gamma(a+x)}{\Gamma(b+x)} e^{ixt} \, dx$$

for real t, where $\operatorname{Im}(a) \neq 0$ (so that the integrand has no singularity on the path of integration) and, for convergence, we require *either* (i) $\operatorname{Re}(a - b) < 0$ when $t \neq \pm 2\pi n$ $(n = 0, 1, 2, \dots)$ or $\operatorname{sgn}(t) = \operatorname{sgn}\{\operatorname{Im}(a)\}$ *or* (ii) $\operatorname{Re}(a - b) < -1$

when $t = \pm 2\pi n$ $(n = 0, 1, 2, \ldots)$ with $\mathrm{sgn}(t) \neq \mathrm{sgn}\{\mathrm{Im}(a)\}$ *or* (iii) $t \neq 0$ when $a - b = -1$. If we take $\mathrm{Im}(a) < 0$, the above integral can be written as

$$I = \pi \int_{-\infty}^{\infty} \frac{1}{\Gamma(b+x)\Gamma(1-a-x)} \frac{e^{ixt}}{\sin \pi(a+x)} \, dx$$

$$= 2\pi i \sum_{m=0}^{\infty} \int_{-\infty}^{\infty} \frac{e^{ixt}}{\Gamma(b+x)\Gamma(1-a-x)} e^{-(2m+1)\pi i(a+x)} \, dx,$$

where the integrals inside the summation sign are of the form (3.3.42), with t replaced by $(2m + 1)\pi - t$. Hence $I = 0$ if $t \leq 0$ and, when $t > 0$, the only non-zero term is that corresponding to $m = [t/2\pi]$. A similar procedure applies when $\mathrm{Im}(a) > 0$ and we obtain the result

$$\int_{-\infty}^{\infty} \frac{\Gamma(a+x)}{\Gamma(b+x)} e^{ixt} \, dx = \begin{cases} 0 & \mathrm{sgn}(t)=\mathrm{sgn}\{\mathrm{Im(a)}\} \\ \pm J & \begin{cases} t \geq 0, & \mathrm{Im}(a) < 0 \\ t \leq 0, & \mathrm{Im}(a) > 0, \end{cases} \end{cases} \tag{3.3.50}$$

where

$$J = \frac{2\pi i}{\Gamma(b-a)} (2 \cos \Theta)^{b-a-1} \exp\{i(b-a-1)\Theta - iat\},$$

$$\Theta = \frac{\pi - t}{2} + \pi \left[\frac{t}{2\pi} \right]$$

and $[x]$ denotes the greatest integer not exceeding x. As a particular case we have the result

$$\int_{-\infty}^{\infty} \frac{\Gamma(a+x)}{\Gamma(b+x)} \, dx = 0 \tag{3.3.51}$$

if $\mathrm{Im}(a) \neq 0$ and $\mathrm{Re}(a - b) < -1$.

A similar procedure can be employed to show that, when $\mathrm{Im}(a) \neq 0$ and $\mathrm{Im}(b) \neq 0$, and *either* (i) t is not an odd multiple of π and $\mathrm{Re}(a + b) < 1$ *or* (ii) t is an odd multiple of π when $\mathrm{sgn}(t) = \mathrm{sgn}\{\mathrm{Im}(b)\}$ or $\mathrm{sgn}(t) = -\mathrm{sgn}\{\mathrm{Im}(a)\}$ and $\mathrm{Re}(a + b) < 0$ *or* (iii) t is an odd multiple of π when $\mathrm{sgn}(t) = \mathrm{sgn}\{\mathrm{Im}(a)\}$, $\mathrm{sgn}(t) = -\mathrm{sgn}\{\mathrm{Im}(b)\}$ and $\mathrm{Re}(a + b) < 1$, then

$$\int_{-\infty}^{\infty} \Gamma(a+x)\Gamma(b-x)e^{ixt} \, dx$$

$$= \frac{2\pi i \Gamma(a+b)}{\left| 2 \cos \frac{1}{2} t \right|^{a+b}} e^{\frac{1}{2}i(b-a)t} \left\{ A(+b)e^{\pi i \Psi} - A(-a)e^{-\pi i \Psi} \right\}, \tag{3.3.52}$$

where

$$\Psi = (a+b) \left[\frac{\pi + t}{2\pi} \right]$$

and

$$A(z) = \begin{cases} 0 & \text{sgn}(\pi - t) = \text{sgn}\{\text{Im}(z)\} \\ 1 & t \le \pi \quad \text{and Im}(z) < 0 \\ -1 & t \ge \pi \quad \text{and Im}(z) > 0. \end{cases}$$

A particular case of (3.3.52), when a and b are not real and $\text{Re}(a + b) < 1$, is given by

$$\int_{-\infty}^{\infty} \Gamma(a + x)\Gamma(b - x)\, dx = \begin{cases} 0 & \text{sgn}\{\text{Im}(a)\} \ne \text{sgn}\{\text{Im}(b)\} \\ \pm \dfrac{\pi i \Gamma(a + b)}{2^{a+b-1}} & \begin{cases} \text{Im}(a, b) < 0 \\ \text{Im}(a, b) > 0. \end{cases} \end{cases}$$

$$(3.3.53)$$

Finally, we mention the particular evaluations which can be determined by the same method

$$\int_{-\infty}^{\infty} \frac{\Gamma(d + x)}{\Gamma(a + x)\Gamma(b - x)\Gamma(c + x)}\, dx = \frac{\pi e^{\pm \frac{1}{2}\pi i(d-c)}}{\Gamma(a - d)\Gamma\left(\frac{1}{2}a + \frac{1}{2}b\right)\Gamma\left(\frac{1}{2}c - \frac{1}{2}d + \frac{1}{2}\right)},$$

$$(3.3.54)$$

when $\text{Im}(d) \ne 0$, $a + 1 = b + c + d$, $\text{Re}(a + b + c - d) > 1$ with the upper or lower sign chosen according as $\text{Im}(d)$ is positive or negative, respectively, and

$$\int_{-\infty}^{\infty} \frac{\Gamma(c + x)\Gamma(d + x)}{\Gamma(a + x)\Gamma(b + x)}\, dx = \begin{cases} 0 & \text{sgn}\{\text{Im}(c)\} = \text{sgn}\{\text{Im}(d)\} \\ \pm J & \begin{cases} \text{Im}(c) > 0, \ \text{Im}(d) < 0 \\ \text{Im}(c) < 0, \ \text{Im}(d) > 0, \end{cases} \end{cases} \quad (3.3.55)$$

when $\text{Im}(c, d) \ne 0$ and $\text{Re}(a + b - c - d) > 1$, where

$$J = \frac{2\pi^2 i}{\sin \pi(c - d)} \frac{\Gamma(a + b - c - d - 1)}{\Gamma(a - c)\Gamma(a - d)\Gamma(b - c)\Gamma(b - d)}.$$

3.3.6 Barnes' Lemmas

We now discuss two lemmas due to Barnes (1908, 1910) which concern more involved evaluations of Mellin-Barnes integrals containing only gamma functions. The first of these lemmas (which is the integral analogue of Gauss' theorem in (3.3.39)) is given by

Lemma 3.5. *If the contour of integration is indented (if necessary) to separate the poles of $\Gamma(s + a)\Gamma(s + b)$ from those of $\Gamma(c - s)\Gamma(d - s)$, then*

$$\frac{1}{2\pi i} \int_{-\infty i}^{\infty i} \Gamma(s + a)\Gamma(s + b)\Gamma(c - s)\Gamma(d - s)\, ds$$

$$= \frac{\Gamma(a + c)\Gamma(a + d)\Gamma(b + c)\Gamma(b + d)}{\Gamma(a + b + c + d)}. \quad (3.3.56)$$

This can be established first as in §3.3.4 by choosing the functions

$$f(x) = \frac{x^b}{(1+x)^{b+d}}, \quad g(x) = \frac{x^{c-1}}{(1+x)^{a+c}} \quad (x \geq 0),$$

for which

$$M[f; s] = \frac{\Gamma(s+b)\Gamma(d-s)}{\Gamma(b+d)}, \quad M[g; s] = \frac{\Gamma(s+c-1)\Gamma(a+1-s)}{\Gamma(a+c)}$$

when $\mathrm{Re}(s+b) > 0$, $\mathrm{Re}(d-s) > 0$ for $M[f; s]$ and $\mathrm{Re}(1-s) < \mathrm{Re}(c)$, $\mathrm{Re}(1-s) > -\mathrm{Re}(a)$ for $M[g; s]$. Then, from Parseval's formula (3.1.11), we obtain

$$I = \frac{1}{2\pi i} \int_{k-\infty i}^{k+\infty i} \Gamma(s+a)\Gamma(s+b)\Gamma(c-s)\Gamma(d-s) \, ds$$

$$= \Gamma(a+c)\Gamma(b+d) \int_0^\infty \frac{x^{b+c-1}}{(1+x)^{a+b+c+d}} \, dx$$

$$= \frac{\Gamma(a+c)\Gamma(b+d)\Gamma(b+c)\Gamma(a+d)}{\Gamma(a+b+c+d)},$$

provided the integration path $\mathrm{Re}(s) = k$ lies in the common strip of analyticity

$$\max\{-\mathrm{Re}(a), -\mathrm{Re}(b)\} < \mathrm{Re}(s) < \max\{\mathrm{Re}(c), \mathrm{Re}(d)\}.$$

These conditions can be relaxed if the result is established by displacement of the contour and evaluation of residues, as described in Whittaker & Watson (1965, p. 289) and Slater (1966, p. 109). If we complete the contour of integration in (3.3.56) by a circular arc of radius R on the right of the imaginary axis then, provided the arc does not pass through a neighbourhood of a pole, the modulus of the integrand on the imaginary axis and on this arc is, by Stirling's formula (2.1.8), controlled by $O(s^{a+b+c+d-2} e^{-2\pi |\mathrm{Im}(s)|})$ as $R \to \infty$. The contribution round the arc therefore tends to zero as $R \to \infty$ when $\mathrm{Re}(a+b+c+d) < 1$. We then find, upon evaluation of the residues at the poles of $\Gamma(c-s)\Gamma(d-s)$ situated at $s = c+n$ and $s = d+n$ for nonnegative integer n, that the integral I is given by

$$\frac{\pi}{\sin \pi(c-d)} \left\{ \sum_{n=0}^\infty \frac{\Gamma(a+d+n)\Gamma(b+d+n)}{n!\,\Gamma(1+d-c+n)} \right.$$

$$\left. - \sum_{n=0}^\infty \frac{\Gamma(a+c+n)\Gamma(b+c+n)}{n!\,\Gamma(1+c-d+n)} \right\}$$

$$= \frac{\pi}{\sin \pi(c-d)} \left\{ \frac{\Gamma(a+d)\Gamma(b+d)}{\Gamma(1+d-c)} {}_2F_1(a+d, b+d; 1+d-c; 1) \right.$$

$$\left. - \frac{\Gamma(a+c)\Gamma(b+c)}{\Gamma(1+c-d)} {}_2F_1(a+c, b+c; 1+c-d; 1) \right\}$$

$$= \frac{\pi \Gamma(1 - a - b - c - d)}{\sin \pi (c - d)} \left\{ \frac{\Gamma(a + d)\Gamma(b + d)}{\Gamma(1 - a - c)\Gamma(1 - b - c)} \right.$$

$$\left. - \frac{\Gamma(a + c)\Gamma(b + c)}{\Gamma(1 - a - d)\Gamma(1 - b - d)} \right\}$$

$$= \frac{\Gamma(a + c)\Gamma(a + d)\Gamma(b + c)\Gamma(b + d)}{\Gamma(a + b + c + d)}$$

$$\times \left\{ \frac{\sin \pi (a + c) \sin \pi (b + c) - \sin \pi (a + d) \sin \pi (b + d)}{\sin \pi (c - d) \sin \pi (a + b + c + d)} \right\}$$

upon use of the Gauss summation formula in (3.3.39). Simple trigonometric identities show that

$$\sin \pi (a + c) \sin \pi (b + c) - \sin \pi (a + d) \sin \pi (b + d)$$

$$= \tfrac{1}{2} \{ \cos \pi (a + b + 2d) - \cos \pi (a + b + 2c) \}$$

$$= \sin \pi (c - d) \sin \pi (a + b + c + d),$$

whence

$$I = \frac{\Gamma(a + c)\Gamma(a + d)\Gamma(b + c)\Gamma(b + d)}{\Gamma(a + b + c + d)}.$$

The above result has been obtained subject to the condition $\mathrm{Re}(a+b+c+d) < 1$; but by analytic continuation the result will hold in the domain of the parameters for which both sides of the equation are analytic functions of a, b, c and d. Hence it is true for all values of the parameters for which none of the poles of $\Gamma(s+a)\Gamma(s+b)$ coincides with any poles of $\Gamma(c - s)\Gamma(d - s)$. In addition, if we write $s + k$, $a - k$, $b - k$, $c + k$, $d + k$ in place of s, a, b, c, d we see that the result is still true when the path of integration is $\mathrm{Re}(s) = k$, where k is any real constant.

Barnes' second lemma states that:

Lemma 3.6. *If the contour of integration is indented (if necessary) to separate the poles of $\Gamma(s + a)\Gamma(s + b)\Gamma(s + c)$ from those of $\Gamma(d - s)\Gamma(-s)$, then*

$$I = \frac{1}{2\pi i} \int_{-\infty i}^{\infty i} \frac{\Gamma(s + a)\Gamma(s + b)\Gamma(s + c)\Gamma(d - s)\Gamma(-s)}{\Gamma(e + s)} \, ds$$

$$= \frac{\Gamma(a)\Gamma(b)\Gamma(c)\Gamma(a + d)\Gamma(b + d)\Gamma(c + d)}{\Gamma(e - a)\Gamma(e - b)\Gamma(e - c)}, \tag{3.3.57}$$

when

$$a + b + c + d = e.$$

A proof of this lemma using a more involved extension of the Parseval formula to double integrals is given in Titchmarsh (1975, p. 195). An alternative approach is to observe that use of the Gauss summation formula (3.3.39) shows that I can be written as

$$\frac{\Gamma(c)}{\Gamma(e)}\frac{1}{2\pi i}\int_{-\infty i}^{\infty i}\Gamma(s+a)\Gamma(s+b)\Gamma(d-s)\Gamma(-s)_2F_1(e-c,-s;e;1)\,ds$$

$$=\frac{\Gamma(c)}{\Gamma(e)}\sum_{n=0}^{\infty}\frac{(e-c)_n}{n!\,(e)_n}\frac{1}{2\pi i}\int_{-\infty i}^{\infty i}\Gamma(s+a)\Gamma(s+b)\Gamma(n-s)\Gamma(d-s)\,ds,$$

provided $\mathrm{Re}(s+c)>0$ to justify the interchange in the order of integration and summation. The integral appearing inside the summation can now be evaluated by Barnes' first lemma (3.3.56) to yield

$$I=\frac{\Gamma(a)\Gamma(b)\Gamma(c)\Gamma(a+d)\Gamma(b+d)}{\Gamma(e)\Gamma(a+b+d)}\sum_{n=0}^{\infty}\frac{(e-c)_n(a)_n(b)_n}{n!\,(e)_n(a+b+d)_n}$$

$$=\frac{\Gamma(a)\Gamma(b)\Gamma(c)\Gamma(a+d)\Gamma(b+d)}{\Gamma(e)\Gamma(a+b+d)}\,_3F_2(a,b,e-c;a+b+d,e;1)$$

provided $\mathrm{Re}(c+d)>0$ for convergence† of the generalised hypergeometric function. If we now take $a+b+c+d=e$, the $_3F_2(1)$ contracts to a $_2F_1(1)$ series, which is then summable by (3.3.39), and we obtain the result stated in (3.3.57). The restrictions $\mathrm{Re}(s+c)>0$ and $\mathrm{Re}(c+d)>0$ can be relaxed by analytic continuation to enable the contour of integration to be indented (if necessary) to separate the poles of $\Gamma(s+a)\Gamma(s+b)\Gamma(s+c)$ from those of $\Gamma(d-s)\Gamma(-s)$. This will be possible provided none of $a,b,c,a+d,b+d,c+d$ equals a nonpositive integer.

Special cases of other integrals involving gamma functions are listed in Slater (1966, pp. 112–114). All these integrals, including those appearing in Barnes' lemmas and in (3.3.36) and (3.3.40), are special cases of the more general integral of the type

$$\frac{1}{2\pi i}\int_{-\infty i}^{\infty i}\frac{\prod_{j=1}^{m}\Gamma(a_j+s)\prod_{j=1}^{p}\Gamma(c_j-s)}{\prod_{j=1}^{n}\Gamma(b_j+s)\prod_{j=1}^{q}\Gamma(d_j-s)}\,ds$$

which, subject to certain restrictions on the parameters and on the integers m,n,p,q, can be expressed as sums of generalised hypergeometric functions of argument equal to ±1; for details, see Slater (1966, p. 133).

3.4 Mellin-Barnes Integral Representations

We conclude this chapter with a selection of standard Mellin-Barnes integral representations for some of the more familiar special functions. We begin with the integrals of this type for the confluent and ordinary hypergeometric functions, which were first given by Barnes (1908).

---

† The sum $_3F_2(\alpha,\beta,\gamma;\delta,\epsilon;1)$ is absolutely convergent when $\mathrm{Re}(\delta+\epsilon-\alpha-\beta-\gamma)>0$.

3.4.1 The Confluent Hypergeometric Functions

The confluent hypergeometric function $_1F_1(a; b; z)$ is defined by the absolutely convergent series†

$$_1F_1(a; b; z) = \sum_{n=0}^{\infty} \frac{(a)_n}{(b)_n} \frac{z^n}{n!} \qquad (|z| < \infty),$$

where $(a)_n = \Gamma(a + n)/\Gamma(a)$ and we suppose that $b \neq 0, -1, -2, \ldots$. Let us consider the integral

$$I = \frac{1}{2\pi i} \int_{c-\infty i}^{c+\infty i} \frac{\Gamma(-s)\Gamma(s + a)}{\Gamma(s + b)} z^s \, ds. \qquad (3.4.1)$$

The integration path can pass to infinity parallel to the imaginary axis with any finite value of $c = \mathrm{Re}(s)$, provided the contour can be indented to separate the poles of $\Gamma(-s)$ at $s = n$ from the poles of $\Gamma(s+a)$ at $s = -a-n$, for nonnegative integer n; see Fig. 3.2. From Stirling's formula in (2.1.8) the modulus of the integrand is

$$O\left(|t|^{-\sigma+a-b-\frac{1}{2}} e^{-\theta t - \frac{1}{2}\pi|t|}\right)$$

as $t \to \pm\infty$, where $s = \sigma + it$ and $\theta = \arg z$. It then follows that the above integral converges absolutely in the sector $|\arg z| < \frac{1}{2}\pi$; see also Rule 1 in §2.4.

Now consider the rectangle with vertices at $c \pm iT$ and $N + \frac{1}{2} \pm iT$, where $T > |\mathrm{Im}(a)|$ and N is a positive integer which, when $\mathrm{Re}(a) < 0$, exceeds $-\mathrm{Re}(a) - \frac{1}{2}$; cf. Fig. 3.2(b). Because of the exponential decay of the integrand it is seen that if N is kept fixed and $T \to \infty$, the integrals along the upper and lower sides of the rectangle both vanish. For the integral I_N, say, taken along the third side we have,

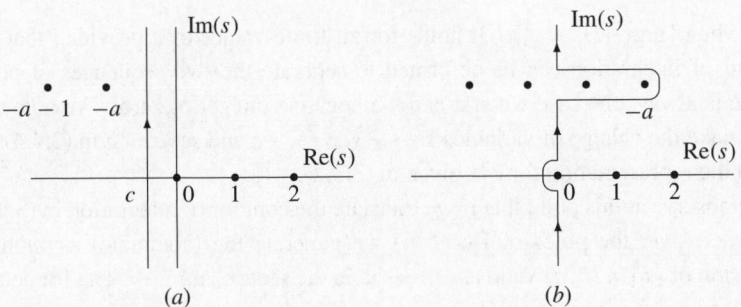

Fig. 3.2. Paths of integration when (a) $\mathrm{Re}(a) > 0$ and (b) $\mathrm{Re}(a) \leq 0$.

† When a equals zero or a negative integer $_1F_1(a; b; z)$ reduces to a polynomial in z.

with $s = N + \frac{1}{2} + it$,

$$|I_N| \leq \frac{1}{2}|z|^{N+\frac{1}{2}} \int_{-\infty}^{\infty} \left| \frac{\Gamma(a + N + \frac{1}{2} + it)}{\Gamma(b + N + \frac{1}{2} + it)} \right| \frac{e^{-\theta t} \operatorname{sech} \pi t}{|\Gamma(N + \frac{3}{2} + it)|} \, dt,$$

where we have used the reflection formula for the gamma function. Using the result in (5.1.3) and the fact† that for large N

$$\left| \frac{\Gamma(a + N + \frac{1}{2} + it)}{\Gamma(b + N + \frac{1}{2} + it)} \right| \leq C|(N + it)^{a-b}| \leq C' N^{\operatorname{Re}(a-b)} \lambda_{a-b}(t) \qquad (3.4.2)$$

uniformly in $t \in (-\infty, \infty)$, where C, C' denote positive assignable constants and

$$\lambda_\nu(t) = \begin{cases} (1 + t^2)^{\frac{1}{2}\operatorname{Re}(\nu)} & \operatorname{Re}(\nu) > 0 \\ 1 & \operatorname{Re}(\nu) \leq 0, \end{cases} \qquad (3.4.3)$$

we find

$$|I_N| \leq \frac{C'|z|^{N+\frac{1}{2}} N^{\operatorname{Re}(a-b)}}{2\Gamma(N + \frac{3}{2})} \int_{-\infty}^{\infty} \lambda_{a-b}(t) \frac{e^{-\theta t}}{(\cosh \pi t)^{\frac{1}{2}}} \, dt.$$

Since the last integral converges absolutely when $|\arg z| < \frac{1}{2}\pi$, we see that $|I_N| \to 0$ as $N \to \infty$.

From Cauchy's theorem it then follows that I is given by the sum of residues of the integrand at the poles of $\Gamma(-s)$, so that

$$I = \sum_{n=0}^{\infty} \frac{\Gamma(a + n)}{\Gamma(b + n)} \frac{(-z)^n}{n!}.$$

Thus, replacing z by $-z$, we obtain Barnes' integral representation

$$\frac{\Gamma(a)}{\Gamma(b)} {}_1F_1(a; b; z) = \frac{1}{2\pi i} \int_{c-\infty i}^{c+\infty i} \frac{\Gamma(-s)\Gamma(s + a)}{\Gamma(s + b)} (-z)^s \, ds \qquad (3.4.4)$$

valid when $|\arg(-z)| < \frac{1}{2}\pi$. It holds for all finite values of c provided that the contour of integration can be deformed to separate the two sequences of poles (which is always the case when a is not a negative integer or zero). We note that if we make the change of variables $s \to -s$, $z \to -z$ and set $a = b$ in (3.4.4) we obtain the representation for e^{-z} given in (3.3.2).

We observe at this point that if we translate the contour of integration in (3.4.4) to the left over the poles of $\Gamma(s + a)$ we generate the (algebraic) asymptotic expansion of ${}_1F_1(a; b; z)$ valid as $|z| \to \infty$ in the sector $|\arg z| < \frac{1}{2}\pi$; for details

---

† In deriving the second of these inequalities, we have employed the fact that for $\nu = \alpha + i\beta$, where α and β are real, $N \geq 1$ and $t \in (-\infty, \infty)$,

$$|(1 + it/N)^\nu| \leq |1 + it/N|^\alpha e^{\frac{1}{2}\pi|\beta|} \leq \lambda_\nu(t) e^{\frac{1}{2}\pi|\beta|}.$$

of this process see Chapter 5. Then, upon evaluation of the residues at the poles $s = -a - k, k = 0, 1, 2, \ldots, N - 1$, we find after a little rearrangement

$$_1F_1(a; b; z) = \frac{\Gamma(b)}{\Gamma(b - a)}(-z)^{-a} \sum_{k=0}^{N-1} \frac{(a)_k(1 + a - b)_k}{k!}(-z)^{-k} + R_N(z),$$

where the remainder after N terms satisfies $|R_N(z)| = O(z^{-N})$ as $|z| \to \infty$ in $|\arg z| < \frac{1}{2}\pi$. This fact illustrates the power of the Mellin-Barnes approach: the series expansion and asymptotic structure of a function are contained in the nature and distribution of the singularities of the integrand. Displacement of the path of integration in one direction produces the convergent (ascending) power series expansion, while displacement in the opposite direction yields the (descending) power series or asymptotic expansion.

We remark that (3.4.4) defines $_1F_1(a; b; z)$ only in the sector $|\arg(-z)| < \frac{1}{2}\pi$. As with the function e^{-z} in (3.3.4), we can analytically continue this definition by writing (for convenience, we have replaced the variable s by $-s$)

$$\frac{\Gamma(a)}{\Gamma(b)}\,_1F_1(a; b; z) = \frac{1}{2\pi i}\int_C \frac{\Gamma(s)\Gamma(a - s)}{\Gamma(b - s)}(-z)^{-s}ds, \qquad (3.4.5)$$

where C denotes a loop encircling the poles of $\Gamma(s)$ (in the positive sense) with endpoints at infinity in $\text{Re}(s) < 0$; cf. Fig. 3.1. This representation takes advantage of the $\exp\{s \log s\}$ decay contained in the gamma function (see §2.4) and accordingly defines $_1F_1(a; b; z)$ *without restriction* on arg z.

The second type of confluent hypergeometric function, denoted by $U(a; b; z)$, is defined by

$$U(a; b; z) = \frac{\Gamma(1 - b)}{\Gamma(1 + a - b)}\,_1F_1(a; b; z)$$

$$+ \frac{\Gamma(b - 1)}{\Gamma(a)}z^{1-b}\,_1F_1(1 + a - b; 2 - b; z). \qquad (3.4.6)$$

The function $U(a; b; z)$, unlike $_1F_1(a; b; z)$, is in general a multi-valued function of z; we shall take as its principal branch that which lies in the z plane cut along the negative real axis. This function is analytic in the cut plane for all complex values of a, b and z (even when b is a negative integer or zero). Let us consider the integral

$$I = \frac{1}{2\pi i}\int_{c-\infty i}^{c+\infty i} \Gamma(-s)\Gamma(s + a)\Gamma(s + 1 + a - b)z^{-s}ds,$$

where the path of integration can be indented (if necessary) to separate the poles of $\Gamma(-s)$ from those of $\Gamma(s + a)\Gamma(s + 1 + a - b)$ (which is always possible provided $a \neq 0, -1, -2, \ldots$ and $a - b \neq -1, -2, \ldots$). Reference to Rule 1 in §2.4 shows that the above integral is defined for $|\arg z| < \frac{3}{2}\pi$. Translation of the integration contour to the left over the poles of $\Gamma(s + a)\Gamma(s + 1 + a - b)$ then produces a *convergent* expansion in ascending powers of z proportional to z^a times the right-hand side of (3.4.6). Thus we find the representation

$$\Gamma(a)\Gamma(1 + a - b)z^a\, U(a; b; z)$$

$$= \frac{1}{2\pi i} \int_{c-\infty i}^{c+\infty i} \Gamma(-s)\Gamma(s + a)\Gamma(s + 1 + a - b)z^{-s}ds \qquad (3.4.7)$$

valid for $|\arg z| < \frac{3}{2}\pi$. Again, we observe that displacement of the integration path to the right over the poles of $\Gamma(-s)$ generates the (algebraic) asymptotic expansion given by

$$U(a; b; z) \sim z^{-a} \sum_{k=0}^{\infty} \frac{(a)_k(1 + a - b)_k}{k!}(-z)^{-k}$$

valid as $|z| \to \infty$ in the sector $|\arg z| < \frac{3}{2}\pi$. A representation valid without restriction on $\arg z$ can also be obtained by replacement of the rectilinear integration path in (3.4.7) by an appropriate loop C as in (3.4.5).

3.4.2 The Gauss Hypergeometric Function

The Gauss hypergeometric function $_2F_1(a, b; c; z)$ is defined by

$$\frac{\Gamma(a)\Gamma(b)}{\Gamma(c)}\,_2F_1(a, b; c; z) = \sum_{n=0}^{\infty} \frac{\Gamma(a + n)\Gamma(b + n)}{\Gamma(c + n)\, n!}\, z^n \quad (|z| < 1)$$

when neither of the parameters a and b is a negative integer or zero. To obtain a Mellin-Barnes integral representation for this function we consider the integral

$$I = \frac{1}{2\pi i} \int_{-\infty i}^{\infty i} \frac{\Gamma(-s)\Gamma(s + a)\Gamma(s + b)}{\Gamma(s + c)}(-z)^s\, ds,$$

where the integration path is indented (if necessary) to separate the poles of $\Gamma(-s)$ from those of $\Gamma(s + a)\Gamma(s + b)$ (which is always possible when neither a nor b is a negative integer or zero). We remark that the integration path can pass to infinity parallel to the imaginary axis with any finite value of $\mathrm{Re}(s)$; we take the limits $\pm\infty i$ for convenience.

The procedure for evaluating the above integral closely follows that for $_1F_1(a; b; z)$. Application of Rule 1 in §2.4 shows that the integral converges absolutely provided $|\arg(-z)| < \pi$. We now apply Cauchy's theorem to the above integral taken round the rectangle with vertices at $\pm iT$, $N + \frac{1}{2} \pm iT$, where $T > \max\{|\mathrm{Im}(a)|, |\mathrm{Im}(b)|\}$ and N is a positive integer exceeding $-\mathrm{Re}(a) - \frac{1}{2}$ and $-\mathrm{Re}(b) - \frac{1}{2}$. Since the integrand is $O\big(|t|^{a+b-c-1}e^{-t\arg(-z)-\pi|t|}\big)$ as $t \to \pm\infty$ by Stirling's formula in (2.1.8), the integrals along the upper and lower sides of the rectangle both vanish as $T \to \infty$. The modulus of the contribution from the third side satisfies

$$|I_N| \leq \tfrac{1}{2}|z|^{N+\frac{1}{2}} \int_{-\infty}^{\infty} M(t) \frac{e^{-\theta t}}{\cosh \pi t} \, dt,$$

where (compare (3.4.2))

$$M(t) = \left| \frac{\Gamma(a + N + \frac{1}{2} + it)\Gamma(b + N + \frac{1}{2} + it)}{\Gamma(c + N + \frac{1}{2} + it)\Gamma(N + \frac{3}{2} + it)} \right|$$

$$\leq C N^{\mathrm{Re}(a+b-c-1)} \lambda_{a-c}(t)\lambda_{b-1}(t)$$

as $N \to \infty$ and $\lambda_\nu(t)$ is defined in (3.4.3). The last integral is absolutely convergent when $|\arg(-z)| < \pi$ and hence $|I_N| = O(|z|^{N+\frac{1}{2}} N^{\mathrm{Re}(a+b-c-1)})$. If $|z| < 1$, we then see that $|I_N| \to 0$ as $N \to \infty$.

We therefore obtain Barnes' integral representation

$$\frac{\Gamma(a)\Gamma(b)}{\Gamma(c)} {}_2F_1(a, b; c; z) = \frac{1}{2\pi i} \int_{-\infty i}^{\infty i} \frac{\Gamma(-s)\Gamma(s + a)\Gamma(s + b)}{\Gamma(s + c)} (-z)^s \, ds.$$

$$(3.4.8)$$

In establishing this result it was necessary to suppose that $|z| < 1$. By the principle of analytic continuation the integral (3.4.8) converges and defines an analytic function of z which is holomorphic in the sector $|\arg(-z)| < \pi$. Hence (3.4.8) defines ${}_2F_1(a, b; c; z)$ not only inside the unit circle but also in the entire z plane cut along the real z axis from 0 to ∞, provided the parameters are such that the integration path can be drawn to separate the poles of $\Gamma(-s)$ from the poles of $\Gamma(s + a)\Gamma(s + b)$.

If the path of integration is displaced to the left we can proceed to evaluate the integral in (3.4.8) in a similar fashion. In this case we encounter the poles of $\Gamma(s+a)\Gamma(s+b)$ which, provided $a - b$ is not an integer,[†] will correspond to two sequences of simple poles with residues

$$(-)^n \frac{\Gamma(b - a - n)\Gamma(a + n)}{n!\,\Gamma(c - a - n)} (-z)^{-a-n}$$

$$= \frac{\Gamma(a)\Gamma(b - a)}{\Gamma(c - a)} (-z)^{-a} \frac{(1 + a - c)_n (a)_n}{(1 + a - b)_n \, n!} z^{-n}$$

at the points $s = -a - n$; the residues at $s = -b - n$ are obtained by interchanging a and b. Provided $|z| > 1$, the integral round the translated path vanishes as the

---

[†] When $a - b$ is an integer some of the poles become double poles and the residues then involve terms in $\log(-z)$; see Abramowitz & Stegun (1965, p. 560).

path moves to infinity and we obtain the result

$$
\begin{aligned}
{}_2F_1(a, b; c; z) &= \frac{\Gamma(c)\Gamma(b-a)}{\Gamma(b)\Gamma(c-a)}(-z)^{-a}{}_2F_1(a, 1+a-c; 1+a-b; z^{-1}) \\
&+ \frac{\Gamma(c)\Gamma(a-b)}{\Gamma(a)\Gamma(c-b)}(-z)^{-b}{}_2F_1(b, 1+b-c; 1+b-a; z^{-1}).
\end{aligned}
$$

$$(3.4.9)$$

Analytic continuation enables us to remove the restriction $|z| > 1$ to establish that (3.4.9) holds for $|\arg(-z)| < \pi$ and $a - b$ not equal to an integer. This result describes the behaviour of the hypergeometric function for large $|z|$.

3.4.3 Some Special Functions

The confluent hypergeometric functions include as particular cases many of the commonly used special functions, such as the incomplete gamma functions, Bessel and the parabolic cylinder functions. As a consequence, it is possible to write down Mellin-Barnes integral representations for these functions from the representations given in (3.4.4) and (3.4.7). For ease of reference, we present in Table 3.1 a short list of some familiar special functions and their corresponding expressions in terms of confluent hypergeometric functions.

To illustrate, we take the incomplete gamma functions which are defined when $|\arg z| < \pi$ by

$$
\gamma(a, z) = \int_0^z t^{a-1}e^{-t}\, dt \quad (\mathrm{Re}(a) > 0), \qquad \Gamma(a, z) = \int_z^\infty t^{a-1}e^{-t}\, dt,
$$

Table 3.1. *Some familiar special functions*

Special function	Confluent hypergeometric representation
$J_\nu(z)$	$e^{-iz}\left(\frac{1}{2}z\right)^\nu (\Gamma(1+\nu))^{-1}{}_1F_1(\nu + \frac{1}{2}; 2\nu + 1; 2iz)$
$I_\nu(z)$	$e^{-z}\left(\frac{1}{2}z\right)^\nu (\Gamma(1+\nu))^{-1}{}_1F_1(\nu + \frac{1}{2}; 2\nu + 1; 2z)$
$M_{k,m}(z)$	$z^{m+\frac{1}{2}}e^{-\frac{1}{2}z}{}_1F_1\left(\frac{1}{2} + m - k; 1 + 2m; z\right)$
$\gamma(a, z)$	$a^{-1}z^a e^{-z}{}_1F_1(1; a + 1; z)$
$\Gamma(a, z)$	$e^{-z}\, U(1 - a; 1 - a; z)$
$\mathrm{erfc}(z)$	$\pi^{-\frac{1}{2}}e^{-z^2}\, U\left(\frac{1}{2}; \frac{1}{2}; z^2\right)$
$D_\nu(z)$	$2^{\frac{1}{2}\nu}e^{-z^2/4}\, U\left(-\frac{1}{2}\nu; \frac{1}{2}; \frac{1}{2}z^2\right)$
$K_\nu(z)$	$\pi^{\frac{1}{2}}e^{-z}(2z)^\nu\, U\left(\nu + \frac{1}{2}; 2\nu + 1; 2z\right)$
$H_\nu^{(1,2)}(z)$	$2\pi^{-\frac{1}{2}}e^{\mp\pi i\left(\nu+\frac{1}{2}\right)\pm iz}(2z)^\nu\, U\left(\nu + \frac{1}{2}; 2\nu + 1; \mp 2iz\right)$
$W_{k,m}(z)$	$z^{m+\frac{1}{2}}e^{-\frac{1}{2}z}\, U\left(\frac{1}{2} + m - k; 1 + 2m; z\right)$

and which satisfy the relation $\gamma(a, z) + \Gamma(a, z) = \Gamma(a)$. From Table 3.1, (3.4.4) and (3.4.7) we then find the Mellin-Barnes integral representations

$$\gamma(a, z) = \frac{1}{2\pi i} \int_{c-\infty i}^{c+\infty i} \frac{\Gamma(-s)}{s+a} z^{s+a} \, ds, \qquad (3.4.10)$$

where $|\arg z| < \frac{1}{2}\pi$ and the contour separates the pole at $s = -a$ from those of $\Gamma(-s)$ $(a \neq 0, -1, -2, \dots)$, and

$$\Gamma(a, z) = -\frac{z^{a-1}e^{-z}}{\Gamma(1-a)} \frac{1}{2\pi i} \int_{-c-\infty i}^{-c+\infty i} \Gamma(s+1-a) \frac{\pi z^{-s}}{\sin \pi s} \, ds \qquad (3.4.11)$$

valid in the sector $|\arg z| < \frac{3}{2}\pi$. Provided a is not a positive integer, the path of integration (which is indented if necessary) has $0 < c < 1$ and passes to the left of the poles situated at $s = 0, 1, 2, \dots$ but to the right of those of $\Gamma(1 - a + s)$ and at $s = -1, -2, \dots$.

An alternative integral representation for $\Gamma(a, z)$, valid in the narrower sector $|\arg z| < \frac{1}{2}\pi$, can be obtained from (3.4.10). If we displace the integration path to the left over the pole at $s = -a$, we find

$$\gamma(a, z) = \Gamma(a) + \frac{1}{2\pi i} \int_{c-\infty i}^{c+\infty i} \frac{\Gamma(-s)}{s+a} z^{s+a} \, ds,$$

where $c < \min\{0, -\text{Re}(a)\}$. Identifying the second term on the right of this last expression as $-\Gamma(a, z)$ and making the change of variable $u = -a - s$, we then find

$$\Gamma(a, z) = \frac{1}{2\pi i} \int_{c-\infty i}^{c+\infty i} \Gamma(a+u) \frac{z^{-u}}{u} \, du \qquad (3.4.12)$$

valid for $|\arg z| < \frac{1}{2}\pi$. The path of integration is either the vertical line $\text{Re}(s) = c$, with $c > \max\{0, -\text{Re}(a)\}$, or with c any finite value provided the contour is indented to lie to the right of all the poles of the integrand.

Mellin-Barnes integral representations for the Bessel function $J_\nu(z)$ and the Weber function $D_\nu(z)$ are similarly found from Table 3.1 to be

$$J_\nu(x) = \pi^{-\frac{1}{2}} e^{ix} (2x)^\nu \frac{1}{2\pi i} \int_{c-\infty i}^{c+\infty i} \frac{\Gamma(-s)\Gamma(s+\nu+\frac{1}{2})}{\Gamma(s+2\nu+1)} (2ix)^s \, ds, \qquad (3.4.13)$$

when $x > 0$ and $\nu + \frac{1}{2} \neq 0, -1, -2, \dots$, and

$$D_\nu(z) = \frac{z^\nu e^{-\frac{1}{4}z^2}}{\Gamma(-\nu)} \frac{1}{2\pi i} \int_{c-\infty i}^{c+\infty i} \Gamma(-s)\Gamma(2s - \nu)(2z^2)^{-s} \, ds, \qquad (3.4.14)$$

valid when $|\arg z| < \frac{3}{4}\pi$, provided ν is not equal to a positive integer or zero. In each of these integrals the path of integration separates the poles of $\Gamma(-s)$ from those of the other gamma functions. We note that (3.4.13) converges only for $x > 0$. Use of an appropriate loop C, as in (3.4.5), enables a representation for $J_\nu(z)$ without restriction on arg z to be obtained in the form

$$J_\nu(z) = \pi^{-\frac{1}{2}} e^{iz} (2z)^\nu \frac{1}{2\pi i} \int_C \frac{\Gamma(s)\Gamma(\nu + \frac{1}{2} - s)}{\Gamma(2\nu + 1 - s)} (2iz)^{-s} \, ds, \qquad (3.4.15)$$

where C is a loop encircling the poles of $\Gamma(s)$ (in the positive sense) with endpoints at infinity in $\text{Re}(s) < 0$; see the discussion surrounding (3.4.22).

Mellin-Barnes integral representations for the Whittaker functions are given by [see also Slater (1960, p. 51); Whittaker & Watson (1965, p. 343)]

$$M_{k,m}(z) = \frac{z^{m+\frac{1}{2}} e^{-\frac{1}{2}z} \Gamma(1 + 2m)}{\Gamma(\frac{1}{2} + m - k)}$$

$$\times \frac{1}{2\pi i} \int_{c-\infty i}^{c+\infty i} \frac{\Gamma(-s)\Gamma(s + \frac{1}{2} + m - k)}{\Gamma(s + 2m + 1)} (-z)^s \, ds, \qquad (3.4.16)$$

for $|\arg z| < \frac{1}{2}\pi$ and $1 + 2m \neq 0, -1, -2, \ldots$ and $k - m - \frac{1}{2} \neq 0, 1, 2, \ldots$, and

$$\Gamma\left(\tfrac{1}{2} + m - k\right)\Gamma\left(\tfrac{1}{2} - m - k\right) W_{k,m}(z)$$

$$= z^k e^{-\frac{1}{2}z} \frac{1}{2\pi i} \int_{c-\infty i}^{c+\infty i} \Gamma(-s)\Gamma\left(s + \tfrac{1}{2} + m - k\right)\Gamma\left(s + \tfrac{1}{2} - m - k\right) z^{-s} \, ds$$

$$(3.4.17)$$

for $|\arg z| < \frac{3}{2}\pi$ and when neither of the numbers $k \pm m - \frac{1}{2}$ is a positive integer or zero. The quantity c in (3.4.16) and (3.4.17) can be any finite value provided the path of integration can be indented to separate the sequences of poles.

It is also possible to represent Bessel functions by integrals in which the exponential factor is not present. From (3.3.34), with $a = 0$ and $b = -\nu$, we obtain the representation of the modified Bessel function of the second kind

$$K_\nu(z) = \tfrac{1}{2}\left(\tfrac{1}{2}z\right)^\nu \frac{1}{2\pi i} \int_{c-\infty i}^{c+\infty i} \Gamma(s)\Gamma(s - \nu)\left(\tfrac{1}{2}z\right)^{-2s} \, ds \qquad (3.4.18)$$

valid in $|\arg z| < \pi$, where the contour can be taken to be the vertical line $\text{Re}(s) = c$ with $c > \max\{0, \text{Re}(\nu)\}$. Alternatively, c can be any finite value provided the contour is indented to lie to the right of the poles of $\Gamma(s)\Gamma(s - \nu)$. From the relation between $K_\nu(z)$ and the Hankel function [Watson (1966, p. 78)]

$$K_\nu(z) = \tfrac{1}{2}\pi i e^{\frac{1}{2}\pi i\nu} H_\nu^{(1)}(iz),$$

we then find the representations

$$\pi e^{\frac{1}{2}(\nu+1)\pi i} H_\nu^{(1)}(z) = \frac{1}{2\pi i} \int_{c-\infty i}^{c+\infty i} \Gamma(s)\Gamma(s - \nu)\left(-\tfrac{1}{2}iz\right)^{\nu-2s} ds, \qquad (3.4.19)$$

$$\pi e^{-\frac{1}{2}(\nu+1)\pi i} H_\nu^{(2)}(z) = \frac{1}{2\pi i} \int_{c-\infty i}^{c+\infty i} \Gamma(s)\Gamma(s - \nu)\left(\tfrac{1}{2}iz\right)^{\nu-2s} ds \qquad (3.4.20)$$

valid in the sectors $|\arg(\mp iz)| < \frac{1}{2}\pi$, respectively, where $c > \max\{0, \mathrm{Re}(\nu)\}$ in both integrals. Representations valid for all $\arg z$ can be obtained by bending back the paths of integration in (3.4.19) and (3.4.20) into a loop C encircling the poles of $\Gamma(s)$ with endpoints at infinity in $\mathrm{Re}(s) < 0$.

Since

$$J_\nu(z) = \tfrac{1}{2}\{H_\nu^{(1)}(z) + H_\nu^{(2)}(z)\},$$

we find from (3.4.19) and (3.4.20) the result

$$J_\nu(z) = \frac{1}{2\pi i} \int_{-\infty i}^{\infty i} \frac{\Gamma(s)}{\Gamma(1+\nu-s)} \left(\tfrac{1}{2}z\right)^{\nu-2s} ds, \qquad (3.4.21)$$

where the path is suitably indented at $s = 0$ to lie to the right of the poles of $\Gamma(s)$. As it stands, this representation is rather limited since it defines $J_\nu(z)$ only for $z > 0$ and $\mathrm{Re}(\nu) > 0$. To see this we apply Stirling's formula (2.1.8) to show that the modulus of the integrand has the behaviour $O(|t|^{2\sigma - \mathrm{Re}(\nu) - 1} e^{2\theta t})$ as $t \to \pm\infty$, where we have put $s = \sigma + it$ and $\theta = \arg z$. On the path of integration as $s \to \pm\infty i$, we see that the integral is absolutely convergent† only when $\arg z = 0$ and $\mathrm{Re}(\nu) > 0$, as stated. If, however, we bend back the path of integration into a loop C that encircles the poles of $\Gamma(s)$ (in the positive sense) with endpoints at infinity in $\mathrm{Re}(s) < 0$ (cf. Fig. 3.1), these restrictions on the validity of (3.4.21) can be removed. If C approaches ∞ in the directions $\pm\theta_0$ say, where $\frac{1}{2}\pi < \theta_0 \le \pi$, then the modulus of the logarithm of the integrand on C behaves like $2|s|\cos\theta_0 \log|s| + O(|s|) \to -\infty$ as $|s| \to \infty$. Consequently, we find that the integral

$$J_\nu(z) = \frac{1}{2\pi i} \int_C \frac{\Gamma(s)}{\Gamma(1+\nu-s)} \left(\tfrac{1}{2}z\right)^{\nu-2s} ds \qquad (3.4.22)$$

defines $J_\nu(z)$ *without restriction* on $\arg z$ and ν, since on C full advantage is being taken of the decay of the integrand in $\mathrm{Re}(s) < 0$.

Alternative representations for the Hankel functions which are valid in the wider sectors $|\arg(\mp iz)| < \frac{3}{2}\pi$ (cf. Rule 1 in §2.4) are given in Watson (1966, p. 193) in the form

$$H_\nu^{(1)}(z) = -\frac{\cos\pi\nu}{\pi^{5/2}} e^{i(z-\pi\nu)}(2z)^\nu$$

$$\times \int_{-\infty i}^{\infty i} \Gamma(s)\Gamma(s-2\nu)\Gamma\left(\nu + \tfrac{1}{2} - s\right)(-2iz)^{-s} ds, \qquad (3.4.23)$$

$$H_\nu^{(2)}(z) = \frac{\cos\pi\nu}{\pi^{5/2}} e^{-i(z-\pi\nu)}(2z)^\nu \int_{-\infty i}^{\infty i} \Gamma(s)\Gamma(s-2\nu)\Gamma\left(\nu + \tfrac{1}{2} - s\right)(2iz)^{-s} ds, \qquad (3.4.24)$$

---

† We remark that displacement of the contour to coincide with the vertical line $\mathrm{Re}(s) = c$ requires $0 < 2c < \mathrm{Re}(\nu)$ for absolute convergence and $\mathrm{Re}(\nu) \le 2c \le \mathrm{Re}(\nu) + 1$ for non-absolute convergence; cf. §2.4 and also Titchmarsh (1975, pp. 197–198).

where the integration path separates the poles of $\Gamma(s)\Gamma(s - 2v)$ from those of $\Gamma(v + \frac{1}{2} - s)$, provided $2v$ is not an odd integer. These formulas can be verified by displacement of the path of integration and evaluation of the residues of the poles of $\Gamma(s)\Gamma(s - 2v)$.

It is possible to establish in a similar manner the Mellin-Barnes integral representation for the product of Bessel functions [Watson (1966, p. 436)]

$$J_\mu(x)J_v(x) = \frac{1}{2\pi i} \int_{-\infty i}^{\infty i} \frac{\Gamma(-s)\Gamma(2s + \mu + v + 1)\left(\frac{1}{2}x\right)^{\mu+v+2s}}{\Gamma(s + \mu + 1)\Gamma(s + v + 1)\Gamma(s + \mu + v + 1)}\, ds$$

(3.4.25)

for $x > 0$, in which the contour separates the poles of $\Gamma(-s)$ from those of $\Gamma(2s + \mu + v + 1)$. Other representations of this type are given by [see Titchmarsh (1975, §7.10); Erdélyi (1954, pp. 332–334)]

$$K_\mu(z)K_v(z) = \frac{1}{2\pi i} \int_{c-\infty i}^{c+\infty i} \frac{1}{4\Gamma(2s)}\Gamma\left(s + \tfrac{1}{2}\mu + \tfrac{1}{2}v\right)\Gamma\left(s + \tfrac{1}{2}\mu - \tfrac{1}{2}v\right)$$

$$\times \Gamma\left(s - \tfrac{1}{2}\mu + \tfrac{1}{2}v\right)\Gamma\left(s - \tfrac{1}{2}\mu - \tfrac{1}{2}v\right)\left(\tfrac{1}{2}z\right)^{-2s}ds \qquad (3.4.26)$$

when $|\arg z| < \frac{1}{2}\pi$ and $2c > |\text{Re}(\mu)| + |\text{Re}(v)|$, together with

$$J_v(z)K_v(z) = \frac{1}{2\pi i} \int_{c-\infty i}^{c+\infty i} \frac{\Gamma(s)\Gamma(2s - v)}{2\Gamma(1 + v - s)}\left(\tfrac{1}{4}z^2\right)^{v-2s}ds \qquad (3.4.27)$$

when $|\arg z| < \frac{1}{4}\pi$ and $c > \max\{0, \frac{1}{2}\text{Re}(v)\}$, and

$$I_v(z)K_v(z) = \frac{1}{2\pi i} \int_{c-\infty i}^{c+\infty i} \frac{\Gamma(s)\Gamma(s + v)\Gamma\left(\frac{1}{2} - s\right)}{2\pi^{\frac{1}{2}}\Gamma(1 + v - s)}z^{-2s}\, ds \qquad (3.4.28)$$

when $|\arg z| < \frac{1}{2}\pi$ and $\max\{0, -\text{Re}(v)\} < c < \frac{1}{2}$.

Special cases of (3.4.25) are given by

$$\sin x\, J_v(x) = \frac{1}{2\pi i} \int_{c-\infty i}^{c+\infty i} \frac{2^{v-1}\Gamma\left(\frac{1}{2}s + \frac{1}{2}v + \frac{1}{2}\right)\Gamma\left(\frac{1}{2} - s\right)}{\Gamma(1 + v - s)\Gamma\left(1 - \frac{1}{2}v - \frac{1}{2}s\right)}x^{-s}\, ds, \qquad (3.4.29)$$

$$\cos x\, J_v(x) = \frac{1}{2\pi i} \int_{c-\infty i}^{c+\infty i} \frac{2^{v-1}\Gamma\left(\frac{1}{2}s + \frac{1}{2}v\right)\Gamma\left(\frac{1}{2} - s\right)}{\Gamma(1 + v - s)\Gamma\left(\frac{1}{2} - \frac{1}{2}v - \frac{1}{2}s\right)}x^{-s}\, ds \qquad (3.4.30)$$

when $x > 0$ and $-1 - \text{Re}(v) < c < \frac{1}{2}$ for (3.4.29) and $-\text{Re}(v) < c < \frac{1}{2}$ for (3.4.30).

4

Applications of Mellin Transforms

4.1 Transformation of Series

A problem of frequent occurrence is the summation of series which become very slowly convergent as a parameter or variable approaches a certain limiting value. In many cases, the terms in such series oscillate in sign and decay in absolute value so slowly that evaluation of the sum from the series as it stands is impracticable. There are several well-known methods available for transforming such series into rapidly convergent, or asymptotic, sums. One well-established approach is to convert the series into a contour integral which is then evaluated by use of Cauchy's theorem. Another method makes use of the Poisson summation formula which involves Fourier cosine transforms – see §8.1.1 for an example of this procedure. A method equivalent to the Poisson summation formula is based on the Mellin transform and is described below.

4.1.1 The Mellin Transform Method

Consider the sum $S(x)$ defined by

$$S(x) = \sum_{n=1}^{\infty} f(nx),$$

where we suppose that the function $f(x)$ is locally integrable on $(0, \infty)$, is $O(x^{-a})$ as $x \to 0+$ and $O(x^{-b})$, where $b > 1$, as $x \to \infty$. If $a < b$, the Mellin inversion theorem (see (3.1.5)) shows that

$$f(x) = \frac{1}{2\pi i} \int_{c-\infty i}^{c+\infty i} F(s) x^{-s} \, ds \quad (a < c < b),$$

where

$$F(s) = \int_0^\infty x^{s-1} f(x)\,dx$$

denotes the Mellin transform of $f(x)$ defined in the strip $a < \operatorname{Re}(s) < b$. Then

$$\sum_{n=1}^\infty f(nx) = \frac{1}{2\pi i} \int_{c-\infty i}^{c+\infty i} F(s) \sum_{n=1}^\infty (nx)^{-s}\,ds$$

$$= \frac{1}{2\pi i} \int_{c-\infty i}^{c+\infty i} F(s)\zeta(s)x^{-s}\,ds, \qquad (4.1.1)$$

where $\max\{1, a\} < c < b$ and $\zeta(s)$ denotes the Riemann zeta function. The inversion of the order of summation and integration is justified by absolute convergence provided $\max\{1, a\} < \operatorname{Re}(s) < b$. Suitable manipulation of the path of integration, so as to enclose certain poles of the integrand, followed by use of the properties of $\zeta(s)$ and Cauchy's theorem, can then lead to an alternative representation involving rapidly convergent (or, in some cases, asymptotic) sums. This depends on the structure of the singularities of the analytic continuation of $F(s)$ outside the strip $a < \operatorname{Re}(s) < b$; see §3.2. For example, if $f(x)$ satisfies the conditions of Lemma 3.3, then the analytic continuation of $F(s)$ in $\operatorname{Re}(s) < a$ possesses a sequence of simple poles, with the first pole satisfying $\operatorname{Re}(s) = a$.

To be more specific, let us suppose that $f(x)$ is bounded at the origin, so that $f(x) = f(0+) + O(x^\alpha)$, where $\alpha > 0$, as $x \to 0+$. In this case, $F(s)$ is regular in the strip $0 < \operatorname{Re}(s) < b$ and has, in general, a simple pole at $s = 0$, with residue $f(0+)$. Then, if it is permissible to displace the contour in (4.1.1) first over the simple pole of $\zeta(s)$ at $s = 1$ and then over the pole at $s = 0$, we find consecutively

$$\sum_{n=1}^\infty f(nx) - \frac{F(1)}{x} = \frac{1}{2\pi i} \int_{c-\infty i}^{c+\infty i} F(s)\zeta(s)x^{-s}\,ds \quad (0 < c < 1) \qquad (4.1.2)$$

$$= -\tfrac{1}{2} f(0+) + \frac{1}{2\pi i} \int_{c-\infty i}^{c+\infty i} F(s)\zeta(s)x^{-s}\,ds, \qquad (4.1.3)$$

where $-\alpha < c < 0$ in the last integral and we have used the result $\zeta(0) = -\tfrac{1}{2}$.

Further manipulation of the contour now depends on the detailed structure of the singularities of $F(s)$. In some cases, it may happen that the remaining simple poles of $F(s)$ coincide with the 'trivial zeros' of $\zeta(s)$ at $s = -2, -4, \ldots$, with the result that further displacement of the contour will not produce any additional algebraic contributions. This is usually the signal that the integral in (4.1.3) yields an exponentially small contribution to the sum $S(x)$. In this case, evaluation of the integral then proceeds by use of the functional relation for $\zeta(s)$ given by [Whittaker & Watson (1965, p. 269)]

$$\zeta(s) = 2^s \pi^{s-1} \zeta(1-s)\Gamma(1-s)\sin\tfrac{1}{2}\pi s, \qquad (4.1.4)$$

followed by appropriate manipulation of the resulting integrand. We give examples of this procedure, both in the general case and in the above 'exponentially small' case, in the following sections.

If we assume that the function $f(x)$ admits an expansion as $x \to 0+$ of the form

$$f(x) \sim f(0+) + \sum_{k=1}^{\infty} A_k x^{\alpha_k},$$

where $\text{Re}(\alpha_k)$ increase monotonically to ∞ as $k \to \infty$, then it follows from Lemma 3.3 and the paragraph following (3.2.4) that $F(s)$ has, in addition to the simple pole at $s = 0$, simple poles at $s = -\alpha_k$ with residue A_k. Upon noting that $F(1) = \int_0^{\infty} f(x)\,dx$, we then obtain from (4.1.3) the expansion

$$\sum_{n=0}^{\infty} f(nx) \sim \frac{1}{x} \int_0^{\infty} f(x)\,dx + \tfrac{1}{2} f(0+) + \sum_{k=1}^{\infty} A_k \zeta(-\alpha_k) x^{\alpha_k} \qquad (4.1.5)$$

as $x \to 0+$, which is a form of the Euler-Maclaurin summation formula.

We remark that further results on the summation of series can be derived from simple properties of the zeta function. For example, we have the results valid when $\text{Re}(s) > 1$

$$\sum_{n=1}^{\infty} (-)^{n-1} n^{-s} = (1 - 2^{1-s}) \zeta(s),$$

$$\sum_{n=0}^{\infty} (2n + 1)^{-s} = (1 - 2^{-s}) \zeta(s).$$

It then follows from (4.1.1) that

$$\sum_{n=1}^{\infty} (-)^{n-1} f(nx) = \frac{1}{2\pi i} \int_{c-\infty i}^{c+\infty i} F(s)(1 - 2^{1-s}) \zeta(s) x^{-s}\,ds \qquad (4.1.6)$$

and

$$\sum_{n=0}^{\infty} f((2n + 1)x) = \frac{1}{2\pi i} \int_{c-\infty i}^{c+\infty i} F(s)(1 - 2^{-s}) \zeta(s) x^{-s}\,ds, \qquad (4.1.7)$$

where in both cases $\max\{1, a\} < c < b$.

To conclude this introductory section we record two results of general interest. The first concerns the Möbius inversion formula for the sum $S(x)$ in (4.1.1). Under modest hypotheses on the functions $S(x)$ and $f(x)$, we have the inversion pair

$$S(x) = \sum_{n=1}^{\infty} f(nx), \qquad f(x) = \sum_{n=1}^{\infty} \hat{\mu}(n) S(nx) \quad (x > 0), \qquad (4.1.8)$$

where $\hat{\mu}(n)$ denotes the Möbius function defined by

$$\hat{\mu}(n) = \begin{cases} 1 & n = 1 \\ (-)^r & \text{if } n \text{ has } r \text{ distinct prime factors} \\ 0 & \text{otherwise.} \end{cases} \tag{4.1.9}$$

This result can be established by taking the Mellin transform of both sides of the above relations to obtain (when $\text{Re}(s) > 1$)

$$M[S; s] = \zeta(s)F(s), \qquad F(s) = M[S; s] \sum_{n=1}^{\infty} \hat{\mu}(n)n^{-s},$$

where $M[S; s]$ is the Mellin transform of $S(x)$. These last equations are seen to be equivalent to the statement of the well-known definition [Titchmarsh (1986, p. 3)]

$$\frac{1}{\zeta(s)} = \sum_{n=1}^{\infty} \frac{\hat{\mu}(n)}{n^s} \qquad (\text{Re}(s) > 1). \tag{4.1.10}$$

The second result is a consequence of (4.1.2). Application of the Mellin inversion formula to (4.1.2) immediately yields the formula

$$\zeta(s) = \frac{1}{F(s)} \int_0^{\infty} x^{s-1} \left\{ \sum_{n=1}^{\infty} f(nx) - \frac{F(1)}{x} \right\} dx, \tag{4.1.11}$$

valid when $0 < \text{Re}(s) < 1$. This is known as Müntz's formula [Titchmarsh (1986, p. 29)] and expresses $\zeta(s)$ in the critical strip in terms of an expansion of arbitrary functions subject to the above restrictions.

4.1.2 The Poisson-Jacobi Formula

The archetypal example of series transformations is the familiar Poisson-Jacobi formula for the sum

$$S(a) = \sum_{n=1}^{\infty} e^{-an^2}.$$

The transformation formula is expressed in the form

$$S(a) = \frac{1}{2}\sqrt{\frac{\pi}{a}} - \frac{1}{2} + \sqrt{\frac{\pi}{a}} \sum_{n=1}^{\infty} e^{-\pi^2 n^2/a}, \tag{4.1.12}$$

where the parameter a satisfies $\text{Re}(a) > 0$. It is seen that the convergence of $S(a)$ as $a \to 0$ in $\text{Re}(a) > 0$ becomes very slow, while the sum on the right-hand side of (4.1.12) is rapidly convergent in this limit. A variety of different proofs of this well-known result can be found in the literature — see, for example, Whittaker & Watson (1965, p. 124); Titchmarsh (1975, p. 60). We shall indulge ourselves here in supplying a proof that relies on Mellin transforms, since it provides a simple illustration of the treatment of the 'exponentially small' case mentioned above; see also §8.1 for a generalisation of this transformation.

We have $f(x) = e^{-x^2}$, which has the Mellin transform $F(s) = \frac{1}{2}\Gamma(\frac{1}{2}s)$ valid in the half-plane $\text{Re}(s) > 0$. Thus we obtain from (4.1.1)

$$S(a) = \frac{1}{4\pi i} \int_{c-\infty i}^{c+\infty i} \Gamma(\tfrac{1}{2}s)\zeta(s)a^{-\frac{1}{2}s}\,ds \qquad (c > 1); \tag{4.1.13}$$

compare (8.1.29). Since $|\zeta(s)| \leq \sum_{n=1}^{\infty} n^{-c} = O(1)$ on the path of integration, the convergence of (4.1.13) is controlled by the factor $\Gamma(\frac{1}{2}s)$. It follows by comparison with the integral for e^{-z} in (3.3.2) (or by Rule 1 in §2.4) that (4.1.13) defines $S(a)$ in the sector $|\arg a| < \frac{1}{2}\pi$.

The integrand in (4.1.13) has simple poles at $s = 1$ and $s = 0$ with the remaining poles of $\Gamma(\frac{1}{2}s)$ at $s = -2, -4, \ldots$ being cancelled by the 'trivial zeros' of $\zeta(2s)$, since $\zeta(-2n) = 0$ for $n = 1, 2, \ldots$. We now consider the integral taken round the rectangular contour with vertices at $c \pm iT$ and $-c' \pm iT$, where $c' > 0$ and both c, c' are supposed finite. The contribution from the upper and lower sides, given by $s = \sigma \pm iT$, $-c' \leq \sigma \leq c$, vanishes as $T \to \infty$, since the modulus of the integrand on these paths is $O(T^{\frac{1}{2}\sigma - \frac{1}{2} + \mu(\sigma)} \log T \, e^{-\frac{1}{2}\Delta T})$, where $\Delta = \frac{1}{2}\pi - |\arg a| < \frac{1}{2}\pi$. This follows from Stirling's formula in (2.1.8), which shows that for finite σ

$$\Gamma(\sigma \pm it) = O\big(t^{\sigma - \frac{1}{2}} e^{-\frac{1}{2}\pi t}\big) \qquad (t \to \infty) \tag{4.1.14}$$

and from the well-known behaviour of $\zeta(s)$ [Titchmarsh (1986, p. 95); Ivić (1985, p. 25)]

$$|\zeta(\sigma \pm it)| = O\big(t^{\mu(\sigma)} \log^A t\big) \qquad (t \to \infty), \tag{4.1.15}$$

where†

$$\mu(\sigma) = \begin{cases} 0 & (\sigma > 1) \\ \frac{1}{2} - \frac{1}{2}\sigma & (0 \leq \sigma \leq 1) \\ \frac{1}{2} - \sigma & (\sigma < 0) \end{cases}$$

and $A = 1$ when $0 \leq \sigma \leq 1$, $A = 0$ otherwise.‡ We may therefore displace the contour to the left over the poles at $s = 1$ and $s = 0$. Upon noting that the residues are given by $(\pi/a)^{\frac{1}{2}}$ and -1, respectively, we consequently find

$$S(a) - \frac{1}{2}\sqrt{\frac{\pi}{a}} + \frac{1}{2} = \frac{1}{2\pi i} \int_{c-\infty i}^{c+\infty i} \Gamma(-s)\zeta(-2s)a^s\,ds, \tag{4.1.16}$$

---

† If we could assume the truth of the Lindelöf hypothesis (which is equivalent to the statement that $\zeta(\frac{1}{2} + it) = O(t^\epsilon)$ for every positive ϵ), we would have $\mu(\sigma) = \frac{1}{2} - \sigma$ for $\sigma \leq \frac{1}{2}$, $\mu(\sigma) = 0$ for $\sigma > \frac{1}{2}$.

‡ An alternative way of expressing this is to say that $|\zeta(\sigma \pm it)| = O(t^{\mu(\sigma)+\epsilon})$ as $t \to +\infty$ for every positive ϵ.

where $c > 0$ and, for convenience, we have made the change of variable $s \to -2s$. It is seen that the poles of $\Gamma(-s)$ at $s = 1, 2, \ldots$ are cancelled by the 'trivial zeros' of $\zeta(-2s)$. This has the consequence that the integrand is holomorphic in $\mathrm{Re}(s) > 0$, so that further displacement of the contour can produce no additional algebraic terms in the expansion of $S(a)$.

The functional relation for $\zeta(s)$ in (4.1.4) can now be employed to convert the argument of the zeta function (which has negative real part) in the above integral into one with real part greater than unity. Use of the reflection and duplication formulas for the gamma function in (2.1.20) and (2.2.23) shows that

$$\Gamma(-s)\zeta(-2s) = (2\pi)^{-2s}\zeta(1+2s)\frac{\Gamma(1+2s)}{\Gamma(1+s)} = \pi^{-2s-\frac{1}{2}}\zeta(1+2s)\Gamma\left(s+\tfrac{1}{2}\right).$$

It follows that the integral on the right-hand side of (4.1.16) can be written as

$$\sqrt{\frac{\pi}{a}}\frac{1}{2\pi i}\int_{c-\infty i}^{c+\infty i}\zeta(1+2s)\Gamma\left(s+\tfrac{1}{2}\right)(\pi^2/a)^{-s-\frac{1}{2}}ds \qquad (c > 0). \qquad (4.1.17)$$

From the behaviour of the gamma function and the fact that $\zeta(1+2s)$ is $O(1)$ as $s \to \infty$ in $\mathrm{Re}(s) > 0$, the contour cannot be closed in the right-hand half-plane; it therefore follows that, as there are no poles to the right of the contour, the contribution made by the above integral must be exponentially small. If we now expand $\zeta(1+2s)$ and reverse the order of summation and integration (which is permissible since $\mathrm{Re}(1+2s) > 1$ on the path of integration), the integral in (4.1.17) then becomes

$$\sum_{n=1}^{\infty}\frac{1}{2\pi i}\int_{c-\infty i}^{c+\infty i}\Gamma\left(s+\tfrac{1}{2}\right)\left(\pi^2 n^2/a\right)^{-s-\frac{1}{2}}ds = \sum_{n=1}^{\infty}e^{-\pi^2 n^2/a},$$

when $\mathrm{Re}(a) > 0$ by (3.3.2), thereby establishing the Poisson-Jacobi formula (4.1.12).

4.2 Examples

In the next two sections we give some examples of the above Mellin transform technique for the transformation of slowly convergent series to illustrate the general procedure. The first set of examples concerns some standard sums while the second set in §4.3 deals with more arcane number-theoretic functions.

4.2.1 An Infinite Series

Our first example concerns the evaluation of the function $S(z)$ defined by the sum

$$S(z) = \sum_{n=1}^{\infty}\frac{n^{\alpha-1}}{n+z} \qquad (0 < \alpha < 1)$$

as $z \to \infty$ in $|\arg z| < \pi$. With $f(x) = x^{\alpha-1}/(1+x)$, the Mellin transform is given by

$$F(s) = \int_0^\infty \frac{x^{s+\alpha-2}}{1+x}\,dx = -\frac{\pi}{\sin \pi(s+\alpha)} \quad (1 < \alpha + \mathrm{Re}(s) < 2)$$

by (3.3.8). Then, from (4.1.1), we find

$$S(z) = z^{\alpha-2}\sum_{n=1}^\infty f(n/z) = -\frac{1}{2\pi i}\int_{c-\infty i}^{c+\infty i} \frac{\pi \zeta(s)}{\sin \pi(s+\alpha)}z^{s+\alpha-2}\,ds, \qquad (4.2.1)$$

where $1 < c < 2 - \alpha$.

The integrand has simple poles at $s = 1$ and at integer values of $s + \alpha$. Let $N \geq 2$ be an integer and let $c' = N + \alpha - \frac{3}{2}$. We consider the integral taken round the rectangular contour with vertices at $c \pm iT$ and $-c' \pm iT$, so that the side in $\mathrm{Re}(s) < 0$ parallel to the imaginary axis passes midway between the poles at $s = 2 - \alpha - N$ and $s = 1 - \alpha - N$. The contribution from the upper and lower sides $s = \sigma \pm iT$, $-c' \leq \sigma \leq c$, vanishes as $T \to \infty$ provided $|\arg z| < \pi$, since, from (4.1.14) and (4.1.15), the modulus of the integrand is controlled by $O(T^{\mu(\sigma)}\log T\, e^{-\Delta T})$, where $\Delta = \pi - |\arg z|$. Displacement of the contour to the left then yields

$$S(z) = \frac{\pi z^{\alpha-1}}{\sin \pi \alpha} + \sum_{k=0}^{N-1}\frac{(-)^k}{z^{1+k}}\zeta(1-\alpha-k) + R_N(z),$$

where the remainder integral $R_N(z)$ is given by

$$R_N(z) = -\frac{1}{2\pi i}\int_{-c'-\infty i}^{-c'+\infty i}\frac{\pi \zeta(s)}{\sin \pi(s+\alpha)}z^{s+\alpha-2}\,ds.$$

If we set $s = -c' + it$ and use (4.1.4), together with the fact that

$$|\zeta(\sigma+it)| \leq \zeta(\sigma) \qquad (\sigma > 1), \qquad (4.2.2)$$

we find

$$|R_N(z)| \leq 2|z|^{-N-\frac{1}{2}}\zeta(N+\alpha-\tfrac{1}{2})\int_{-\infty}^\infty g(t)\frac{e^{-\theta t}}{\cosh \pi t}\,dt,$$

where

$$g(t) = (2\pi)^{-N-\alpha-\frac{1}{2}}|\Gamma(N+\alpha-\tfrac{1}{2}+it)\sin \tfrac{1}{2}\pi(N+\alpha-\tfrac{3}{2}+it)|$$

and $\theta = \arg z$. Since $g(t) = O(|t|^{N+\alpha-1})$ as $t \to \pm\infty$, by (4.1.14), it follows that the above integral is independent of $|z|$ and converges when $|\arg z| < \pi$. Hence $|R_N(z)| = O(|z|^{-N-\frac{1}{2}})$ and we obtain the asymptotic expansion

$$S(z) \sim \frac{\pi z^{\alpha-1}}{\sin \pi \alpha} + \sum_{k=0}^\infty \frac{(-)^k}{z^{1+k}}\zeta(1-\alpha-k) \qquad (4.2.3)$$

as $z \to \infty$ in $|\arg z| < \pi$.

We note that the conditions imposed on α are unnecessarily restrictive. Extension of the expansion (4.2.3) to complex values of α satisfying $\mathrm{Re}(\alpha) < 1$ is straightforward. In this case, the relation (4.2.1) holds in the strip $\max\{1, 1 - \mathrm{Re}(\alpha)\} < c < 2 - \mathrm{Re}(\alpha)$. The analysis proceeds as above, with the exception that now a double pole at $s = 1$ will arise whenever $\alpha = -m, m = 0, 1, 2, \dots$. The residue at this double pole can be evaluated by observing that in the neighbourhood of $s = 1$ we have

$$\frac{\pi \zeta(s)}{\sin \pi (s - m)} z^{s-m-2} = \frac{(-z)^{-m-1}}{\delta} \left(\frac{1}{\delta} + \gamma + \cdots \right) (1 + \delta \log z + \cdots)$$

$$= (-z)^{-m-1} \left\{ \frac{1}{\delta^2} + \frac{\gamma + \log z}{\delta} + \cdots \right\},$$

where $\gamma = 0.577215\dots$ is Euler's constant and $\delta = s - 1$. Thus we find that, provided α does not equal a nonpositive integer, (4.2.3) applies when $\mathrm{Re}(\alpha) < 1$, and that when $\alpha = -m$

$$S(z) \sim (-)^m z^{-m-1} (\gamma + \log z) + \sum_{k=0}^{\infty} {}' \frac{(-)^k}{z^{1+k}} \zeta(1 + m - k) \tag{4.2.4}$$

as $z \to \infty$ in $|\arg z| < \pi$, where the prime on the summation sign indicates the omission of the term corresponding to $k = m$.

It is worth noting that when $\alpha = 0$ we have the result

$$\sum_{n=1}^{\infty} \frac{1}{n(n+z)} = \frac{1}{z} \{ \psi(z) + \gamma + z^{-1} \},$$

where $\psi(z)$ is the logarithmic derivative of the gamma function. Use of the values $\zeta(-2k) = 0$, $\zeta(1 - 2k) = -B_{2k}/(2k)$ $(k = 1, 2, \dots)$ shows that (4.2.4) correctly reduces to the expansion

$$\psi(z) \sim \log z - \frac{1}{2z} - \sum_{k=1}^{\infty} \frac{B_{2k}}{2k \, z^{2k}}$$

as $z \to \infty$ in $|\arg z| < \pi$. We also remark that differentiation of both sides of (4.2.3) with respect to α yields the expansion, when $0 < \alpha < 1$,

$$\sum_{n=1}^{\infty} \frac{n^{\alpha-1} \log n}{n+z} \sim \frac{\pi z^{\alpha-1}}{\sin \pi \alpha} (\log z - \pi \cot \pi \alpha) + \sum_{k=0}^{\infty} \frac{(-)^{k-1}}{z^{1+k}} \zeta'(1 - \alpha - k)$$

$$\tag{4.2.5}$$

as $z \to \infty$ in $|\arg z| < \pi$. This last result† is obtained in Evgrafov (1961, p. 163) by the Euler-Maclaurin summation formula.

---

† There is a misprint in Eq. (11) of this reference.

4.2.2 A Smoothed Dirichlet Series

We consider the evaluation of the sum

$$\sum_{n=1}^{\infty} n^{-w} e^{-an}$$

for real values of $w > 0$ (although the analysis is easily extended to arbitrary complex w) as $a \to 0$ in the sector $|\arg a| < \frac{1}{2}\pi$. With $f(x) = x^{-w} e^{-x}$, the Mellin transform is given by

$$F(s) = \int_0^{\infty} x^{s-w-1} e^{-x}\, dx = \Gamma(s - w) \qquad (\mathrm{Re}(s) > w)$$

so that, from (4.1.1), we find

$$\sum_{n=1}^{\infty} n^{-w} e^{-an} = \frac{1}{2\pi i} \int_{c-\infty i}^{c+\infty i} \Gamma(s - w)\zeta(s) a^{w-s}\, ds, \qquad (4.2.6)$$

where $c > \max\{1, w\}$.

The integrand possesses simple poles at $s = 1$ and $s = w - k$ ($k = 0, 1, 2, \ldots$), except if w equals a positive integer m when the pole at $s = 1$ is double. As in the preceding example, we consider the integral taken round the rectangular contour with vertices at $c \pm iT$, $-c' \pm iT$, where $c' > 0$. The contribution from the upper and lower sides $s = \sigma \pm iT$, $-c' \le \sigma \le c$, vanishes as $T \to \infty$ provided $|\arg a| < \frac{1}{2}\pi$, since, from (4.1.14) and (4.1.15), the modulus of the integrand is controlled by $O(T^{\sigma-w+\mu(\sigma)-\frac{1}{2}} \log T\, e^{-\Delta T})$, where $\Delta = \frac{1}{2}\pi - |\arg a|$. Upon noting that the residue at the double pole at $s = 1$ (when $w = m$) is given by $(-a)^{m-1}\{\gamma - \log a + \psi(m)\}/\Gamma(m)$, where γ is Euler's constant and $\psi(m)$ denotes the logarithmic derivative of the gamma function, we can then displace the contour to the left over the poles to obtain

$$\sum_{n=1}^{\infty} n^{-w} e^{-an} = J(a, w) + \sum_{k=0}^{N-1}{}' \frac{(-)^k}{k!} \zeta(w - k) a^k + R_N. \qquad (4.2.7)$$

Here

$$J(a, w) = \begin{cases} \Gamma(1 - w) a^{w-1} & (w \ne m) \\[2mm] \dfrac{(-a)^{m-1}}{\Gamma(m)}\{\gamma - \log a + \psi(m)\} & (w = m), \end{cases} \qquad (4.2.8)$$

N is a positive integer such that $N > w + \frac{1}{2}$ and the prime on the sum over k denotes the omission of the term corresponding to $k = m - 1$ when $w = m$.

The remainder integral R_N is

$$R_N = \frac{1}{2\pi i} \int_{C_N} \Gamma(s - w)\zeta(s) a^{w-s}\, ds, \qquad (4.2.9)$$

where C_N denotes the displaced contour $\text{Re}(s) = -N + w + \frac{1}{2}$. We let $s = -N + w + \frac{1}{2} + it$ and put $N - \frac{1}{2} - w = M + \delta$, where $M = [N - \frac{1}{2} - w]$ denotes the integer part and δ the fractional part $(0 \leq \delta < 1)$. Then, when $0 < w \leq \frac{1}{2}$ we have $M = N - 1$ with $0 \leq \delta < \frac{1}{2}$, while when $\frac{1}{2} < w < N - \frac{1}{2}$ we have $0 \leq M \leq N - 2$. By repeated application of the result $\Gamma(z+1) = z\Gamma(z)$ it follows that

$$\left| \frac{\Gamma(N + \frac{1}{2} - w + it)}{\Gamma(N + \frac{1}{2} + it)} \right| = P(t) \left| \frac{\Gamma(1 + \delta + it)}{\Gamma(\frac{1}{2} + it)} \right|,$$

where†

$$P(t) = \frac{\prod_{r=1}^{M} \{(\delta + r)^2 + t^2\}^{\frac{1}{2}}}{\prod_{r=1}^{N} \{(r - \frac{1}{2})^2 + t^2\}^{\frac{1}{2}}} < (\tfrac{1}{4} + t^2)^{-\frac{1}{2}} \leq 2$$

for $-\infty < t < \infty$. Then by use (4.1.4) and (4.2.2) we find

$$|R_N| \leq \frac{(|a|/2\pi)^{N - \frac{1}{2}}}{(2\pi)^{1-w}} \int_{-\infty}^{\infty} |\zeta(N + \tfrac{1}{2} - w + it)| \left| \frac{\Gamma(N + \frac{1}{2} - w + it)}{\Gamma(N + \frac{1}{2} + it)} \right|$$

$$\times \left| \sin \tfrac{1}{2}\pi(-N + w + \tfrac{1}{2} + it) \right| \frac{e^{\psi t}}{\cosh \pi t} \, dt$$

$$\leq 2 \frac{(|a|/2\pi)^{N - \frac{1}{2}}}{(2\pi)^{1-w}} \zeta(N + \tfrac{1}{2} - w) \int_{-\infty}^{\infty} \left| \frac{\Gamma(1 + \delta + it)}{\Gamma(\frac{1}{2} + it)} \right| \frac{\cosh \frac{1}{2}\pi t}{\cosh \pi t} e^{\psi t} \, dt,$$

where $\psi = \arg a$. The ratio of gamma functions in the above integrand is $O(|t|^{\frac{1}{2}+\delta})$ as $t \to \pm\infty$, so that the integral is absolutely convergent and is $O(1)$ when $|\arg a| < \frac{1}{2}\pi$.

Then, since $\zeta(N + \frac{1}{2} - w) = O(1)$ for large N, we see that

$$|R_N| = O\left((|a|/2\pi)^{N - \frac{1}{2}}\right) \qquad (N \to \infty).$$

Hence, $|R_N| \to 0$ as $N \to \infty$ provided $|a|/2\pi < 1$ and $|\arg a| < \frac{1}{2}\pi$, and we therefore obtain the convergent expansion‡

$$\sum_{n=1}^{\infty} n^{-w} e^{-an} = J(a, w) + \sum_{k=1}^{\infty} {}' \frac{(-)^k}{k!} \zeta(w - k) a^k \qquad (|a| < 2\pi). \qquad (4.2.10)$$

That the sum over k in (4.2.10) is absolutely convergent when $|a| < 2\pi$ can also be seen from (4.1.4), since the behaviour of the late terms is controlled by $(|a|/2\pi)^k \Gamma(k + 1 - w)/\Gamma(k + 1) \sim k^{-w}(|a|/2\pi)^k$ as $k \to \infty$.

---

† This bound can be sharpened to $\pi^{\frac{1}{2}}/\Gamma(N - M + \frac{1}{2})$.

‡ This sum can be expressed alternatively in terms of Lerch's transcendent Φ [Erdélyi (1954, §1.11)] in the form $e^{-a}\Phi(e^{-a}, w, 1)$.

The sector of validity $|\arg a| < \frac{1}{2}\pi$ in (4.2.10) can be extended to $|\arg a| \leq \frac{1}{2}\pi$, but at the expense of dealing with conditionally convergent integrals. To see this, we put $a = \pm ix$ (with $x > 0$) in (4.2.6) and make the temporary restriction $w > 1$, to find

$$\sum_{n=1}^{\infty} \frac{\cos nx}{n^w} = \frac{1}{2\pi i} \int_{c-\infty i}^{c+\infty i} \Gamma(s-w)\zeta(s)x^{w-s} \cos \tfrac{1}{2}\pi(s-w)\,ds; \qquad (4.2.11)$$

an analogous expression holds for the sum involving $\sin nx$. Due to the presence of the cosine term, the integrand now no longer decays exponentially as $\mathrm{Im}(s) \to \pm\infty$, with the consequence that the integral (4.2.11) is only conditionally convergent† for $\max\{1, w\} < c < w + \frac{1}{2}$. In general, the integrand in this case has a simple pole at $s = 1$ and a sequence of poles at $s = w - 2k$, $(k = 0, 1, 2, \ldots)$, except for integer values of w, when the sequence of poles is finite when w is even (due to $\zeta(-2m) = 0$, $m = 1, 2, \ldots$), or the pole at $s = 1$ is double when w is odd.

We then deduce from (4.2.10) and (4.2.8) that

$$\sum_{n=1}^{\infty} \frac{\cos nx}{n^w} = \hat{J}(x, w) + \sum_{k=0}^{\infty}{}' \frac{(-)^k}{(2k)!}\zeta(w - 2k)x^{2k}, \qquad (4.2.12)$$

when $0 < x < 2\pi$, where

$$\hat{J}(x, w) = \begin{cases} \Gamma(1-w)x^{w-1}\cos\tfrac{1}{2}\pi(1-w) & (w \neq 2m+1) \\[2mm] \dfrac{(-)^m x^{2m}}{(2m)!}\{\gamma - \log x + \psi(2m+1)\} & (w = 2m+1), \end{cases} \qquad (4.2.13)$$

and the prime indicates the omission of the term corresponding to $k = m$ when $w = 2m + 1$ $(m = 0, 1, 2, \ldots)$. We remark that the modulus of the integrand of the associated remainder integral R_N, taken along the path $\mathrm{Re}(s) = -2N + w + 1$, is no longer exponentially decaying as $|t| \to \infty$, but is algebraic given by $(x/2\pi)^{2N-1}O(|t|^{-w})$. The remainder integral consequently converges absolutely when $w > 1$. The temporary restriction $w > 1$ can now be replaced by $w > 0$ by appeal to analytic continuation, since both sides of (4.2.12) are analytic functions of w when $\mathrm{Re}(w) > 0$. Continuous extension also shows that the open interval $0 < x < 2\pi$ can be replaced by the closed interval $0 \leq x \leq 2\pi$ when $w > 1$. See also the papers by Boersma (1975) and Grossman (1997) for a discussion of sums of this type.

---

† We note that the Mellin transform of the associated function $x^{-w}\cos x$ is also conditionally convergent.

The special cases of (4.2.12) corresponding to $w = 1$ and $w = 2$ yield the standard results [Abramowitz & Stegun (1965, p. 1005)]

$$\sum_{n=1}^{\infty} \frac{\cos nx}{n} = -\log x + \sum_{k=1}^{\infty} \frac{\zeta(2k)}{k}(x/2\pi)^{2k} = -\log\left(2\sin\tfrac{1}{2}x\right) \quad (0 < x < 2\pi)$$

and

$$\sum_{n=1}^{\infty} \frac{\cos nx}{n^2} = \tfrac{1}{6}\pi^2 - \tfrac{1}{2}\pi x + \tfrac{1}{4}x^2 \quad (0 \le x \le 2\pi),$$

where we have used the facts that $\psi(1) = -\gamma$ and that, when $w = 2$, the sum over k in (4.2.12) consists of only the terms corresponding to $k = 0$ and $k = 1$, since $\zeta(2 - 2k) = 0$ for $k \ge 2$.

The situation corresponding to $0 < w < 1$ is more interesting†, since the sum on the left-hand side of (4.2.12) is divergent when $x = 0$ (and $x = 2\pi$). Thus, for example, if $w = \tfrac{1}{2}$ we find upon application of (4.1.4) the formula, valid when $0 < x < 2\pi$,

$$\sum_{n=1}^{\infty} \frac{\cos nx}{n^{1/2}} = \left(\frac{\pi}{2x}\right)^{1/2} + \pi^{-\frac{1}{2}} \sum_{k=0}^{\infty} \frac{\Gamma\left(2k + \tfrac{1}{2}\right)}{(2k)!} \zeta\left(2k + \tfrac{1}{2}\right) \left(\frac{x}{2\pi}\right)^{2k}, \quad (4.2.14)$$

which readily permits the evaluation of the slowly convergent sum on the left-hand side in the limit $x \to 0+$.

4.2.3 A Finite Sum

Finite sums can also be dealt with by the same procedure. Suppose we have the sum

$$\sum_{n=1}^{N} \frac{\left(1 - n^\alpha x^\alpha\right)^{\nu-\frac{1}{2}}}{n^\beta} \quad (\nu > \tfrac{1}{2}),$$

where $x > 0$, $\alpha, \beta > 0$ and N is the largest integer such that $xN \le 1$. Then as $x \to 0+$, we find $N \to \infty$ and direct evaluation of the sum in this limit will involve a large number of terms. However, with this example, a complication presents itself in the form of the resulting expansion (without further manipulation) *not* being asymptotic. The particular case $\alpha = \beta = \tfrac{2}{3}$ and $\nu = 1$ has been discussed by Macfarlane (1949) [see also Davies (1978, p. 215)], although this complication was not correctly identified in these references.

We let

$$f(x) = \begin{cases} x^{-\beta}(1 - x^\alpha)^{\nu-\frac{1}{2}} & (0 < x < 1) \\ 0 & (x > 1), \end{cases}$$

---

† The sums involving $\sin(x\sqrt{n})/n$ and $\cos(x\sqrt{n})/n$ have been evaluated by Boersma (1995); see also Chapman (1995) and Glasser (1995).

so that the Mellin transform $F(s)$ of $f(x)$ is given by

$$F(s) = \int_0^1 x^{s-\beta-1}(1-x^\alpha)^{\nu-\frac{1}{2}}dx = \frac{1}{\alpha}\int_0^1 y^{(s-\beta)/\alpha-1}(1-y)^{\nu-\frac{1}{2}}dy$$

$$= \frac{\Gamma\left(\dfrac{s-\beta}{\alpha}\right)\Gamma(\nu+\frac{1}{2})}{\alpha\Gamma\left(\dfrac{s-\beta}{\alpha}+\nu+\frac{1}{2}\right)} \qquad (\mathrm{Re}(s) > \beta),$$

where the integral has been evaluated in terms of the beta function in (3.3.8). It then follows from (4.1.1) that

$$\sum_{n=1}^N \frac{(1-n^\alpha x^\alpha)^{\nu-\frac{1}{2}}}{n^\beta} = x^\beta \sum_{n=1}^\infty f(nx)$$

$$= \frac{1}{2\pi i}\int_{c-\infty i}^{c+\infty i} \frac{\Gamma\left(\dfrac{s-\beta}{\alpha}\right)\Gamma(\nu+\frac{1}{2})}{\alpha\Gamma\left(\dfrac{s-\beta}{\alpha}+\nu+\frac{1}{2}\right)}\zeta(s)\,x^{\beta-s}\,ds,$$

(4.2.15)

where $c > \max\{1, \beta\}$. The above integrand has, in general, simple poles at $s = 1$ and $s = \beta - \alpha k$ for nonnegative integer k, although, according to the values of α, β and ν, it can happen that either $s = 1$ is a double pole or that some of the poles of the infinite sequence become ordinary points.

We shall consider two specific cases. First, let $\alpha = 2$ and $\beta = 0$ so that

$$\sum_{n=1}^N (1-n^2 x^2)^{\nu-\frac{1}{2}} = \frac{1}{2\pi i}\int_{c-\infty i}^{c+\infty i} \frac{\Gamma(\frac{1}{2}s)\Gamma(\nu+\frac{1}{2})}{2\Gamma(\frac{1}{2}s+\nu+\frac{1}{2})}\zeta(s)\,x^{-s}\,ds \quad (c > 1).$$

From (4.1.14) and (4.1.15), the modulus of the integrand on $s = \sigma \pm it$ is $O(t^{-\nu-\frac{1}{2}+\mu(\sigma)}\log t)$ as $t \to \infty$. As a consequence, the contour can be displaced to the left only up to the vertical line $\mathrm{Re}(s) = -\nu$, where the displaced integral ceases to be conditionally convergent. Since the only poles of the integrand are at $s = 0$ and $s = 1$ (the remaining poles of $\Gamma(\frac{1}{2}s)$ are cancelled by the 'trivial zeros' of the zeta function), we therefore obtain

$$\sum_{n=1}^N (1-n^2 x^2)^{\nu-\frac{1}{2}} = \frac{\Gamma(\nu+\frac{1}{2})}{\Gamma(\nu+1)}\frac{\pi^{\frac{1}{2}}}{2x} - \frac{1}{2} + \frac{1}{2}\Gamma(\nu+\frac{1}{2})I(x), \qquad (4.2.16)$$

where, with the change of variable $s \to -s$,

$$I(x) = \frac{1}{2\pi i}\int_{c-\infty i}^{c+\infty i} \frac{\Gamma(-\frac{1}{2}s)}{\Gamma(-\frac{1}{2}s+\nu+\frac{1}{2})}\zeta(-s)x^s\,ds \qquad (0 < c < \nu).$$

Use of (4.1.4) and the duplication formula for the gamma function in (2.2.23), followed by expansion of $\zeta(1+s)$, then shows that

$$
I(x) = \frac{1}{2\pi i} \int_{c-\infty i}^{c+\infty i} \frac{\zeta(1+s)}{\pi^{\frac{1}{2}}} \frac{\Gamma\left(\frac{1}{2}s+\frac{1}{2}\right)}{\Gamma\left(-\frac{1}{2}s+\nu+\frac{1}{2}\right)} \left(\frac{x}{\pi}\right)^s ds
$$

$$
= \pi^{-\frac{1}{2}} \sum_{k=1}^{\infty} k^{-1} \frac{1}{2\pi i} \int_{c-\infty i}^{c+\infty i} \frac{\Gamma\left(\frac{1}{2}s+\frac{1}{2}\right)}{\Gamma\left(-\frac{1}{2}s+\nu+\frac{1}{2}\right)} \left(\frac{x}{\pi k}\right)^s ds \quad (0 < c < \nu)
$$

$$
= 2\pi^{\frac{1}{2}-\nu} x^{\nu-1} \sum_{k=1}^{\infty} k^{-\nu} J_\nu(2\pi k/x), \tag{4.2.17}
$$

where the last integral has been evaluated in terms of the Bessel function J_ν by means of the result [cf. (3.4.21); see also Watson (1966, p. 192)]

$$
\left(\tfrac{1}{2}x\right)^{1-\nu} J_\nu(x) = \frac{1}{2\pi i} \int_{c-\infty i}^{c+\infty i} \frac{\Gamma\left(\frac{1}{2}s+\frac{1}{2}\right)}{2\Gamma\left(\nu+\frac{1}{2}-\frac{1}{2}s\right)} \left(\tfrac{1}{2}x\right)^{-s} ds,
$$

where $x > 0$ and $0 < c < \nu$.

It is seen that the sum of Bessel functions in (4.2.17) does not yield (without further manipulation) an asymptotic series in the limit $x \to 0$, with the result that we are unable to obtain an asymptotic expansion from (4.2.16). However, an estimate for the contribution made by this sum can be found by employing the asymptotic representation of $J_\nu(z)$ as $z \to +\infty$ given by

$$
J_\nu(z) = \sqrt{\frac{2}{\pi z}} \left\{ \cos\left(z - \tfrac{1}{2}\pi\nu - \tfrac{1}{4}\pi\right) + O(z^{-1}) \right\}.
$$

Then we find

$$
|I(x)| \leq 2\pi^{-\frac{1}{2}-\nu} x^{\nu-\frac{1}{2}} \sum_{k=1}^{\infty} \left\{ \frac{\left|\cos\left(2\pi k/x - \tfrac{1}{2}\pi\nu - \tfrac{1}{4}\pi\right)\right|}{k^{\nu+\frac{1}{2}}} + O\left(x k^{-\nu-\frac{3}{2}}\right) \right\}
$$

$$
\leq 2\pi^{-\frac{1}{2}-\nu} x^{\nu-\frac{1}{2}} \zeta\left(\nu+\tfrac{1}{2}\right)\{1 + O(x)\}
$$

so that, when $\nu > \frac{1}{2}$,

$$
\sum_{n=1}^{N} (1 - n^2 x^2)^{\nu-\frac{1}{2}} = \frac{\Gamma\left(\nu+\frac{1}{2}\right)}{\Gamma(\nu+1)} \frac{\pi^{\frac{1}{2}}}{2x} - \frac{1}{2} + O(x^{\nu-\frac{1}{2}}) \quad (x \to 0). \tag{4.2.18}
$$

The same complication is thrown into even sharper focus in our second example where we take $\alpha = \beta = \frac{2}{3}$ and $\nu = 1$. In this case we have from (4.2.15)

$$S(x) = \sum_{n=1}^{N} \frac{\left(1 - n^{2/3}x^{2/3}\right)^{1/2}}{n^{2/3}}$$

$$= \frac{3\pi^{\frac{1}{2}}}{8\pi i} \int_{c-\infty i}^{c+\infty i} \frac{\Gamma\left(\frac{3}{2}s - 1\right)}{\Gamma\left(\frac{3}{2}s + \frac{1}{2}\right)} \zeta(s)\, x^{\frac{2}{3}-s}\, ds \quad (c > 1), \qquad (4.2.19)$$

where the integrand possesses poles at $s = 1$ and at $s = \frac{2}{3} - \frac{2}{3}k$ for nonnegative integer values of k, with the values $k = 4, 7, 10, \ldots$ deleted. However, due to the algebraic growth of the modulus of the integrand given by $O(t^{-\frac{3}{2}+\mu(\sigma)} \log t)$ as $t \to \infty$ (when $s = \sigma \pm it$), the contour can pass to infinity along a vertical path only when $\mathrm{Re}(s) > -1$. Consequently, without modification, the contour in (4.2.19) *cannot* be displaced in the usual manner over an arbitrary number of poles of the infinite sequence to yield an asymptotic expansion.

To establish that simple evaluation of the residues does not lead to a full description of the asymptotic expansion of $S(x)$, we follow the approach described in Titchmarsh (1986, p. 315) and first displace the contour in (4.2.19) to the imaginary axis (indented at $s = 0$). The portion of the path lying between $\pm iT$, where $T > 0$ is an arbitrary fixed value to be suitably chosen, is then deformed into the rectangular contour with vertices at $\pm iT$, $-c' \pm iT$, where $c' = \frac{2}{3}M - 1$ for $M = 2, 3, \ldots$, as depicted in Fig. 4.1. Application of Cauchy's theorem then shows that†

$$S(x) = \frac{3\pi}{4x^{1/3}} - \frac{1}{2\sqrt{\pi}} \sum_{k=0}^{M-1} \frac{\Gamma\left(k - \frac{1}{2}\right)}{k!} \zeta\left(\frac{2}{3} - \frac{2}{3}k\right) x^{\frac{2}{3}k} + R_M(x), \qquad (4.2.20)$$

where $R_M(x)$ is the integral along the path $ABCDEF$.

We only give an outline of the order estimate for $R_M(x)$. It is not difficult to establish that the contribution to $R_M(x)$ from the sides ABC and DEF is $O(T^{-1}x^{\frac{2}{3}}) + O(T^{\frac{2}{3}M-2}x^{\frac{2}{3}M-\frac{1}{3}})$. To estimate the contribution from the side CD we first separate off the interval $(-t_0, t_0)$, where $t_0 \ll T$ with $t_0 = O(1)$, to find that the contribution from this part of the path is $O(x^{\frac{2}{3}M-\frac{1}{3}})$. Over the interval (t_0, T), we use the functional relation in (4.1.4) to express $\zeta(s)$ in terms of $\zeta(1-s)$ and Stirling's formula (2.1.8) to obtain the approximate behaviour of the integrand in the form (omitting numerical factors)

$$x^{\frac{2}{3}M-\frac{1}{3}}t^{\frac{2}{3}M-2} \exp\{it[1 - \log(xt/2\pi)]\}\{1 + O(t^{-1})\}.$$

---

† We observe that the terms corresponding to $k = 1 + 3m$ $(m = 1, 2, \ldots)$ make no contribution.

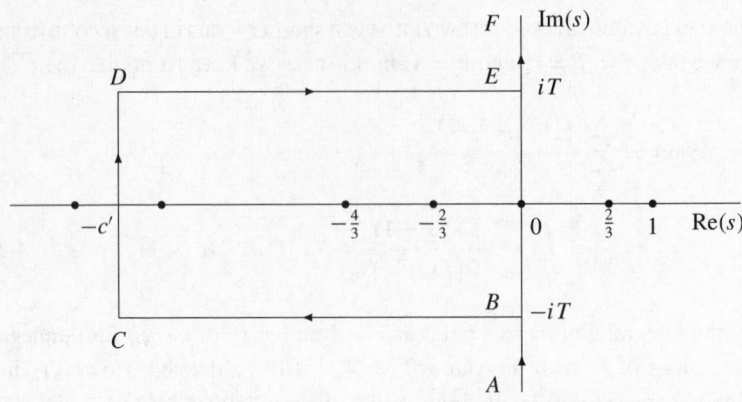

Fig. 4.1. The deformed path of integration for $R_M(x)$. The heavy points represent poles and $c' = \frac{2}{3}M - 1$, $M = 2, 3, \ldots$.

Then, upon defining the new variable $\tau = xt/2\pi$, this part of the integral yields

$$x^{\frac{2}{3}M - \frac{1}{3}}\left\{ \int_{t_0}^{T} t^{\frac{2}{3}M - 2} \exp\{it[1 - \log(xt/2\pi)]\}\, dt + O\left(T^{\frac{2}{3}M - 2}\right) \right\}$$

$$= (2\pi)^{\frac{2}{3}M - 1} x^{\frac{2}{3}} \int_{\tau_0}^{xT/2\pi} \tau^{\frac{2}{3}M - 2} e^{2\pi i f(\tau)/x}\, d\tau + O\left(T^{\frac{2}{3}M - 2} x^{\frac{2}{3}M - \frac{1}{3}}\right),$$

where $f(\tau) = \tau - \tau \log \tau$ and $\tau_0 = x t_0/2\pi = O(x)$. For $x \to 0$, the integrand has a point of stationary phase at $\tau = 1$, where $f'(\tau) = 0$. If we now choose $T = 2\pi/x$, so that the upper limit of integration coincides with this point, we find by the method of stationary phase [see, for example, Olver (1974, p. 97); Wong (1989, p. 76)] that the modulus of the above integral is $O(x^{\frac{1}{2}})$ as $x \to 0$. It then follows that the contribution from the interval (t_0, T) yields the order estimate $x^{\frac{2}{3}} O(x^{\frac{1}{2}}) + O(x^{\frac{5}{3}}) = O(x^{\frac{7}{6}})$. A similar result holds for the interval $(-T, -t_0)$ and, since the contribution from ABC and DEF is $O(x^{\frac{5}{3}})$ when $T = O(x^{-1})$, we consequently obtain the order estimate

$$R_M(x) = O\left(x^{\frac{7}{6}}\right) + O\left(x^{\frac{2}{3}M - \frac{1}{3}}\right) \qquad (x \to 0).$$

Thus, the contribution from $R_M(x)$ (when $M \geq 3$) is $O(x^{\frac{7}{6}})$ as $x \to 0$, and so (4.2.20) is not an asymptotic expansion. We remark, however, that the constant implied in the O symbol does depend on M with the result that $R_M(x) \to \infty$ as $M \to \infty$. (This fact can also be seen from (4.2.20) since the sum over k is clearly divergent as $M \to \infty$.) To verify the above conclusion, we show in Table 4.1 the numerical values of the error term $R_M(x)$ in (4.2.20) for different values of x when $M = 6$. The last column indicates that $|R_M(x)|$ indeed scales approximately like $x^{\frac{7}{6}}$ as $x \to 0$, as predicted.

Table 4.1. *Values of the sum $S(x)$ and the error $|R_M(x)|$ for different x when $M = 6$*

x	$S(x)$	$|R_M(x)|$	$|R_M(x)|/x^{7/6}$
1×10^{-1}	2.6718053586	1.1673×10^{-2}	0.1713
5×10^{-2}	3.9772343344	5.1747×10^{-3}	0.1705
1×10^{-2}	8.4997627388	7.8858×10^{-4}	0.1699
5×10^{-3}	11.3385038142	3.5111×10^{-4}	0.1698
1×10^{-3}	21.1168124272	5.3681×10^{-5}	0.1697
1×10^{-4}	48.3156257669	3.6569×10^{-6}	0.1697

This example is considered in more depth in §5.6.3, where it is shown how one can circumvent the apparent obstacle of not being able to displace the contour beyond the vertical line $\text{Re}(s) = -1$. It will be found that the term $R_M(x)$ yields an additional asymptotic expansion consisting of an infinite number of exponentially oscillatory terms.

4.3 Number-Theoretic Examples

A wide range of number-theoretic sums and combinatorial problems arising in the analysis of algorithms in computing science has been treated extensively in a series of papers by Flajolet *et al.* (1985, 1994, 1995) and Flajolet & Sedgewick (1995); see also the monograph by Sedgewick & Flajolet (1996). The approach adopted throughout these papers is that of the Mellin transform, which is ideally suited for problems of this type. The first example we describe concerns harmonic sums and the asymptotic evaluation of the nth harmonic number, which is discussed in more detail in Flajolet *et al.* (1985, 1995). The next two examples in §§4.3.2–3 have been drawn from a recent article by Ninham *et al.* (1992) on the application of number-theoretic functions to mathematical physics. The final section dealing with some other number-theoretic sums has been taken from Titchmarsh (1986) and Hardy & Littlewood (1918).

4.3.1 A Harmonic Sum

A harmonic sum is a sum of the type

$$\xi(x) = \sum_k \lambda_k f(\mu_k x), \tag{4.3.1}$$

where $f(x)$ is called the base function, with λ_k, μ_k being known as the amplitudes and frequencies, respectively. If $\Xi(s)$ and $F(s)$ denote the Mellin transforms of $\xi(x)$ and $f(x)$, then we find formally

$$\Xi(s) = \omega(s)F(s), \qquad \omega(s) = \sum_k \lambda_k \mu_k^{-s} \tag{4.3.2}$$

and by the Mellin inversion formula (3.1.5)

$$\xi(x) = \frac{1}{2\pi i} \int_{c-\infty i}^{c+\infty i} \omega(s)F(s)x^{-s}\,ds. \qquad (4.3.3)$$

The transformed equation (4.3.2) and the above integral representation for $\xi(x)$ will be valid provided s lies in the intersection of the strip of analyticity of $F(s)$ and the domain of absolute convergence of $\omega(s)$.

As a simple example, consider the nth harmonic number given by

$$\theta_n = \sum_{k=1}^{n} \frac{1}{k}.$$

This can be extended to a continuous function of x by introducing

$$h(x) = \sum_{k=1}^{\infty} \left\{ \frac{1}{k} - \frac{1}{x+k} \right\} = \sum_{k=1}^{\infty} \frac{1}{k} \frac{(x/k)}{(1+x/k)},$$

so that $h(n) = \theta_n$. This is seen to be a harmonic sum with the base function $f(x) = x/(1+x)$ and $\lambda_k = \mu_k = k^{-1}$. The Mellin transform of $f(x)$ is $F(s) = -\pi \operatorname{cosec} \pi s$ in $-1 < \operatorname{Re}(s) < 0$ and $\omega(s) = \zeta(1-s)$ in $\operatorname{Re}(s) < 0$, so that the Mellin transform of $h(x)$ is given by $H(s) = -\pi \operatorname{cosec} \pi s\, \zeta(1-s)$ in the common strip $-1 < \operatorname{Re}(s) < 0$. Then we obtain

$$h(x) = -\frac{1}{2\pi i} \int_{-c-\infty i}^{-c+\infty i} \frac{\pi}{\sin \pi s} \zeta(1-s)x^{-s}\,ds \qquad (0 < c < 1).$$

Since $\zeta(s) \sim 1/(s-1)+\gamma$ as $s \to 1$, where γ is Euler's constant, the integrand has a double pole at $s = 0$ with residue $\log x + \gamma$, together with a sequence of simple poles at $s = 1, 2, \dots$ on the right of the contour. Displacement of the path of integration to the right then generates the expansion

$$h(x) = \log x + \gamma + \frac{1}{2x} + \sum_{k=2}^{N-1} \frac{(-)^{k-1}B_k}{kx^k} + R_N$$

for integer $N \geq 2$, where we have used the facts that $\zeta(0) = -\frac{1}{2}$ and $\zeta(1-k) = -B_k/k$ in terms of the Bernoulli numbers B_k. The remainder R_N is given by the integral taken along the displaced contour

$$R_N = i \int_{-c+N-\infty i}^{-c+N+\infty i} \frac{\Gamma(s)\zeta(s)}{2\sin\frac{1}{2}\pi s}(2\pi x)^{-s}\,ds,$$

upon use of (4.1.4), to express $\zeta(1-s)$ in terms of $\zeta(s)$, and the reflection formula for the gamma function. Then, by means of the bounds $|\Gamma(\sigma + it)| \leq \Gamma(\sigma)$ $(\sigma > 0)$, $|\zeta(\sigma + it)| \leq \zeta(\sigma)$ $(\sigma > 1)$ and the first inequality in (2.5.12), we find

$$|R_N| \leq \Gamma(N-c)\zeta(N-c)(2\pi x)^{-N+c}\operatorname{cosec} \pi c \int_{-\infty}^{\infty} \frac{dt}{2\cosh\frac{1}{2}\pi t}$$

when $0 < c < 1$, where the last integral equals unity by (2.5.13). Hence, $R_N = O(x^{-N+c})$ as $x \to \infty$ for fixed N; but, since $|R_N| \to \infty$ as $N \to \infty$ (for fixed x), we conclude that the above expansion for $h(x)$ is asymptotic. Recalling that the B_k with odd index $k \geq 3$ are zero, we then obtain the well-known expansion

$$\theta_n \sim \log n + \gamma + \frac{1}{2n} - \sum_{k=1}^{\infty} \frac{B_{2k}}{2k\, n^{2k}} \qquad (4.3.4)$$

as $n \to \infty$.

Other interesting examples arising in analysis algorithms in computing science are discussed in Flajolet $et\ al.$ (1985, 1995) where, in particular, the asymptotic expansion of sums such as $\sum_{k=0}^{\infty} f(2^{\pm k}x)$ are obtained as $x \to 0$ or $x \to \infty$, respectively. As a typical illustration of this type of sum we mention the evaluation of

$$\sum_{k=0}^{\infty} (-)^k e^{-2^k x} \qquad (x > 0).$$

From (4.3.2) and (4.3.3), with $\lambda_k = (-)^k$, $\mu_k = 2^k$ and $f(x) = e^{-x}$, we find that $\omega(s) = \sum_{k=0}^{\infty} (-)^k 2^{-ks} = (1 + 2^{-s})^{-1}$ (provided $\mathrm{Re}(s) > 0$) and hence that

$$\sum_{k=0}^{\infty} (-)^k e^{-2^k x} = \frac{1}{2\pi i} \int_{c-\infty i}^{c+\infty i} \frac{\Gamma(s)}{1 + 2^{-s}} x^{-s}\, ds \qquad (c > 0).$$

The integrand has simple poles at $s = 0, -1, -2, \ldots$ and on the imaginary axis at $s = \pm\chi_k$ with residue equal to $\Gamma(\pm\chi_k) x^{\mp\chi_k} / \log 2$, where $\chi_k = (2k+1)\pi i / \log 2$ ($k = 0, 1, 2, \ldots$). Because of the exponential decay of the gamma function for large $|\mathrm{Im}(s)|$, we are free to displace the contour to the left to obtain the (convergent) result

$$\sum_{k=0}^{\infty} (-)^k e^{-2^k x} = \sum_{k=0}^{\infty} \frac{(-)^k}{k!} \frac{x^k}{1 + 2^k} + \frac{2}{\log 2} \mathrm{Re} \sum_{k=0}^{\infty} \Gamma(\chi_k) x^{-\chi_k},$$

from which it is possible to obtain the behaviour as $x \to 0+$. The second sum on the right-hand side possesses terms that decay very rapidly with k (cf. §4.3.3) and represents a fluctuating contribution of magnitude controlled by the leading term $(2/\log 2)|\Gamma(\pi i / \log 2)| = 2\{\log 2\ \sinh(\pi^2/\log 2)\}^{-\frac{1}{2}} \simeq 2.75 \times 10^{-3}$.

A result equivalent to this has been given by Keating & Reade (2000) using the Poisson summation formula in their analysis of the summability of alternating gap series. We note that the sum $\sum_{n=1}^{\infty} \exp(-2^k x)$ yields the Mellin-Barnes integral above with the factor $1 + 2^{-s}$ in the denominator replaced by $1 - 2^{-s}$. The poles on the imaginary axis are now situated at $s = \pm 2\pi i k / \log 2$ ($k = 0, 1, 2, \ldots$), with the result that $s = 0$ becomes a double pole, thereby producing a term in $\log x$.

4.3.2 Euler's Product

Our next number-theoretic example concerns Euler's product defined by

$$F_E(z) = \prod_{n=1}^{\infty} (1 - z^n) \qquad (0 \le z \le 1),$$

which plays a central role in the theory of partitions and elliptic modular functions [Ninham et al. (1992, pp. 464–465)]. This function clearly satisfies $F_E(0) = 1$ and $F_E(1) = 0$, but the accurate determination of $F_E(x)$ as $x \to 1-$ requires the evaluation of many terms in the product; see also Harsoyo & Temme (1982).

The first step in the process is to obtain a Mellin-Barnes integral representation for the function $h(e^{-x})$ given by

$$h(e^{-x}) = -\log F_E(e^{-x}) = -\sum_{n=1}^{\infty} \log(1 - e^{-nx}),$$

which enables the sum on the right-hand side to be easily evaluated as $x \to 0+$. We note that, in terms of the variable x, the limiting situation now corresponds to $x \to 0+$. We achieve this by putting $f(x) = \log(1 - e^{-x})$, which possesses the Mellin transform given by

$$F(s) = \int_0^{\infty} x^{s-1} \log(1 - e^{-x})\, dx = -\frac{1}{s} \int_0^{\infty} \frac{x^s}{e^x - 1}\, dx \quad (\mathrm{Re}(s) > 0)$$

$$= -\zeta(s+1)\Gamma(s)$$

upon integration by parts and use of a standard integral representation for the Riemann zeta function; see Abramowitz & Stegun (1965, p. 807). By the Mellin inversion formula in (3.1.5) we then find

$$-\log(1 - e^{-x}) = \frac{1}{2\pi i} \int_{c-\infty i}^{c+\infty i} \zeta(s+1)\Gamma(s) x^{-s}\, ds \qquad (c > 0).$$

We remark that $f(x) \sim \log x$ as $x \to 0+$ and so does not satisfy the conditions in (4.1.1). Nevertheless, the weak nature of this divergence does not prevent a similar process from being employed. If we replace x by nx, where n is a positive integer, and sum over n, we then obtain

$$h(e^{-x}) = -\sum_{n=1}^{\infty} \log(1 - e^{-nx}) = \frac{1}{2\pi i} \int_{c-\infty i}^{c+\infty i} \zeta(s)\zeta(s+1)\Gamma(s) x^{-s}\, ds,$$

$$(4.3.5)$$

where $c > 1$ and the reversal of the order of summation and integration is justified by absolute convergence.

The integrand in (4.3.5) has simple poles at $s = \pm 1$ and a double pole at $s = 0$. The remaining poles of $\Gamma(s)$ at $s = -2, -3, \ldots$ are cancelled by the 'trivial zeros' of one or other of the zeta functions. We now displace the integration path to the left of $s = -1$. Since, from (4.1.14) and (4.1.15), the modulus of the integrand (when $s = \sigma \pm it$) is

$$O\left(t^{\mu(\sigma)+\mu(\sigma+1)+\sigma-\frac{1}{2}}(\log t)^2 e^{-\frac{1}{2}\pi t}\right) \qquad (t \to \infty),$$

the contributions from the upper and lower sides of the rectangular contour, formed by the displacement of the path of integration, vanish. Observing that the residue at the double pole is $\zeta'(0) - \zeta(0) \log x$, we therefore find

$$h(e^{-x}) = -\zeta(0)\zeta(-1)x - \zeta(0) \log x + \zeta'(0) + \frac{\zeta(2)}{x} + I_R,$$

where

$$I_R = \frac{1}{2\pi i} \int_{c-\infty i}^{c+\infty i} \zeta(-s)\zeta(1-s)\Gamma(-s)x^s \, ds \qquad (c > 1)$$

and we have made the change of variable $s \to -s$. Use of (4.1.4) and the reflection formula for the gamma function shows that

$$I_R = \frac{1}{2\pi i} \int_{c-\infty i}^{c+\infty i} \zeta(s)\zeta(1+s)\Gamma(s)(4\pi^2/x)^{-s} \, ds \qquad (c > 1)$$

$$= h(e^{-4\pi^2/x}).$$

Then with the well-known values of the Riemann zeta function $\zeta(-1) = -\frac{1}{12}$, $\zeta(0) = -\frac{1}{2}$, $\zeta'(0) = -\frac{1}{2}\log 2\pi$ and $\zeta(2) = \frac{1}{6}\pi^2$, we deduce that

$$h(e^{-x}) = -\frac{x}{24} + \frac{1}{2}\log\left(\frac{x}{2\pi}\right) + \frac{\pi^2}{6x} + h(e^{-4\pi^2/x}).$$

Exponentiation of both sides of the above equation finally yields the result

$$F_E(e^{-x}) = \sqrt{\frac{2\pi}{x}} \exp\left(\frac{x}{24} - \frac{\pi^2}{6x}\right) F_E(e^{-4\pi^2/x}), \qquad (4.3.6)$$

from which the value of $F_E(x)$ as $x \to 0+$ can now be easily computed.

We note that this formula can be rewritten in a different form to reveal a symmetry under inversion. If we define $G(x) = x^{\frac{1}{4}}e^{-\pi x/12}F_E(e^{-2\pi x})$, then we have

$$G(x) = G(1/x).$$

4.3.3 Ramanujan's Function

Our next example concerns the function $g(x)$, defined for $x > 0$ by the series

$$g(x) = \sum_{n=-\infty}^{\infty} \left\{\exp(-xe^n) - \exp(-e^n)\right\}; \qquad (4.3.7)$$

see Ninham *et al.* (1992, pp. 461–464). This function was considered by Ramanujan in connection with a fallacious proof of the prime number theorem

[see Hardy (1940, p. 38)]. The series converges uniformly for all finite closed
subintervals of $(0, \infty)$, so that $g(x)$ is continuous for $x > 0$. We note that $g(x)$
satisfies the same functional equation as $\log x$, namely $g(ex) = g(x) + g(e)$,
with $g(1) = 0$. This fact suggests that the function $g(x)$ should contain a $\log x$
dependence.

A different way of understanding qualitatively the growth of $g(x)$ as $x \to 0+$
can be seen from the behaviour of the terms in (4.3.7) as a function of n; see
Fig. 4.2. As $x \to 0+$, it is found that the terms are approximately equal to unity
for $1 \lesssim n \lesssim n^*$, where n^* corresponds to the value of n at which the argument of
the first exponential becomes $O(1)$; that is, roughly speaking, when $xe^{n^*} \sim 1$, or
$n^* \sim -\log x$. Since there are $O(n^*)$ terms which each contribute approximately
unity to the sum, this means that $g(x)$ will diverge like $-\log x$ as $x \to 0+$ and we
are again in a situation which will involve the evaluation of many terms in this limit.

Use of the Mellin transform of e^{-x}, given by $\Gamma(s)$ for $\text{Re}(s) > 0$, enables us to
deduce that

$$g(x) = \sum_{n=-\infty}^{\infty} \frac{1}{2\pi i} \int_{c-\infty i}^{c+\infty i} e^{-ns}(x^{-s} - 1)\Gamma(s)\,ds \qquad (c > 0).$$

To deal with the inversion of the order of summation and integration, we write
$g(x) = g_1(x) + g_2(x)$, where

$$g_1(x) = \sum_{n=0}^{\infty} \frac{1}{2\pi i} \int_{c-\infty i}^{c+\infty i} e^{-ns}(x^{-s} - 1)\Gamma(s)\,ds$$

and

$$g_2(x) = \sum_{n=1}^{\infty} \frac{1}{2\pi i} \int_{c-\infty i}^{c+\infty i} e^{ns}(x^{-s} - 1)\Gamma(s)\,ds.$$

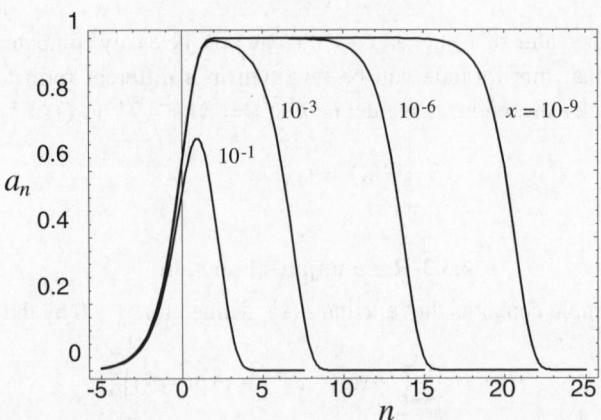

Fig. 4.2. The behaviour of $a_n = \exp(-xe^n) - \exp(-e^n)$ as a function of n (regarded as a
continuous variable) for different values of x in the limit $x \to 0+$.

Then the series defining $g_1(x)$ converges uniformly with respect to s on the path of integration and accordingly we can make the interchange to find

$$g_1(x) = \frac{1}{2\pi i} \int_{c-\infty i}^{c+\infty i} \frac{(x^{-s} - 1)\Gamma(s)}{1 - e^{-s}} \, ds \qquad (c > 0).$$

This interchange is not permissible in the series defining $g_2(x)$ as it stands. Noting that the integrand has no pole at $s = 0$ (since the pole of $\Gamma(s)$ is cancelled by the zero of $x^{-s} - 1$) and that the modulus of the integrand is controlled by $e^{-\frac{1}{2}\pi|t|}$ as $t = \text{Im}(s) \to \pm\infty$, we can displace the contour to the left to obtain

$$g_2(x) = \sum_{n=1}^{\infty} \frac{1}{2\pi i} \int_{c-\infty i}^{c+\infty i} e^{ns}(x^{-s} - 1)\Gamma(s) \, ds \qquad (-1 < c < 0).$$

On the new contour the sum converges uniformly so that the interchange can be carried out to yield

$$g_2(x) = -\frac{1}{2\pi i} \int_{c-\infty i}^{c+\infty i} \frac{(x^{-s} - 1)\Gamma(s)}{1 - e^{-s}} \, ds \qquad (-1 < c < 0).$$

We therefore deduce that

$$g(x) = \frac{1}{2\pi i} \int_{C} \frac{(x^{-s} - 1)\Gamma(s)}{1 - e^{-s}} \, ds,$$

where the integration contour C denotes a path that encloses the imaginary axis in the positive sense; see Fig. 4.3. Evaluation of the residues at the enclosed simple poles at $s = 0$ and $s = \pm 2\pi n i$ ($n = 1, 2, \ldots$), combined with use of the conjugacy property of the gamma function, then yields the final result

$$g(x) = -\log x + 2\text{Re} \sum_{n=1}^{\infty} \Gamma(2\pi i n)(x^{-2\pi i n} - 1). \qquad (4.3.8)$$

Since $|\Gamma(iy)| = (\pi/y \sinh(\pi y))^{\frac{1}{2}}$ for $y > 0$, the absolute value of the coefficients in this expansion is given by

$$|\Gamma(2\pi i n)| = \{2n \sinh(2\pi^2 n)\}^{-\frac{1}{2}}.$$

As a result of this rapid exponential decay in n, computation of $g(x)$ in the limit $x \to 0+$ can now be readily achieved.

The function $g(x)$ coincides with $-\log x$ at isolated points where $x^{-2\pi i n} - 1 = 0$ for all n, that is, when $x = e^m$ ($m = 0, \pm 1, \pm 2, \ldots$). In addition to the above zeros of $g(x) + \log x$, there is also another infinite sequence of zeros. This can be more readily perceived by putting $x = e^t$, so that (4.3.8) becomes

$$g(e^t) + t = 2\text{Re} \sum_{n=1}^{\infty} \Gamma(2\pi i n)(e^{-2\pi i n t} - 1), \qquad (4.3.9)$$

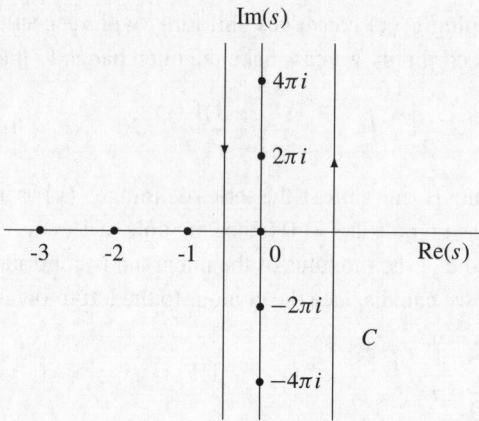

Fig. 4.3. The contour C enclosing the poles on the imaginary s axis.

which shows that $g(e^t) + t$ is a periodic function of t with unit period[†]. The behaviour of the right-hand side of (4.3.9) in the interval $0 \le t \le 1$ is shown in Fig. 4.4. Due to the rapid decay of the coefficients, the additional zero t_0 in the interval $0 \le t \le 1$ can be approximated by setting the first term in the sum in (4.3.9) to zero to find

$$\mathrm{Re}(\Gamma(2\pi i))(\cos 2\pi t_0 - 1) + \mathrm{Im}(\Gamma(2\pi i)) \sin 2\pi t_0 \simeq 0;$$

this yields

$$\tan \pi t_0 \simeq A, \quad A = \frac{\mathrm{Im}(\Gamma(2\pi i))}{\mathrm{Re}(\Gamma(2\pi i))} = 3.97388,$$

whence $t_0 \simeq 0.42153$.

In Ninham et al. (1992, pp. 461–462) it is shown that the same analysis can also be applied to the more general sum given by

$$g(x) = \sum_{n=-\infty}^{\infty} \{h(xe^n) - h(e^n)\},$$

where the function $h(x)$ is continuous for $0 \le x < \infty$ and satisfies the restrictions $h(x) = h(0+) + O(x^\alpha)$ as $x \to 0+$ and $h(x) = O(x^{-\alpha})$ as $x \to \infty$ for some $\alpha > 0$. In this case we have the result

$$g(x) = -h(0+) \log x + {\sum_{n=-\infty}^{\infty}}' H(2\pi in)(x^{-2\pi in} - 1), \tag{4.3.10}$$

[†] This also follows from the functional equation $g(ex) = g(x) + g(e) = g(x) - 1$ upon putting $x = e^t$. We note that if we put $g(e^t) = -1/y(t)$, then (4.3.9) yields the recurrence relation $y(t+1) = y(t)/(1 + y(t))$. The solution of this nonlinear difference equation is simple for integer values of t, but possesses an irregular structure for noninteger values of t.

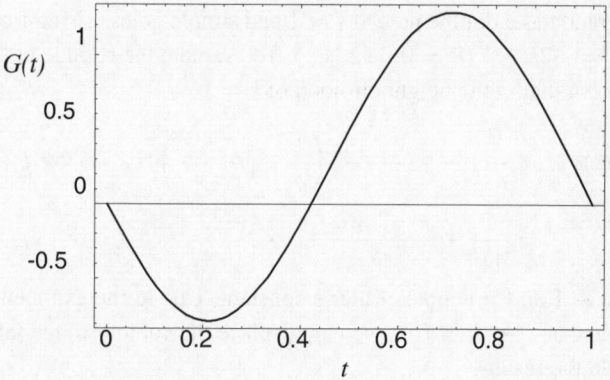

Fig. 4.4. The behaviour of $G(t) = (g(e^t) + t) \times 10^4$ in the interval $0 \leq t \leq 1$.

where the prime denotes the omission of the term corresponding to $n = 0$ and $H(s)$ denotes the Mellin transform of $h(x)$, holomorphic in the strip $0 < \mathrm{Re}(s) < \alpha$, with a simple pole at $s = 0$ of residue $h(0+)$.

4.3.4 Some Other Number-Theoretic Sums

The final number-theoretic example we consider in this section concerns the behaviour as $a \to 0$ of the harmonic sum (see §4.3.1)

$$\sum_{n=1}^{\infty} \lambda_n e^{-an}$$

for suitably defined coefficients λ_n when $\mathrm{Re}(a) > 0$. If $\omega(s) = \sum_{n=1}^{\infty} \lambda_n n^{-s}$ is absolutely convergent for $\mathrm{Re}(s) > \sigma_0$, then by (4.3.2) the Mellin transform of the above sum is $\omega(s)\Gamma(s)$ in $\mathrm{Re}(s) > \max\{0, \sigma_0\}$. Then by (4.3.3) we have

$$\sum_{n=1}^{\infty} \lambda_n e^{-an} = \frac{1}{2\pi i} \int_{c-\infty i}^{c+\infty i} \omega(s)\Gamma(s)a^{-s} ds,$$

where $c > \max\{0, \sigma_0\}$.

If, for example, we let $\lambda_n = d(n)$, where $d(n)$ denotes the number of divisors of n (including 1 and n itself), then [Titchmarsh (1986, p. 4)]

$$\omega(s) = \sum_{n=1}^{\infty} d(n)n^{-s} = \zeta^2(s) \qquad (\mathrm{Re}(s) > 1)$$

and so we have for the following divisor sum the result, when $|\arg a| < \frac{1}{2}\pi$,

$$\sum_{n=1}^{\infty} d(n)e^{-an} = \frac{1}{2\pi i} \int_{c-\infty i}^{c+\infty i} \Gamma(s)\zeta^2(s)a^{-s} ds \qquad (c > 1). \qquad (4.3.11)$$

The integrand has a double pole at $s = 1$ and simple poles arising from $\Gamma(s)$ at $s = 0$ and $s = -2k - 1$ ($k = 0, 1, 2, \ldots$). To evaluate the residue at the double pole we observe that in the neighbourhood of $s = 1$

$$\Gamma(s)\zeta^2(s)a^{-s} = \frac{1}{a}(1 - \gamma\delta + \cdots)\left(\frac{1}{\delta} + \gamma + \cdots\right)^2(1 - \delta\log a + \cdots)$$

$$= \frac{1}{a\delta^2} + \frac{\gamma - \log a}{a\delta} + \cdots,$$

where $\delta = s - 1$ and γ denotes Euler's constant. Due to the exponential decay of the integrand as $s \to \sigma \pm i\infty$, we can displace the contour to the left over the poles, to yield the result

$$\sum_{n=1}^{\infty} d(n)e^{-an} = \frac{\gamma - \log a}{a} + \frac{1}{4} - \sum_{k=1}^{N-1} \frac{\zeta^2(-2k-1)}{(2k+1)!}a^{2k+1} + R_N, \qquad (4.3.12)$$

where N is a positive integer and the remainder integral

$$R_N = \frac{1}{2\pi i}\int_{-2N-\infty i}^{-2N+\infty i} \Gamma(s)\zeta^2(s)a^{-s}ds \qquad (|\arg a| < \tfrac{1}{2}\pi).$$

From (4.1.4) the above integrand may be expressed in the form

$$\Gamma(s)\zeta^2(s) = (2\pi)^{2s-1}\Gamma(1-s)\zeta^2(1-s)\tan\tfrac{1}{2}\pi s.$$

If we let $s = -2N + it$, $\psi = \arg a$ and use the facts that $|\tanh\tfrac{1}{2}\pi t| \leq 1$ and

$$|\zeta(\sigma + it)| \leq \zeta(\sigma) \qquad (\sigma > 1),$$

we find

$$|R_N| \leq \frac{|a|^{2N}}{(2\pi)^{4N+2}}\int_{-\infty}^{\infty} |\Gamma(2N+1+it)\zeta^2(2N+1+it)|\tanh(\tfrac{1}{2}\pi|t|)\,e^{\psi t}\,dt$$

$$\leq \frac{|a|^{2N}}{(2\pi)^{4N+2}}\zeta^2(2N+1)\int_{-\infty}^{\infty} |\Gamma(2N+1+it)|\,e^{\psi t}\,dt$$

$$= |a|^{2N}O(e^{-2N}(N/2\pi^2)^{2N+1}\sec^{2N+\frac{3}{2}}\psi) \qquad (N \to \infty)$$

upon application of Lemma 2.6. Hence, $|R_N| = O(|a|^{2N})$ as $a \to 0$ for fixed N when $|\arg a| < \tfrac{1}{2}\pi$; but, since $|R_N| \to \infty$ as $N \to \infty$ (for fixed $|a|$), we conclude that (4.3.12) yields an asymptotic expansion. Thus we obtain [Titchmarsh (1986, p. 163)]

$$\sum_{n=1}^{\infty} d(n)e^{-an} \sim \frac{\gamma - \log a}{a} + \frac{1}{4} - \frac{1}{\pi^2}\sum_{k=1}^{\infty} \zeta^2(2k+2)\Gamma(2k+2)(a/4\pi^2)^{2k+1},$$

$$(4.3.13)$$

valid as $a \to 0$ in $|\arg a| < \tfrac{1}{2}\pi$.

We remark that the sum in (4.3.11) for $a \to 0+$ can be viewed as an exponentially-smoothed analogue of the well-known Dirichlet divisor sum given by [Titchmarsh (1986, p. 312)]

$$D(x) = \sum_{n \le x} d(n) = \frac{1}{2\pi i} \int_{c-\infty i}^{c+\infty i} \zeta^2(s) x^s \frac{ds}{s} \qquad (c > 1).$$

Since, with $s = \sigma \pm it$, the modulus of the integrand is $O(t^{-1})$ when $\sigma > 1$ and $O(t^{-\sigma}(\log t)^2)$ when $0 \le \sigma \le 1$ as $t \to \infty$, the contour can be displaced to the left over the double pole at $s = 1$, but *not* over the simple pole at $s = 0$. Thus, upon evaluation of the residue at $s = 1$, we find

$$D(x) = x \log x + (2\gamma - 1)x + \Delta(x),$$

where $\Delta(x)$ denotes the displaced integral taken along the vertical line $\mathrm{Re}(s) = c$, with $0 < c < 1$. By a process of path deformation similar to that employed in §4.2.3, it is found† in the limit $x \to +\infty$ [Titchmarsh (1986, p. 315)] that $\Delta(x) = O(x^{\frac{1}{3}+\epsilon})$ for every positive ϵ.

As a second example, we consider the harmonic sum with $\lambda_n = \hat{\mu}(n)/n$ given by

$$\mathfrak{S}(a) = a^{\frac{1}{2}}\pi^{\frac{1}{4}} \sum_{n=1}^{\infty} \frac{\hat{\mu}(n)}{n} e^{-\pi a^2/n^2} = a^{\frac{1}{2}}\pi^{\frac{1}{4}} \sum_{n=1}^{\infty} \frac{\hat{\mu}(n)}{n}(e^{-\pi a^2/n^2} - 1),$$

where $a > 0$ and $\hat{\mu}(n)$ denotes the Möbius function‡ in (4.1.9). This sum has been considered in a famous memoir by Hardy & Littlewood (1918, §2.5). Straightforward evaluation shows that

$$\mathfrak{S}(a) = a^{\frac{1}{2}}\pi^{\frac{1}{4}} \sum_{k=1}^{\infty} \frac{(-)^k (\pi a^2)^k}{k!} \sum_{n=1}^{\infty} \frac{\hat{\mu}(n)}{n^{2k+1}} = a^{\frac{1}{2}}\pi^{\frac{1}{4}} \sum_{k=1}^{\infty} \frac{(-)^k (\pi a^2)^k}{k! \, \zeta(2k+1)} \qquad (4.3.14)$$

by (4.1.10). Upon noting that the Mellin transform $F(s) = M[e^{-x} - 1; s] = \Gamma(s)$ in $-1 < \mathrm{Re}(s) < 0$ and that

$$\omega(s) = \sum_{n=1}^{\infty} \frac{\hat{\mu}(n)}{n^{1-2s}} = \frac{1}{\zeta(1-2s)} \qquad (\mathrm{Re}(s) < 0),$$

we obtain from (4.3.3) that

$$\mathfrak{S}(a) = \frac{1}{2\pi i} \int_{-c-\infty i}^{-c+\infty i} \frac{\Gamma(s)}{\zeta(1-2s)} (a\pi^{\frac{1}{2}})^{\frac{1}{2}-2s} ds, \qquad (4.3.15)$$

where $0 < c < 1$, so that the path of integration lies in the intersection of the strip of analyticity of $F(s)$ and the half-plane of absolute convergence of $\omega(s)$. Use of

† More recent estimates have slightly improved on this result; see Titchmarsh (1986, p. 325).
‡ We make use of the result $\sum_{n=1}^{\infty} \hat{\mu}(n)/n = 0$.

the functional relation (4.1.4) and the duplication formula for the gamma function (2.2.23) shows that this last integral can be written as

$$\mathfrak{S}(a) = \frac{1}{2\pi i} \int_{-c-\infty i}^{-c+\infty i} \frac{\Gamma\left(\frac{1}{2}-s\right)}{\zeta(2s)} \left(\pi^{\frac{1}{2}}/a\right)^{2s-\frac{1}{2}} ds \quad (0 < c < 1).$$

The usual method of displacement of the contour to the right over a set of poles of the integrand now faces a major problem: namely, the nature of the complex zeros of the Riemann zeta function (which are known to lie in the critical strip $0 < \mathrm{Re}(s) < 1$). If we assume that (a) we can displace the path past the vertical line $\mathrm{Re}(s) = \frac{1}{4}$, (b) the complex zeros are all simple and (c) the resulting sum of the residues is convergent, then we find

$$\mathfrak{S}(a) = -\frac{1}{2} \sum \left(\pi^{\frac{1}{2}}/a\right)^{\rho-\frac{1}{2}} \frac{\Gamma\left(\frac{1}{2}-\frac{1}{2}\rho\right)}{\zeta'(\rho)}$$
$$+ \frac{1}{2\pi i} \int_{c'-\infty i}^{c'+\infty i} \frac{\Gamma\left(\frac{1}{2}-s\right)}{\zeta(2s)} \left(\pi^{\frac{1}{2}}/a\right)^{2s-\frac{1}{2}} ds,$$

where $\frac{1}{2} < c' < \frac{3}{2}$ and the sum is taken over all the complex zeros $s = \rho$ of $\zeta(s)$. With the change of variable $u = \frac{1}{2} - s$, the last integral† can be written as

$$\frac{1}{2\pi i} \int_{-c-\infty i}^{-c+\infty i} \frac{\Gamma(u)}{\zeta(1-2u)} \left(\pi^{\frac{1}{2}}/a\right)^{\frac{1}{2}-2u} du \quad (0 < c < 1)$$

which is seen to equal $\mathfrak{S}(1/a)$ by (4.3.15). Hence, *on the above assumptions*, we obtain the relation

$$\mathfrak{S}(1/a) = \mathfrak{S}(a) + \frac{1}{2} \sum \left(\pi^{\frac{1}{2}}/a\right)^{\rho-\frac{1}{2}} \frac{\Gamma\left(\frac{1}{2}-\frac{1}{2}\rho\right)}{\zeta'(\rho)}. \tag{4.3.16}$$

Now let $a \to 0$; then by (4.3.14) $\mathfrak{S}(a) = O(a^{\frac{5}{2}})$. If we assume the Riemann hypothesis (i.e., that the complex zeros satisfy $\mathrm{Re}(\rho) = \frac{1}{2}$) the right-hand side of (4.3.16) is $O(1)$ in this limit. This would indicate that $\mathfrak{S}(1/a) = O(1)$ as $a \to 0$, and hence that

$$\sum_{n=1}^{\infty} \frac{\hat{\mu}(n)}{n} e^{-\pi a^2/n^2} = O(a^{-\frac{1}{2}})$$

as $a \to +\infty$.

---

† This can also be seen by displacement of the contour to the right over the simple poles at $s = k + \frac{1}{2}$, $k = 1, 2, \ldots$ and comparison with (4.3.14).

As a third example, we note that if $\lambda_n = \Lambda(n)$, where $\Lambda(n) = \log p$ if n is a prime p, or a power of p, and zero otherwise, then [Titchmarsh (1986, p. 4)]

$$\omega(s) = \sum_{n=1}^{\infty} \frac{\Lambda(n)}{n^s} = -\frac{\zeta'(s)}{\zeta(s)} \qquad (\text{Re}(s) > 1).$$

Then we have from (4.3.3)

$$\sum_{n=1}^{\infty} \Lambda(n) e^{-an} = -\frac{1}{2\pi i} \int_{c-\infty i}^{c+\infty i} \Gamma(s) \frac{\zeta'(s)}{\zeta(s)} a^{-s} ds$$

provided $\text{Re}(a) > 0$ and $c > 1$. It is shown in Hardy & Littlewood (1918, §2.2), by displacement of the contour to the left over the poles of the subject of integration, that

$$\sum_{n=1}^{\infty} (\Lambda(n) - 1) e^{-an} = \Psi(a) - \sum \Gamma(\rho) a^{-\rho},$$

where the sum is over the complex zeros of $\zeta(s)$ and the function $\Psi(a)$ has the form $\Psi_1(a) + a^2 \log(1/a) \Psi_2(a)$, with $\Psi_1(a)$ and $\Psi_2(a)$ being power series convergent in $|a| < 2\pi$. Then on the assumption of the Riemann hypothesis it follows that

$$\sum_{n=1}^{\infty} (\Lambda(n) - 1) e^{-an} = O(a^{-\frac{1}{2}})$$

as $a \to 0+$.

To close this section, we mention an interesting closed-form evaluation of an inverse Mellin transform involving the Riemann zeta function which has been given in Flajolet *et al.* (1994). In the harmonic sum in (4.3.1), we take $f(x) = 1 - x$ in $0 < x < 1$, $f(x) = 0$ in $x > 1$, with the Mellin transform given by $F(s) = 1/s(s + 1)$, and put $\lambda_k = 1$, $\mu_k = k$, $x = n^{-1}$ (where n is a positive integer). Then, we find from (4.3.3) the Perron formula (cf. (3.3.29))

$$\sum_{1 \le k \le n} \left(1 - \frac{k}{n}\right) = \tfrac{1}{2}(n - 1) = \frac{1}{2\pi i} \int_{c-\infty i}^{c+\infty i} \frac{\zeta(s) n^s}{s(s + 1)} ds,$$

where $c > 1$. Displacement of the contour to the left over the simple poles at $s = 1$ and $s = 0$ then yields, successively, the astonishing results†

$$\frac{1}{2\pi i} \int_{c-\infty i}^{c+\infty i} \frac{\zeta(s) n^s}{s(s + 1)} ds = -\tfrac{1}{2} \qquad (0 < c < 1),$$

$$\frac{1}{2\pi i} \int_{-c-\infty i}^{-c+\infty i} \frac{\zeta(s) n^s}{s(s + 1)} ds = 0 \qquad (0 < c < 1).$$

† We note that in the second integral the convergence is absolute only when $0 < c < \frac{1}{2} -$ cf. (4.1.15).

This last result has been exploited by Flajolet *et al.* (1994) in the derivation of the asymptotics of the sum–of–digits sum by a Mellin integral approach.

4.4 Solution of Differential Equations

Although the subjects of this and the following sections are not directly connected with asymptotics, it is felt that a chapter on applications would not be complete without a mention of the use of the Mellin transform in the solution of certain types of differential, integral and difference equations. It is not our intention here to give an exhaustive survey of these important areas of applied mathematics, but merely to give examples which illustrate clearly the utility of the Mellin transform. For further information, the interested reader is referred to specialist texts on integral transforms, such as Sneddon (1972) and Davies (1978).

4.4.1 Potential Problems in Wedge-Shaped Regions

In plane polar r, θ coordinates, the Laplacian operator ∇^2 is

$$\nabla^2 f(r, \theta) = \frac{\partial^2 f}{\partial r^2} + \frac{1}{r} \frac{\partial f}{\partial r} + \frac{1}{r^2} \frac{\partial^2 f}{\partial \theta^2} = \frac{1}{r^2} \left(r \frac{\partial}{\partial r} \right)^2 f + \frac{1}{r^2} \frac{\partial^2 f}{\partial \theta^2}.$$

We let $F(s, \theta)$ denote the Mellin transform of $f(r, \theta)$ with respect to the variable r, so that

$$F(s, \theta) = \int_0^\infty r^{s-1} f(r, \theta) \, dr.$$

Provided $f(r, \theta) \sim r^{-a}$ as $r \to 0$ and $f(r, \theta) \sim r^{-b}$ as $r \to \infty$, $F(s, \theta)$ is defined in the strip of analyticity $a < \mathrm{Re}(s) < b$; see §3.1.1. From the differential properties of the Mellin transform in (3.1.10) combined with (3.1.7b), we then find that†

$$M\left[\left(r \frac{\partial}{\partial r} \right)^2 f; s \right] = s^2 F(s, \theta)$$

in $a < \mathrm{Re}(s) < b$. Hence in the strip of analyticity we obtain the simple result

$$M[r^2 \nabla^2 f; s] = \left(\frac{d^2}{d\theta^2} + s^2 \right) F(s, \theta). \tag{4.4.1}$$

Consider the boundary value problem relating to the determination of the steady-state temperature distribution inside an infinite wedge given by

$$\nabla^2 u = 0 \qquad (0 \le r < \infty, \quad -\alpha \le \theta \le \alpha) \tag{4.4.2}$$

---

† It can also be shown that $M[(r\partial/\partial r)^2 f; s] = s^2 F(s, \theta) + [r^{s+1} \partial f/\partial r - s r^s f]_0^\infty$ by integration by parts. In the strip $a < \mathrm{Re}(s) < b$ the integrated terms vanish.

(where $\alpha < \pi$) subject to the boundary conditions

$$u(r, \pm\alpha) = \begin{cases} T_0 & (0 \leq r < a) \\ 0 & (r > a). \end{cases} \tag{4.4.3}$$

If we suppose that $u(r, \theta)$ is bounded as $r \to 0$ and that $u(r, \theta) \sim r^{-b}$ as $r \to \infty$, the Mellin transform, $U(s, \theta)$, with respect to r exists in the strip $0 < \mathrm{Re}(s) < b$. From (4.4.1), the differential equation (4.4.2) and boundary conditions (4.4.3) transform to

$$\frac{d^2U}{d\theta^2} + s^2 U = 0, \qquad U(s, \pm\alpha) = T_0 \frac{a^s}{s},$$

for which the solution is readily shown to be

$$U(s, \theta) = T_0 \frac{a^s \cos s\theta}{s \cos \alpha s}.$$

Application of the Mellin inversion formula in (3.1.5) then shows that in the sector $|\theta| \leq \alpha$

$$u(r, \theta) = \frac{T_0}{2\pi i} \int_{c-\infty i}^{c+\infty i} \frac{(r/a)^{-s} \cos s\theta}{s \cos \alpha s} \, ds \qquad (0 < c < \beta)$$

where $\beta = \pi/2\alpha$. The evaluation of this integral can be carried out either by displacement of the integration path over the poles of the integrand in the usual way, or by making use of the result in (3.3.17). If we express the term $\cos s\theta$ as a sum of exponentials, make the change of variable $s = \beta w$ and let $z_{\pm} = (r/a)^\beta e^{\pm i\beta\theta}$, we find

$$u(r, \theta) = \frac{T_0}{4\pi i} \int_{c'-\infty i}^{c'+\infty i} \frac{z_+^{-w} + z_-^{-w}}{w \cos \frac{1}{2}\pi w} \, dw \qquad (0 < c' < 1)$$

$$= \frac{2T_0}{\pi} \mathrm{Re}(\mathrm{arccot}\, z_+).$$

Use of the standard identities $\mathrm{arccot}\, z = \arctan(1/z)$ and $\mathrm{arccot}\, z = \frac{1}{2}\pi - \arctan z$ $(\mathrm{Re}(z) \geq 0)$, together with the result [Abramowitz & Stegun (1965, p. 81)]

$$\mathrm{Re}(\arctan z) = \frac{1}{2} \arctan\left(\frac{2\mathrm{Re}(z)}{1 - |z|^2}\right),$$

when $|z| < 1$, then shows that

$$u(r, \theta) = T_0\left\{\frac{1}{2} + \frac{1}{\pi} \arctan\left(\frac{a^{2\beta} - r^{2\beta}}{2(ar)^\beta \cos \beta\theta}\right)\right\}$$

valid for $0 \leq r < \infty$, $-\alpha \leq \theta \leq \alpha$. From this we see that as $r \to \infty$ the solution $u(r, \theta) = O(r^{-\beta})$, so that the strip of analyticity of the Mellin transform $U(s, \theta)$ corresponds to $0 < \mathrm{Re}(s) < \beta$. In particular, we have for $-\alpha \leq \theta \leq \alpha$ that $u(a, \theta) = \frac{1}{2}T_0$ and $u(r, \theta) \gtrless \frac{1}{2}T_0$ when $r \lessgtr a$.

We now consider the angular domain $0 \le \theta \le \alpha$ and let the boundary $\theta = 0$ be held at zero temperature while the other boundary at $\theta = \alpha$ is maintained as in (4.4.3). The solution for the Mellin transform in this case is

$$U(s, \theta) = T_0 \frac{a^s \sin s\theta}{s \sin \alpha s},$$

and hence

$$u(r, \theta) = \frac{T_0}{2\pi i} \int_{c-\infty i}^{c+\infty i} \frac{(r/a)^{-s} \sin s\theta}{s \sin \alpha s} \, ds \qquad (0 < c < 2\beta).$$

Evaluation of this integral as described above and use of the result (3.3.16) shows that†

$$u(r, \theta) = \frac{T_0}{\pi} \operatorname{Im} \left(\log[1 + (a/r)^{2\beta} e^{2i\beta\theta}] \right)$$

$$= T_0 \left\{ \tfrac{1}{2} - \frac{1}{\pi} \arctan \left(\frac{r^{2\beta} + a^{2\beta} \cos 2\beta\theta}{a^{2\beta} \sin 2\beta\theta} \right) \right\}$$

for $0 \le r < \infty, 0 \le \theta \le \alpha$.

The solution of the above boundary value problem with the more general boundary conditions

$$u(r, \alpha) = g(r), \qquad u(r, -\alpha) = h(r)$$

is given by

$$U(s, \theta) = G(s) \frac{\sin s(\alpha + \theta)}{\sin 2\alpha s} + H(s) \frac{\sin s(\alpha - \theta)}{\sin 2\alpha s},$$

where $G(s)$ and $H(s)$ denote the Mellin transforms of the boundary functions $g(r)$ and $h(r)$, respectively. These transforms are assumed to be such that the strip of analyticity of $U(s, \theta)$ is contained in the strips of analyticity of $G(s)$ and $H(s)$. The inversion of $U(s, \theta)$ in this general case can be achieved by employing the Parseval formula in (3.1.13). If we let $J_\alpha(x, \phi) = M^{-1}[\sin s\phi / \sin 2\alpha s; s]$, it readily follows from (3.3.12) by letting $s \to s/\beta$, where $\beta = \pi/2\alpha$, that

$$J_\alpha(x, \phi) = \frac{\beta}{\pi} \frac{x^\beta \sin \beta\phi}{1 + 2x^\beta \cos \beta\phi + x^{2\beta}} \qquad (|\phi| < 2\alpha).$$

† This result and that in the previous example were communicated to us by J. Boersma.

Then we have

$$u(r,\theta) = \int_0^\infty g(u) J_\alpha(r/u, \alpha + \theta) \frac{du}{u} + \int_0^\infty h(u) J_\alpha(r/u, \alpha - \theta) \frac{du}{u}$$

$$= \frac{\beta r^\beta}{\pi} \cos \beta\theta \left\{ \int_0^\infty \frac{g(u) u^{\beta-1}}{u^{2\beta} - 2(ru)^\beta \sin \beta\theta + r^{2\beta}} du \right.$$

$$\left. + \int_0^\infty \frac{h(u) u^{\beta-1}}{u^{2\beta} + 2(ru)^\beta \sin \beta\theta + r^{2\beta}} du \right\}$$

when $-\alpha < \theta < \alpha$.

Thus, for example, if we let $\alpha = \frac{1}{2}\pi$ (so that $\beta = 1$) and take $g(r) = h(r) = e^{-r}$ as boundary functions, the variation of temperature on axis $(\theta = 0)$ is given by

$$u(r, 0) = \frac{2r}{\pi} \int_0^\infty \frac{e^{-u} \, du}{u^2 + r^2} = -\frac{2}{\pi} \operatorname{Im}[e^{ir} E_1(ir)],$$

where $E_1(x)$ denotes the exponential integral.

4.4.2 Ordinary Differential Equations

The Mellin transform can be used to solve ordinary differential equations of the type

$$L_x(y) \equiv x^m G(\Theta) y(x) - H(\Theta) y(x) = 0, \qquad (4.4.4)$$

where G and H are polynomials of their argument of degree n and p, respectively, and $\Theta \equiv xd/dx$. The order of the differential equation is $\max\{n, p\}$. If we make the transformation $z = x^m$, so that the operator $\Theta = m\Theta'$, where $\Theta' = zd/dz$, we find

$$zG(m\Theta')y(z) - H(m\Theta')y(z) = 0. \qquad (4.4.5)$$

Introduction of the Mellin transform

$$Y(s) = \int_0^\infty z^{s-1} y(z) \, dz$$

combined with use of (3.1.10) and (3.1.7b) then shows that the differential equation (4.4.5) transforms into the simple two-term recurrence relation for $Y(s)$

$$G(-ms - m)Y(s + 1) = H(-ms)Y(s). \qquad (4.4.6)$$

Since G and H are polynomials, they can be expressed in factored form and a solution of (4.4.6) is readily obtained in terms of gamma functions. It then follows that a solution of (4.4.4) is found by inverting the Mellin transform $Y(s)$ to yield

$$y(x) = \frac{1}{2\pi i} \int_{c-\infty i}^{c+\infty i} Y(s) x^{-ms} \, ds, \qquad (4.4.7)$$

where c is chosen so that the integration path lies in the strip of analyticity of $Y(s)$.

To illustrate this method of solution, let us first consider the Bessel equation

$$x^2\frac{d^2y}{dx^2} + x\frac{dy}{dx} + (x^2 - v^2)y = 0. \qquad (4.4.8)$$

This can be rewritten in the form

$$(\Theta^2 - v^2)y + x^2 y = 0,$$

so that $G \equiv 1$, $H \equiv v^2 - \Theta^2$ and $m = 2$. From (4.4.6) we obtain the equation satisfied by the Mellin transform $Y(s)$ given by

$$Y(s+1) = (v^2 - 4s^2)Y(s) = 4(\tfrac{1}{2}v + s)(\tfrac{1}{2}v - s)Y(s).$$

The general solution of this recurrence relation is seen to be

$$Y(s) = C\, 2^{2s}\, \frac{\Gamma(s + \tfrac{1}{2}v)}{\Gamma(1 + \tfrac{1}{2}v - s)}\, p(s),$$

where C is an arbitrary constant and $p(s)$ denotes a periodic function of unit period. A solution of (4.4.8) is consequently

$$y(x) = \frac{C}{2\pi i} \int_{c-\infty i}^{c+\infty i} \frac{\Gamma(s + \tfrac{1}{2}v)}{\Gamma(1 + \tfrac{1}{2}v - s)}\, p(s)(\tfrac{1}{2}x)^{-2s}\, ds, \qquad (4.4.9)$$

provided $p(s)$ is chosen in such a manner that the integral converges for some c in the strip of analyticity.

Different choices of $p(s)$ subject to the above restriction then produce different solutions of the differential equation. The simplest choice in this case corresponds to $p(s) = 1$, with the result that the integral (4.4.9) then converges† only for $x > 0$ when $0 < 2c < 1$, and we find

$$y(x) = \frac{C}{2\pi i} \int_{c-\infty i}^{c+\infty i} \frac{\Gamma(s + \tfrac{1}{2}v)}{\Gamma(1 + \tfrac{1}{2}v - s)}(\tfrac{1}{2}x)^{-2s}\, ds.$$

Comparison with (3.4.21) shows that, upon replacement of the variable of integration s by $s - \tfrac{1}{2}v$, this solution corresponds to the Bessel function $J_v(x)$ when $C = 1$.

If, on the other hand, we choose $p(s) = \cot \pi(s - \tfrac{1}{2}v)$ we will generate a different solution. This function has unit period and does not affect the convergence properties of (4.4.9), since $\cot \pi(s - \tfrac{1}{2}v) \sim \mp i$ as $\mathrm{Im}(s) \to \pm\infty$. Then a second solution of Bessel's equation is given by

$$y(x) = \frac{C}{2\pi i} \int_{c-\infty i}^{c+\infty i} \frac{\Gamma(s + \tfrac{1}{2}v)}{\Gamma(1 + \tfrac{1}{2}v - s)}(\tfrac{1}{2}x)^{-2s} \cot \pi\left(s - \tfrac{1}{2}v\right) ds$$

$$= \frac{\pi^{-1}C}{2\pi i} \int_{c-\infty i}^{c+\infty i} \Gamma(s + \tfrac{1}{2}v)\Gamma(s - \tfrac{1}{2}v)(\tfrac{1}{2}x)^{-2s} \cos \pi\left(s - \tfrac{1}{2}v\right) ds,$$

† This follows from the fact that the ratio of gamma functions in the integrand is $O(|t|^{2\sigma - 1})$ as $t \to \pm\infty$ (where $s = \sigma + it$); see also the discussion surrounding (3.4.21).

where $|\text{Re}(v)| < 2c < 1$. From the Appendix, this is seen to be the Bessel function $Y_v(x)$ when $C = 1$.

In a similar fashion, we can consider the modified Bessel equation

$$x^2 \frac{d^2 y}{dx^2} + x \frac{dy}{dx} - (x^2 + v^2) y = 0. \tag{4.4.10}$$

This yields the transformed equation

$$Y(s + 1) = 4\left(s + \tfrac{1}{2}v\right)\left(s - \tfrac{1}{2}v\right) Y(s),$$

for which the solution is given by

$$Y(s) = C \, 2^{2s} \Gamma\left(s + \tfrac{1}{2}v\right) \Gamma\left(s - \tfrac{1}{2}v\right) p(s).$$

The solutions of (4.4.10) can then be expressed in the form

$$y(x) = \frac{C}{2\pi i} \int_{c-\infty i}^{c+\infty i} \Gamma\left(s + \tfrac{1}{2}v\right) \Gamma\left(s - \tfrac{1}{2}v\right) p(s) \left(\tfrac{1}{2}x\right)^{-2s} ds$$

upon appropriate choice of c and the periodic function $p(s)$.

If we again choose $p(s) = 1$, with $2c > |\text{Re}(v)|$, the resulting integral converges for $|\arg x| < \tfrac{1}{2}\pi$ by Rule 1 in §2.4. Straightforward displacement of the contour and evaluation of the residues at the poles $s = \pm\tfrac{1}{2}v - k$ for nonnegative integer k (assuming v is not equal to an integer) yields

$$y(x) = \frac{\pi C}{\sin \pi v} \left\{ \sum_{k=0}^{\infty} \frac{\left(\tfrac{1}{2}x\right)^{2k-v}}{k! \, \Gamma(1 - v + k)} - \sum_{k=0}^{\infty} \frac{\left(\tfrac{1}{2}x\right)^{2k+v}}{k! \, \Gamma(1 + v + k)} \right\}$$

$$= \frac{\pi C}{\sin \pi v} \{ I_{-v}(x) - I_v(x) \},$$

where $I_v(x)$ is the modified Bessel function of the first kind. Hence we can identify our solution $y(x)$ as the modified Bessel function of the second kind, $K_v(x)$ when $C = \tfrac{1}{2}$.

Another example is furnished by the Gauss hypergeometric equation

$$x(1 - x) \frac{d^2 y}{dx^2} + [c - (a + b + 1)x] \frac{dy}{dx} - aby = 0,$$

where we suppose, for convenience, that a and b are positive. This equation can be written in the form (4.4.4) as

$$x(\Theta^2 + (a + b)\Theta + ab)y - \Theta(\Theta + c - 1)y = 0,$$

so that we obtain the Mellin transformed equation given by

$$Y(s + 1) = \frac{-s(c - 1 - s)}{(a - 1 - s)(b - 1 - s)} Y(s).$$

This last equation has the solution

$$Y(s) = C \frac{\Gamma(s)\Gamma(a-s)\Gamma(b-s)}{\Gamma(c-s)} (-)^s p(s),$$

where C is an arbitrary constant and $p(s)$ a periodic function with unit period.

If we let $p(s) = 1$, we find the solution

$$y(x) = \frac{C}{2\pi i} \int_{c'-\infty i}^{c'+\infty i} \frac{\Gamma(s)\Gamma(a-s)\Gamma(b-s)}{\Gamma(c-s)} (-x)^{-s} \, ds$$

provided $0 < c' < \min\{a, b\}$. By making the change of variable $s \to -s$ and comparison with (3.4.8), we see that this solution is a multiple of the hypergeometric function $_2F_1(a, b; c; x)$. This integral representation is readily continued analytically by suitable indentation of the integration path to deal with general (nonpositive integer) values of a and b.

4.4.3 Inverse Mellin Transform Solutions

We observe that the solutions in the above examples are expressed as inverse Mellin transforms. An alternative, but entirely equivalent, procedure is to substitute an inverse Mellin transform directly in the differential equation (4.4.4) and solve for the inverse transform. To illustrate this approach, let us consider the Hermite equation

$$\frac{d^2 y}{dx^2} - x \frac{dy}{dx} + vy = 0 \tag{4.4.11}$$

and look for a solution in the form

$$y(x) = \frac{1}{2\pi i} \int_{-\infty i}^{\infty i} h(s)x^{2s} \, ds. \tag{4.4.12}$$

The function $h(s)$ is a meromorphic function of s to be determined and the path of integration can be suitably indented (if necessary) to separate the poles of $h(s)$. The choice x^{2s} in the integrand (rather than x^{-s}) has been made for convenience in the solution of the resulting difference equation defining $h(s)$.

Formal substitution of (4.4.12) into the differential equation (4.4.11) gives

$$\frac{1}{2\pi i} \int_{-\infty i}^{\infty i} h(s)x^{2s}\{2s(2s-1)x^{-2} - 2s + v\} \, ds = 0.$$

If we now replace s by $s+1$ in the first part of the above integral, we find

$$\frac{1}{2\pi i} \int_{-1-\infty i}^{-1+\infty i} h(s+1)2(s+\tfrac{1}{2})(s+1)x^{2s} \, ds$$

$$- \frac{1}{2\pi i} \int_{-\infty i}^{\infty i} h(s)(s - \tfrac{1}{2}v)x^{2s} \, ds = 0.$$

A solution of (4.4.11) will be obtained provided we choose the function $h(s)$ to satisfy the first-order difference equation

$$h(s+1) = \frac{(s - \frac{1}{2}v)}{2(s + \frac{1}{2})(s + 1)} h(s)$$

and to be such that

$$\frac{1}{2\pi i} \left\{ \int_{-1-\infty i}^{-1+\infty i} - \int_{-\infty i}^{\infty i} \right\} h(s)(s - \frac{1}{2}v) x^{2s} \, ds = 0.$$

The function $h(s)$ takes the form

$$h(s) = C \, 2^{-s} \frac{\Gamma(s - \frac{1}{2}v)}{\Gamma(s + \frac{1}{2})\Gamma(s + 1)} p(s) = C\pi^{-\frac{1}{2}} 2^s \frac{\Gamma(s - \frac{1}{2}v)}{\Gamma(2s + 1)} p(s),$$

so that (4.4.12) represents a solution provided

$$\frac{1}{2\pi i} \left\{ \int_{-1-\infty i}^{-1+\infty i} - \int_{-\infty i}^{\infty i} \right\} \frac{\Gamma(s - \frac{1}{2}v + 1)}{\Gamma(2s + 1)} (2x^2)^s p(s) \, ds = 0 \qquad (4.4.13)$$

upon appropriate choice† of the periodic function $p(s)$.

Let us suppose temporarily that $\mathrm{Re}(v) < 0$. Let $p(s)$ be given by

$$p(s) = -\frac{\pi}{\sin 2\pi s}$$

and suppose the path of integration in (4.4.12) to be indented to the left at $s = 0$ to separate the poles of $\Gamma(s - \frac{1}{2}v)$ from those of $-\pi/(\Gamma(2s + 1)\sin 2\pi s) = \Gamma(-2s)$ at $s = 0, \frac{1}{2}, 1, \ldots$. Then the condition (4.4.13) becomes

$$\frac{1}{2\pi i} \left\{ \int_{-1-\infty i}^{-1+\infty i} - \int_{-\infty i}^{\infty i} \right\} \Gamma(s - \frac{1}{2}v + 1)\Gamma(-2s)(2x^2)^s \, ds = 0 \qquad (4.4.14)$$

and we see that in the strip between the indented contours the integrand possesses no poles. From Stirling's formula (2.1.8), the integrand in the strip (where $s = \sigma + it$) is

$$O\left(|t|^{-\sigma - \frac{1}{2}v} e^{-\frac{3}{2}\pi|t| \mp 2|t| \arg x} \right)$$

as $t \to \pm\infty$. It follows that, provided $|\arg x| < \frac{3}{4}\pi$, the contribution from the ends of the strip vanishes as $|t| \to \infty$ and that by Cauchy's theorem the integral (4.4.14) vanishes.

A solution of (4.4.11) is consequently given by

$$y(x) = \frac{\pi^{-\frac{1}{2}} C}{2\pi i} \int_{-\infty i}^{\infty i} \Gamma(s - \frac{1}{2}v)\Gamma(-2s)(2x^2)^s \, ds$$

---

† We remark that the choice $p(s) = 1$ is not possible in this case since the integrals in (4.4.13) would then not converge. This can be seen from the fact that, with $s = \sigma + it$, the ratio of gamma functions in the integrand is $O(|t|^{-\sigma - v/2} e^{\pi|t|/2})$ as $t \to \pm\infty$.

when $|\arg x| < \frac{3}{4}\pi$, $\mathrm{Re}(v) < 0$ and where the integration path is indented at $s = 0$ to lie to the left of the poles of $\Gamma(-2s)$. A simple change of variable $s \to \frac{1}{2}v - s$ and comparison with (3.4.14) shows that, when $C = 2^{-\frac{1}{2}v}\pi^{\frac{1}{2}}/\Gamma(-v)$, the solution $y(x)$ is given by

$$y(x) = e^{\frac{1}{4}x^2}D_v(x) = He_v(x),$$

where $He_v(x)$ is the Hermite function; see Abramowitz & Stegun (1965, p. 780).

If we close the contour to the right and evaluate the residues at the poles of $\Gamma(-2s)$ we generate the convergent expansion

$$He_v(x) = \frac{1}{2}\pi^{-\frac{1}{2}}C \sum_{k=0}^{\infty} \frac{(-)^k}{k!}\Gamma\left(\tfrac{1}{2}k - \tfrac{1}{2}v\right)(2x^2)^{\frac{1}{2}k}$$

$$= \frac{1}{2}\pi^{-\frac{1}{2}}C\left\{\sum_{k=0}^{\infty} \frac{\Gamma\left(k - \tfrac{1}{2}v\right)}{(2k)!}(2x^2)^k - \sum_{k=0}^{\infty} \frac{\Gamma\left(k + \tfrac{1}{2} - \tfrac{1}{2}v\right)}{(2k+1)!}(2x^2)^{k+\frac{1}{2}}\right\}$$

$$= \frac{2^{\frac{1}{2}v}\pi^{\frac{1}{2}}}{\Gamma\left(\tfrac{1}{2} - \tfrac{1}{2}v\right)}{}_1F_1\left(-\tfrac{1}{2}v; \tfrac{1}{2}; \tfrac{1}{2}x^2\right) - \frac{2^{\frac{1}{2}v+\frac{1}{2}}\pi^{\frac{1}{2}}}{\Gamma\left(-\tfrac{1}{2}v\right)}x\,{}_1F_1\left(\tfrac{1}{2} - \tfrac{1}{2}v; \tfrac{3}{2}; \tfrac{1}{2}x^2\right).$$

The restriction on v may be removed by appeal to analytic continuation. If v is a positive integer or zero $He_v(x)$ is a polynomial of degree v.

This procedure can be extended to obtain solutions of the nth-order differential equation

$$y^{(n)} - \sum_{r=0}^{p} a_r x^r y^{(r)} = 0, \qquad (4.4.15)$$

where n and p are integers satisfying $n > p \geq 0$ and a_r are constants. This equation is a generalisation of the Hermite equation (4.4.11) corresponding to $n = 2$, $p = 1$ and has been discussed at length in Paris & Wood (1986, Ch. 3). Applications with $n = 3, 4$ and $p = 1$ arise in hydrodynamic stability theory, while cases with $n = 4, 6$ and $p = 2$ arise in magnetohydrodynamic stability theory and also in deficiency index theory [Wood (1971)]. If we make use of the result

$$x^r \frac{d^r}{dx^r} = \prod_{k=0}^{r-1}(\Theta - k) \qquad (r = 1, 2, \ldots),$$

where $\Theta \equiv xd/dx$, the differential equation (4.4.15) can be written in the form

$$y^{(n)} - \prod_{r=1}^{p}(\Theta + \beta_r)y = 0. \qquad (4.4.16)$$

The parameters β_r ($1 \leq r \leq p$) are defined in terms of the coefficients a_r by

$$a_0 + \sum_{r=1}^{p} a_r \prod_{k=0}^{r-1} (\Theta - k) = \prod_{r=1}^{p} (\Theta + \beta_r).$$

We now seek a solution in the form

$$y(x) = \frac{1}{2\pi i} \int_{-\infty i}^{\infty i} h(s) x^{ns} \, ds. \tag{4.4.17}$$

Substitution of $y(x)$ into (4.4.16) yields

$$\frac{1}{2\pi i} \int_{-\infty i}^{\infty i} h(s) x^{ns} \left\{ ns(ns - 1) \ldots (ns - n + 1) x^{-n} - \prod_{r=1}^{p} (ns + \beta_r) \right\} ds = 0,$$

so that replacement of s by $s + 1$ in the first part of the integrand gives a solution if

$$h(s + 1) = n^{p-n} \frac{\prod_{r=1}^{p} (s + \beta_r/n)}{\prod_{r=1}^{n} (s + r/n)} h(s)$$

and

$$\frac{1}{2\pi i} \left\{ \int_{-1-\infty i}^{-1+\infty i} - \int_{-\infty i}^{\infty i} \right\} h(s) \prod_{r=1}^{p} \left(s + \frac{\beta_r}{n} \right) x^{ns} \, ds = 0.$$

A solution of the recurrence relation for $h(s)$ is given by

$$h(s) = Cn^{(p-n)s} \frac{\prod_{r=1}^{p} \Gamma(s + \beta_r/n)}{\prod_{r=1}^{n} \Gamma(s + r/n)} p(s) = n^{ps} \frac{\prod_{r=1}^{p} \Gamma(s + \beta_r/n)}{\Gamma(ns + 1)} p(s),$$

where $p(s)$ is a periodic function of unit period and we have employed the multiplication formula for the gamma function

$$\Gamma(nz) = (2\pi)^{\frac{1}{2}(1-n)} n^{nz-\frac{1}{2}} \prod_{r=0}^{n-1} \Gamma\left(z + \frac{r}{n} \right)$$

with the constant C chosen equal to $(2\pi)^{\frac{1}{2}(n-1)} n^{-\frac{1}{2}}$.

If, for example, we make the choice

$$p(s) = \frac{\pi n (-)^{ns+1}}{\sin \pi ns}$$

and let the integration path in (4.4.17) be suitably indented so as to separate the poles of $\Gamma(s + \beta_r/n)$ from those of $\Gamma(-ns)$, we see that in the strip between the indented contours the integrand of

$$\frac{1}{2\pi i} \left\{ \int_{-1-\infty i}^{-1+\infty i} - \int_{-\infty i}^{\infty i} \right\} \prod_{r=1}^{p} \Gamma\left(s + \frac{\beta_r}{n} + 1 \right) \Gamma(-ns)(-n^{p/n} x)^{ns} \, ds$$

has no poles. As $\text{Im}(s) = t \to \pm\infty$ in the strip, the integrand is dominated by $O(e^{-\frac{1}{2}(n+p)\pi|t| \mp n|t| \arg(-x)})$ so that the contribution from the ends of the strip vanishes as $|t| \to \infty$ provided $|\arg(-x)| < \frac{1}{2}\pi(1 + p/n)$. A solution of (4.4.15) can consequently be found in the form

$$y(x) = \frac{n}{2\pi i} \int_{-\infty i}^{\infty i} \Gamma(-ns) \prod_{r=1}^{p} \Gamma\left(s + \frac{\beta_r}{n}\right) (-n^{p/n} x)^{ns} \, ds$$

$$= \frac{1}{2\pi i} \int_{-\infty i}^{\infty i} \Gamma(-s) \prod_{r=1}^{p} \Gamma\left(\frac{s}{n} + \frac{\beta_r}{n}\right) (-n^{p/n} x)^s \, ds$$

in $|\arg(-x)| < \frac{1}{2}\pi(1 + p/n)$, where the contour is indented at $s = 0$ to lie to the right of all the poles of $\Gamma((s + \beta_r)/n)$ but to the left of those of $\Gamma(-s)$. Assuming this separation of poles to be possible†, we may swing round the contour into the infinite semi-circle in the right half-plane to find the absolutely convergent expansion (when $p < n$)

$$y(x) = \sum_{k=0}^{\infty} \frac{(n^{p/n} x)^k}{k!} \prod_{r=1}^{p} \Gamma\left(\frac{k}{n} + \frac{\beta_r}{n}\right) \qquad (|x| < \infty).$$

The validity of this solution (an integral function of x) for all values of arg x follows by analytic continuation.

Other choices of $p(s)$ lead to different solutions of (4.4.15). For example, when the order n of the differential equation is even, the choices

$$p(s) = \frac{(-)^{\frac{1}{2}ns}}{\sin \frac{1}{2}\pi ns} \quad \text{and} \quad p(s) = \frac{(-)^{\frac{1}{2}ns}}{\cos \frac{1}{2}\pi ns}$$

yield solutions which are respectively even and odd functions of x. These solutions, together with other solutions possessing certain asymptotic properties as $x \to \infty$, are discussed fully in Paris (1980); see also Paris & Wood (1986, Ch. 3).

4.5 Solution of Integral Equations

The solution of certain integral equations can be conveniently achieved by means of Mellin transforms. The types of equation we consider to illustrate this application are the Fredholm equations of the first and second kinds (taken over the semi-infinite interval $0 \le x < \infty$)

$$\int_0^\infty k(x, t) f(t) \, dt = g(x) \tag{4.5.1}$$

and

$$f(x) = g(x) + \lambda \int_0^\infty k(x, t) f(t) \, dt. \tag{4.5.2}$$

† This is always possible except when the β_r ($1 \le r \le p$) take on negative integer values.

In these equations the unknown function is $f(x)$, with the function $g(x)$, the kernel $k(x, t)$ and the parameter λ being specified. In what follows, we shall be concerned only with the particular solution of (4.5.2) where the parameter λ is supposed not to coincide with an eigenvalue of the linear integral operator

$$\mathcal{K}f = \int_0^\infty k(x, t) f(t) \, dt.$$

For such values of λ a solution of (4.5.2) can be found only if the given function $g(x)$ is orthogonal to all the eigenfunctions of the adjoint of $\mathcal{K}$. The reader interested in a more general treatment should consult specialist texts† on integral equations, for example Courant & Hilbert (1953, Ch. 3), Tricomi (1957), Pogorzelski (1966) and Moiseiwitsch (1977).

4.5.1 Kernels of the Form $k(xt)$

To illustrate the use of the Mellin transform in the solution of integral equations we shall suppose in this section that the kernel is of the special form $k(xt)$. We first consider the equation

$$\int_0^\infty k(xt) f(t) \, dt = g(x),$$

where $g(x)$ must be in the range of the operator $\mathcal{K}$ to have a solution. If we take the Mellin transform of both sides, using the Parseval formula (3.1.12), we obtain formally

$$G(s) = M \left[\int_0^\infty k(xt) f(t) \, dt; s \right] = K(s) F(1 - s), \qquad (4.5.3)$$

where $F(s)$, $G(s)$ are the Mellin transforms of $f(x)$, $g(x)$ and $K(s) = M[k(x); s]$. It is assumed here that the transforms in (4.5.3) have a common strip of analyticity; for a discussion of the procedure when there is no common strip, see Morse & Feshbach (1953, Ch. 8). If we replace s by $1 - s$ we find

$$F(s) = G(1 - s)/K(1 - s),$$

so that by the Mellin inversion formula in (3.1.5) the solution is given by

$$f(x) = \frac{1}{2\pi i} \int_{c-\infty i}^{c+\infty i} \frac{G(1 - s)}{K(1 - s)} x^{-s} \, ds. \qquad (4.5.4)$$

The solution $f(x)$ can also be expressed in the alternative 'resolvent' form by using the Parseval formula (3.1.12) in reverse. Thus, if the inverse Mellin transform

---

† Other treatments of integral equations are included in Whittaker & Watson (1965, Ch. 11), Morse & Feshbach (1953, Ch. 8), Titchmarsh (1975, Ch. 11) and Ledermann (1990).

of $1/K(1-s)$ is denoted by $L(x)$, that is if $L(x) = M^{-1}[1/K(1-s); x]$ (assuming it exists), then we find†

$$f(x) = \int_0^\infty L(xt)g(t)\, dt. \qquad (4.5.5)$$

In a similar manner, the Fredholm equation of the second kind with kernel $k(xt)$ (also known as Fox's equation)

$$f(x) = g(x) + \lambda \int_0^\infty k(xt)f(t)\, dt$$

yields upon taking the Mellin transform

$$F(s) = G(s) + \lambda K(s)F(1-s).$$

Replacing s by $1-s$ to produce

$$F(1-s) = G(1-s) + \lambda K(1-s)F(s),$$

we find, upon elimination of $F(1-s)$, that

$$\{1 - \lambda^2 K(s)K(1-s)\}F(s) = G(s) + \lambda K(s)G(1-s).$$

The solution $f(x)$ can then be expressed by the Mellin inversion formula as

$$f(x) = \frac{1}{2\pi i}\int_{c-\infty i}^{c+\infty i} \frac{G(s) + \lambda K(s)G(1-s)}{1 - \lambda^2 K(s)K(1-s)} x^{-s}\, ds. \qquad (4.5.6)$$

Conditions for the existence of $f(x)$ as an L^2 solution are given in Titchmarsh (1975, p. 333).

Example 1. Consider the integral equation

$$\int_0^\infty \frac{f(t)}{(1+x^2t^2)^2}\, dt = x^{-1}\{1 + (m/x)\}e^{-m/x} \qquad (x > 0, \quad m > 0).$$

From the Appendix, we have

$$k(x) = (1+x^2)^{-2}, \qquad K(s) = \tfrac{1}{2}\Gamma\left(\tfrac{1}{2}s\right)\Gamma\left(2 - \tfrac{1}{2}s\right) \quad (0 < \mathrm{Re}(s) < 4)$$

and

$$G(s) = \int_0^\infty x^{s-2}\{1 + (m/x)\}e^{-m/x}\, dx = \int_0^\infty u^{-s}(1 + mu)e^{-mu}\, du$$

$$= m^{s-1}\Gamma(1-s)(2-s) \qquad (\mathrm{Re}(s) < 1),$$

---

† We remark that the resolvent kernel $L(xt) = k(xt)$ if $K(s)K(1-s) = 1$. This is the necessary and sufficient condition for $k(xt)$ to be a Fourier kernel.

so that we have a common strip of analyticity given by $0 < \text{Re}(s) < 1$. Then from (4.5.4) the solution is given by

$$f(x) = \frac{1}{2\pi i} \int_{c-\infty i}^{c+\infty i} \frac{2\Gamma(s)(s+1)}{\Gamma(\frac{1}{2} - \frac{1}{2}s)\Gamma(\frac{1}{2}s + \frac{3}{2})} (mx)^{-s}\, ds \qquad (c > 0)$$

$$= \frac{4/\pi}{2\pi i} \int_{c-\infty i}^{c+\infty i} \Gamma(s) \cos \tfrac{1}{2}\pi s\, (mx)^{-s} ds = \frac{4}{\pi} \cos mx$$

by (3.3.6).

Example 2. Consider the Fredholm equation of the second kind

$$f(x) = g(x) + \lambda \int_0^\infty \sqrt{\frac{2}{\pi}} \cos xt\, f(t)\, dt \quad (x > 0).$$

In this case we have

$$k(x) = \sqrt{\frac{2}{\pi}} \cos x, \qquad K(s) = \sqrt{\frac{2}{\pi}} \Gamma(s) \cos \tfrac{1}{2}\pi s \quad (0 < \text{Re}(s) < 1),$$

so that

$$K(s)K(1-s) = \frac{2}{\pi} \Gamma(s)\Gamma(1-s) \cos \tfrac{1}{2}\pi s\, \sin \tfrac{1}{2}\pi s = 1.$$

Then, assuming $g(x)$ is such that its Mellin transform $G(s)$ has a strip of analyticity that overlaps with the strip $0 < \text{Re}(s) < 1$, we find from (4.5.6), when $\lambda^2 \neq 1$, the solution

$$f(x) = \frac{1}{2\pi i} \int_{c-\infty i}^{c+\infty i} \{G(s) + \lambda\sqrt{\frac{2}{\pi}} \Gamma(s) \cos \tfrac{1}{2}\pi s\, G(1-s)\} \frac{x^{-s}}{1 - \lambda^2}\, ds$$

$$= \frac{g(x)}{1 - \lambda^2} + \sqrt{\frac{2}{\pi}} \frac{\lambda}{1 - \lambda^2} \frac{1}{2\pi i} \int_{c-\infty i}^{c+\infty i} \Gamma(s) \cos \tfrac{1}{2}\pi s\, G(1-s) x^{-s}\, ds$$

$$= \frac{g(x)}{1 - \lambda^2} + \frac{\lambda}{1 - \lambda^2} \int_0^\infty \sqrt{\frac{2}{\pi}} \cos xt\, g(t)\, dt$$

upon use of the Parseval formula (3.1.12). This example may also be dealt with by Fourier cosine transformation.

Example 3. In the integral equation given in Titchmarsh (1975, p. 334)

$$f(x) = g(x) + \frac{1}{\sqrt{\pi}} \int_0^\infty e^{-xt} f(t)\, dt \qquad (x > 0)$$

let

$$g(x) = \frac{\log(1 + x)}{x} - \frac{\pi}{x} H(x - 1),$$

where $H(x - 1)$ denotes the Heaviside step function. We have

$$k(x) = \frac{e^{-x}}{\sqrt{\pi}}, \qquad K(s) = \frac{1}{\sqrt{\pi}}\Gamma(s) \quad (\text{Re}(s) > 0)$$

and

$$G(s) = M[g(x); s] = \int_0^\infty x^{s-2} \log(1 + x)\, dx - \pi \int_1^\infty x^{s-2}\, dx$$

$$= \frac{1}{1 - s} \int_0^\infty \frac{x^{s-1}}{1 + x}\, dx - \frac{\pi}{1 - s}$$

$$= \frac{\pi}{1 - s}(\operatorname{cosec} \pi s - 1) \qquad (0 < \text{Re}(s) < 1)$$

upon integrating by parts and using the Mellin transform of $(1+x)^{-1}$. The common strip of analyticity in this case is seen to correspond to $0 < \text{Re}(s) < 1$. Then, since

$$K(s)K(1 - s) = \operatorname{cosec} \pi s,$$

and

$$G(s) + K(s)G(1 - s) = \left\{\frac{\pi}{1 - s} + \pi^{\frac{1}{2}}\Gamma(s - 1)\right\}(\operatorname{cosec} \pi s - 1),$$

we find from (4.5.6) that

$$f(x) = \frac{1}{2\pi i} \int_{c-\infty i}^{c+\infty i} \left\{\frac{\pi}{s - 1} - \pi^{\frac{1}{2}}\Gamma(s - 1)\right\} x^{-s}\, ds \quad (0 < c < 1)$$

$$= -\frac{\pi}{x} + \frac{1}{2\pi i} \int_{c-\infty i}^{c+\infty i} \frac{\pi}{s - 1} x^{-s}\, ds + \frac{\pi^{\frac{1}{2}}}{x}(1 - e^{-x}) \quad (c > 1)$$

by displacement of the path across the pole at $s = 1$ in the first part of the integral and evaluation of the second part by an obvious modification of (3.3.2). The remaining integral can be evaluated by means of the discontinuous integral in (3.3.20) to yield the solution

$$f(x) = \frac{\pi^{\frac{1}{2}}}{x}(1 - e^{-x}) - \frac{\pi}{x}H(x - 1).$$

4.5.2 Kernels of the Form $k(x/t)$

We now suppose the kernel $k(x, t)$ in (4.5.1) and (4.5.2) to be of the form $k(x/t)$ and consider the integral equation

$$\int_0^\infty k(x/t)f(t)\frac{dt}{t} = g(x). \tag{4.5.7}$$

Use of the Parseval formula (3.1.13) shows that the Mellin transform of this equation is given by

$$K(s)F(s) = G(s),$$

provided the Mellin transforms of $F(s)$, $G(s)$ and $K(s)$ have a common strip of analyticity. The solution $f(x)$ is therefore given formally by

$$f(x) = \frac{1}{2\pi i} \int_{c-\infty i}^{c+\infty i} \frac{G(s)}{K(s)} x^{-s} \, ds. \tag{4.5.8}$$

For the Fredholm equation of the second kind with kernel $k(x/t)$

$$f(x) = g(x) + \lambda \int_0^\infty k(x/t) f(t) \frac{dt}{t}, \tag{4.5.9}$$

we have (assuming a common strip of analyticity) the resulting transformed equation given by

$$F(s) = G(s) + \lambda K(s) F(s), \tag{4.5.10}$$

so that

$$f(x) = \frac{1}{2\pi i} \int_{c-\infty i}^{c+\infty i} \frac{G(s)}{1 - \lambda K(s)} x^{-s} \, ds. \tag{4.5.11}$$

Conditions for the solution $f(x)$ to be L^2 are given in Titchmarsh (1975, p. 304).

Example 4. In the integral equation of the first kind

$$\int_0^\infty e^{-x/t} f(t) \frac{dt}{t} = e^{-a\sqrt{x}} \qquad (x > 0, \quad a > 0),$$

we have

$$k(x) = e^{-x}, \qquad K(s) = \Gamma(s) \quad (\text{Re}(s) > 0)$$

and

$$G(s) = M[e^{-a\sqrt{x}}; s] = 2a^{-2s} \Gamma(2s) \qquad (\text{Re}(s) > 0).$$

In this example both transforms are analytic in the right half-plane $\text{Re}(s) > 0$. Then, from (4.5.8), we find the solution

$$f(x) = \frac{1}{\pi i} \int_{c-\infty i}^{c+\infty i} \frac{\Gamma(2s)}{\Gamma(s)} (a^2 x)^{-s} \, ds \quad (c > 0)$$

$$= \frac{1}{2\pi i} \int_{c-\infty i}^{c+\infty i} \pi^{-\frac{1}{2}} \Gamma\left(s + \tfrac{1}{2}\right) \left(\tfrac{1}{4} a^2 x\right)^{-s} \, ds = \tfrac{1}{2} a \sqrt{\frac{x}{\pi}} e^{-\frac{1}{4} a^2 x}$$

by (3.3.2).

Example 5. Consider the integral equation

$$\int_x^1 \frac{t/x}{\sqrt{t^2 - x^2}} f(t) \, dt = g(x) \qquad (0 < x < 1),$$

where the kernel is given by $k(x) = x^{-1}/\sqrt{1 - x^2}$. This equation can be expressed in the form (4.5.7) by letting

$$f^*(x) = f(x)H(1 - x), \quad g^*(x) = g(x)H(1 - x), \quad k^*(x) = k(x)H(1 - x),$$

where $H(1 - x)$ is the Heaviside step function, to yield the equivalent form

$$\int_0^\infty k^*(x/t) f^*(t) \frac{dt}{t} = g^*(x).$$

Now

$$K(s) = M[k^*(x); s] = \int_0^1 \frac{x^{s-2}}{\sqrt{1 - x^2}} dx$$

$$= \frac{1}{2}\pi^{\frac{1}{2}} \frac{\Gamma\left(\frac{1}{2}s - \frac{1}{2}\right)}{\Gamma\left(\frac{1}{2}s\right)} \quad (\mathrm{Re}(s) > 1),$$

so that from (4.5.8) we have the solution

$$f(x) = \frac{1}{2\pi i} \int_{c-\infty i}^{c+\infty i} \frac{2}{\sqrt{\pi}} \frac{\Gamma\left(\frac{1}{2}s\right)}{\Gamma\left(\frac{1}{2}s - \frac{1}{2}\right)} G(s) x^{-s} ds.$$

We wish to employ the Parseval formula (3.1.13) in reverse to express the solution $f(x)$ in resolvent form as an integral involving $g(x)$, as was done in Example 2. However this cannot be done without some rearrangement of the above integrand, since the inverse Mellin transform of the ratio of gamma functions $\Gamma\left(\frac{1}{2}s\right)/\Gamma\left(\frac{1}{2}s - \frac{1}{2}\right)$ does not exist†. To overcome this difficulty, we rearrange the integrand by introducing the factor $\frac{1}{2}s - \frac{1}{2}$ into both the numerator and denominator to find

$$f(x) = \frac{1}{2\pi i} \int_{c-\infty i}^{c+\infty i} \frac{s - 1}{\sqrt{\pi}} G(s) \frac{\Gamma\left(\frac{1}{2}s\right)}{\Gamma\left(\frac{1}{2}s + \frac{1}{2}\right)} x^{-s} ds.$$

The inverse of the gamma function ratio now exists and is given by

$$L(x) = M^{-1}\left[\Gamma\left(\tfrac{1}{2}s\right)/\Gamma\left(\tfrac{1}{2}s + \tfrac{1}{2}\right); x\right] = \frac{2}{\sqrt{\pi(1 - x^2)}} H(1 - x),$$

since

$$\frac{2}{\sqrt{\pi}} \int_0^1 \frac{x^{s-1}}{\sqrt{1 - x^2}} dx = \frac{1}{\sqrt{\pi}} \int_0^1 \frac{u^{\frac{1}{2}s-1}}{(1 - u)^{\frac{1}{2}}} du = \frac{\Gamma\left(\frac{1}{2}s\right)}{\Gamma\left(\frac{1}{2}s + \frac{1}{2}\right)}$$

by (3.3.8). From (3.1.10) the inverse of $(s - 1)G(s)$ is

$$M^{-1}[(s - 1)G(s); x] = -\left(x\frac{d}{dx} + 1\right) g^*(x) = -\frac{d}{dx}(xg^*(x)),$$

† This can be seen from the fact that this ratio is $O(|t|^{\frac{1}{2}})$ as $t \to \pm\infty$ (where $s = \sigma + it$).

so that by application of (3.1.13) we obtain the solution in the form

$$f(x) = -\frac{1}{\sqrt{\pi}} \int_0^\infty L(x/t) \frac{d}{dt}(tg^*(t)) \frac{dt}{t}$$

$$= -\frac{2}{\pi} \int_x^1 \frac{1}{\sqrt{t^2 - x^2}} \frac{d}{dt}(tg(t)) \, dt.$$

In Sneddon (1972, p. 280) this result is generalised by showing that the solution of the integral equation

$$\int_x^1 \frac{T_n(t/x)}{\sqrt{t^2 - x^2}} f(t) \, dt = g(x) \quad (0 < x < 1),$$

where $T_n(x) = \cos(n \arccos x)$ denotes the Chebyshev polynomial, is given by

$$f(x) = -\frac{2}{\pi} \int_x^1 \frac{T_{n-1}(t/x)}{\sqrt{t^2 - x^2}} t^{1-n} \frac{d}{dt}(t^n g(t)) \, dt.$$

Example 6. As a final example let us consider the Fredholm equation of the second kind

$$f(x) = e^{-ax} + \frac{\lambda}{\pi} \int_0^\infty \frac{f(t)}{x + t} \, dt \quad (x > 0),$$

where we restrict the parameter λ to the range $0 < \lambda < 1$ and suppose that $a > 0$. We have

$$k(x) = \frac{1}{1 + x}, \qquad K(s) = \frac{\pi}{\sin \pi s} \quad (0 < \text{Re}(s) < 1),$$

$$g(x) = e^{-ax}, \qquad G(s) = a^{-s} \Gamma(s) \quad (\text{Re}(s) > 0),$$

so that, in the common strip of analyticity $0 < \text{Re}(s) < 1$, we find from (4.5.10) that

$$F(s) = \frac{a^{-s} \Gamma(s) \sin \pi s}{\sin \pi s - \lambda}.$$

The solution $f(x)$ is given by (4.5.11) as

$$f(x) = \frac{1}{2\pi i} \int_{c-\infty i}^{c+\infty i} \frac{\Gamma(s) \sin \pi s}{\sin \pi s - \lambda} (ax)^{-s} \, ds \quad (0 < c < \beta), \tag{4.5.12}$$

where $\beta = (1/\pi) \arcsin \lambda$.

For $0 < \lambda < 1$, the integrand has simple poles at the zeros of $\sin \pi s - \lambda$; that is, at the points

$$s_n = \beta - 2n, \quad -\beta - 2n + 1 \quad (n = 1, 2, \ldots)$$

on the left of the contour of integration. Displacement of the contour to the left and evaluation of the residues at the poles at s_n yields the (absolutely convergent) solution

$$f(x) = \frac{1}{\sqrt{1-\lambda^2}} \left\{ \sum_{n=1}^{\infty} \frac{(ax)^{2n-\beta}}{\Gamma(2n-\beta+1)} - \sum_{n=1}^{\infty} \frac{(ax)^{2n-1+\beta}}{\Gamma(2n+\beta)} \right\},$$

which can be expressed in terms of two $_1F_2$ hypergeometric functions. The asymptotic behaviour of $f(x)$ can be obtained by displacement of the contour to the right over the poles at $s = \beta, 1-\beta, \dots$ to find as $ax \to \infty$

$$f(x) \sim -\frac{1}{\sqrt{1-\lambda^2}} \left\{ \frac{(ax)^{-\beta}}{\Gamma(1-\beta)} - \frac{(ax)^{-1+\beta}}{\Gamma(\beta)} + \cdots \right\}.$$

An alternative representation of this solution can be found by writing (4.5.12) as

$$f(x) = e^{-ax} + \frac{1}{2\pi i} \int_{c-\infty i}^{c+\infty i} \frac{\lambda \Gamma(s)}{\sin \pi s - \lambda} (ax)^{-s} \, ds \quad (0 < c < \beta).$$

With $\lambda = \sin \pi \beta$ (where $0 < \beta < \frac{1}{2}$), we have the inverse transform†

$$M^{-1} \left[\frac{\sin \pi \beta}{\sin \pi s - \sin \pi \beta}; x \right] = \frac{\tan \pi \beta}{\pi} \frac{(x^{2-\beta} - x^{1+\beta})}{1 - x^2}.$$

Hence, by Parseval's formula (3.1.13), the solution can also be written in the form

$$f(x) = e^{-ax} + \frac{\tan \pi \beta}{\pi} \int_0^{\infty} L(x/t) e^{-at} \frac{dt}{t},$$

where the resolvent kernel $L(x/t)$ is given by

$$L(x) = \frac{x^{2-\beta}}{1 - x^2} (1 - x^{2\beta-1}).$$

4.6 Solution of Difference Equations

Difference equations have many applications in both pure and applied mathematics and are associated with an extensive literature. Recurrence relations, for instance, for which each term of a certain series is obtained from a number of preceding terms, can be viewed as cases of equations of this type. As in the analogous theory of ordinary differential equations, the simplest difference equations correspond to those with constant coefficients. When $n = 2$, the homogeneous second-order

---

† This inverse transform can be found either by residue evaluation as above or through use of the identity $\sin \pi \beta / (\sin \pi s - \sin \pi \beta) = \frac{1}{2} \tan \pi \beta \{ \cot \frac{1}{2} \pi (s - \beta + 2) - \cot \frac{1}{2} \pi (s + \beta + 1) \}$ combined with the result $M^{-1}[\cot \frac{1}{2} \pi (s + \alpha); x] = (2/\pi) x^{\alpha}/(1 - x^2)$; see the Appendix. There is a misprint in the corresponding formula given in Titchmarsh (1975, p. 310).

difference equation with constant coefficients is

$$a_2 y(x + 2) + a_1 y(x + 1) + a_0 y(x) = 0; \qquad (4.6.1)$$

this case is representative since the method of solution when $n = 2$ applies to equations of any order. It is then well known that the solution $y(x)$ of (4.6.1) is given by

$$y(x) = \begin{cases} \lambda_1^x p_1(x) + \lambda_2^x p_2(x) & (\lambda_1 \neq \lambda_2) \\ \lambda_1^x \{p_1(x) + x p_2(x)\} & (\lambda_1 = \lambda_2), \end{cases}$$

where $p_1(x)$, $p_2(x)$ are periodic functions of unit period and λ_1, λ_2 are the roots of the characteristic equation

$$a_2 \lambda^2 + a_1 \lambda + a_0 = 0. \qquad (4.6.2)$$

If x is restricted to nonnegative integer values (as in recurrence relations, for example) then we can replace the periodic functions $p_1(x)$ and $p_2(x)$ by constants A and B. Then when $x = 0, 1, 2, \ldots$, the solution of (4.6.1) is

$$y(x) = \begin{cases} A\lambda_1^x + B\lambda_2^x & (\lambda_1 \neq \lambda_2) \\ \lambda_1^x (A + Bx) & (\lambda_1 = \lambda_2). \end{cases}$$

A classical example of a second-order recurrence relation is produced by the Fibonacci sequence $0, 1, 1, 2, 3, 5, 8, 13, 21, \ldots$, in which each term after the second is the sum of the two preceding terms; that is, for nonnegative integer values of x the terms in the sequence are defined by the recurrence relation

$$y(x + 2) - y(x + 1) - y(x) = 0.$$

In this case the roots of the characteristic equation $\lambda^2 - \lambda - 1 = 0$ are $\lambda_{1,2} = \frac{1}{2}(1 \pm \sqrt{5})$. Since $y(0) = 0$, $y(1) = 1$ it follows that the constants $A = -B = 1/\sqrt{5}$, so that the general term of the Fibonacci sequence is therefore

$$y(x) = \frac{1}{\sqrt{5}} \left\{ \left(\frac{1 + \sqrt{5}}{2} \right)^x - \left(\frac{1 - \sqrt{5}}{2} \right)^x \right\},$$

a result attributed to Binet.

In this section we shall be concerned with the use of the Mellin transform in the solution of the homogeneous second-order difference equation with linear coefficients given by

$$(a_2 x + b_2) y(x + 2) + (a_1 x + b_1) y(x + 1) + (a_0 x + b_0) y(x) = 0 \qquad (4.6.3)$$

in which it will be assumed that the roots of the characteristic equation (4.6.2) are unequal. This equation is called the *hypergeometric difference equation* because, as we shall see, its solutions can be expressed in terms of the hypergeometric function

$$_2F_1(\alpha, \beta; \gamma; z) = \sum_{n=0}^{\infty} \frac{(\alpha)_n (\beta)_n}{(\gamma)_n \, n!} z^n, \tag{4.6.4}$$

where $(\alpha)_n = \Gamma(\alpha + n)/\Gamma(\alpha)$.

4.6.1 Solution by Mellin Transforms

We shall employ the Mellin transform to determine solutions of the second-order difference equation (4.6.3) in the form of integrals taken round certain paths in the complex plane. Before discussing this more general case, we first illustrate the method by means of the simpler and very well-known first-order difference equation

$$y(x + 1) = xy(x) \tag{4.6.5}$$

which is satisfied by the gamma function $\Gamma(x)$. We look for a solution in the form

$$y(x) = \int_C t^{x-1} f(t) \, dt, \tag{4.6.6}$$

where the function $f(t)$ and the contour of integration C in the complex t plane are to be determined. We note that when C coincides with the positive real axis this form is then the usual Mellin transform of $f(t)$. Substitution of this integral for $y(x)$ in (4.6.5) yields

$$\int_C \{t^x - xt^{x-1}\} f(t) \, dt = 0,$$

or, upon integration by parts of the second term,

$$\int_C t^x \{f(t) + f'(t)\} \, dt - [t^x f(t)]_C = 0.$$

A solution is therefore obtained if $f(t) + f'(t) = 0$, that is if $f(t) = e^{-t}$, and if C is chosen such that $[t^x e^{-t}]_C = 0$. When $\text{Re}(x) > 0$, we can take the path from 0 to ∞ along any ray in the right half-plane, and in particular along the positive real axis. Then a solution of (4.6.5) is given by

$$y_1(x) = p(x) \int_0^{\infty} t^{x-1} e^{-t} \, dt \quad (\text{Re}(x) > 0),$$

where $p(x)$ is a periodic function of unit period. If we choose $p(x) = 1$ then we have the well-known solution $y_1(x) = \Gamma(x)$. By analytic continuation this solution is then valid throughout the x plane except at the poles on the negative real axis.

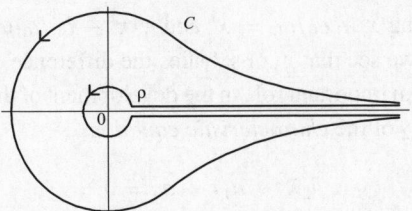

Fig. 4.5. The loop C and the deformed contour, with a circular indentation of radius ρ about the origin, when $\text{Re}(x) > 0$.

A second solution of (4.6.5), which places no restriction on x, can be found by choosing C to be a loop starting and ending at $+\infty$ which encircles the origin $t = 0$ in the positive sense, namely

$$y_2(x) = \int_{\infty}^{(0+)} t^{x-1} e^{-t} \, dt.$$

To evaluate this integral, we introduce a branch cut along the positive real axis and shrink the loop contour onto the upper and lower sides of the positive real axis between (ρ, ∞) together with a small circular contour of radius ρ centred at $t = 0$; see Fig. 4.5.

We take the branch of t^{x-1} specified by $0 < \arg t < 2\pi$, so that t^{x-1} is real for real values of x on the upper side of the cut. Then, with $\arg t = 0$ on the upper side between ∞ and ρ, and $\arg t = 2\pi$ on the lower side between ρ and ∞, we find

$$\int_{\infty}^{(0+)} t^{x-1} e^{-t} \, dt = \int_{\infty}^{\rho} t^{x-1} e^{-t} \, dt + e^{2\pi i x} \int_{\rho}^{\infty} t^{x-1} e^{-t} \, dt$$

$$+ i \int_{0}^{2\pi} (\rho e^{i\theta})^x e^{-\rho e^{i\theta}} \, d\theta.$$

Provided $\text{Re}(x) > 0$, the last integral vanishes as $\rho \to 0$ and accordingly we have

$$\int_{\infty}^{(0+)} t^{x-1} e^{-t} \, dt = (e^{2\pi i x} - 1) \int_{0}^{\infty} t^{x-1} e^{-t} \, dt$$

$$= (e^{2\pi i x} - 1)\Gamma(x).$$

We note that the right-hand side of this last expression is of the form $p(x)\Gamma(x)$, where $p(x) = e^{2\pi i x} - 1$ is a periodic function of unit period. This form is equivalent to Hankel's contour integral for $\Gamma(x)$; see Whittaker & Watson (1965, p. 244).

4.6.2 The Hypergeometric Difference Equation

We now apply the Mellin transform method to the solution of the second-order homogeneous equation in (4.6.3). We note that without loss of generality the coefficient b_0 can be set to zero to yield

$$(a_2 x + b_2)y(x + 2) + (a_1 x + b_1)y(x + 1) + a_0 x y(x) = 0. \qquad (4.6.7)$$

This follows by letting $x + b_0/a_0 = x'$ and $y(x' - b_0/a_0) = g(x')$, say; then, dropping the prime, we see that $g(x)$ satisfies the difference equation of the form (4.6.3) with $b_0 = 0$. An important role in the development of the solutions is played by the roots λ_1 and λ_2 of the *characteristic equation*

$$a_2\lambda^2 + a_1\lambda + a_0 = 0.$$

We shall consider only the general case in which λ_1 and λ_2 are finite, distinct from each other and different from zero; in terms of the coefficients, this means that none of the three quantities a_2, a_0 and $a_1^2 - 4a_0a_2$ is zero.

In the discussion of the solutions we shall find it more convenient to adopt the canonical form (4.6.7) (when $\lambda_1 \neq \lambda_2$)

$$(x + \beta_1 + \beta_2 + 2)y(x + 2) - [(\lambda_1 + \lambda_2)(x + 1) + \beta_1\lambda_2 + \beta_2\lambda_1]y(x + 1)$$

$$+ \lambda_1\lambda_2 xy(x) = 0, \tag{4.6.8}$$

where the constants β_1 and β_2 are related to the coefficients a_r and b_r in a straightforward manner. Then, substitution of the integral

$$y(x) = \frac{1}{2\pi i} \int_C t^{x-1} f(t)\, dt$$

into the left hand-side of (4.6.8) yields

$$\frac{1}{2\pi i} \int_C t^{x-1} \left\{ x[t^2 - (\lambda_1 + \lambda_2)t + \lambda_1\lambda_2] + (\beta_1 + \beta_2 + 2)t^2 \right.$$

$$\left. -(\lambda_1 + \lambda_2 + \beta_1\lambda_2 + \beta_2\lambda_1)t \right\} f(t)\, dt.$$

If we use the integration by parts result

$$\int_C x t^{x+k-1} f(t)\, dt = [t^{x+k} f(t)]_C - \int_C t^x \frac{d}{dt}\{t^k f(t)\}\, dt$$

successively for $k = 0$, 1 and 2 in the first part of the above integral, we find that the left hand-side of (4.6.8) becomes

$$\frac{1}{2\pi i} \int_C t^x \left\{ -\frac{d}{dt}[(t - \lambda_1)(t - \lambda_2) f(t)] + [(\beta_1 + \beta_2 + 2)t \right.$$

$$\left. - (\beta_1 + 1)\lambda_2 - (\beta_2 + 1)\lambda_1] f(t) \right\} dt$$

$$+ \frac{1}{2\pi i}[t^x (t - \lambda_1)(t - \lambda_2) f(t)]_C.$$

A solution of (4.6.8) will consequently be obtained if the integral vanishes identically and if, in addition, the path of integration C is chosen so that the integrated part is zero. Thus we choose the function $f(t)$ to satisfy the differential equation

$$\frac{d}{dt}[(t - \lambda_1)(t - \lambda_2)f(t)] = \{(\beta_1 + \beta_2 + 2)t - (\beta_1 + 1)\lambda_2 - (\beta_2 + 1)\lambda_1\}f(t).$$

After a little simplification this equation reduces to

$$\frac{f'(t)}{f(t)} = \frac{\beta_1}{t - \lambda_1} + \frac{\beta_2}{t - \lambda_2},$$

with the solution

$$f(t) = (t - \lambda_1)^{\beta_1}(t - \lambda_2)^{\beta_2}.$$

Then a solution of (4.6.8) is given by the integral

$$y(x) = \frac{1}{2\pi i}\int_C t^{x-1}(t - \lambda_1)^{\beta_1}(t - \lambda_2)^{\beta_2}\, dt \qquad (4.6.9)$$

provided C is chosen so that

$$\left[t^x(t - \lambda_1)^{\beta_1+1}(t - \lambda_2)^{\beta_2+1}\right]_C = 0. \qquad (4.6.10)$$

The expression (4.6.10) vanishes at $t = 0$ if $\mathrm{Re}(x) > 0$, at $t = \lambda_{1,2}$ if $\mathrm{Re}(\beta_{1,2}) > -1$ and at $t = \infty$ if $\mathrm{Re}(x + \beta_1 + \beta_2 + 2) < 0$. Consequently solutions can be found by taking contours of integration associated with any pair of the values $t = 0, \lambda_1, \lambda_2, \infty$. If $\mathrm{Re}(x) > 0$, we can take the solutions†

$$\left.\begin{aligned}
y_1(x) &= \frac{1}{2\pi i}\int_0^{(\lambda_1+)} t^{x-1}(t - \lambda_1)^{\beta_1}(t - \lambda_2)^{\beta_2}\, dt, \\
y_2(x) &= \frac{1}{2\pi i}\int_0^{(\lambda_2+)} t^{x-1}(t - \lambda_1)^{\beta_1}(t - \lambda_2)^{\beta_2}\, dt,
\end{aligned}\right\} \qquad (4.6.11)$$

where the loops start from the origin and encircle the points λ_1 and λ_2 in the positive sense, respectively, as shown in Fig. 4.6(a). These solutions can be shown to form a fundamental set.

Provided $\mathrm{Re}(\beta_1) > -1$, we can shrink the loop in the solution $y_1(x)$ onto the ray joining the origin to the point λ_1 together with a vanishingly small circular indentation surrounding the point $t = \lambda_1$. Then as t describes this deformed loop in the positive sense starting from $t = 0$, $\arg(t - \lambda_1)$ increases from $\arg \lambda_1 - \pi$ to $\arg \lambda_1 + \pi$, while $\arg t = \arg \lambda_1$ on the line segment. Then we find‡

$$y_1(x) = \frac{(e^{-\pi i\beta_1} - e^{\pi i\beta_1})}{2\pi i}\int_0^{\lambda_1} t^{x-1}(\lambda_1 - t)^{\beta_1}(t - \lambda_2)^{\beta_2}\, dt$$

---

† If $\arg \lambda_1 = \arg \lambda_2$ (but $|\lambda_1| \neq |\lambda_2|$) the loops in $y_1(x)$ and $y_2(x)$ must be deformed so that the points λ_2 and λ_1, respectively, are not enclosed.

‡ If β_1 is an integer the loop integral $y_1(x)$ vanishes. In this case a solution can be taken as the integral between 0 and λ_1.

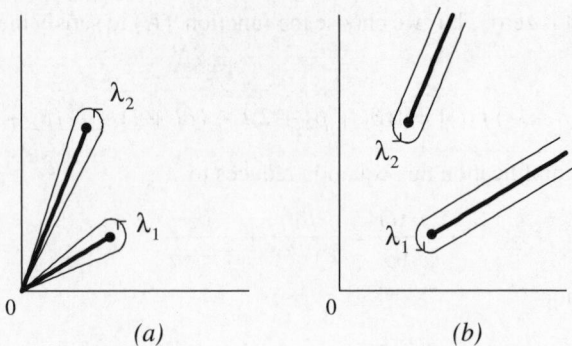

Fig. 4.6. The loop contours in the t plane for the solutions $y_r(x)$ when (a) $r = 1, 2$ and (b) $r = 3, 4$.

$$= -\frac{\sin \pi \beta_1}{\pi} \lambda_1^{x+\beta_1} (-\lambda_2)^{\beta_2} \int_0^1 \tau^{x-1}(1-\tau)^{\beta_1}(1-k\tau)^{\beta_2} \, d\tau,$$

where we have made the change of variable $t = \lambda_1 \tau$ and put $k = \lambda_1/\lambda_2$. From Abramowitz & Stegun (1965, p. 558) we identify this last integral as a hypergeometric function, so that

$$y_1(x) = -\frac{\sin \pi \beta_1}{\pi} \lambda_1^{x+\beta_1} (-\lambda_2)^{\beta_2} \, B(x, \beta_1 + 1) {}_2F_1(x, -\beta_2; x + \beta_1 + 1; k),$$

$$(4.6.12)$$

where $B(x, y) = \Gamma(x)\Gamma(y)/\Gamma(x + y)$ denotes the beta function; see (3.3.8). The treatment of the solution $y_2(x)$ is similar and can be obtained from (4.6.12) by interchanging λ_1, β_1 and λ_2, β_2, and replacing k by k^{-1}.

Provided $\mathrm{Re}(x + \beta_1 + \beta_2) < 0$, another fundamental set of solutions† is given by

$$\left.\begin{aligned} y_3(x) &= \frac{1}{2\pi i} \int_\infty^{(\lambda_1+)} t^{x-1}(t - \lambda_1)^{\beta_1}(t - \lambda_2)^{\beta_2} \, dt, \\[2ex] y_4(x) &= \frac{1}{2\pi i} \int_\infty^{(\lambda_2+)} t^{x-1}(t - \lambda_1)^{\beta_1}(t - \lambda_2)^{\beta_2} \, dt, \end{aligned}\right\} \qquad (4.6.13)$$

where the loops start and end at $\infty \exp(i \arg \lambda_{1,2})$ and encircle the points $t = \lambda_{1,2}$ in the positive sense, respectively; see Fig. 4.6(b). As for the solution $y_1(x)$, the loop in $y_3(x)$ can, when $\mathrm{Re}(\beta_1) > -1$, be shrunk onto the ray $\arg t = \arg \lambda_1$ between λ_1 and ∞ together with a vanishingly small circular contour around the point $t = \lambda_1$. On this path we let $\arg(t - \lambda_1)$ increase from $\arg \lambda_1$ to $\arg \lambda_1 + 2\pi$

† Other solutions can be found by taking loops connecting the points λ_1, λ_2 and $0, \infty$. We do not consider these solutions; they can, of course, be expressed as linear combinations of the solutions in (4.6.11) and (4.6.13).

and arg $t = \arg \lambda_1$ on the line segment. Then

$$y_3(x) = \frac{(e^{2\pi i \beta_1} - 1)}{2\pi i} \int_{\lambda_1}^{\infty} t^{x-1} (t - \lambda_1)^{\beta_1} (t - \lambda_2)^{\beta_2} \, dt$$

$$= \frac{(e^{2\pi i \beta_1} - 1)}{2\pi i} \lambda_1^{x+\beta_1+\beta_2} \int_0^1 \tau^{-x-\beta_1-\beta_2-1} (1 - \tau)^{\beta_1} (1 - k^{-1}\tau)^{\beta_2} \, d\tau,$$

where we have made the change of variable $t = \lambda_1/\tau$. Evaluation of the integral leads to the solution

$$y_3(x) = \frac{(e^{2\pi i \beta_1} - 1)}{2\pi i} \lambda_1^{x+\beta_1+\beta_2} \, B(\beta_1 + 1, -x - \beta_1 - \beta_2)$$

$$\times \, _2F_1(-\beta_2, -x - \beta_1 - \beta_2; 1 - x - \beta_2; k^{-1}). \qquad (4.6.14)$$

The solution $y_4(x)$ is obtained from the above by interchanging λ_1, β_1 and λ_2, β_2, and replacing k^{-1} by k.

If $|\lambda_1| < |\lambda_2|$ (i.e., $|k| < 1$), the hypergeometric series in $y_1(x)$ and $y_4(x)$ converge for all values of x except at $x = -\beta_1 - n$ and $x = -\beta_2 + n$ ($n = 1, 2, \ldots$), respectively, whereas the hypergeometric series in $y_2(x)$ and $y_3(x)$ are divergent. However, the only points where the solutions $y_1(x)$ and $y_4(x)$ are not analytic are $x = 0, -1, -2, \ldots$, where $B(x, \beta_1 + 1)$ and $B(x, \beta_2 + 1)$ have simple poles. If $|k| > 1$, then we use the solutions $y_2(x)$ and $y_3(x)$, where the only points of non-analyticity are $x = -\beta_1 - \beta_2 + n$ ($n = 0, 1, 2, \ldots$) corresponding to the simple poles of $B(\beta_1 + 1, -x - \beta_1 - \beta_2)$ and $B(\beta_2 + 1, -x - \beta_1 - \beta_2)$. For $|k| = 1$, $k \neq 1$ the hypergeometric series in (4.6.4) converges if $\mathrm{Re}(\gamma - \alpha - \beta) > -1$ and diverges if $\mathrm{Re}(\gamma - \alpha - \beta) \leq -1$. Hence, if λ_1, λ_2 are such that $|k| = 1$ (but not $k = 1$, since by hypothesis $\lambda_1 \neq \lambda_2$) the solutions $y_1(x)$, $y_2(x)$, $y_3(x)$ and $y_4(x)$ in (4.6.12) and (4.6.14) converge when $\mathrm{Re}(\beta_1 + \beta_2) > -2$.

Alternative forms of solution can be obtained by use of the familiar transformation properties of the hypergeometric function listed in Abramowitz & Stegun (1965, p. 559). For example, from the result

$$_2F_1(\alpha, \beta; \gamma; z) = (1 - z)^{-\alpha} \, _2F_1\left(\alpha, \gamma - \beta; \gamma; \frac{z}{z - 1}\right), \qquad (4.6.15)$$

we find that the solution $y_3(x)$ in (4.6.7) can also be expressed as

$$y_3(x) = \frac{(e^{2\pi i \beta_1} - 1)}{2\pi i} \lambda_1^{x+\beta_1} (\lambda_1 - \lambda_2)^{\beta_2} \, B(\beta_1 + 1, -x - \beta_1 - \beta_2)$$

$$\times \, _2F_1\left(\beta_1 + 1, -\beta_2; 1 - x - \beta_2; \frac{1}{1 - k}\right).$$

The solutions of (4.6.3) in the special cases when the roots of the characteristic equation are equal, or one root is either zero or infinite, require special treatment and have been discussed at great length in Batchelder (1967).

4.6.3 Solution of the Inhomogeneous First-Order Equation

We briefly discuss the solution of the inhomogeneous first-order difference equation

$$y(x + \omega) - y(x) = \omega\phi(x), \tag{4.6.16}$$

where, for simplicity in presentation, we restrict ω to be positive and $\phi(x)$ is a given function. A formal solution of this equation, which may be readily verified by substitution, is given in the form

$$y(x) = A - \omega \sum_{k=0}^{\infty} \phi(x + \omega k),$$

with A denoting a constant independent of x. For example, if $\phi(x) = e^{-ax}$ $(a > 0)$ we obtain the solution

$$y(x) = A - \omega \sum_{k=0}^{\infty} e^{-a(x+\omega k)} = A - \frac{\omega e^{-ax}}{1 - e^{-a\omega}}.$$

In general, however, the convergence of the sum cannot be guaranteed. One way of overcoming this difficulty is to consider the difference equation (4.6.16) with the function $\phi(x)$ replaced by another function $\phi(x; \mu)$ (with $\mu > 0$) and to write the constant A as

$$A = \int_c^{\infty} \phi(t; \mu) \, dt,$$

where $c \geq 0$ is another constant. Then, assuming that $\phi(x; \mu)$ is such that (i) $\phi(x; \mu) \to \phi(x)$ as $\mu \to 0$ and (ii) the integral for A and the infinite sum both converge, we find that the principal solution of the difference equation

$$y(x + \omega) - y(x) = \omega\phi(x; \mu) \tag{4.6.17}$$

is given by

$$y(x; \mu) = \int_c^{\infty} \phi(t; \mu) \, dt - \omega \sum_{k=0}^{\infty} \phi(x + \omega k; \mu). \tag{4.6.18}$$

If we let $\mu \to 0$, the difference equation (4.6.17) becomes the difference equation (4.6.16) with the principal solution

$$y(x) = \lim_{\mu \to 0} y(x; \mu).$$

This method depends on the difference of the integral and the infinite series having a limit as $\mu \to 0$. Although each separately may diverge when $\mu = 0$, the choice of the function $\phi(x; \mu)$ has to be made so that in this difference the divergent parts cancel. A convenient choice for $\phi(x; \mu)$ in many cases can be taken as $\phi(x; \mu) = \phi(x)e^{-\mu x}$.

To illustrate this method let $\phi(x) = 1$ in (4.6.16). Then taking $\phi(x; \mu) = e^{-\mu x}$, we find

$$
\begin{aligned}
y(x) &= \lim_{\mu \to 0} \left\{ \int_c^\infty e^{-\mu t} \, dt - \omega \sum_{k=0}^\infty e^{-\mu(x+\omega k)} \right\} \\
&= \lim_{\mu \to 0} \left\{ \frac{e^{-\mu c}}{\mu} - \frac{\omega e^{-\mu x}}{1 - e^{-\mu \omega}} \right\} \\
&= x - c - \tfrac{1}{2} \omega.
\end{aligned}
\tag{4.6.19}
$$

If we take $\omega = 1$ and set the constant $c = 0$, we have the principal solution

$$
y(x) = x - \tfrac{1}{2} = B_1(x),
$$

which is the Bernoulli polynomial of order $n = 1$. It is a well-known result that the Bernoulli polynomial $B_n(x)$ satisfies the difference equation

$$
y(x+1) - y(x) = nx^{n-1};
\tag{4.6.20}
$$

see, for example, Abramowitz & Stegun (1965, p. 804). It follows from this fact that (4.6.16) can be solved in terms of Bernoulli polynomials when $\phi(x)$ is a polynomial in x.

By a detailed argument (which we do not reproduce here) using Cauchy's theorem to express the series in (4.6.18) as a contour integral followed by deformation of contours, it is shown in Milne-Thomson (1933, pp. 222–230) that the principal solution of (4.6.16) can be written as the Mellin-type integral

$$
y(x) = \frac{1}{2\pi i} \int_{-a-\infty i}^{-a+\infty i} f(x + \omega s) \left(\frac{\pi}{\sin \pi s} \right)^2 ds,
\tag{4.6.21}
$$

where $0 < a < 1$. Here the function $f(x)$ is defined by

$$
f(x) = \int_c^x \phi(t) \, dt,
$$

where c is a constant and it is supposed that (i) $\phi(x)$ is holomorphic in the half-plane $\mathrm{Re}(x) \geq a$, (ii) for $\mathrm{Re}(x) \geq a$ there exist positive constants C and κ such that

$$
|\phi(x)| < Ce^{(\kappa+\epsilon)|x|}
$$

for arbitrarily small positive ϵ and (iii) ω satisfies the condition $0 < \omega < 2\pi/\kappa$. The choice of functions which satisfy these conditions includes all integral functions of order 1 and, in particular, $\log x$. Then, for example, for the difference equation (4.6.20) with $\phi(x) = nx^{n-1}$ and $c = 0$, we obtain $f(x) = x^n$ which leads to the integral representation for the Bernoulli polynomials

$$
B_n(x) = \frac{1}{2\pi i} \int_{-a-\infty i}^{-a+\infty i} (s + x)^n \left(\frac{\pi}{\sin \pi s} \right)^2 ds \quad (0 < a < 1).
\tag{4.6.22}
$$

For the difference equation (4.6.16) with $\phi(x) = e^{mx}$ $(m > 0)$, we have $\kappa = m$ and

$$f(x) = \int_c^x e^{mt}\, dt = \frac{1}{m}\{e^{mx} - e^{mc}\}.$$

Then, for $0 < \omega < 2\pi/\kappa$, a solution of the difference equation is given by

$$y(x) = \frac{1}{2\pi i} \int_{-a-\infty i}^{-a+\infty i} f(x + \omega s) \left(\frac{\pi}{\sin \pi s}\right)^2 ds \qquad (0 < a < 1).$$

Displacing the path of integration to the right over the double pole at $s = 0$ with residue ωe^{mx}, followed by the change of variable $s \to 1 + s$, we find

$$y(x) = -\omega e^{mx} + \frac{1}{2\pi i} \int_{-a+1-\infty i}^{-a+1+\infty i} f(x + \omega s) \left(\frac{\pi}{\sin \pi s}\right)^2 ds$$

$$= -\omega e^{mx} + \frac{1}{2\pi i} \int_{-a-\infty i}^{-a+\infty i} f(x + \omega s + \omega) \left(\frac{\pi}{\sin \pi s}\right)^2 ds.$$

Now it is easily seen that

$$f(x + \omega s + \omega) = e^{m\omega} f(x + \omega s) - \frac{e^{mc}}{m}(1 - e^{m\omega}),$$

so that $y(x)$ can be rewritten as

$$y(x) = \frac{\omega e^{mx}}{e^{m\omega} - 1} - \frac{e^{mc}}{m} \frac{1}{2\pi i} \int_{-a-\infty i}^{-a+\infty i} \left(\frac{\pi}{\sin \pi s}\right)^2 ds.$$

This last integral can be readily integrated directly to yield

$$\frac{1}{2\pi i} \int_{-a-\infty i}^{-a+\infty i} \left(\frac{\pi}{\sin \pi s}\right)^2 ds = \left[\frac{1}{e^{-2\pi i s} - 1}\right]_{-a-\infty i}^{-a+\infty i} = 1;$$

cf. also (3.3.18) in the limit $x \to 1$. Then the solution of (4.6.16) in this case is

$$y(x) = \frac{\omega e^{mx}}{e^{m\omega} - 1} - \frac{e^{mc}}{m} \qquad (0 < \omega < 2\pi/m),$$

which is equivalent to that in (4.6.19) with μ replaced by $-m$.

Finally, we mention that a discussion of the difference equation (4.6.16) with $\phi(x) = \log x$ (for which a solution is $y(x) = \log \Gamma(x)$) based on the integral (4.6.21) has been given in Lawless (1995).

4.7 Convergent Inverse Factorial Series

An inverse factorial series is an expansion of the type

$$\Omega(z) = \sum_{n=0}^{\infty} \frac{a_n\, n!}{z(z+1) \cdots (z+n)}, \qquad (4.7.1)$$

where the coefficients a_n are such that the series converges for sufficiently large $\mathrm{Re}(z)$. We give here only a summary discussion of the convergence of such series: for a detailed treatment see Milne-Thomson (1933, Ch. 10). In general the series in (4.7.1) converges in a right half-plane†. We have the results that if the series converges for $z = z_0$, then it converges for all z satisfying $\mathrm{Re}(z) > \mathrm{Re}(z_0)$ and converges absolutely for $\mathrm{Re}(z) > \mathrm{Re}(z_0) + 1$. If the series for $\Omega(z)$ converges absolutely for $z = z_0$, then it converges absolutely when $\mathrm{Re}(z) > \mathrm{Re}(z_0)$. We mention that absolutely convergent inverse factorial expansions for the Bessel functions of the form (4.7.1) have been considered in Dunster & Lutz (1991).

The domains of convergence and absolute convergence are half-planes which, in general, do not coincide. We denote the abscissa of convergence by $\mathrm{Re}(z) = \lambda$ and that of absolute convergence by $\mathrm{Re}(z) = \mu$. Then the series for $\Omega(z)$ converges when $\mathrm{Re}(z) > \lambda$ and converges absolutely when $\mathrm{Re}(z) > \mu$. It can be established that

$$0 \leq \mu - \lambda \leq 1$$

and that in the strip $\lambda < \mathrm{Re}(z) < \mu$ the series converges, but not absolutely. Thus, if $\lambda = +\infty$ the series (4.7.1) is divergent for all z; if $\lambda = -\infty$ the series converges in the whole z plane. A result due to Landau enables the abscissa of convergence to be determined as follows: if

$$\alpha = \limsup_{N \to \infty} \frac{\log |\sum_{n=0}^{N} a_n|}{\log N}, \quad \beta = \limsup_{N \to \infty} \frac{\log |\sum_{n=N}^{\infty} a_n|}{\log N}, \qquad (4.7.2)$$

then the abscissa of convergence λ of (4.7.1) is given by‡ α if $\lambda \geq 0$ and β if $\lambda < 0$. It can also be established that if the series for $\Omega(z)$ is convergent at the point $z = z_0$, then the series is uniformly convergent in the sector $-\frac{1}{2}\pi + \epsilon \leq \arg(z - z_0) \leq \frac{1}{2}\pi - \epsilon$, where ϵ is an arbitrarily small positive quantity. This domain of uniform convergence can be extended further to the half-plane $\mathrm{Re}(z) \geq \mathrm{Re}(z_0) + \epsilon$. In addition we note that the expansion of a function as an inverse factorial series is unique and that, as a consequence, an inverse factorial series cannot vanish identically unless all the coefficients a_n vanish.

The function $\Omega(z)$ can be written in the form

$$\Omega(z) = \Gamma(z) \sum_{n=0}^{\infty} \frac{a_n \, n!}{\Gamma(z + n + 1)} = \sum_{n=0}^{\infty} a_n \, B(z, n + 1),$$

where $B(x, y)$ is the beta function given in (3.3.8), which shows that the poles of $\Omega(z)$ are those of $\Gamma(z)$ at the points $z = 0, -1, -2, \ldots$. If we replace the beta

---

† It is of interest to note that the main convergence properties of the associated factorial series $\sum_{n=0}^{\infty}(-)^n a_n z(z-1)\cdots(z-n)/n!$ are the same as those of (4.7.1).
‡ A proof of this result is given in Milne-Thomson (1933, pp. 279–283).

function by its well-known integral representation in (3.3.8)

$$B(x, y) = \int_0^1 t^{x-1}(1 - t)^{y-1}\, dt \quad (\mathrm{Re}(x, y) > 0),$$

we have, when $\mathrm{Re}(z) > 0$,

$$\Omega(z) = \sum_{n=0}^{\infty} a_n \int_0^1 t^{z-1}(1 - t)^n\, dt. \tag{4.7.3}$$

The *generating function* $f(t)$ defined by

$$t^{z-1} f(t) = \sum_{n=0}^{\infty} a_n t^{z-1}(1 - t)^n \quad (0 \le t \le 1), \tag{4.7.4}$$

converges uniformly in the interval $0 \le t \le 1$, provided $\mathrm{Re}(z) > \max\{1, \lambda + 2\}$, where λ is the abscissa of convergence of $\Omega(z)$. To show this, let $\mathrm{Re}(z) = x$, where x satisfies this last condition. Then it follows that the series for $\Omega(z)$ converges when $z = x - 2$ and hence that the nth term of the series tends to zero as $n \to \infty$; that is

$$\lim_{n \to \infty} \frac{|a_n|\, n!}{(x - 1)x \cdots (x + n - 2)} = \lim_{n \to \infty} |a_n| \Big/ \binom{x + n - 2}{n} = 0.$$

Thus, given an $\epsilon > 0$, we can find an $n = n_0$ such that

$$|a_n| < \epsilon \binom{x + n - 2}{n} \quad (n \ge n_0).$$

We therefore have that

$$\left| \sum_{n=n_0}^{\infty} a_n t^{z-1}(1 - t)^n \right| < \epsilon t^{x-1} \sum_{n=n_0}^{\infty} \binom{x + n - 2}{n}(1 - t)^n$$

$$< \epsilon t^{x-1}[1 - (1 - t)]^{1-x} = \epsilon,$$

which proves the uniform convergence of (4.7.4).

Then, provided $\mathrm{Re}(z) > \max\{1, \lambda + 2\}$, we can interchange the order of summation and integration in (4.7.3) to obtain the representation

$$\Omega(z) = \int_0^1 t^{z-1} f(t)\, dt, \tag{4.7.5}$$

which we recognise as a finite Mellin transform. When the function $\Omega(z)$ is given, the generating function $f(t)$ can be obtained by Mellin inversion to yield

$$f(t) = \frac{1}{2\pi i} \int_{c-\infty i}^{c+\infty i} t^{-s} \Omega(s)\, ds, \tag{4.7.6}$$

where the constant $c > \mu$, the abscissa of absolute convergence of the inverse factorial series for $\Omega(z)$.

The following examples are taken from Milne-Thomson (1933, pp. 290–291).

Example 1. The function

$$\Omega(z) = \sum_{n=1}^{\infty} \frac{(n-1)!}{z(z+1)\cdots(z+n)}$$

is associated with the coefficients $a_n = n^{-1}$. Since $\sum_{n=1}^{N} n^{-1} \sim \gamma + \log N$ for large N, we find from (4.7.2) that the abscissa of convergence λ is given by

$$\lambda = \limsup_{N \to \infty} \frac{\log(\gamma + \log N)}{\log N} = 0.$$

To determine the abscissa of absolute convergence we consider the ratio of the $(n+1)$th to the nth terms of the series given by

$$\frac{n}{z+n+1} = 1 - \frac{z+1}{n} + O(n^{-2}) \qquad (n \to \infty),$$

and employ Weierstrass' test†. This shows that the series for $\Omega(z)$ is absolutely convergent when $\mathrm{Re}(z) > 0$; hence in this case we have $\mu = \lambda = 0$.

The generating function $f(t)$ is given by

$$f(t) = \sum_{n=1}^{\infty} \frac{(1-t)^n}{n} = -\log t.$$

Hence, from (4.7.5) when $\mathrm{Re}(z) > 0$, we find

$$\Omega(z) = -\int_0^1 t^{z-1} \log t \, dt = z^{-1} \int_0^1 t^{z-1} \, dt = z^{-2}.$$

Example 2. The function $\Omega(z) = (z-a)^{-1}$ is associated with the generating function

$$f(t) = \frac{1}{2\pi i} \int_{c-\infty i}^{c+\infty i} \frac{t^{-s}}{s-a} \, dt = t^{-a} \qquad (c > \mathrm{Re}(a)).$$

Application of the binomial theorem shows that

$$t^{-a} = [1 - (1-t)]^{-a} = 1 + \sum_{n=1}^{\infty} \binom{a+n-1}{n} (1-t)^n.$$

Hence, the coefficients a_n in the inverse factorial series expansion of $(z-a)^{-1}$ are given by $a_0 = 1$, $a_n = a(a+1)\cdots(a+n-1)/n!$ $(n \geq 1)$, and so we obtain Waring's formula

$$\frac{1}{z-a} = \frac{1}{z} + \frac{a}{z(z+1)} + \frac{a(a+1)}{z(z+1)(z+2)} + \frac{a(a+1)(a+2)}{z(z+1)(z+2)(z+3)} + \cdots.$$

† Weierstrass' test states that if $u_{n+1}/u_n = 1 - An^{-1} + O(n^{-B})$, where A is independent of n and $B > 1$, the series $\sum u_n$ is absolutely convergent if, and only if, $\mathrm{Re}(A) > 1$; see, for example, Knopp (1956, p. 133).

By Weierstrass' test, the abscissa of absolute convergence is easily seen to be $\mu = \text{Re}(a)$; since $(z - a)^{-1}$ has a pole at $z = a$, it follows that the abscissa of convergence is also $\lambda = \text{Re}(a)$. We remark that the series on the right-hand side can be written alternatively as $z^{-1} {}_2F_1(1, a; z + 1; 1)$, whereupon the sum on the left-hand side can be recovered using Gauss' summation theorem in (3.3.39).

5

Asymptotic Expansions

5.1 Algebraic Asymptotic Expansions

The standard Mellin-Barnes integral representation of a function $f(z)$ in some sector S, with vertex at $z = 0$, is given by

$$f(z) = \frac{1}{2\pi i} \int_{c-\infty i}^{c+\infty i} g(s) z^{-s}\, ds, \qquad (5.1.1)$$

where $g(s)$ usually consists of gamma functions and possibly trigonometric functions. The path of integration, which may be indented if necessary, is taken parallel to the imaginary s axis; in some cases, however, it is found expedient to take as integration contour a loop C with endpoints at infinity in an appropriate half-plane.

The method of determination of the asymptotic expansion of $f(z)$ for large $|z|$ from the above integral is a well-known and powerful technique. Suitable displacement of the path parallel to the imaginary axis over a subset of the poles of $g(s)$ then produces expansions in either ascending or descending powers of the variable z. The expansion in descending powers corresponds to the asymptotic expansion of $f(z)$ valid as $|z| \to \infty$ in S in the Poincaré sense. The form of this expansion must, by the nature of the integral (5.1.1), be of algebraic type.† Another powerful feature of this approach is that the remainder term which results when the path is displaced over a finite number of poles is simply given by the integral (5.1.1) taken over the displaced path. It is then usually a relatively straightforward matter to obtain a bound for the remainder to establish the nature of the expansion.

We consider four examples of this well-known procedure which illustrate the most important and peculiarly useful features of the Mellin transform.

---

† The terms *algebraic* and *exponential* expansions refer to expansions in which the controlling behaviour is either an algebraic power of the variable z or contains an exponential function of z.

5.1.1 The Exponential Integral $E_1(z)$

The exponential integral $E_1(z)$ is defined for $|\arg z| < \pi$ by

$$e^z E_1(z) = e^z \int_z^\infty \frac{e^{-t}}{t}\, dt,$$

where the path of integration excludes the origin $t = 0$ and does not cross the negative real axis, and elsewhere by analytic continuation. In the sector $|\arg z| < \frac{1}{2}\pi$, this definition is equivalent to

$$e^z E_1(z) = \int_0^\infty \frac{e^{-z\tau}}{1+\tau}\, d\tau,$$

where the integration path is the positive real axis. From Parseval's formula in (3.1.11) and the fact that the Mellin transforms of e^{-x} and $(1+x)^{-1}$ are $\Gamma(s)$ and $\pi/\sin \pi s$, it then follows that

$$e^z E_1(z) = \frac{1}{2\pi i} \int_{c-\infty i}^{c+\infty i} \Gamma(s) \frac{\pi z^{-s}}{\sin \pi s}\, ds \qquad (0 < c < 1). \tag{5.1.2}$$

With $s = \sigma + it$, Stirling's formula in (2.1.8) shows that the modulus of the integrand has the behaviour

$$|z|^{-\sigma} O(|t|^{\sigma - \frac{1}{2}} e^{-\Delta|t|}), \qquad \Delta = \tfrac{3}{2}\pi \mp \arg z$$

as $t \to \pm\infty$, so that (5.1.2) defines $e^z E_1(z)$ in the sector $|\arg z| < \frac{3}{2}\pi$; cf. also Rule 1 in §2.4. Because of this exponential decay as $t \to \pm\infty$, we can displace the path of integration either to the right or left over the two subsets of poles situated at $s = k + 1$ (simple) and $s = -k$ (double), where k is a nonnegative integer. The residues at the double poles can be determined as the coefficient of δ^{-1} in the expansion of

$$\frac{\pi^2 z^{-s}}{\Gamma(1-s)\sin^2 \pi s} = \frac{z^k}{k!} \frac{\pi^2}{\sin^2 \pi \delta} \frac{(1 - \delta \log z + \cdots)}{(1 - \delta \psi(k+1) + \cdots)}$$

$$= \frac{z^k}{k!} \left\{ \frac{1}{\delta^2} + \frac{1}{\delta}[\psi(k+1) - \log z] + \cdots \right\},$$

where $\delta = s + k$ and ψ denotes the logarithmic derivative of the gamma function. Hence, upon displacement of the path to the left over the first M poles, we obtain

$$e^z E_1(z) = \sum_{k=0}^{M-1} \frac{z^k}{k!} \{\psi(k+1) - \log z\} + R_M(z),$$

where the remainder $R_M(z)$ is given by

$$R_M(z) = \frac{1}{2\pi i} \int_{c-M-\infty i}^{c-M+\infty i} \frac{\pi^2 z^{-s}}{\Gamma(1-s)\sin^2 \pi s}\, ds \qquad (0 < c < 1).$$

For simplicity in the estimation of the remainder term, we choose $c = \frac{1}{2}$ and let $s = -M + \frac{1}{2} + it$. Then, with $\theta = \arg z$, we have

$$|R_M(z)| \leq \tfrac{1}{2}\pi|z|^{M-\frac{1}{2}} \int_{-\infty}^{\infty} \frac{1}{|\Gamma(M + \frac{1}{2} + it)|} \frac{e^{\theta t}}{\cosh^2 \pi t}\, dt$$

$$\leq \frac{\pi|z|^{M-\frac{1}{2}}}{2\Gamma(M + \frac{1}{2})} \int_{-\infty}^{\infty} \frac{e^{\theta t}}{(\cosh \pi t)^{\frac{3}{2}}}\, dt,$$

since, by repeated use of $\Gamma(z + 1) = z\Gamma(z)$ and the fact that $|\Gamma(\frac{1}{2} + it)| = (\pi/\cosh \pi t)^{\frac{1}{2}}$,

$$\frac{1}{|\Gamma(M + \frac{1}{2} + it)|} \leq \frac{\pi^{\frac{1}{2}}}{\Gamma(M + \frac{1}{2})} \frac{1}{|\Gamma(\frac{1}{2} + it)|} = \frac{(\cosh \pi t)^{\frac{1}{2}}}{\Gamma(M + \frac{1}{2})} \tag{5.1.3}$$

for $M \geq 1$ and $t \in (-\infty, \infty)$. This last integral is independent of $|z|$ and converges when $|\arg z| < \frac{3}{2}\pi$, so that

$$|R_M(z)| = \frac{1}{\Gamma(M + \frac{1}{2})} O\big(|z|^{M-\frac{1}{2}}\big).$$

Hence $|R_M(z)| \to 0$ as $M \to \infty$ for fixed $|z|$ and we consequently obtain the convergent series expansion

$$e^z E_1(z) = \sum_{n=0}^{\infty} \frac{z^n}{n!}\{\psi(n + 1) - \log z\},$$

whence

$$E_1(z) + \log z = e^{-z} \sum_{n=0}^{\infty} \frac{z^n}{n!}\psi(n + 1). \tag{5.1.4}$$

This result holds for all $\arg z$ by analytic continuation when an appropriate branch is chosen for $\log z$.

If the path in (5.1.2) is displaced a finite distance to the right to coincide with the vertical line $\mathrm{Re}(s) = M + \frac{1}{2}$, say, we find (remembering that a multiplicative factor of -1 arises since we are closing the contour in the negative direction)

$$e^z E_1(z) = \sum_{k=0}^{M-1} \frac{(-)^k k!}{z^{k+1}} + R_M(z), \tag{5.1.5}$$

where the modulus of the remainder in this case is given by

$$|R_M(z)| = \left| \frac{1}{2\pi i} \int_{M+\frac{1}{2}-\infty i}^{M+\frac{1}{2}+\infty i} \Gamma(s) \frac{\pi z^{-s}}{\sin \pi s}\, ds \right|$$

$$\leq \tfrac{1}{2}|z|^{-M-\frac{1}{2}} \int_{-\infty}^{\infty} |\Gamma(M + \tfrac{1}{2} + it)| \frac{e^{\theta t}}{\cosh \pi t}\, dt, \tag{5.1.6}$$

where $\theta = \arg z$. Since $|\Gamma(M + \frac{1}{2} + it)| = O(|t|^M e^{-\frac{1}{2}\pi|t|})$ as $t \to \pm\infty$ and the last integral is absolutely convergent when $|\theta| < \frac{3}{2}\pi$, we consequently find that $|R_M(z)| = O(|z|^{-M-\frac{1}{2}})$ in the sector $|\arg z| < \frac{3}{2}\pi$. The constant implied in the O symbol, however, depends on M and results in the divergence of $|R_M(z)|$ as $M \to \infty$; see §5.2. We therefore find the asymptotic expansion

$$E_1(z) \sim e^{-z} \sum_{k=0}^{\infty} \frac{(-)^k k!}{z^{k+1}} \tag{5.1.7}$$

valid as $|z| \to \infty$ in $|\arg z| < \frac{3}{2}\pi$.

5.1.2 The Parabolic Cylinder Function $D_\nu(z)$

The Weber parabolic cylinder function $D_\nu(z)$ has, for $\nu \neq 0, 1, 2, \ldots$, the Mellin-Barnes integral representation

$$D_\nu(z) = \frac{z^\nu e^{-\frac{1}{4}z^2}}{\Gamma(-\nu)} I(z), \tag{5.1.8}$$

where

$$I(z) = \frac{1}{2\pi i} \int_{-\infty i}^{\infty i} \Gamma(-s)\Gamma(2s - \nu)(2z^2)^{-s} \, ds \tag{5.1.9}$$

and the path of integration is indented to separate the poles of $\Gamma(-s)$ from those of $\Gamma(2s - \nu)$; see (3.4.14). Application of Rule 1 in §2.4 shows that $I(z)$ is defined by (5.1.9) in the sector $|\arg z| < \frac{3}{4}\pi$. Because the integrand decays exponentially like

$$|z|^{-2\sigma} O(|t|^{\sigma - \text{Re}(\nu) - 1} e^{-2\Delta|t|}), \qquad \Delta = \frac{3}{4}\pi \mp |\arg z| \tag{5.1.10}$$

as $t \to \pm\infty$, where $s = \sigma + it$, the integration path may be displaced either to the right or left over the subsets of poles of the integrand at $s = k$ and $s = \frac{1}{2}\nu - \frac{1}{2}k$ ($k = 0, 1, 2, \ldots$).

Closing the contour to the left, which is permissible on account of the asymptotic form of the integrand, we obtain the convergent (ascending) series expansion

$$I(z) = 2^{-\frac{1}{2}\nu-1} z^{-\nu} \sum_{n=0}^{\infty} \frac{(-)^n}{n!} \Gamma\left(\frac{1}{2}n - \frac{1}{2}\nu\right)(2z^2)^{\frac{1}{2}n}$$

$$= 2^{-\frac{1}{2}\nu-1} z^{-\nu} \pi^{\frac{1}{2}} \left\{ \sum_{n=0}^{\infty} \frac{\Gamma\left(n - \frac{1}{2}\nu\right)}{n! \, \Gamma\left(n + \frac{1}{2}\right)} \left(\frac{1}{2}z^2\right)^n \right.$$

$$\left. - 2^{-\frac{1}{2}} z \sum_{n=0}^{\infty} \frac{\Gamma\left(n + \frac{1}{2} - \frac{1}{2}\nu\right)}{n! \, \Gamma\left(n + \frac{3}{2}\right)} \left(\frac{1}{2}z^2\right)^n \right\}$$

upon separating the sum into even and odd k and using the duplication formula for the gamma function in (2.2.23). After a little rearrangement, this yields the standard definition in terms of the confluent hypergeometric function [Whittaker & Watson (1965, p. 347)]

$$
D_\nu(z) = e^{-\frac{1}{4}z^2} \pi^{\frac{1}{2}} \left\{ \frac{2^{\frac{1}{2}\nu}}{\Gamma\left(\frac{1}{2} - \frac{1}{2}\nu\right)} {}_1F_1\left(-\frac{1}{2}\nu; \frac{1}{2}; \frac{1}{2}z^2\right) \right.
$$

$$
\left. - \frac{2^{\frac{1}{2}\nu+\frac{1}{2}}z}{\Gamma\left(-\frac{1}{2}\nu\right)} {}_1F_1\left(\frac{1}{2} - \frac{1}{2}\nu; \frac{3}{2}; \frac{1}{2}z^2\right) \right\}. \tag{5.1.11}
$$

The algebraic asymptotic expansion (the descending series) of $I(z)$ can be determined by displacing the contour of integration to the right over the first M poles of the sequence $s = 0, 1, 2, \ldots$ to coincide† with the vertical line $\mathrm{Re}(s) = M - \frac{1}{2}$. Noting that the residue of $\Gamma(-s)$ at $s = k$ is $(-)^{k-1}/k!$, we therefore find

$$
I(z) = \sum_{k=0}^{M-1} \frac{(-)^k}{k!} \Gamma(2k - \nu)(2z^2)^{-k} + R_M(z), \tag{5.1.12}
$$

where the remainder term is given by

$$
R_M(z) = \frac{1}{2\pi i} \int_{M-\frac{1}{2}-\infty i}^{M-\frac{1}{2}+\infty i} \Gamma(-s)\Gamma(2s - \nu)(2z^2)^{-s} \, ds.
$$

If we suppose that M is chosen such that $2M - 1 > \mathrm{Re}(\nu)$ we can take the path in this integral to be the vertical line $s = M - \frac{1}{2} + it$ (without any indentation) to find, with $\theta = \arg z$,

$$
|R_M(z)| \le \frac{1}{2}(2|z|^2)^{-M+\frac{1}{2}} \int_{-\infty}^{\infty} \left| \frac{\Gamma(2M - 1 - \nu + 2it)}{\Gamma(M + \frac{1}{2} + it)} \right| \frac{e^{2\theta t}}{\cosh \pi t} \, dt.
$$

The modulus of the last integrand is $O(|t|^{M-\mathrm{Re}(\nu)-\frac{3}{2}} e^{-2\Delta|t|})$ as $t \to \pm\infty$, where Δ is defined in (5.1.10), so that the integral converges in the sector $|\arg z| < \frac{3}{4}\pi$. Hence $|R_M(z)| = O(|z|^{-2M+1})$ in this sector, where the constant implied in the O symbol is independent of $|z|$ but diverges‡ as $M \to \infty$.

We therefore obtain the well-known asymptotic expansion given by [Whittaker & Watson (1965, p. 347)]

$$
D_\nu(z) \sim \frac{z^\nu e^{-\frac{1}{4}z^2}}{\Gamma(-\nu)} \sum_{k=0}^{\infty} \frac{(-)^k}{k!} \Gamma(2k - \nu)(2z^2)^{-k} \tag{5.1.13}
$$

as $|z| \to \infty$ in the sector $|\arg z| < \frac{3}{4}\pi$.

---

† The choice $\mathrm{Re}(s) = M - \frac{1}{2}$ is made for convenience; we are, in fact, free to choose $\mathrm{Re}(s) = M - c$ with $0 < c < 1$.

‡ This automatically follows from the fact that the sum on the right-hand side of (5.1.12) diverges as $M \to \infty$.

5.1.3 A Bessel Function Integral

Consider the behaviour of the integral

$$F_\nu(x) = \int_0^\infty \frac{J_\nu(xt)}{1+t}\,dt \qquad (\nu > -1)$$

in the limits $x \to 0+$ and $x \to +\infty$, where $J_\nu(t)$ is the Bessel function of the first kind of order ν; a generalisation of this integral has been discussed in Watson (1966, §13.6). From the Appendix, we have the Mellin transforms

$$M[J_\nu(t); s] = 2^{s-1} \frac{\Gamma\left(\frac{1}{2}\nu + \frac{1}{2}s\right)}{\Gamma\left(1 + \frac{1}{2}\nu - \frac{1}{2}s\right)} \qquad \left(-\nu < \mathrm{Re}(s) < \tfrac{3}{2}\right),$$

$$M\left[(1+t)^{-1}; 1-s\right] = \frac{\pi}{\sin \pi s} \qquad (0 < \mathrm{Re}(s) < 1).$$

Use of Parseval's formula in (3.1.11) then shows that $F_\nu(x)$ can be expressed as the Mellin-Barnes integral

$$F_\nu(x) = \frac{1}{2\pi i} \int_{c-\infty i}^{c+\infty i} 2^{s-1} \frac{\Gamma\left(\frac{1}{2}\nu + \frac{1}{2}s\right)}{\Gamma\left(1 + \frac{1}{2}\nu - \frac{1}{2}s\right)} \frac{\pi x^{-s}}{\sin \pi s}\,ds, \qquad (5.1.14)$$

where $\max\{0, -\nu\} < c < 1$.

From Stirling's formula (2.1.8), the modulus of the integrand is seen to possess the controlling behaviour $x^{-\sigma} O(|t|^{\sigma-1} e^{-\pi|t|})$ as $t \to \pm\infty$, where $s = \sigma + it$, so that we are free to displace the integration contour in (5.1.14) both to the right and to the left. Provided $\nu \neq 0, 1, 2, \ldots$, the integrand has two infinite sequences of simple poles on the left of the contour at $s = -k$ and $s = -\nu - 2k$ ($k = 0, 1, 2, \ldots$). Displacement of the integration path to the left over these poles then yields the *convergent* expansion

$$F_\nu(x) = \frac{1}{2} \sum_{k=0}^\infty \left(-\tfrac{1}{2}x\right)^k \frac{\Gamma\left(\frac{1}{2}\nu - \frac{1}{2}k\right)}{\Gamma\left(1 + \frac{1}{2}\nu + \frac{1}{2}k\right)} - \frac{\pi}{\sin \pi \nu} \left(\tfrac{1}{2}x\right)^\nu \sum_{k=0}^\infty \frac{\left(-\frac{1}{4}x^2\right)^k}{k!\,\Gamma(1+\nu+k)}$$

$$= \frac{\pi}{\sin \pi \nu} \left\{\mathbf{J}_\nu(x) - J_\nu(x)\right\} \qquad (5.1.15)$$

upon identification of the first series in terms of the Anger function $\mathbf{J}_\nu(x)$ [Watson (1966, p. 309)] and the second series as a Bessel function of the first kind.

When ν is a nonnegative integer, the poles of the integrand in (5.1.14) on the left of the integration contour are no longer all simple. We consider only the case $\nu = 0$ in detail; the situation when ν equals a positive integer can be dealt with in a similar manner. When $\nu = 0$, we have simple poles at $s = -1, -3, \ldots$ and double poles at $s = 0, -2, -4, \ldots$. The residues at the double poles are given by the coefficient of δ^{-1} in the expansion of

$$\frac{2^{s-1}x^{-s}}{\Gamma^2(1-\frac{1}{2}s)}\frac{\pi^2}{\sin \pi s \, \sin \frac{1}{2}\pi s}$$

$$= \frac{1}{2}\left(-\frac{1}{4}x^2\right)^k \frac{\left(\frac{1}{2}x\right)^{-\delta}}{\Gamma^2(k+1-\frac{1}{2}\delta)}\frac{\pi^2}{\sin \pi \delta \, \sin \frac{1}{2}\pi \delta}$$

$$= \frac{\left(-\frac{1}{4}x^2\right)^k}{(k!)^2}\left\{\frac{1}{\delta^2} - \frac{1}{\delta}\left[\log \left(\frac{1}{2}x\right) - \psi(k+1)\right] + \cdots\right\},$$

where $s = -2k + \delta$ and ψ denotes the logarithmic derivative of the gamma function. Then we find the convergent expansion

$$F_0(x) = \frac{1}{4}\pi x \sum_{k=0}^{\infty} \frac{\left(-\frac{1}{4}x^2\right)^k}{\Gamma^2\left(k+\frac{3}{2}\right)} - \sum_{k=0}^{\infty} \frac{\left(-\frac{1}{4}x^2\right)^k}{(k!)^2}\left\{\log \left(\frac{1}{2}x\right) - \psi(k+1)\right\}$$

$$= \frac{1}{2}\pi\{\mathbf{H}_0(x) - Y_0(x)\} \tag{5.1.16}$$

upon identification of the first series as a Struve function $\mathbf{H}_0(x)$ and the second as a Bessel function of the second kind.

The behaviour of $F_\nu(x)$ for large x is obtained by displacement of the integration path to the right over the simple poles at $s = 1, 2, \ldots, M-1$, where M is a positive integer, to yield the expansion

$$F_\nu(x) = \frac{1}{2}\sum_{k=1}^{M-1} \frac{(-)^{k-1}\Gamma\left(\frac{1}{2}\nu + \frac{1}{2}k\right)}{\Gamma\left(1+\frac{1}{2}\nu - \frac{1}{2}k\right)}\left(\frac{1}{2}x\right)^{-k} + R_M(x),$$

where the remainder term $R_M(x)$ is

$$R_M(x) = \frac{1}{2\pi i}\int_{-c+M-\infty i}^{-c+M+\infty i} 2^{s-1}\frac{\Gamma\left(\frac{1}{2}\nu + \frac{1}{2}s\right)}{\Gamma\left(1+\frac{1}{2}\nu - \frac{1}{2}s\right)}\frac{\pi x^{-s}}{\sin \pi s}\,ds$$

with $0 < c < 1$. The last integral is absolutely convergent by virtue of (2.1.8) and has the order estimate $O(x^{-M+c})$. Thus we find the asymptotic expansion as $x \to +\infty$ given by

$$F_\nu(x) \sim \frac{1}{2}\sum_{k=1}^{\infty} \frac{(-)^{k-1}\Gamma\left(\frac{1}{2}\nu + \frac{1}{2}k\right)}{\Gamma\left(1+\frac{1}{2}\nu - \frac{1}{2}k\right)}\left(\frac{1}{2}x\right)^{-k} \tag{5.1.17}$$

valid for $\nu > -1$. The leading behaviour of $F_\nu(x)$ as $x \to 0+$ and $x \to +\infty$ then follows from (5.1.15)–(5.1.17).

5.1.4 The Mittag-Leffler Function $\mathcal{E}_a(z)$

The final example we consider in this section is the Mittag-Leffler function $\mathcal{E}_a(z)$ defined by the Maclaurin series, for $a > 0$,

$$\mathcal{E}_a(z) = \sum_{n=0}^{\infty} \frac{z^n}{\Gamma(1+an)} \qquad (|z| < \infty). \qquad (5.1.18)$$

This function is an integral function of z (of order a^{-1}) which reduces to e^z when $a = 1$. The determination of the asymptotics of $\mathcal{E}_a(z)$ in the sector $-\pi < \arg z \leq \pi$ is considerably more subtle than the preceding examples of this section and is well worth careful study. The analysis we shall present is partly based on Barnes (1906, pp. 285–289).

Since the residue of $\cot \pi s$ at any integer n is $1/\pi$, we have

$$\mathcal{E}_a(z) = \frac{1}{2\pi i} \int_C \frac{z^{-s}}{\Gamma(1-as)} \pi \cot \pi s \, ds, \qquad (5.1.19)$$

where C denotes a loop described in the positive sense enclosing the poles at $s = 0, -1, -2, \dots$ and with endpoints at infinity in $\mathrm{Re}(s) < 0$; see Fig. 5.1. From Stirling's formula (2.1.8), the logarithm of the modulus of the integrand for large $|s|$ is

$$aR \cos \theta \, \log(aR) + AR + O(\log R),$$

where $A = \sin \theta \, \arg z + a|\sin \theta|(\pi - |\theta|) - (a + \log |z|) \cos \theta$, with $s = Re^{i\theta}$. Hence, if C approaches infinity in the directions $\pm \theta_0$ say, where $\frac{1}{2}\pi < \theta_0 \leq \pi$, the logarithm of the integrand is controlled by $aR \cos \theta_0 \log(aR) \to -\infty$ as $R \to \infty$ independently of the value of $\arg z$. Consequently (5.1.19) defines $\mathcal{E}_a(z)$ without restriction on $\arg z$. Note, however, that it is *not* possible at this stage to straighten C into the vertical line $\mathrm{Re}(s) = c$, $0 < c < 1$, since the logarithm of the modulus of the integrand would then have the controlling behaviour $(\frac{1}{2}\pi a \pm |\arg z|)R$ as $\mathrm{Im}(s) \to \pm\infty$.

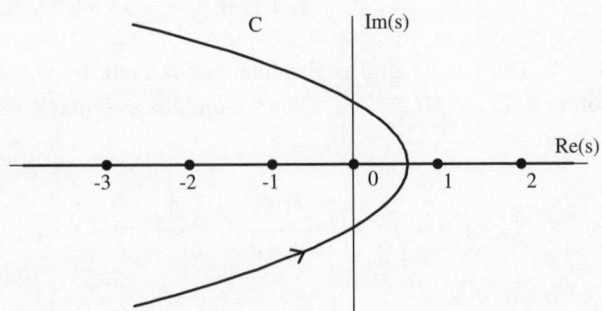

Fig. 5.1. The contour of integration for the Mittag-Leffler function $\mathcal{E}_a(z)$.

In order to extract the exponential dependence from $\mathcal{E}_a(z)$ we shall make use of the trigonometric identity† for arbitrary ν

$$\frac{\sin \pi \nu s}{\sin \pi s} = 2 \sum_{r=0}^{N-1} \cos(\nu - 2r - 1)\pi s + \frac{\sin \pi (\nu - 2N)s}{\sin \pi s}, \qquad (5.1.20)$$

where $N = 1, 2, \ldots$. Then, we write (5.1.19) as

$$\mathcal{E}_a(z) = \frac{1}{2\pi i} \int_C z^{-s} \Gamma(as) \sin \pi as \cot \pi s \, ds,$$

where, with $\nu = a-1$, the trigonometric functions appearing in the above integrand can be rearranged in the form

$$\sin \pi as \cot \pi s = \cos \pi as + \frac{\sin \pi (a-1)s}{\sin \pi s}$$

$$= \cos \pi as + 2 \sum_{r=1}^{N} \cos(a - 2r)\pi s + \frac{\sin(a - 2N - 1)\pi s}{\sin \pi s}.$$

We consequently find that

$$\mathcal{E}_a(z) = \frac{1}{2\pi i} \int_C z^{-s} \Gamma(as) \left\{ \cos \pi as + 2 \sum_{r=1}^{N} \cos(a - 2r)\pi s \right\} ds + J(z)$$

$$= \frac{1}{a} \sum_{r=-N}^{N} \exp\{z^{1/a} e^{2\pi i r/a}\} + J(z), \qquad (5.1.21)$$

where

$$J(z) = \frac{1}{2\pi i} \int_C z^{-s} \Gamma(as) \frac{\sin(a - 2N - 1)\pi s}{\sin \pi s} \, ds. \qquad (5.1.22)$$

The integrals in (5.1.21) have been evaluated by writing the cosines in terms of exponentials and use of the result given in (3.3.3).

To determine the asymptotics of the integral $J(z)$, we now straighten the contour C into the vertical line $\text{Re}(s) = c$, so that

$$J(z) = \frac{1}{2\pi i} \int_{c-\infty i}^{c+\infty i} z^{-s} \Gamma(as) \frac{\sin(a - 2N - 1)\pi s}{\sin \pi s} \, ds, \qquad (5.1.23)$$

where $0 < c < 1$. The controlling behaviour of the integrand is given by

$$O\left(|t|^{a\sigma - \frac{1}{2}} \exp\left\{ (|a - 2N - 1| - \tfrac{1}{2}a - 1)\pi |t| + t \arg z \right\} \right)$$

† This expansion follows from the recursion relation $T_\nu(s) = 2\cos(\nu - 1)\pi s + T_{\nu-2}(s)$, where $T_\nu(s) = \sin \pi \nu s / \sin \pi s$ and ν is an arbitrary parameter; see also Gradshteyn & Rhyzhik (1980, p. 29, Eq. 1.341(3)).

as $t \to \pm\infty$, with $s = \sigma + it$. Thus, the sectors of convergence of (5.1.23) are given by

$$|\arg z| < \begin{cases} (2N + 2 - \frac{1}{2}a)\pi & (a > 2N + 1) \\ (\frac{3}{2}a - 2N)\pi & (a < 2N + 1). \end{cases} \tag{5.1.24}$$

Because of the exponential decay of the integrand, we can displace the path in (5.1.23) to the right, over the poles at $s = 1, 2, \ldots$, to coincide with the vertical line $\mathrm{Re}(s) = M - \frac{1}{2}$, where M is a positive integer. We therefore find

$$J(z) = -\frac{1}{\pi} \sum_{k=1}^{M-1} (-z)^{-k} \Gamma(ak) \sin \pi(a-1)k + O\left(|z|^{-M+\frac{1}{2}}\right)$$

$$= -\sum_{k=1}^{M-1} \frac{z^{-k}}{\Gamma(1 - ak)} + O\left(|z|^{-M+\frac{1}{2}}\right) \tag{5.1.25}$$

as $|z| \to \infty$ in (5.1.24). Note that for integer a the above sum vanishes and we have $J(z) = O(|z|^{-M+\frac{1}{2}})$ for arbitrary M; see below.

We now choose $N = [\frac{1}{2}a - \frac{1}{2}]$ where the square brackets denote the nearest integer part.† It is then not difficult to see that, with this choice of N, the integral $J(z)$ converges in (at least) the whole plane (i.e., the sector $-\pi < \arg z \le \pi$) for $a \ge 1$, while for $0 < a < 1$ (when $N = 0$) it converges only in the sector $|\arg z| < \frac{3}{2}\pi a$. Hence, for $a \ge 1$, we have the expansion

$$\mathcal{E}_a(z) \sim \frac{1}{a} \sum_{r=-N}^{N} \exp\{z^{1/a} e^{2\pi i r/a}\} - \sum_{k=1}^{\infty} \frac{z^{-k}}{\Gamma(1 - ak)}$$

as $|z| \to \infty$ in $-\pi < \arg z \le \pi$.

To deal with the case $0 < a < 1$ and z not contained in the sector $|\arg z| < \frac{3}{2}\pi a$, we consider the representation for $\mathcal{E}_a(-z)$ given by the integral

$$\mathcal{E}_a(-z) = \frac{1}{2\pi i} \int_{c-\infty i}^{c+\infty i} \frac{\pi z^{-s}}{\Gamma(1 - as)} \frac{ds}{\sin \pi s} \quad (0 < c < 1).$$

This integrand has the controlling behaviour $O(|t|^{a\sigma - \frac{1}{2}} e^{(\frac{1}{2}a-1)\pi |t|} e^{t \arg z})$ as $t \to \pm\infty$, so that the sector of convergence is $|\arg z| < \frac{1}{2}\pi(2 - a)$. Displacement of the path of integration to the right then yields

$$\mathcal{E}_a(-z) = -\sum_{k=1}^{M-1} \frac{(-z)^{-k}}{\Gamma(1 - ak)} + O\left(|z|^{-M+\frac{1}{2}}\right).$$

Changing z into $-z$, we have, when $0 < a < 2$, the asymptotic expansion given in (5.1.25) valid as $|z| \to \infty$ in the sector $\frac{1}{2}\pi a < \arg z < 2\pi - \frac{1}{2}\pi a$.

---

† The nearest integer part is defined by $[x] = N$ when x is in the interval $(N - \frac{1}{2}, N + \frac{1}{2}]$. Thus $[\frac{1}{2}a - \frac{1}{2}] = N$ when $2N < a \le 2N + 2$.

Collecting together these results we therefore find that, when $0 < a < 2$,

$$\mathcal{E}_a(z) \sim \begin{cases} \dfrac{1}{a} \exp\{z^{1/a}\} - \displaystyle\sum_{k=1}^{\infty} \dfrac{z^{-k}}{\Gamma(1-ak)} & (|\arg z| < \tfrac{3}{2}\pi a) \\[4mm] -\displaystyle\sum_{k=1}^{\infty} \dfrac{z^{-k}}{\Gamma(1-ak)} & (|\arg(-z)| < \tfrac{1}{2}\pi(2-a)) \end{cases}$$

(5.1.26)

and, when $a > 2$,

$$\mathcal{E}_a(z) \sim \frac{1}{a} \sum_{r=-N}^{N} \exp\left\{z^{1/a} e^{2\pi i r/a}\right\} - \sum_{k=1}^{\infty} \frac{z^{-k}}{\Gamma(1-ak)} \qquad (-\pi < \arg z \le \pi)$$

(5.1.27)

where $N = [\tfrac{1}{2}a - \tfrac{1}{2}]$.

For integer values of a we have either $a = 2N + 2$ or $a = 2N + 1$. In this case the integral $J(z)$ can be evaluated exactly to find from (5.1.22) and (3.3.3) that

$$J(z) = \begin{cases} \dfrac{1}{a} \exp\{-z^{1/a}\} & (a = 2N + 2) \\[3mm] 0 & (a = 2N + 1). \end{cases}$$

Thus, when a is an integer, we obtain the *exact* result

$$\mathcal{E}_a(z) = \frac{1}{a} \sum_{r=-N}^{N'} \exp\left\{z^{1/a} e^{2\pi i r/a}\right\},$$

(5.1.28)

where $N' = N + 1$ or N according as a is even or odd, respectively. For example, when $a = 1, 2, 3, 4$ we have the sums

$$\mathcal{E}_1(z) = e^z, \qquad \mathcal{E}_2(z) = \cosh z^{\frac{1}{2}},$$

$$\mathcal{E}_3(z) = \tfrac{1}{3} e^{z^{1/3}} + \tfrac{2}{3} e^{-\frac{1}{2}z^{1/3}} \cos\left(\tfrac{\sqrt{3}}{2} z^{\frac{1}{3}}\right),$$

$$\mathcal{E}_4(z) = \tfrac{1}{2} \cosh z^{\frac{1}{4}} + \tfrac{1}{2} \cos z^{\frac{1}{4}}.$$

It should be remarked that the expansion (5.1.26) is valid in the Poincaré sense. In the sector $|\arg z| < \tfrac{1}{2}\pi a$, $\mathcal{E}_a(z)$ is exponentially large as $|z| \to \infty$ while in the common sectors $\tfrac{1}{2}\pi a < |\arg z| < \tfrac{3}{2}\pi a$, the exponential term is recessive. More detailed analysis shows that the rays $\arg z = \pm\pi a$ are Stokes lines, where the exponential term is maximally subdominant with respect to the algebraic expansion. In the neighbourhood of these rays the exponential term is multiplied by a factor (the Stokes multiplier) which undergoes a smooth but rapid change from unity to zero as one crosses $\arg z = \pm\pi a$ in the sense of increasing $|\arg z|$. This is the Stokes phenomenon which is discussed in detail in Chapter 6. Taking account of the exponentially small contribution in $\tfrac{1}{2}\pi a < |\arg z| < \pi a$, the expansion of $\mathcal{E}_a(z)$ when $0 < a < 2$ is then given in the complete sense by the composite expansion (5.1.26a) in the sector $|\arg z| < \pi a$ and by the algebraic expansion (5.1.26b) in the sector $|\arg(-z)| < \pi(1 - \frac{1}{2}a)$.

When $a = 2$, $\mathcal{E}_a(z) = \cosh z^{\frac{1}{2}}$ is exponentially large for all $\arg z$ except in the neighbourhood of the negative real axis where it is oscillatory. Finally, for $a > 2$, the expansion of $\mathcal{E}_a(z)$ is exponentially large throughout the z plane. In the expansion (5.1.27) we note that exponentially small contributions are included: these correspond to values of r satisfying $|(\arg z) \pm 2\pi r| > \frac{1}{2}\pi a$. The dominant terms in the expansions (5.1.26) and (5.1.27) can also be deduced from (2.3.10)–(2.3.12), where the parameters $\kappa = a$, $h = a^{-a}$ and $\vartheta = 0$.

We note that the generalisation of the Mittag-Leffler function given by

$$\mathcal{E}_{a,b}(z) = \sum_{n=0}^{\infty} \frac{z^n}{\Gamma(an + b)} \qquad (|z| < \infty),$$

where $a > 0$ and b is such that $an + b \neq 0, -1, -2, \ldots,$ can be handled in a similar manner. The dominant expansion of $\mathcal{E}_{a,b}(z)$ follows from §2.3; see also Erdélyi (1953, Vol. 3, p. 210).

5.2 Remainder Integrals

In this section we show how to obtain estimates for the remainder terms in the expansions that are generated in the process of path displacement in Mellin-Barnes integrals. We shall illustrate this procedure by means of the exponential integral $E_1(z)$ and the confluent hypergeometric function. The analysis of the remainder integrals in these cases relies heavily on certain bounds for the gamma function which are given in §2.1.3. We also discuss the numerical computation of the remainder integral (for the particular case of $E_1(z)$) and consider the variation of its rate of convergence as a function of the truncation index M.

5.2.1 Error Bounds

Let us take as our first example the exponential integral $E_1(z)$ discussed in §5.1.1. From (5.1.5) and (5.1.6) we have

$$e^z E_1(z) = \sum_{k=0}^{M-1} \frac{(-)^k k!}{z^{k+1}} + R_M(z), \tag{5.2.1}$$

where M denotes a positive integer and the remainder is given by

$$R_M(z) = \frac{1}{2\pi i} \int_{M+\frac{1}{2}-\infty i}^{M+\frac{1}{2}+\infty i} \Gamma(s) \frac{\pi z^{-s}}{\sin \pi s} \, ds. \tag{5.2.2}$$

As we saw in §5.1, $|R_M(z)| = O(|z|^{-M-\frac{1}{2}})$ in the sector $|\arg z| < \frac{3}{2}\pi$, which is sufficient to establish the asymptotic nature of the expansion (5.2.1) in this sector. We now determine estimates for the constant implied in the O symbol.

In the sector $|\arg z| < \pi$, we can use the inequality $|\Gamma(x+iy)| \leq \Gamma(x) \, (x > 0)$ to find (with $\theta = \arg z$)

$$|R_M(z)| \leq \tfrac{1}{2}|z|^{-M-\frac{1}{2}} \int_{-\infty}^{\infty} \left|\Gamma\left(M + \tfrac{1}{2} + it\right)\right| \frac{e^{\theta t}}{\cosh \pi t} \, dt \qquad (5.2.3)$$

$$\leq |z|^{-M-\frac{1}{2}} \Gamma\left(M + \tfrac{1}{2}\right) \int_{0}^{\infty} \frac{\cosh \theta t}{\cosh \pi t} \, dt \qquad (|\theta| < \pi)$$

$$= \tfrac{1}{2}|z|^{-M-\frac{1}{2}} \Gamma\left(M + \tfrac{1}{2}\right) \sec \tfrac{1}{2}\theta \qquad (5.2.4)$$

by (2.5.13). This shows that the error bound for $|R_M(z)|$ in $|\arg z| < \pi$ possesses the same structure as the terms in the asymptotic expansion and diverges factorially as $M \to \infty$ like $\Gamma(M + \tfrac{1}{2})$ in this sector.

In the adjacent sectors $\pi \leq |\arg z| < \tfrac{3}{2}\pi$, we can no longer use the above simple inequality for the gamma function and instead we resort to the bound given in (2.1.21), namely

$$|\Gamma(w)| < \Gamma(x)(|w|/x)^{x-\frac{1}{2}} e^{-|t|\phi(t)} \exp\left\{\tfrac{1}{6}|w|^{-1}\right\}, \qquad (5.2.5)$$

where $w = x + it$ (with $x > 0$) and $\phi(t) = \arctan(|t|/x)$. Then, since $\operatorname{sech} \pi t \leq 2e^{-\pi|t|}$ for $t \in (-\infty, \infty)$, we find from (5.2.3), upon letting $x = M + \tfrac{1}{2}$,

$$|R_M(z)| < |z|^{-M-\frac{1}{2}} \Gamma\left(M + \tfrac{3}{2}\right) \exp\left\{\tfrac{1}{6}M^{-1}\right\} H_M(\theta) \qquad (5.2.6)$$

valid in $|\arg z| < \tfrac{3}{2}\pi$, where

$$H_M(\theta) = \frac{1}{M + \frac{1}{2}} \int_{-\infty}^{\infty} e^{-(\pi+\phi(t))|t|+\theta t} \left\{1 + \left(t/(M + \tfrac{1}{2})\right)^2\right\}^{M/2} dt$$

$$= \int_{0}^{\infty} (e^{-\Delta_+ \tau} + e^{-\Delta_- \tau}) e^{\left(M+\frac{1}{2}\right)\psi(\tau)} (1 + \tau^2)^{M/2} d\tau. \qquad (5.2.7)$$

Here we have defined the new variable $\tau = t/(M + \tfrac{1}{2})$, used the relation $\arctan \tau = \tfrac{1}{2}\pi - \arctan(1/\tau)$, and defined the quantities $\Delta_\pm = (\tfrac{3}{2}\pi \pm \theta)(M + \tfrac{1}{2})$ and $\psi(\tau) = \tau \arctan(1/\tau)$.

A crude bound for $H_M(\theta)$ can be derived by using the fact that $0 \leq \psi(\tau) < 1$ for $\tau \in [0, \infty)$, where the limit 1 is approached as $\tau \to \infty$. Then

$$H_M(\theta) < e^{M+\frac{1}{2}} \int_{0}^{\infty} (e^{-\Delta_+ \tau} + e^{-\Delta_- \tau})(1 + \tau^2)^{M/2} d\tau$$

$$< e^{M+\frac{1}{2}} K(0), \qquad (5.2.8)$$

where

$$K(x) = J(\Delta_+ + x) + J(\Delta_- + x)$$

with

$$J(x) = \frac{e^x}{x^{M+1}}\Gamma(M+1,x).$$

In deriving this result we have made use of the estimate

$$\int_0^\infty e^{-x\tau}(1+\tau^2)^{\alpha/2}\,d\tau < \int_0^\infty e^{-x\tau}(1+\tau)^\alpha\,d\tau$$

$$= \frac{e^x}{x^{\alpha+1}}\Gamma(\alpha+1,x) \qquad (x>0, \quad \alpha>0) \tag{5.2.9}$$

in terms of the incomplete gamma function; see, for example, Abramowitz & Stegun (1965, p. 260). A sharper bound can be obtained if we use the inequality

$$e^{a\psi(\tau)} \le 1 + e^a(1-e^{-\lambda\tau}) \qquad (a\ge 0)$$

in (5.2.7) for suitable $\lambda \ge \lambda(a)$. The values of $\lambda(a)$, obtained by direct computation, for different $a = M + \frac{1}{2}$ are summarised in Table 5.1; see Fig. 5.2 for the particular case $M = 5$. This then yields the improved bound

$$H_M(\theta) < K(0) + e^{M+\frac{1}{2}}\{K(0) - K(\lambda)\}. \tag{5.2.10}$$

Numerical results illustrating the accuracy of the bounds in (5.2.4) and (5.2.6) are shown in Table 5.2. The modulus of the remainder $|R_M(z)|$ is computed from (5.2.2) for $M = 5$ and $M = 10$ when $|z| = 20$ and different values of the phase θ. The values of the error bound when $\theta < \pi$ and $\theta = \pi$ are obtained from (5.2.4) and (5.2.6), respectively, where $H_M(\theta)$ has been computed from (5.2.7). It will be observed that these bounds are quite sharp but, of course, (5.2.4) becomes useless as $\theta \to \pm\pi$. The bound for the integral $H_M(\theta)$ in (5.2.8) is, however, considerably less realistic (approximately out by two or three orders of magnitude for the values used in Table 5.2). This loss of sharpness stems from the use of the simple bound $\psi(\tau) \le 1$ in the integral (5.2.7). The bound in (5.2.10) is found to yield an order of magnitude improvement over that in (5.2.8).

Table 5.1. *Values of* $\lambda(a)$
for different $a = M + \frac{1}{2}$

M	$\lambda(a)$	M	$\lambda(a)$
1	0.705	6	0.532
2	0.725	7	0.495
3	0.684	8	0.464
4	0.629	9	0.438
5	0.576	10	0.416

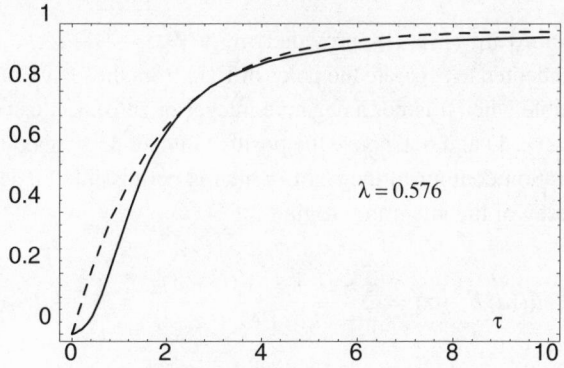

Fig. 5.2. Comparison of $(e^{a\psi(\tau)} - 1)e^{-a}$ (solid curve) with $1 - e^{-\lambda\tau}$ (dashed curve) when $a = M + \frac{1}{2}$ for $M = 5$ and $\lambda = 0.576$.

Table 5.2. *Comparison of* $|R_M(z)|$ *with*
the error bounds when $z = 20e^{i\theta}$

	$\|z\| = 20 \quad M = 5$	
θ/π	$\|R_M(z)\|$	Bound
0	1.45450×10^{-6}	1.82878×10^{-6}
0.25	1.52885×10^{-6}	1.97946×10^{-6}
0.50	1.77739×10^{-6}	2.58629×10^{-6}
0.75	2.26548×10^{-6}	4.77884×10^{-6}
1.00	2.79526×10^{-6}	1.17184×10^{-5}
	$\|z\| = 20 \quad M = 10$	
θ/π	$\|R_M(z)\|$	Bound
0	1.15585×10^{-8}	1.23735×10^{-8}
0.25	1.23668×10^{-8}	1.33930×10^{-8}
0.50	1.53695×10^{-8}	1.74987×10^{-8}
0.75	2.33743×10^{-8}	3.23334×10^{-8}
1.00	4.71485×10^{-8}	1.06200×10^{-7}

As a second example, we consider the confluent hypergeometric function† of the first kind which has the integral representation [see (3.4.4)]

$$\frac{\Gamma(a)}{\Gamma(b)} {}_1F_1(a; b; -z) = \sum_{n=0}^{\infty} \frac{\Gamma(a+n)}{\Gamma(b+n)} \frac{(-z)^n}{n!} \qquad (|z| < \infty)$$

$$= \frac{1}{2\pi i} \int_{-\infty i}^{\infty i} \Gamma(s) \frac{\Gamma(a-s)}{\Gamma(b-s)} z^{-s} \, ds$$

† We remark that the exponential integral can also be expressed in terms of a confluent hypergeometric function by $e^z E_1(z) = U(1; 1; z)$.

valid in the sector $|\arg z| < \frac{1}{2}\pi$, provided† a, $b \neq 0, -1, -2, \ldots$. The path of integration is indented to separate the poles of $\Gamma(s)$ from those of $\Gamma(a-s)$ (which is always possible when a is not a negative integer or zero). Let us consider only real values of a (< 1) and b. Choose the positive integer $M \geq \max\{1, b-a\}$ and shift the integration contour to the right (which is permissible on account of the exponential decay of the integrand) to find

$$\frac{\Gamma(a)}{\Gamma(b)}\,_1F_1(a;b;-z) = \sum_{k=0}^{M} \frac{(-)^k}{k!} \frac{\Gamma(k+a)}{\Gamma(b-a-k)} z^{-k-a} + R_M(z),$$

where

$$R_M(z) = \frac{1}{2\pi i} \int_{C_M} \Gamma(s) \frac{\Gamma(a-s)}{\Gamma(b-s)} z^{-s}\, ds$$

and C_M denotes the vertical line $\mathrm{Re}(s) = a + M + \frac{1}{2}$. Then

$$|R_M(z)| \leq \frac{|z|^{-a-M-\frac{1}{2}}}{2\pi} \int_{-\infty}^{\infty} \left| \frac{\Gamma\left(a+M+\frac{1}{2}+it\right)\Gamma\left(-M-\frac{1}{2}-it\right)}{\Gamma\left(b-a-M-\frac{1}{2}-it\right)} \right| e^{\theta t}\, dt,$$

where we have again put $\theta = \arg z$.

Now, for $t \in (-\infty, \infty)$, we have

$$\left|\Gamma\left(-M-\frac{1}{2}-it\right)\right| = \frac{\pi}{\cosh \pi t} \frac{1}{\left|\Gamma\left(M+\frac{3}{2}+it\right)\right|} \leq \frac{2\pi e^{-\pi|t|}}{\left|\Gamma\left(M+\frac{3}{2}+it\right)\right|},$$

from (2.1.16) and (2.1.17)

$$\left| \frac{\Gamma\left(a+M+\frac{1}{2}+it\right)}{\Gamma\left(M+\frac{3}{2}+it\right)} \right| \leq \frac{N(a)}{\left(M+\frac{1}{2}\right)^{1-a}},$$

where $N(a) = 1$ when $a \leq 0$ and $N(a) = 1 + 1/(M+\frac{1}{2})$ when $0 < a < 1$, and

$$\frac{1}{\left|\Gamma\left(b-a-M-\frac{1}{2}-it\right)\right|} \leq \frac{e^{\pi|t|}}{\pi}\left|\Gamma\left(a-b+M+\frac{3}{2}+it\right)\right|.$$

Substitution of these bounds in the integral for $|R_M(z)|$ then yields

$$|R_M(z)| \leq |z|^{-a-M-\frac{1}{2}} \frac{N(a)}{\pi\left(M+\frac{1}{2}\right)^{1-a}} \int_{-\infty}^{\infty} e^{\theta t}\left|\Gamma\left(a-b+M+\frac{3}{2}+it\right)\right| dt.$$

† If a is a negative integer $_1F_1(a;b;-z)$ reduces to a polynomial.

Use of (5.2.5), with $\chi = a - b + M + 1 \geq 1$, finally leads to the bound

$$|R_M(z)| < |z|^{-a-M-\frac{1}{2}} \frac{N(a)}{\pi \left(M + \frac{1}{2}\right)^{1-a}} \Gamma\left(\chi + \tfrac{1}{2}\right) \exp\left\{\tfrac{1}{6}\chi^{-1}\right\}$$

$$\times \int_{-\infty}^{\infty} e^{\theta t - |t|\phi(t)} \left\{1 + \left(t/\left(\chi + \tfrac{1}{2}\right)\right)^2\right\}^{\chi/2} dt$$

$$= |z|^{-a-M-\frac{1}{2}} \frac{N(a)}{\pi \left(M + \frac{1}{2}\right)^{1-a}} \Gamma\left(\chi + \tfrac{3}{2}\right) \exp\left\{\tfrac{1}{6}\chi^{-1}\right\} H_\chi(\theta), \quad (5.2.11)$$

where $\phi(t)$ is now given by $\arctan\left(|t|/\left(\chi + \tfrac{1}{2}\right)\right)$ and $H_\chi(\theta)$ is defined in (5.2.7) with $\Delta_\pm = \left(\tfrac{1}{2}\pi \pm \theta\right)\left(\chi + \tfrac{1}{2}\right)$.

A result equivalent to this has been given by Riekstiņš (1983), who employed a simpler bound for $H_\chi(\theta)$ based on the estimate† for the integral in (5.2.9) (when $\tfrac{1}{2}\alpha$ is a positive integer)

$$\int_0^\infty e^{-x\tau}(1 + \tau^2)^{\alpha/2}\, d\tau < \frac{\Gamma(\alpha + 1)}{x^{\alpha+1}}(1 + x^2)^{\alpha/2}.$$

5.2.2 Numerical Evaluation

We now consider direct *numerical evaluation* of the remainder integrals and, in particular, investigate their rate of convergence as a function of the truncation index M. We return to the simple exponential integral $e^z E_1(z)$ with the expansion given by [see (5.2.1)]

$$e^z E_1(z) = \sum_{k=0}^{M-1} \frac{(-)^k k!}{z^{k+1}} + R_M(z). \quad (5.2.12)$$

The remainder $R_M(z)$ in (5.2.2) takes the form

$$R_M(z) = \tfrac{1}{2}(-)^M z^{-M-\frac{1}{2}} \int_{-\infty}^{\infty} \Gamma\left(M + \tfrac{1}{2} + it\right) \frac{z^{-it}}{\cosh \pi t}\, dt, \quad (5.2.13)$$

where we have set $s = M + \tfrac{1}{2} + it$.

From Stirling's formula (2.1.8), the modulus of the integrand possesses the behaviour $|t|^M e^{-\frac{3}{2}\pi|t|+\theta t}$ as $t \to \pm\infty$. The decay of the integrand is consequently exponential, although when $\theta = \arg z \to \pm\frac{3}{2}\pi$ this decay becomes prohibitively slow as $t \to \pm\infty$, respectively. In Fig. 5.3 we show the behaviour of

$$f(t) = \frac{\mathrm{Re}\left(\Gamma\left(M + \tfrac{1}{2} + it\right)\right)}{\Gamma\left(M + \tfrac{1}{2}\right)}$$

as a function of t for different values of M; the behaviour of the imaginary part is similar. It is found that the real and imaginary parts of $\Gamma\left(M + \tfrac{1}{2} + it\right)$ decay more

---

† This estimate is obtained by expanding the factor $(1 + \tau^2)^{\alpha/2}$ by the binomial theorem and integrating term by term.

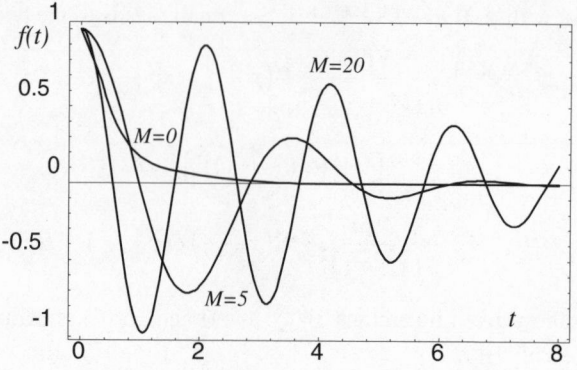

Fig. 5.3. Graph of $f(t)$ for different M.

Table 5.3. *Computation of $e^z E_1(z)$ when $z = 5$*

M	$S_M(z)$	$R_M(z)$	$S_M(z) + R_M(z)$	t^*
0	0	0.17042 21763	0.17042 21763	3.27
1	0.20000 00000	−0.02957 78237	0.17042 21763	3.54
2	0.16000 00000	0.01042 21763	0.17042 21763	3.86
3	0.17600 00000	−0.00557 78237	0.17042 21763	4.24
4	0.16640 00000	0.00402 21763	0.17042 21763	4.67
5	0.17408 00000	−0.00365 78237	0.17042 21763	5.14
6	0.16640 00000	0.00402 21763	0.17042 21763	5.67
7	0.17561 60000	−0.00519 38237	0.17042 21763	6.23
8	0.16271 36000	0.00770 85763	0.17042 21763	6.82
9	0.18335 74400	−0.01293 52637	0.17042 21763	7.45
10	0.14619 85280	0.02422 36483	0.17042 21763	8.10

slowly and oscillate more rapidly with *increasing M*. The remainder integral in (5.2.13) has been computed numerically using *Mathematica* for different values of M and z. We show in Table 5.3 the results of such computations for the particular case when $z = 5$. The second and third columns show the values of the finite sum on the right-hand side of (5.2.12) (which we denote by $S_M(z)$) and $R_M(z)$. In each case, the sum of these two quantities equals the value of $e^5 E_1(5)$ to ten decimal places. Optimal truncation is given by the value of M which corresponds to the least value of $|R_M(z)|$ and is seen to be $M = 5$. It is immediately apparent that it does not matter what particular value of M – whether sub- or post-optimal – is chosen to compute the right-hand side of (5.2.12). This is, of course, as it must be and is a consequence of our using 'exact' numerical evaluation of the remainder $R_M(z)$.

To give a measure of the rate of convergence of the integral in (5.2.13), we also show in the last column of Table 5.3 the value of t^* for which $|\Gamma(M + \frac{1}{2} + it)|/\cosh \pi t \leq 10^{-6}$ when $|t| \geq t^*$. It is seen that t^* increases as the truncation index M increases. Thus, *we conclude that there is no special significance with regard to the convergence of the remainder integral $R_M(z)$ at optimal truncation.*

Table 5.4. *Values of t^* for different M, θ and α*

	$M = 2$				$M = 4$			
	θ/π				θ/π			
α/π	0	0.5	1.0	1.5	0	0.5	1.0	1.5
0.5	3.86	6.07	13.10	∞	4.67	7.54	17.07	∞
0.6	3.68	5.35	9.14	22.68	4.32	6.40	11.19	28.11
0.7	3.95	5.39	8.05	14.01	4.51	6.27	9.54	16.76
	$M = 6$				$M = 8$			
	θ/π				θ/π			
α/π	0	0.5	1.0	1.5	0	0.5	1.0	1.5
0.5	5.67	9.28	21.61	∞	6.82	11.26	26.61	∞
0.6	5.11	7.63	13.47	33.79	6.03	9.01	15.94	39.64
0.7	5.21	7.28	11.17	19.63	6.03	8.41	12.91	22.60

Finally, we briefly consider complex z when the phase $0 \leq \theta < \frac{3}{2}\pi$; negative values of the phase can be dealt with in an analogous fashion. In this case, the remainder term becomes

$$R_M(z) = \frac{1}{2}(-)^M z^{-M-\frac{1}{2}} \int_{-\infty}^{\infty} \Gamma(M + \frac{1}{2} + it)|z|^{-it} \frac{e^{\theta t}}{\cosh \pi t} \, dt. \qquad (5.2.14)$$

The integral along $[0, \infty)$ now decays more slowly due to the presence of the factor $e^{\theta t}$; the decay along $(-\infty, 0]$, of course, is accelerated by the presence of this factor. The convergence of the integral over the upper half of the path can be ameliorated by allowing this part of the path to be inclined at an obtuse angle α, say, to the positive real axis (i.e., so that $s = M + \frac{1}{2} + te^{i\alpha}$, $t \geq 0$) to make use of the $\exp\{\mathrm{Re}(s) \log |s|\}$ decay† in $\mathrm{Re}(s) < 0$ contained in $\Gamma(s)$.

In Table 5.4 we show, for different truncation index M and inclination angle α, the values of t^* for which the quantity

$$\left| \Gamma(M + \tfrac{1}{2} + te^{i\alpha}) \frac{e^{\theta t \sin \alpha}}{\cos(\pi t e^{i\alpha})} \right| \leq 10^{-6}$$

when $t \geq t^*$. This gives an indication of the rate of convergence of the modified integral in (5.2.14) taken over the part of the path situated in $\mathrm{Im}(s) \geq 0$.

5.3 Saddle-Point Approximation of Integrals

The procedure outlined in §5.1 for the determination of algebraic expansions is standard. In the following sections, we show how the Mellin-Barnes approach can be extended to functions whose expansion is exponential or oscillatory in character.

---

† This would also enable us to extend the sector of validity beyond $\theta = \frac{3}{2}\pi$.

We begin this extension of the theory by first considering an approximation to such integrals. In some applications it is sufficient to have only an estimate for the leading behaviour of a function as a variable or parameter becomes large. One of the most commonly used techniques for the asymptotic evaluation of Laplace-type integrals is the saddle point method; see, for example, Copson (1965, Ch. 7); Olver (1974, p. 121); Wong (1989, p. 84). In this section we show, by means of examples, how the saddle point method can be applied to certain types of Mellin-Barnes integrals.

5.3.1 An Integral Due to Heading and Whipple

The first example we consider is an integral investigated by Heading & Whipple (1952) that arose in a study of the oblique reflection of long wavelength radio waves from the ionosphere. Let $I(z)$ denote the integral

$$I(z) = \frac{1}{2\pi i} \int_C z^{-s} \prod_{j=1}^{4} \Gamma(s + a_j)\, ds, \qquad (5.3.1)$$

where the a_j are arbitrary constants. The contour of integration C passes to the right (in the positive sense) of all the poles of the integrand situated at $s = -a_j - k$, $k = 0, 1, 2, \ldots$ $(1 \leq j \leq 4)$, with endpoints at infinity in the third and fourth quadrants of the s plane as shown in Fig. 5.4(a). From Rule 2 in §2.4, it is seen that the above integral defines the function $I(z)$ for *all complex values of* z, since we are taking advantage of the $\exp\{\mathrm{Re}(s)\log|s|\}$ decay in $\mathrm{Re}(s) < 0$ contained in the gamma functions. We remark in passing that, if the path in (5.3.1) were taken parallel to the imaginary s axis with $\mathrm{Re}(s) > \max_{1 \leq j \leq 4}\{-\mathrm{Re}\,(a_j)\}$, Rule 1 in §2.4 shows that the resulting integral would then define $I(z)$ only in the sector $|\arg z| < 2\pi$.

It is required to obtain the behaviour of $I(z)$ for large (complex) values of z. To apply the saddle point method to (5.3.1) we observe that, since there are no poles on the right of C, we are free to displace the contour as far to the right as we please (but with the endpoints at infinity still in $\mathrm{Re}(s) < 0$). Then, on the

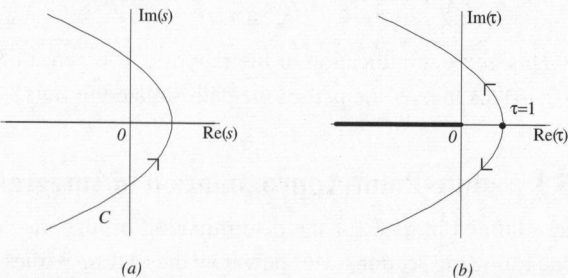

(a)   (b)

Fig. 5.4. (a) The path of integration C. (b) The path of steepest descent in the τ plane through the saddle point $\tau = 1$ when $\arg z = 0$. The τ plane is cut along the negative real axis and the arrows denote the direction of steepest descent.

new, displaced contour (which we still call C), we can approximate the gamma functions in the integrand by means of Stirling's formula (2.1.8) (since $|s|$ is now everywhere large on C)

$$\Gamma(s+a) \sim (2\pi)^{\frac{1}{2}} e^{-s} s^{s+a-\frac{1}{2}} \qquad (|s| \to \infty, \quad |\arg s| < \pi), \qquad (5.3.2)$$

to find

$$I(z) \sim \frac{1}{2\pi i} \int_C (2\pi)^2 s^{\vartheta-2} \exp\{4s \log s - 4s - s \log z\} \, ds,$$

where $\vartheta = \sum_{j=1}^4 a_j$. Introduction of the new variable $s = z^{\frac{1}{4}} \tau$ then shows that $I(z)$ can be cast in the standard form of a generalised Laplace integral

$$I(z) \sim (2\pi)^2 z^{\frac{1}{4}\vartheta-\frac{1}{4}} \frac{1}{2\pi i} \int_{C'} \tau^{\vartheta-2} \exp\{4|z|^{\frac{1}{4}} f(\tau)\} \, d\tau,$$

where

$$f(\tau) = (\tau \log \tau - \tau) e^{i\phi/4},$$

$\phi = \arg z$ and C' denotes the map of the loop C in the τ plane (which corresponds to a rotation of C through an angle $-\phi/4$).

As $z \to \infty$, the exponential factor in the integrand possesses a saddle point at $\tau = 1$, where $f'(\tau) = 0$. The path of steepest descent through the saddle, described by

$$\text{Im}(f(\tau) - f(1)) = 0, \qquad \text{Re}(f(\tau) - f(1)) \leq 0,$$

is illustrated in Fig. 5.4(b) for $\arg z = 0$. It is readily shown that the directions of this path at $\tau = 1$ are given by $\arg(\tau-1) = \pm\frac{1}{2}\pi - \frac{1}{8} \arg z$. For complex values of z, the path of steepest descent through $\tau = 1$ is found to be the topological equivalent (rotated through an angle $-\phi/4$) of that for $\arg z = 0$, provided $|\arg z| < 4\pi$, so that the contour C' can always be deformed to coincide with the path of steepest descent for this range of $\arg z$. Straightforward application of the saddle point approximation then yields

$$I(z) \sim \frac{(2\pi)^2 z^{\frac{1}{4}\vartheta-\frac{1}{4}}}{2\pi i} \exp\{4|z|^{\frac{1}{4}} f(1)\} \sqrt{\frac{2\pi}{4|z|^{\frac{1}{4}}|f''(1)|}} e^{i(\frac{1}{2}\pi-\frac{1}{8}\phi)}$$

$$= 2^{\frac{1}{2}} \pi^{\frac{3}{2}} z^{\frac{1}{4}\vartheta-\frac{3}{8}} \exp\{-4z^{\frac{1}{4}}\} \qquad (5.3.3)$$

as $|z| \to \infty$ in $|\arg z| < 4\pi$, which is the result† obtained by Heading & Whipple (1952).

---

† The extension of this result to the situation where the integrand contains the product of $p \ (\geq 1)$ gamma functions is obvious and we find

$$I(z) \sim (2\pi)^{(p-1)/2} p^{-\frac{1}{2}} z^{(\vartheta+\frac{1}{2}-\frac{1}{2}p)/p} \exp\{-pz^{1/p}\}$$

as $|z| \to \infty$ in $|\arg z| < p\pi$, where now $\vartheta = \sum_{j=1}^p a_j$.

5.3.2 The Bessel Function $J_n(nx)$

Another example of the saddle point approximation applied to Mellin-Barnes integrals is furnished by the familiar integral representation for the Bessel function $J_n(nx)$ given by [cf. Watson (1966, p. 192) and (3.4.22)]

$$J_n(nx) = \frac{1}{2\pi i} \int_C \frac{\Gamma(s)}{\Gamma(n+1-s)} \left(\tfrac{1}{2}nx\right)^{n-2s} ds, \qquad (5.3.4)$$

where, as in the previous example, C denotes a loop that encircles the poles of $\Gamma(s)$ in the positive sense and passes to infinity in the third and fourth quadrants of the s plane. Although this representation holds for arbitrary complex‡ x and n, we shall only consider $x > 0$ and suppose that the order n (not necessarily an integer) is positive and large.

Since there are no poles of the integrand in $\mathrm{Re}(s) > 0$ we can expand the contour C to the right so that $|s|$ is large everywhere on C. With the new variable $s = n\tau$, the integral (5.3.4) can be written as

$$J_n(nx) = \frac{1}{2\pi i} \int_C g(\tau) \exp\{nf(\tau)\} \, d\tau,$$

where

$$g(\tau) = \{\tau(1-\tau)\}^{-\frac{1}{2}}$$

and C now denotes a similar loop in the τ plane. The function $f(\tau)$ is given by

$$f(\tau) = \frac{1}{n} \log \left\{ \frac{\Gamma(n\tau)}{(1-\tau)g(\tau)\Gamma(n-n\tau)} \right\} + (1-2\tau) \log \left(\tfrac{1}{2}nx\right)$$

$$\sim \tau \log \tau - (1-\tau) \log (1-\tau) + (1-2\tau)\{1 + \log \left(\tfrac{1}{2}x\right)\}$$

as $n \to \infty$, provided τ is not in the neighbourhoods of the branch cuts $(-\infty, 0]$ and $[1, +\infty)$ in the τ plane where Stirling's approximation ceases to be valid. To leading order, the exponential factor in the above integral has saddles at the points given by

$$\log \tau + \log (1-\tau) - \log \left(\tfrac{1}{4}x^2\right) = 0;$$

that is, at the points

$$\tau_{1,2} = \tfrac{1}{2}\left(1 \mp \sqrt{1-x^2}\right).$$

We note that for $x > 0$, these saddle points are bounded away from $\tau = 0$ and $\tau = 1$.

The paths of steepest descent associated with the function $f(\tau)$ are illustrated in Fig. 5.5. When $0 < x < 1$, the saddle points (which we label P_1 and P_2,

---

‡ When C is the path $(-\infty i, \infty i)$ (indented at $s = 0$), then (5.3.4) defines $J_n(nx)$ only for positive values of x and n, since the integrand on the contour is $O(|s|^{-n-1})$ as $|s| \to \infty$; see the discussion surrounding (3.4.21).

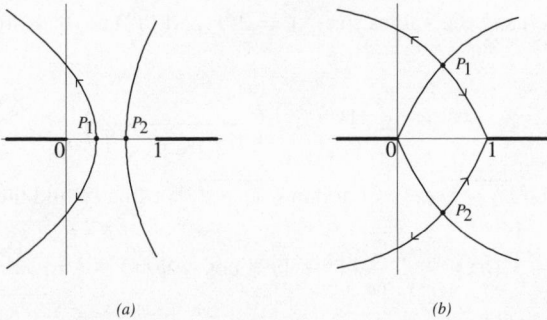

(a)   (b)

Fig. 5.5. The paths of steepest descent and ascent through the saddles P_1 and P_2 when (a) $0 < x < 1$ and (b) $x > 1$. The arrows denote the direction of steepest descent. The τ plane is cut along the negative real axis and along the positive real axis from $\tau \geq 1$.

respectively) are situated on the real axis in the interval $(0, 1)$. The path of steepest descent through P_1 passes to infinity in the third and fourth quadrants (with the directions at the saddle given by $\arg(\tau - \tau_1) = \pm\frac{1}{2}\pi$), while that through P_2 coincides with the interval $(\tau_1, 1)$ of the real axis. The integration contour C can therefore be reconciled with the steepest descent path through P_1. Upon noting that $g(\tau_1) = 2/x$, $f''(\tau_1) \simeq 4(1 - x^2)^{\frac{1}{2}}/x^2$ and

$$e^{nf(\tau_1)} \simeq \left(\frac{x}{1 + \sqrt{1 - x^2}}\right)^n e^{n\sqrt{1-x^2}},$$

we find

$$J_n(nx) \sim \frac{g(\tau_1)}{2\pi i} e^{nf(\tau_1)} \sqrt{\frac{2\pi}{n|f''(\tau_1)|}} e^{\frac{1}{2}\pi i}$$

$$\sim \frac{1}{\sqrt{2\pi n}} \frac{e^{n\sqrt{1-x^2}}}{(1 - x^2)^{\frac{1}{4}}} \left(\frac{x}{1 + \sqrt{1 - x^2}}\right)^n \qquad (5.3.5)$$

as $n \to \infty$ when $0 < x < 1$.

When $x > 1$, the saddles P_1 and P_2 move off the real axis into the complex plane along the line $\mathrm{Re}(\tau) = \frac{1}{2}$. The paths of steepest descent through P_1 and P_2 (with the directions at the saddles points given by $\arg(\tau - \tau_1) = \pm\frac{1}{2}\pi + \frac{1}{4}\pi$ and $\arg(\tau - \tau_2) = \mp\frac{1}{2}\pi - \frac{1}{4}\pi$) now both terminate at the branch point $\tau = 1$ and pass to infinity in the third and fourth quadrants, respectively; see Fig. 5.5(b). In this case, the contour C can be deformed to pass over both saddles and accordingly we obtain

$$J_n(nx) \sim \frac{1}{2\pi i} \sqrt{\frac{2\pi}{n}} \left\{ g(\tau_1) \frac{e^{nf(\tau_1) + \frac{3}{4}\pi i}}{\sqrt{|f''(\tau_1)|}} + g(\tau_2) \frac{e^{nf(\tau_2) + \frac{1}{4}\pi i}}{\sqrt{|f''(\tau_2)|}} \right\}$$

$$\sim \frac{(x^2 - 1)^{-\frac{1}{4}}}{\sqrt{2\pi n}} \left\{ e^{nf(\tau_1) + \frac{1}{4}\pi i} + e^{nf(\tau_2) - \frac{1}{4}\pi i} \right\},$$

where we have used the values $g(\tau_{1,2}) = 2/x$ and $|f''(\tau_{1,2})| \simeq 4(x^2 - 1)^{\frac{1}{2}}/x^2$. Then, since

$$e^{nf(\tau_{1,2})} \simeq e^{\mp in\sqrt{x^2-1}} \left(\frac{x}{1 \mp i\sqrt{x^2 - 1}}\right)^n$$

and $\log\{(1 + i\sqrt{x^2 - 1})/x\} = i \arctan\sqrt{x^2 - 1}$, we finally find the result

$$J_n(nx) \sim \sqrt{\frac{2}{\pi n}}(x^2 - 1)^{-\frac{1}{4}} \cos\left\{n\Psi(x) - \tfrac{1}{4}\pi\right\}, \tag{5.3.6}$$

$$\Psi(x) = \sqrt{x^2 - 1} - \arctan\sqrt{x^2 - 1}$$

as $n \to \infty$ when $x > 1$.

The asymptotic expressions in (5.3.5) and (5.3.6) are the well-known leading approximations for the Bessel function $J_n(nx)$ [see Watson (1966, pp. 227, 234)]. In the particular case $x = 1$, the saddles P_1 and P_2 form a double saddle at $\tau = \frac{1}{2}$ and an approximation for $J_n(n)$ can be derived in an analogous manner. Similarly, uniform approximations in terms of the Airy function valid in the neighbourhood of $x = 1$ can also, if desired, be derived from (5.3.4) by standard techniques; see, for example, Olver (1974, p. 351) or Wong (1989, p. 366). We do not include these details here.

5.3.3 A Gauss Hypergeometric Function

The final example we consider concerns the evaluation of the Gauss hypergeometric function

$$_2F_1(a + cn, 1 - b + c'n; n + 1; x/(1 + x)) \tag{5.3.7}$$

for large positive values of n, where $c + c' = 1$ and $x > 0$. A standard transformation [Abramowitz & Stegun (1965, p. 559, Eq. (15.3.4))] shows that the above hypergeometric function can be written alternatively as

$$(1 + x)^{a+cn} \,_2F_1(a + cn, b + cn; n + 1; -x),$$

which will prove to be a more convenient form when using the Mellin-Barnes approach. The particular case $c = \frac{1}{2}$ is of historical interest since it was this problem, discussed in a fragmentary (posthumous) manuscript by Riemann (1863), which led to the introduction of the saddle point method in the complex plane.† A more modern application of (5.3.7), described by Lighthill (1947) and Cherry (1950), has arisen in aerodynamics where the parameter $c = 1 + \alpha$, $\alpha > 0$ and, to leading order in powers of n^{-1}, a and b are constants independent of n.

---

† It should be pointed out that, in an earlier paper, Stokes (1850) also effectively used the saddle point technique in the complex plane in his asymptotic investigation of a certain integral (now called the Airy function) arising in the theory of the rainbow. For details, see the review paper Paris (1996).

We consider the function

$$G(n; x) = x^{n/2} {}_2F_1(a + cn, b + cn; n + 1; -x)$$

for $n \to \infty$, where, for simplicity, we shall suppose that $c > 0$ with a and b being arbitrary finite parameters. From the representation of the hypergeometric function as a Mellin-Barnes integral [see (3.4.8)], we obtain

$$G(n; x) = \frac{A'(n)}{2\pi i} \int_C \frac{\Gamma(a + cn - s)\Gamma(b + cn - s)}{\Gamma(n + 1 - s)} \Gamma(s) x^{\frac{1}{2}n - s} \, ds, \qquad (5.3.8)$$

where

$$A'(n) = \frac{\Gamma(n + 1)}{\Gamma(a + cn)\Gamma(b + cn)}$$

and C is again a loop enclosing the poles of $\Gamma(s)$ and with endpoints at infinity in the third and fourth quadrants of the s plane. By Rule 2 in §2.4, the integral in (5.3.8) converges when $|x| < 1$. The integrand has additional sequences of poles[†] at $s = cn + a + k$ and $s = cn + b + k, k = 0, 1, 2, \ldots$. Thus, as $n \to \infty$, we are free to deform C far to the right, but not over any of these latter poles.

Introduction of the new variable $s = n\tau$ then enables us to cast (5.3.8) in the form

$$G(n; x) = \frac{A(n)}{2\pi i} \int_C g(\tau) \exp\{nf(\tau)\} \, d\tau, \qquad (5.3.9)$$

where C now denotes a similar loop in the τ plane,

$$g(\tau) = \frac{(c - \tau)^{a+b-1}}{\{\tau(1 - \tau)\}^{\frac{1}{2}}},$$

$$f(\tau) = \frac{1}{n} \log \left\{ \frac{\Gamma(n\tau)\Gamma(a + cn - n\tau)\Gamma(b + cn - n\tau)}{2\pi(1 - \tau)g(\tau)\Gamma(n - n\tau)n^{a+b-1}} \right\}$$

$$+ \left(\tfrac{1}{2} - \tau\right) \log x + (2c - 1)(1 - \log n)$$

and

$$A(n) = 2\pi A'(n) n^{a+b-1+(2c-1)n} e^{(1-2c)n}$$

$$\sim \sqrt{2\pi n} \, c^{1-a-b-2cn} \qquad (n \to \infty).$$

Upon application of Stirling's approximation in (5.3.2), we find that

$$f(\tau) \sim \tau \log \tau - (1 - \tau) \log(1 - \tau) + 2(c - \tau) \log(c - \tau) + \left(\tfrac{1}{2} - \tau\right) \log x$$

as $n \to \infty$, provided τ is not in the neighbourhoods of the branch cuts $(-\infty, 0]$, $[1, +\infty)$ and the cuts parallel to the positive real axis emanating from the

---

[†] In the case of finite n, these poles would yield the algebraic expansion of $G(n; x)$ as $x \to \infty$.

points $c + (a/n)$ and $c + (b/n)$. Then, to leading order, the exponential factor in (5.3.9) has saddles at the points given by

$$\log\{\tau(1 - \tau)\} - 2\log(c - \tau) - \log x = 0;$$

that is, by the roots of the quadratic $\tau(1 - \tau)/(c - \tau)^2 = x$, whence

$$\tau_{1,2} = \frac{1}{1 + x}\left\{cx + \tfrac{1}{2} \mp \sqrt{c(1 - c)x + \tfrac{1}{4}}\right\}. \tag{5.3.10}$$

We consider only Riemann's problem with $c = \tfrac{1}{2}$. In this case the saddles are given by

$$\tau_{1,2} = \tfrac{1}{2}\left(1 \mp \frac{1}{\sqrt{1 + x}}\right).$$

For $x > 0$, we have $0 < \tau_1 < \tfrac{1}{2}$ and $\tfrac{1}{2} < \tau_2 < 1$; hence τ_1 is bounded away from the branch cuts, whereas τ_2 lies on the cut emanating from $c = \tfrac{1}{2}$. The path of steepest descent through τ_1 is similar to that illustrated in Fig. 5.4(b) (where the directions at τ_1 are given by $\arg(\tau - \tau_1) = \pm\tfrac{1}{2}\pi$) and so the contour C can be deformed to pass through τ_1. Then, using the values

$$g(\tau_1) = 2^{2-a-b}x^{-\frac{1}{2}}(1 + x)^{1-\frac{1}{2}a-\frac{1}{2}b}, \quad f''(\tau_1) \simeq 4(1 + x)^{\frac{3}{2}}/x$$

and

$$f(\tau_1) \simeq \tfrac{1}{2}\log\left(\frac{\sqrt{1 + x} - 1}{\sqrt{1 + x} + 1}\right),$$

we arrive at the approximation

$$G(n; x) \sim \frac{A(n)}{2\pi i}g(\tau_1)e^{nf(\tau_1)}\sqrt{\frac{2\pi}{n|f''(\tau_1)|}}\,e^{\frac{1}{2}\pi i}$$

$$\sim 2^n(1 + x)^{\frac{1}{4}-\frac{1}{2}a-\frac{1}{2}b}\left(\frac{\sqrt{1 + x} - 1}{\sqrt{1 + x} + 1}\right)^{n/2} \tag{5.3.11}$$

as $n \to \infty$. This is essentially the result obtained by Riemann, although he employed an Euler integral representation for the hypergeometric function.

From (5.3.10) it is easily shown that, when $0 < c < 1$, only the saddle point at τ_1 is significant (as in Riemann's case), the other saddle at τ_2 lying on the branch cut emanating from c. When $c > 1$, however, both saddles are situated in the domain where Stirling's approximation is valid. In particular, the saddles coalesce when $x = x_c = 1/\{4c(c - 1)\}$ at the point

$$\tau_c = \frac{c(2c - 1)}{1 + 4c(c - 1)}.$$

The situation is then similar to that depicted in Fig. 5.5. For $0 < x < x_c$, only the saddle τ_1 contributes to the leading behaviour of $G(n; x)$, while for $x_c < x < \infty$, the saddles τ_1 and τ_2 move off the real axis into the complex plane and both contribute. In the aerodynamic problem mentioned above, the parameter $c > 1$ and the ranges $0 < x < x_c$ and $x_c < x < \infty$ correspond to subsonic and supersonic flows, respectively. We do not discuss the details of this problem any further: asymptotic expansions for $G(n; x)$ (based on an Euler integral for the hypergeometric function) are given in Cherry (1950), who also derived uniform asymptotic expansions in terms of Bessel functions valid in the neighbourhood of $x = x_c$.

A general and detailed treatment of the expansion of the Gauss hypergeometric function for large parameters can be found in Watson (1918); see also Luke (1969, Ch. 7).

5.4 Exponential Asymptotic Expansions

The standard Mellin-Barnes integral representation defining the function $f(z)$ in some sector S is given by (5.1.1). As we saw in §5.1, suitable displacement of the path of integration over a subset of the poles of $g(s)$ then produces an asymptotic expansion for $f(z)$ valid in the Poincaré sense as $|z| \to \infty$ in S. This procedure, however, can only produce *algebraic*-type expansions: its limitation to the sector S is, in general, a consequence of the existence of one or more subdominant exponential expansions which 'switch' on across Stokes lines situated in the interior of S. For large $|z|$ outside this sector, these exponential expansions then form the dominant asymptotic expansion of $f(z)$.

In this section we consider the situation arising when either the subset of poles is empty, so that displacement of the path of integration does not produce any useful asymptotic information† as $|z| \to \infty$ in S, or we wish to extend the range of validity of the expansion of $f(z)$ beyond the algebraic sector S. When confronted with the determination of the asymptotic expansion of $f(z)$ in these cases it is usual to turn to alternative methods (e.g., the method of steepest descent or a differential equation approach). We show‡ how such exponential expansions can be obtained by a slight, but very important, modification to the integral (5.1.1), which consists essentially of contour deformation combined with use of an inverse factorial expansion of $g(s)$. This is carried out for the exponential integral and the familiar Bessel and Weber parabolic cylinder functions as illustrative examples. We also consider the same procedure applied to a function defined by an infinite sum and, in §5.5, to Faxén's integral where the construction of the asymptotic expansion is more elaborate.

---

† This shows that $f(z)$ tends to zero more rapidly than any algebraic power of z in the sector S. Generally speaking, $f(z)$ will then possess an exponentially small expansion as $|z| \to \infty$ in S.

‡ This method is also described in Riekstiņš (1977, pp. 367–371).

5.4.1 The Exponential Integral $E_1(z)$

In §5.1.1 the behaviour of the exponential integral $E_1(z)$ for $|z| \to \infty$ in the sector $|\arg z| < \frac{3}{2}\pi$ was shown in (5.1.7) to be given by an exponential expansion (containing the factor e^{-z}). To derive this expansion, we considered the function $e^z E_1(z)$ which, since it has an algebraic character in this sector, is then amenable to the standard procedure of evaluation of Mellin-Barnes integrals by path displacement over a certain subset of the poles of the integrand. In the present example, we reconsider the derivation of the expansion of $E_1(z)$ without the device of the inclusion of the factor e^z.

The Mellin transform of $E_1(x)$ for $x > 0$ is given by

$$
\int_0^\infty x^{s-1} \left\{ \int_1^\infty \frac{e^{-xt}}{t} \, dt \right\} dx = \int_1^\infty t^{-1} \left\{ \int_0^\infty x^{s-1} e^{-xt} \, dx \right\} dt
$$

$$
= \Gamma(s) \int_1^\infty t^{-s-1} \, dt = \frac{\Gamma(s)}{s} \quad (\mathrm{Re}(s) > 0),
$$

so that by the Mellin inversion formula (3.1.5)

$$
E_1(z) = \frac{1}{2\pi i} \int_{c-\infty i}^{c+\infty i} z^{-s} \Gamma(s) \frac{ds}{s} \quad (c > 0). \tag{5.4.1}
$$

The sector of convergence for this integral is $|\arg z| < \frac{1}{2}\pi$ (see Rule 1 in §2.4): we note that this is considerably less than the sector in (5.1.7). The integrand has a double pole at $s = 0$ (with residue $-\gamma - \log z$) and a sequence of simple poles at $s = -1, -2, \ldots$. Since the integrand has the controlling behaviour $O(|t|^{\sigma - \frac{3}{2}} e^{\theta t - \frac{1}{2}\pi |t|})$ as $t \to \pm\infty$ (with $s = \sigma + it$ and $\theta = \arg z$), we can displace the integration path to the left to obtain the convergent expansion

$$
E_1(z) + \log z = -\gamma - \sum_{n=1}^\infty \frac{(-)^n z^n}{n \, n!}.
$$

This result holds for all values of $\arg z$ by analytic continuation and is equivalent to (5.1.4).

On the right of the path of integration the integrand is holomorphic. Consequently, displacement of the contour into $\mathrm{Re}(s) > 0$ yields no asymptotic information: this is the signal that the expansion of $E_1(z)$ is exponentially small in $|\arg z| < \frac{1}{2}\pi$ (a fact we already know, of course, from (5.1.7)). To show how we can obtain the exponentially small expansion from (5.4.1), we first displace the path of integration as far to the right as we please so that $|s|$ is everywhere large on the new path C. Then, writing the integrand as

$$
z^{-s} \frac{\Gamma(s)}{s} = z^{-s} \frac{\Gamma^2(s)}{\Gamma(s+1)},
$$

we employ the inverse factorial expansion in (2.2.38)

$$\frac{\Gamma^2(s)}{\Gamma(s+1)} = \sum_{j=0}^{M-1} (-)^j A_j \Gamma(s-1-j) + \rho_M(s) \Gamma(s-1-M) \quad (M=1,2,\dots),$$

where $\rho_M(s) = O(1)$ as $|s| \to \infty$ in $|\arg s| \leq \pi - \epsilon, \epsilon > 0$ and, from (2.2.40), the coefficients $A_j = j!$.

We now substitute this expansion into the integrand of (5.4.1) and integrate term by term, where we suppose that C lies to the right of the point $s = M + 1$. Then we obtain

$$E_1(z) = \sum_{j=0}^{M-1} (-)^j j! \frac{1}{2\pi i} \int_C \Gamma(s-1-j) z^{-s}\, ds + R_M(z)$$

$$= e^{-z} \sum_{j=0}^{M-1} \frac{(-)^j j!}{z^{j+1}} + R_M(z),$$

where we have made use of the result [see §3.3.1]

$$z^{-\alpha} e^{-z} = \frac{1}{2\pi i} \int_C \Gamma(s-\alpha) z^{-s}\, ds \tag{5.4.2}$$

which is valid (i) for $|\arg z| < \frac{1}{2}\pi$ when C is the vertical line $\mathrm{Re}(s) = c > \mathrm{Re}(\alpha)$ and (ii) for all $\arg z$ when C is a loop encircling the point $s = \alpha$ (in the positive sense) with endpoints at infinity in $\mathrm{Re}(s) < 0$. The remainder $R_M(z)$ is given by

$$R_M(z) = \frac{1}{2\pi i} \int_C \rho_M(s) \Gamma(s-1-M) z^{-s}\, ds.$$

Estimates for remainder integrals of this type are discussed in §2.5, where, from Lemma 2.7 and the assumption that $\mathrm{Re}(s) > M + 1$ on C, it follows that

$$|R_M(z)| = O(|z|^{-M-1} e^{-z})$$

as $|z| \to \infty$ in the sector $|\arg z| < \frac{1}{2}\pi$.

Hence, we again obtain the asymptotic expansion

$$E_1(z) \sim e^{-z} \sum_{k=0}^{\infty} \frac{(-)^k k!}{z^{k+1}} \tag{5.4.3}$$

valid as $|z| \to \infty$ in $|\arg z| < \frac{1}{2}\pi$. We remark that the sector of validity of this result is considerably less than that in (5.1.7). It is possible to show that (5.4.3) holds in a wider sector through use of a path in (5.4.1) which, instead of being the vertical line $\mathrm{Re}(s) = c$, is a contour bent backwards to have endpoints at infinity in $\mathrm{Re}(s) < 0$; we do not discuss this modification here.

5.4.2 The Bessel Function $J_\nu(z)$

The second example we consider of the determination of the exponential expansion from a Mellin-Barnes integral is the Bessel function $J_\nu(z)$ for finite ν as $|z| \to \infty$ in $|\arg z| < \frac{1}{2}\pi$. From (3.4.22) we have the Mellin-Barnes integral representation

$$J_\nu(z) = \frac{1}{2\pi i} \int_C \frac{\Gamma(s)}{\Gamma(1 + \nu - s)} \left(\tfrac{1}{2}z\right)^{\nu - 2s} ds, \qquad (5.4.4)$$

where C is a loop that encircles the poles of $\Gamma(s)$ (in the positive sense) with endpoints at infinity in $\mathrm{Re}(s) < 0$ (cf. Fig. 3.1). This integral defines† $J_\nu(z)$ *without restriction* on $\arg z$ and ν. Since there are no poles of the integrand in $\mathrm{Re}(s) > 0$, displacement of the loop to the right yields no useful asymptotic information.

To proceed with the asymptotic expansion of (5.4.4) for large $|z|$, we first employ the reflection formula for the gamma function and write

$$J_\nu(z) = \frac{1}{2\pi i} \left\{ e^{-\frac{1}{2}\pi i \nu} I_+(z) - e^{\frac{1}{2}\pi i \nu} I_-(z) \right\},$$

where

$$I_\pm(z) = \frac{1}{2\pi i} \int_C \Gamma(s)\Gamma(s - \nu) \left(\tfrac{1}{2}z e^{\mp\frac{1}{2}\pi i}\right)^{\nu - 2s} ds.$$

Since there are no poles of the integrand to the right of C, we may expand the loop as far to the right as we please such that $|s|$ is everywhere large on C. In particular, we shall suppose that C encloses the point

$$s = \tfrac{1}{2}\nu + \tfrac{1}{4} + \tfrac{1}{2}M \qquad (5.4.5)$$

for positive integer M. On the expanded loop, we now employ the inverse factorial expansion, given in (2.2.33), for the product of gamma functions appearing in the integrand

$$\Gamma(s)\Gamma(s - \nu) = (2\pi)^{\frac{1}{2}} 2^{1 + \nu - 2s} \left\{ \sum_{j=0}^{M-1} (-)^j c_j \Gamma\left(2s - \nu - \tfrac{1}{2} - j\right) \right.$$

$$\left. + \rho_M(s) \Gamma\left(2s - \nu - \tfrac{1}{2} - M\right) \right\},$$

where $\rho_M(s) = O(1)$ as $|s| \to \infty$ in $|\arg s| \leq \pi - \epsilon$, $\epsilon > 0$ and, from (2.2.35), the coefficients c_j are given by

$$c_j = \frac{(-)^j}{j! 8^j} \prod_{r=1}^{j} \left\{ 4\nu^2 - (2r - 1)^2 \right\}.$$

---

† We observe that if we straighten the contour C to coincide with the imaginary axis (indented at $s = 0$), as in (3.4.21), the resulting integral then defines $J_\nu(z)$ only for $z > 0$ and $\mathrm{Re}(\nu) > 0$.

Substitution of this expansion into the integrand of $I_\pm(z)$, followed by term-by-term integration, then enables us to write

$$I_\pm(z) = (2\pi)^{\frac{1}{2}} \left\{ 2 \sum_{j=0}^{M-1} (-)^j c_j \frac{1}{2\pi i} \int_C \Gamma\left(2s - \nu - \tfrac{1}{2} - j\right) \right.$$

$$\left. \times \left(e^{\mp\frac{1}{2}\pi i} z\right)^{\nu-2s} ds + R_M^\pm(z) \right\} \qquad (5.4.6)$$

$$= (2\pi)^{\frac{1}{2}} \left\{ \sum_{j=0}^{M-1} \frac{(-)^j c_j}{(\mp i z)^{j+\frac{1}{2}}} e^{\pm i z} + R_M^\pm(z) \right\},$$

where we have made use of the result (5.4.2) to evaluate the integrals appearing in the sum over j.

The remainder integrals $R_M^\pm(z)$ are given by

$$R_M^\pm(z) = \frac{1}{\pi i} \int_C \rho_M(s) \Gamma\left(2s - \nu - \tfrac{1}{2} - M\right) \left(z e^{\mp\frac{1}{2}\pi i}\right)^{\nu-2s} ds.$$

From Lemma 2.8 and (5.4.5), it follows that

$$\left| R_M^\pm(z) \right| = O\left(|z|^{-M-\frac{1}{2}} e^{\pm i z}\right)$$

as $|z| \to \infty$ in the sectors $\left| \arg(e^{\mp\frac{1}{2}\pi i} z) \right| < \pi$, respectively. Since the common sector of validity for these bounds is $|\arg z| < \frac{1}{2}\pi$, we finally obtain the asymptotic expansion

$$J_\nu(z) \sim \sqrt{\frac{2}{\pi z}} \sum_{j=0}^{\infty} \frac{(-)^j c_j}{z^j} \cos\left(z + \tfrac{1}{2}\pi j - \tfrac{1}{2}\pi \nu - \tfrac{1}{4}\pi\right) \qquad (5.4.7)$$

valid as $|z| \to \infty$ in $|\arg z| < \frac{1}{2}\pi$.

We can write this expansion in its more familiar form [Abramowitz & Stegun (1965, p. 364)], by separating the above sum into sums over even and odd j, to find

$$J_\nu(z) = \sqrt{\frac{2}{\pi z}} \left\{ P(\nu, z) \cos \chi - Q(\nu, z) \sin \chi \right\},$$

where $\chi = z - \frac{1}{2}\pi\nu - \frac{1}{4}\pi$ and

$$P(\nu, z) \sim \sum_{k=0}^{\infty} \frac{(-)^k}{(2k)!} (8z)^{-2k} \prod_{r=1}^{2k} \{4\nu^2 - (2r-1)^2\},$$

$$Q(\nu, z) \sim \sum_{k=0}^{\infty} \frac{(-)^k}{(2k+1)!} (8z)^{-2k-1} \prod_{r=1}^{2k+1} \{4\nu^2 - (2r-1)^2\}.$$

The expansion (5.4.7) is limited to the sector $|\arg z| < \frac{1}{2}\pi$ because of the presence of Stokes lines on the rays $\arg z = \pm\frac{1}{2}\pi$, where the absolute value of the coefficients multiplying the exponential factors $\exp\{\pm iz\}$, respectively undergo a smooth but rapid decrease to zero (see Chapter 6). In the usual Poincaré sense, however, (5.4.7) holds in the extended sector $|\arg z| < \pi$.

5.4.3 The Parabolic Cylinder Function $D_\nu(z)$

To illustrate how asymptotic information can be obtained from an integral of type (5.1.1) when $\arg z$ lies outside the algebraic sector S, we return to consideration of the Weber parabolic cylinder function $D_\nu(z)$ discussed in §5.1.2. We recall that, for $\nu \neq 0, 1, 2, \ldots$, this function has the Mellin-Barnes integral representation

$$D_\nu(z) = \frac{z^\nu e^{-z^2/4}}{\Gamma(-\nu)} I(z), \qquad (5.4.8)$$

where

$$I(z) = \frac{1}{2\pi i} \int_{-\infty i}^{\infty i} \Gamma(-s)\Gamma(2s - \nu)(2z^2)^{-s}\, ds \qquad (5.4.9)$$

and the path of integration is indented to separate the poles of $\Gamma(-s)$ from those of $\Gamma(2s - \nu)$. Application of Rule 1 in §2.4 shows that $I(z)$ is defined by (5.4.9) in the sector S given by $|\arg z| < \frac{3}{4}\pi$. Displacement of the contour in (5.4.9) to the right over the first M poles of $\Gamma(-s)$ then yields the algebraic expansion

$$I(z) = \sum_{k=0}^{M-1} \frac{(-)^k}{k!}\Gamma(2k - \nu)(2z^2)^{-k} + R_M(z), \qquad (5.4.10)$$

where it was shown at (5.1.12) that $|R_M(z)| = O(|z|^{-2M+1})$ as $|z| \to \infty$ in S. The expansion of $I(z)$ in (5.4.10) is limited to the sector $|\arg z| < \frac{3}{4}\pi$ because of the appearance of a single subdominant exponential expansion across the Stokes lines on $\arg z = \pm\frac{1}{2}\pi$. This latter expansion becomes the dominant expansion in the complementary sector $|\arg(-z)| < \frac{1}{4}\pi$.

To determine the behaviour of $I(z)$ outside the sector S, we first bend back the contour of integration into a loop C, with endpoints at infinity in the directions $\pm\theta_0$, where $\frac{1}{2}\pi < \theta_0 \leq \pi$; see Fig. 5.6. From Stirling's formula (2.1.8) with $s = Re^{i\theta}$, the modulus of the logarithm of the integrand on C behaves like $R\cos\theta_0 \log R + O(R) \to -\infty$ as $R \to \infty$. Accordingly, we find that

$$I(z) = \frac{1}{2\pi i} \int_C \Gamma(-s)\Gamma(2s - \nu)(2z^2)^{-s}\, ds \qquad (5.4.11)$$

without restriction on $\arg z$; see also Rule 2 in §2.4. It is clear that since (5.4.11) defines $I(z)$ completely, it must contain the asymptotic information we seek.

Then we substitute the identity

$$1 = e^{\mp 2\pi i s}(1 \pm 2i e^{\pm\pi i s} \sin \pi s)$$

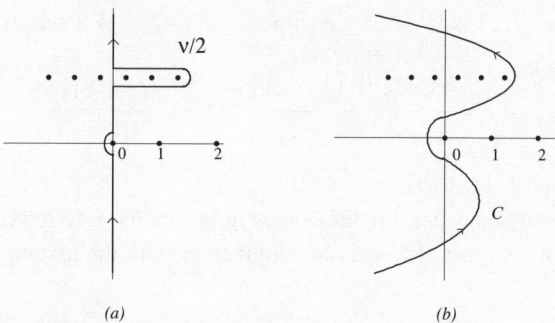

Fig. 5.6. (*a*) The path of integration in (5.4.9) when $\text{Re}(\nu) > 0$ and (*b*) the contour C indented to separate poles of $\Gamma(-s)$ and $\Gamma(2s - \nu)$.

into (5.4.11) to find

$$
I(z) = \frac{1}{2\pi i} \int_C \Gamma(-s)\Gamma(2s - \nu)(2z^2)^{-s} e^{\mp 2\pi i s}\{1 \pm 2i e^{\pm\pi i s} \sin \pi s\}\, ds
$$

$$
= I(ze^{\pm\pi i}) \mp \int_C \frac{\Gamma(2s - \nu)}{\Gamma(s + 1)}(2z^2 e^{\pm\pi i})^{-s}\, ds.
$$

If we define the function $E(z)$ by

$$
E(z) = \int_C \frac{\Gamma(2s - \nu)}{\Gamma(s + 1)}(2z^2)^{-s}\, ds, \tag{5.4.12}
$$

where C embraces the poles of the integrand, we obtain the connection formulas in the form

$$
I(z) = I(ze^{\pm\pi i}) \mp E(ze^{\pm\frac{1}{2}\pi i}). \tag{5.4.13}
$$

We remark that the above decomposition of $I(z)$ has resulted in an integral $E(z)$ whose integrand has no poles to the right of the contour C. It then follows that we can expand the contour C (with endpoints still in $\text{Re}(s) < 0$) in (5.4.12) so that $|s|$ is everywhere large and such that C encloses the point

$$
s = \nu + \tfrac{1}{2} + M \tag{5.4.14}
$$

for positive integer M. On the expanded contour, we now employ the inverse factorial expansion in (2.2.38)

$$
\frac{\Gamma(2s - \nu)}{\Gamma(s + 1)} = \frac{2^{2s-\nu-1}}{\sqrt{\pi}} \frac{\Gamma\left(s - \tfrac{1}{2}\nu\right)\Gamma\left(s - \tfrac{1}{2}\nu + \tfrac{1}{2}\right)}{\Gamma(s + 1)}
$$

$$
= \frac{2^{2s-\nu-1}}{\sqrt{\pi}} \left\{ \sum_{j=0}^{M-1} (-)^j A_j \Gamma\left(s - \nu - \tfrac{1}{2} - j\right) \right.
$$

$$
\left. + \rho_M(s)\Gamma\left(s - \nu - \tfrac{1}{2} - M\right) \right\},
$$

where $\rho_M(s) = O(1)$ as $|s| \to \infty$ in $|\arg s| \leq \pi - \epsilon$, $\epsilon > 0$ and, from (2.2.40),

$$A_j = \frac{\left(1 + \frac{1}{2}\nu\right)_j \left(\frac{1}{2} + \frac{1}{2}\nu\right)_j}{j!} = \frac{2^{-2j}(\nu + 1)_{2j}}{j!}$$

with $(a)_j = \Gamma(a + j)/\Gamma(a)$.

Substitution of this expansion into the integrand of $E(z)$, followed by term-by-term integration and use of (5.4.2) in conjunction with the assumption (5.4.14), then yields

$$E(z) = \frac{2^{-\nu-1}}{\sqrt{\pi}} \sum_{j=0}^{M-1} (-)^j A_j \int_C \Gamma\left(s - \nu - \tfrac{1}{2} - j\right)\left(\tfrac{1}{2}z^2\right)^{-s} ds + R_M(z)$$

$$= (2\pi)^{\frac{1}{2}} i z^{-2\nu-1} e^{-\frac{1}{2}z^2} \sum_{j=0}^{M-1} (-)^j A_j \left(\tfrac{1}{2}z^2\right)^{-j} + R_M(z).$$

The remainder $R_M(z)$ is given by

$$R_M(z) = \frac{2^{-\nu-1}}{\sqrt{\pi}} \int_C \rho_M(s)\Gamma\left(s - \nu - \tfrac{1}{2} - M\right)\left(\tfrac{1}{2}z^2\right)^{-s} ds,$$

where, from (5.4.14), the contour C encloses all the poles of the integrand. Application of Lemma 2.8 shows that

$$|R_M(z)| = O\left(|z|^{-2M-2\mathrm{Re}(\nu)-1} e^{-\frac{1}{2}z^2}\right)$$

as $|z| \to \infty$ in $|\arg z| < \pi$.

Consequently we obtain the exponential asymptotic expansion given by

$$E(z) \sim (2\pi)^{\frac{1}{2}} i z^{-2\nu-1} e^{-\frac{1}{2}z^2} \sum_{j=0}^{\infty} (-)^j A_j \left(\tfrac{1}{2}z^2\right)^{-j} \qquad (5.4.15)$$

valid as $|z| \to \infty$ in $|\arg z| < \pi$. We can now construct the asymptotic expansion of $I(z)$ from (5.4.13) and the algebraic and exponential expansions in (5.4.10) and (5.4.15), respectively. It can be seen that in $|\arg z| < \frac{1}{2}\pi$, the expansion of $I(z)$ is algebraic, while in $|\arg(-z)| < \frac{1}{2}\pi$, $I(z)$ is the sum of an exponential and an algebraic expansion with the positive and negative imaginary axes being Stokes lines. In the sector $\frac{1}{2}\pi < |\arg z| < \frac{3}{4}\pi$, the exponential expansion is recessive so that, in the Poincaré sense, $I(z)$ is algebraic in the wider sector $|\arg z| < \frac{3}{4}\pi$.

The expansion of $D_\nu(z)$ now follows from (5.4.8) and (5.4.13) in the form

$$D_\nu(z) = e^{\mp\pi i\nu} D_\nu(ze^{\pm\pi i}) \mp \frac{z^\nu e^{-z^2/4}}{\Gamma(-\nu)} E\left(ze^{\pm\frac{1}{2}\pi i}\right). \qquad (5.4.16)$$

This then yields the standard asymptotic expansion of $D_\nu(z)$ given by

$$
D_\nu(z) \sim
\begin{cases}
z^\nu e^{-\frac{1}{4}z^2} S_1(z) & (|\arg z| < \frac{1}{2}\pi) \\[2ex]
z^\nu e^{-\frac{1}{4}z^2} S_1(z) + \dfrac{(2\pi)^{\frac{1}{2}}(ze^{\mp\pi i})^{-\nu-1}}{\Gamma(-\nu)} e^{\frac{1}{4}z^2} S_2(z) & (\frac{1}{2}\pi < |\arg z| < \pi),
\end{cases}
$$

$$(5.4.17)$$

where $S_1(z)$ and $S_2(z)$ denote the formal asymptotic sums

$$
S_1(z) = \sum_{k=0}^{\infty} \frac{(-\nu)_{2k}}{k!}(-2z^2)^{-k}, \quad S_2(z) = \sum_{k=0}^{\infty} \frac{(\nu+1)_{2k}}{k!}(2z^2)^{-k}
$$

and the upper or lower sign is chosen according as z lies in the second or third quadrants, respectively. These expansions are valid in the complete sense of Olver (1964), where the sectors of validity are less than those customarily given in the literature [Whittaker & Watson (1965, pp. 347–348); Erdélyi (1953, Vol. 2, p. 123)] for expansions fulfilling the Poincaré condition. The rays $\arg z = \pm\frac{1}{2}\pi$ and $\arg z = \pi$ are Stokes lines for $D_\nu(z)$, where the above expansions exhibit the Stokes phenomenon.

Finally, we note that the standard connection formula [Whittaker & Watson (1965, p. 348)]

$$
D_\nu(z) = e^{-\pi i \nu} D_\nu(ze^{\pi i}) - \frac{i\sqrt{2\pi}}{\Gamma(-\nu)} e^{-\frac{1}{2}\pi i \nu} D_{-\nu-1}\big(ze^{\frac{1}{2}\pi i}\big),
$$

and (5.4.16) enable us to conclude that

$$
i e^{-\frac{1}{2}\pi i \nu} D_{-\nu-1}\big(ze^{\frac{1}{2}\pi i}\big) = \frac{z^\nu e^{-z^2/4}}{\sqrt{2\pi}} E\big(ze^{\frac{1}{2}\pi i}\big).
$$

This yields the alternative Mellin-Barnes integral representation of $D_\nu(z)$, with the exponential factor $\exp(z^2/4)$ instead of $\exp(-z^2/4)$ in (5.4.8), given by

$$
D_\nu(z) = (2\pi)^{\frac{1}{2}} z^{-\nu-1} e^{\frac{1}{4}z^2} \frac{1}{2\pi i} \int_C \frac{\Gamma(2s+\nu+1)}{\Gamma(s+1)} (2z^2)^{-s} \, ds, \quad (5.4.18)
$$

which holds without restriction on $\arg z$ and ν. This result can be easily verified by direct evaluation of the residues of the poles enclosed by C and use of Kummer's theorem for the confluent hypergeometric function.

5.4.4 An Infinite Sum

We consider the sum

$$
S(z) = \sum_{n=1}^{\infty} \frac{1}{n^2(n^2+z^2)^{\frac{1}{2}}}
$$

as $z \to \infty$ in $|\arg z| < \frac{1}{2}\pi$. The function $S(z)$ is seen to possess an infinite sequence of branch points situated on the imaginary z axis at the points $z = \pm ki$ ($k = 1, 2, \dots$), where $S(z)$ becomes infinite.

If we let $f(x) = x^{-2}(1 + x^2)^{-\frac{1}{2}}$, the Mellin transform $F(s)$ is given by

$$F(s) = \int_0^\infty \frac{x^{s-3}}{(1+x^2)^{\frac{1}{2}}}dx = \frac{1}{2}\int_0^\infty \frac{y^{\frac{1}{2}s-2}}{(1+y)^{\frac{1}{2}}}dy$$

$$= \frac{1}{2\sqrt{\pi}}\Gamma\big(\tfrac{1}{2}s - 1\big)\Gamma\big(\tfrac{3}{2} - \tfrac{1}{2}s\big) \qquad (2 < \mathrm{Re}(s) < 3)$$

upon evaluation of the integral as a beta function in (3.3.8). From (4.1.1) we then find that

$$S(z) = \frac{\pi^{-\frac{1}{2}}}{4\pi i}\int_{c-\infty i}^{c+\infty i}\Gamma\big(\tfrac{1}{2}s - 1\big)\Gamma\big(\tfrac{3}{2} - \tfrac{1}{2}s\big)\zeta(s)z^{s-3}\,ds \quad (2 < c < 3).$$

This integral defines $S(z)$ in the sector $|\arg z| < \frac{1}{2}\pi$, since on the path of integration, $s = c + it$, the integrand is $|z|^{c-3}O(|t|^{-\frac{1}{2}}e^{-\Delta|t|})$ as $t \to \pm\infty$, where $\Delta = \frac{1}{2}\pi - |\arg z|$. The integrand has simple poles at $s = 0, 1$ and 2, with the remaining poles of $\Gamma(\frac{1}{2}s - 1)$ being cancelled by the 'trivial zeros' of $\zeta(s)$. In addition, there is an infinite sequence of poles at $s = 3 + 2k$ ($k = 0, 1, 2, \dots$) resulting from $\Gamma(\frac{3}{2} - \frac{1}{2}s)$; see Fig. 5.7. Before proceeding with the determination of the asymptotic expansion of $S(z)$ for large z, we note that if we complete the contour by an infinite semi-circle on the right we obtain the expansion, valid for $|z| < 1$,

$$S(z) = \frac{1}{\sqrt{\pi}}\sum_{k=0}^\infty (-)^k \frac{\Gamma\big(k + \frac{1}{2}\big)}{k!}\zeta(3 + 2k)z^{2k} \qquad (|z| < 1).$$

This provides an alternative representation of $S(z)$ inside the unit circle that yields a more efficient means of computation for small values of z.

Since the integrand is exponentially decaying as $\mathrm{Im}(s) \to \pm\infty$ when $|\arg z| < \frac{1}{2}\pi$, we can displace the contour to the left over the poles at $s = 0, 1, 2$ to find

$$S(z) = \frac{\pi^2}{6z} - \frac{1}{z^2} + \frac{1}{4z^3} - \frac{1}{2\sqrt{\pi}z^3}I(z),$$

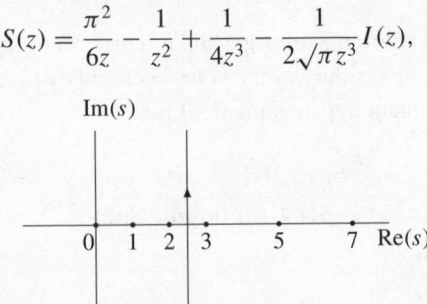

Fig. 5.7. The path of integration for $S(z)$. The heavy points denote poles.

where, provided $|\arg z| < \frac{1}{2}\pi$,

$$I(z) = -\frac{1}{2\pi i} \int_{c-\infty i}^{c+\infty i} \Gamma\left(-\frac{1}{2}s - 1\right)\Gamma\left(\frac{1}{2}s + \frac{3}{2}\right)\zeta(-s)z^{-s}\, ds$$

$$= \frac{1}{2\pi i} \int_{c-\infty i}^{c+\infty i} \zeta(1+s)\frac{\Gamma(s+1)\Gamma\left(\frac{1}{2}s + \frac{3}{2}\right)}{\Gamma\left(\frac{1}{2}s + 2\right)}(2\pi z)^{-s}\, ds \quad (c > 0) \quad (5.4.19)$$

and, for convenience, we have made the change of variable $s \to -s$ and employed the functional relation for $\zeta(s)$ in (4.1.4). The above integrand for $I(z)$ has no poles to the right of the path of integration and, due to the growth of $\Gamma(s+1)$ as $s \to \infty$, cannot be completed by a semi-circular contour on the right. Accordingly, we anticipate that $I(z)$ will be exponentially small as $z \to \infty$ in $|\arg z| < \frac{1}{2}\pi$.

To evaluate $I(z)$ we first move the path far to the right so that $|s|$ is everywhere large on the new path C (this is permissible on account of the exponential decay of the integrand as $\mathrm{Im}(s) \to \pm\infty$). Then we make use of the inverse factorial expansion given in Lemma 2.2 in the form

$$\frac{\Gamma(s+1)\Gamma\left(\frac{1}{2}s + \frac{3}{2}\right)}{\Gamma\left(\frac{1}{2}s + 2\right)} = 2^{\frac{1}{2}}\left\{\sum_{j=0}^{M-1}(-)^j c_j \Gamma\left(s + \frac{1}{2} - j\right) + \rho_M(s)\Gamma\left(s + \frac{1}{2} - M\right)\right\},$$

$$(5.4.20)$$

where M is a positive integer, $c_0 = 1$ and $\rho_M(s) = O(1)$ as $|s| \to \infty$ in $|\arg s| \leq \pi - \epsilon, \epsilon > 0$. The coefficients c_j can be determined by the method described in §2.2.4 (or in §2.2.2 after use of the duplication formula for the gamma function) and the first few are thus found to be

$$c_1 = \frac{9}{8}, \quad c_2 = \frac{345}{128}, \quad c_3 = \frac{9555}{1024}, \quad c_4 = \frac{1371195}{32768}, \quad c_5 = \frac{60259815}{262144}, \ldots.$$

Substitution of (5.4.20) into (5.4.19), where it is supposed that on C the condition $\mathrm{Re}(s) > M - \frac{1}{2}$ is satisfied, and expansion of the zeta function then yields

$$I(z) = \sum_{k=1}^{\infty}\frac{1}{k}\frac{1}{2\pi i}\int_C \frac{\Gamma(s+1)\Gamma\left(\frac{1}{2}s + \frac{3}{2}\right)}{\Gamma\left(\frac{1}{2}s + 2\right)}(2\pi kz)^{-s}\, ds$$

$$= 2(\pi z)^{\frac{1}{2}}\sum_{k=1}^{\infty}k^{-\frac{1}{2}}\left\{\sum_{j=0}^{M-1}\frac{(-)^j c_j}{(2\pi kz)^j}\frac{1}{2\pi i}\int_C \Gamma\left(s + \frac{1}{2} - j\right)\right.$$

$$\left. \times (2\pi kz)^{-s+j-\frac{1}{2}}\, ds + R_M(kz)\right\}$$

$$= 2(\pi z)^{\frac{1}{2}}\sum_{k=1}^{\infty}k^{-\frac{1}{2}}\left\{e^{-2\pi kz}\sum_{j=0}^{M-1}\frac{(-)^j c_j}{(2\pi kz)^j} + R_M(kz)\right\}$$

upon use of (5.4.2). The remainder $R_M(kz)$ is given by

$$R_M(kz) = \frac{1}{2\pi i} \int_C \rho_M(s)\Gamma\left(s + \tfrac{1}{2} - M\right)(2\pi kz)^{-s-\frac{1}{2}} ds,$$

where, from the above assumption that $\text{Re}(s) > M - \frac{1}{2}$, the path C lies to the right of the poles of the integrand. Then, by Lemma 2.7, it follows that

$$|R_M(kz)| = O(e^{-2\pi kz}(kz)^{-M})$$

as $|z| \to \infty$ in $|\arg z| < \frac{1}{2}\pi$.

We therefore finally obtain the asymptotic expansion

$$S(z) \sim \frac{\pi^2}{6z} - \frac{1}{z^2} + \frac{1}{4z^3} - z^{-\frac{5}{2}} \sum_{k=1}^{\infty} \frac{e^{-2\pi kz}}{k^{\frac{1}{2}}} \sum_{j=0}^{\infty} \frac{(-)^j c_j}{(2\pi kz)^j} \tag{5.4.21}$$

valid as $|z| \to \infty$ in $|\arg z| < \frac{1}{2}\pi$. We note that this expansion involves a double sum: the outer sum over k is convergent in $|\arg z| < \frac{1}{2}\pi$, while the inner sum over j is a divergent asymptotic series. We remark that there is an infinite number of exponentially small terms of the type $\exp(-2\pi kz)$: these terms result from the infinite sequence of branch points on the imaginary z axis. An alternative expression for the double sum in (5.4.21) is

$$z^{-\frac{5}{2}} \sum_{j=0}^{\infty} \frac{(-)^j c_j S_j(z)}{(2\pi z)^j},$$

where we have defined the sums

$$S_j(z) = \sum_{k=1}^{\infty} k^{-j-\frac{1}{2}} e^{-2\pi kz} \qquad \left(|\arg z| < \tfrac{1}{2}\pi\right).$$

The expansion (5.4.21) with just the first exponential ($k = 1$) has been stated by Olver (1974, p. 306), where a closed-form representation for the coefficients c_j is given as

$$c_j = (-)^j \frac{\Gamma\left(j + \frac{1}{2}\right)}{\sqrt{\pi}} \sum_{k=0}^{j} (-)^k 2^{k-j} (1+k) \binom{-\frac{1}{2}}{j-k}.$$

5.5 Faxén's Integral

In §5.4 we saw how an exponential asymptotic expansion could be obtained from a Mellin-Barnes integral representation when either the subset of poles on the right of the contour in (5.1.1) is empty or when it is required to extend the range of validity of the asymptotic expansion beyond its algebraic sector. Here we consider a more elaborate situation given by Faxén's integral [Faxén (1921)], which is defined for $0 < a < 1$ and $b > 0$ by

$$\text{Fi}(a, b; z) = \int_0^\infty e^{-t+zt^a} t^{b-1}\, dt$$

$$= \sum_{n=0}^\infty \frac{\Gamma(an + b)}{n!} z^n \qquad (|z| < \infty). \qquad (5.5.1)$$

At this point, it is convenient to introduce the parameters κ, h, ϑ associated with the expansion of $\text{Fi}(a, b; z)$ by (see §2.3)

$$\kappa = 1 - a, \qquad h = a^a, \qquad \vartheta = b - \tfrac{1}{2}, \qquad (5.5.2)$$

and the variable

$$X = \kappa(hz)^{1/\kappa}. \qquad (5.5.3)$$

The Mellin-Barnes integral representation for $\text{Fi}(a, b; z)$ can be derived by replacing the exponential factor $\exp(zt^a)$ in (5.5.1), by means of (3.3.2) and reversal of the order of integration, to obtain

$$\text{Fi}(a, b; z) = \frac{1}{2\pi i} \int_{c-\infty i}^{c+\infty i} \Gamma(s)(ze^{\mp \pi i})^{-s} \left\{ \int_0^\infty e^{-t} t^{-as+b-1}\, dt \right\} ds$$

$$= \frac{1}{2\pi i} \int_{c-\infty i}^{c+\infty i} \Gamma(s)\Gamma(b - as)(ze^{\mp \pi i})^{-s}\, ds, \qquad (5.5.4)$$

where $0 < c < b/a$ so that the path of integration separates the poles of $\Gamma(s)$ from those of $\Gamma(b - as)$. From Rule 1 in §2.4, it is seen that the integral in (5.5.4) converges in the sector $|\arg(-z)| < \frac{1}{2}\pi(1 + a) = \pi - \frac{1}{2}\pi\kappa$. Straightforward displacement of the path to the right over the poles of $\Gamma(b - as)$ situated at $s = (b + k)/a, k = 0, 1, 2, \ldots$, then generates the algebraic expansion

$$\text{Fi}(a, b; z) \sim H(ze^{\mp \pi i}) \qquad (5.5.5)$$

valid as $|z| \to \infty$ in $|\arg(-z)| < \pi - \frac{1}{2}\pi\kappa$, where $H(z)$ denotes the formal asymptotic sum

$$H(z) = \frac{1}{a} z^{-b/a} \sum_{k=0}^\infty \frac{(-)^k}{k!} \Gamma\left(\frac{k + b}{a}\right) z^{-k/a}. \qquad (5.5.6)$$

The upper or lower sign in (5.5.4) and $H(ze^{\mp \pi i})$ is chosen according as $\arg z > 0$ or $\arg z < 0$, respectively.

To determine the exponential expansion associated with $\text{Fi}(a, b; z)$, we first bend back the path of integration in (5.5.4) to have endpoints at infinity in $\mathrm{Re}(s) < 0$ and denote this new contour of integration by C; cf. Fig. 5.6. The resulting representation

$$\mathrm{Fi}(a, b; z) = \frac{1}{2\pi i} \int_C \Gamma(s)\Gamma(b - as)(ze^{\mp\pi i})^{-s}\, ds \qquad (5.5.7)$$

then defines $\mathrm{Fi}(a, b; z)$ without restriction on arg z. This follows from the fact that on C the modulus of the logarithm of the integrand is controlled by $\kappa R \cos \theta_0 \log R$ as $R = |s| \to \infty$, where $\pm\theta_0$ denote the directions in which C passes to infinity with $\frac{1}{2}\pi < \theta_0 \le \pi$; cf. also Rule 2 in §2.4. If we now substitute the identity

$$1 = e^{\mp 2\pi i\omega}(1 \pm 2ie^{\pm\pi i\omega} \sin \pi\omega), \qquad \omega = as - b$$

in the integral in (5.5.7) we obtain

$$\mathrm{Fi}(a, b; z) = e^{\pm 2\pi ib}\mathrm{Fi}(a, b; ze^{\pm 2\pi ia})$$
$$+ \frac{i}{\pi}e^{\pm\pi i\vartheta} \int_C \Gamma(s)\Gamma(b - as) \sin \pi\omega \, (ze^{\mp\pi i\kappa})^{-s}\, ds.$$

Use of the reflection formula for the gamma function then yields the connection formulas in the form

$$\mathrm{Fi}(a, b; z) = e^{\pm 2\pi ib}\mathrm{Fi}(a, b; ze^{\pm 2\pi ia}) + E_{\pm}(z), \qquad (5.5.8)$$

where

$$E_{\pm}(z) = \frac{e^{\pm\pi i\vartheta}}{2\pi i} \int_C \frac{2\pi\Gamma(s)}{\Gamma(1 - b + as)}(ze^{\mp\pi i\kappa})^{-s}\, ds \qquad (5.5.9)$$

for all arg z, provided $0 < a < 1$.

As for the Weber function in the preceding section, there are now no poles of the integrand in $E_{\pm}(z)$ to the right of the contour of integration C, so that we are free to expand C to make $|s|$ everywhere large on the contour. From Lemma 2.2 we have, for $M = 1, 2, \ldots$, the inverse factorial expansion†

$$\frac{\Gamma(s)}{\Gamma(1 - b + as)} = \frac{\kappa(h\kappa^\kappa)^{-s}}{2\pi} \left\{ \sum_{j=0}^{M-1} (-)^j A_j \Gamma(\kappa s + \vartheta - j) \right.$$

$$\left. + \rho_M(s)\Gamma(\kappa s + \vartheta - M) \right\},$$

where $\rho_M(s) = O(1)$ as $|s| \to \infty$ in $|\arg s| \le \pi - \epsilon$, $\epsilon > 0$ and the parameters κ, h and ϑ are defined in (5.5.2). The first few coefficients $A_j = c_j A_0$, where $A_0 = (2\pi)^{\frac{1}{2}}\kappa^{-\frac{1}{2}-\vartheta}a^\vartheta$, can be obtained from §2.2.4. Substitution of this expansion

† The factor κ has been extracted for convenience.

in (5.5.9), together with use of (5.4.2) when it is supposed that C is expanded sufficiently to enclose the point $s = (M - \vartheta)/\kappa$, then yields

$$E_{\pm}(z) = \kappa e^{\pm \pi i \vartheta} \sum_{j=0}^{M-1} (-)^j A_j \frac{1}{2\pi i} \int_C \Gamma(\kappa s + \vartheta - j)(X e^{\mp \pi i})^{-\kappa s}\, ds + R_M^{\pm}(z)$$

$$= X^{\vartheta} e^X \sum_{j=0}^{M-1} A_j X^{-j} + R_M^{\pm}(z), \tag{5.5.10}$$

where X is defined in (5.5.3). The remainder $R_M^{\pm}(z)$ is given by

$$R_M^{\pm}(z) = \frac{\kappa e^{\pm \pi i \vartheta}}{2\pi i} \int_C \rho_M(s) \Gamma(\kappa s + \vartheta - M)(X e^{\mp \pi i})^{-\kappa s}\, ds.$$

From Lemma 2.8, we find that

$$|R_M^{\pm}(z)| = O(X^{\vartheta - M} e^X)$$

as $|z| \to \infty$ in $|\arg(z^{1/\kappa} e^{\mp \pi i})| < \pi$.

A special case of the connection formula (5.5.8) can be verified directly when $a = b = \frac{1}{2}$ (so that $\kappa = \frac{1}{2}$ and $\vartheta = 0$). With the aid of the duplication formula for the gamma function in (2.2.23), we find

$$E_{\pm}(z) = \frac{1}{2\pi i} \int_C \frac{2\pi \Gamma(s)}{\Gamma(\frac{1}{2}s + \frac{1}{2})} (\mp i z)^{-s}\, ds$$

$$= \frac{\pi^{\frac{1}{2}}}{2\pi i} \int_C \Gamma(\tfrac{1}{2}s)(\mp \tfrac{1}{2} i z)^{-s}\, ds = 2\pi^{\frac{1}{2}} e^{\frac{1}{4} z^2}$$

by (3.3.3), so that the connection formula (5.5.8) becomes

$$\mathrm{Fi}\big(\tfrac{1}{2}, \tfrac{1}{2}; z\big) + \mathrm{Fi}\big(\tfrac{1}{2}, \tfrac{1}{2}; z e^{\pm \pi i}\big) = 2\pi^{\frac{1}{2}} e^{\frac{1}{4} z^2}.$$

This is easily seen to be satisfied since, from (5.5.1),

$$\mathrm{Fi}\big(\tfrac{1}{2}, \tfrac{1}{2}; z\big) = 2 \int_0^{\infty} e^{-t^2 + zt}\, dt = \pi^{\frac{1}{2}} e^{\frac{1}{4} z^2} \big\{1 + \mathrm{erf}(\tfrac{1}{2} z)\big\},$$

where erf denotes the error function and we note that $\mathrm{erf}(x e^{\pm \pi i}) = -\mathrm{erf}(x)$.

The connection formula† (5.5.8), combined with the algebraic and exponential expansions in (5.5.5) and (5.5.10), enables us to construct the dominant asymptotic

---

† Note that with the form of the algebraic expansion $H(z)$ in (5.5.6) it follows that $\exp\{\pm 2\pi i b\} H(z \exp\{\pm 2\pi i a\}) = H(z)$.

expansion of $\text{Fi}(a, b; z)$ in the form

$$
\text{Fi}(a, b; z) \sim
\begin{cases}
X^{\vartheta} e^{X} \displaystyle\sum_{j=0}^{\infty} A_j X^{-j} + H(ze^{\mp \pi i}) & \left(|\arg z| < \tfrac{1}{2}\pi\kappa\right) \\[2em]
H(ze^{\mp \pi i}) & \left(|\arg(-z)| < \pi - \tfrac{1}{2}\pi\kappa\right),
\end{cases}
\tag{5.5.11}
$$

where the upper and lower signs are to be chosen according as $\arg z > 0$ or $\arg z < 0$, respectively. This expansion can be seen to agree with the standard result in (2.3.10) applied to the Maclaurin series definition of $\text{Fi}(a, b; z)$ in (5.5.1). The function $\text{Fi}(a, b; z)$ is consequently exponentially large as $|z| \to \infty$ in the sector $|\arg z| < \tfrac{1}{2}\pi\kappa$, while in the sector $|\arg(-z)| < \pi - \tfrac{1}{2}\pi\kappa$, $\text{Fi}(a, b; z)$ is dominated by an algebraic expansion. On the rays $\arg z = \pm\tfrac{1}{2}\pi\kappa$, the exponential term e^{X} is oscillatory, so that these rays are anti-Stokes lines, and the ray $\arg z = 0$ is a Stokes line for the algebraic expansion.

We remark, however, that the expansion in (5.5.11) is valid only in the sense of Poincaré and does not reveal the presence of possible recessive exponential terms between or beyond the rays $\arg z = \pm\tfrac{1}{2}\pi\kappa$. To illustrate this point, we note that repeated application of (5.5.8) yields the result, for $k = 1, 2, \ldots,$

$$
\text{Fi}(a, b; z) = e^{\pm 2\pi i k b}\text{Fi}(a, b; ze^{\pm 2\pi i k a}) + \sum_{r=0}^{k-1} e^{\pm 2\pi i r b} E_{\pm}(ze^{\pm 2\pi i r a}) \tag{5.5.12}
$$

which connects $\text{Fi}(a, b; z)$ with $\text{Fi}(a, b; ze^{\pm 2\pi i k a})$. If we choose $a = b = \tfrac{1}{6}$, so that $\vartheta = -\tfrac{1}{3}$, and select $k = 3$ in (5.5.12), we obtain the identity

$$
\text{Fi}(\tfrac{1}{6}, \tfrac{1}{6}; z) + \text{Fi}(\tfrac{1}{6}, \tfrac{1}{6}; -z) = E_{+}(z) + e^{\frac{1}{3}\pi i} E_{+}(ze^{\frac{1}{3}\pi i}) + e^{\frac{2}{3}\pi i} E_{+}(ze^{\frac{2}{3}\pi i}). \tag{5.5.13}
$$

In this case, use of the standard expansion in (5.5.11) would indicate the presence of only two exponential expansions for the left-hand side of (5.5.13), with the algebraic expansions cancelling exactly. The form given in (5.5.13), however, reveals the presence (in a certain sector) of three exponential expansions, thereby contradicting the prediction of (5.5.11). This example confirms the existence of other recessive exponential terms which are missing from the expansion in (5.5.11). We consider the expansion of a specific case of $\text{Fi}(a, b; z)$ in more detail in §6.2.5.

5.6 Integrals with a 'Contour Barrier'

In the asymptotic evaluation of the Mellin-Barnes integral representation

$$
\frac{1}{2\pi i} \int_{c-\infty i}^{c+\infty i} g(s) z^{-s} \, ds,
$$

one is sometimes confronted with the impossibility of being able to displace the integration path $\text{Re}(s) = c$ beyond a certain point. This will arise, for example, when $|g(s)|$ has algebraic growth like $O(|t|^{a(c-b)})$ as $\text{Im}(s) = t \to \pm\infty$, where $a > 0$ and b is real, so that the integral is no longer convergent along a vertical line with $c > b$. In the process of path displacement, such a situation represents a 'contour barrier' across which it is apparently impossible to penetrate.

An example of this type has already been encountered in the evaluation of a finite sum in §4.2.3. In such cases, one is tempted to abandon the Mellin description and to resort to alternative methods of asymptotic evaluation. Rather than present a general theory, we select three examples which illustrate how this apparent obstacle can be circumvented to extract asymptotic information from the above integral in this special case. In each example we shall find that, in addition to the algebraic expansion that results if we could simply displace the contour over the poles in $\text{Re}(s) > b$, there is an oscillatory contribution to the integral. This is a rather surprising result as on cursory inspection there seems to be nothing in the integrand that might suggest this behaviour.

5.6.1 An Illustrative Example

As a simple example to illustrate the method of dealing with a Mellin-Barnes integral with a contour barrier, we consider

$$I(x) = \frac{1}{2\pi i} \int_{-c-\infty i}^{-c+\infty i} \frac{\pi \, \text{cosec} \frac{1}{2}\pi s}{2\Gamma(\frac{1}{2} - s)} x^{-s} ds, \qquad (5.6.1)$$

where $x > 0$ and† $0 < c < \frac{1}{2}$. From Stirling's formula (5.3.2) with $s = \sigma + it$, the modulus of the integrand is seen to have the controlling behaviour $O(|t|^{\sigma})$ as $t \to \pm\infty$, so that we cannot displace the contour of integration to the right of the imaginary axis. Straightforward displacement of the contour to the left over the poles on the negative axis, however, is permitted and generates the convergent expansion

$$I(x) = \sum_{k=1}^{\infty} \frac{(-)^k x^{2k}}{\Gamma(2k + \frac{1}{2})} \qquad (0 < x < \infty). \qquad (5.6.2)$$

We recognise that $I(x)$ is in fact a generalisation of the Mittag-Leffler function in §5.1.4 so that its asymptotics could also be obtained by application of the theory presented in §2.3.

To show how the asymptotic expansion of $I(x)$ can be obtained from the Mellin-Barnes integral representation in (5.6.1), we need to gain access to the poles situated at $s = 0, 2, 4, \ldots$; but, as this integral stands, such access is denied by virtue of

---

† This choice is quite arbitrary; the modification required in the development of the expansion of the integral for other paths of integration is easily carried out.

the algebraic growth of the integrand to the right of $\text{Re}(s) = 0$. To overcome this difficulty, we write $I(x)$ in the form

$$I(x) = \frac{1}{2\pi i} \int_{-c-\infty i}^{-c+\infty i} \frac{\Gamma\left(s + \frac{1}{2}\right) \cos \pi s}{2 \sin \frac{1}{2}\pi s} x^{-s} \, ds$$

$$= \frac{1}{2\pi i} \int_{-c-\infty i}^{-c+\infty i} \Gamma\left(s + \frac{1}{2}\right)\left\{\frac{1}{2 \sin \frac{1}{2}\pi s} - \sin \frac{1}{2}\pi s\right\} x^{-s} \, ds \qquad (5.6.3)$$

upon a straightforward rearrangement of the ratio of trigonometric functions. With this decomposition we see that $I(x)$ has been split up into the difference of two parts, which we shall call $I_1(x)$ and $I_2(x)$. The first part $I_1(x)$ has an integrand with the controlling behaviour $O(|t|^\sigma e^{-\pi|t|})$ as $t \to \pm\infty$, together with an infinite sequence of poles at $s = 0, 2, 4, \ldots$ on the right of the contour. The second part $I_2(x)$ has an integrand with the controlling behaviour $O(|t|^\sigma)$, but has no poles on the right of the contour. Because of the exponential decay in $|t|$ of the integrand of $I_1(x)$, we can displace its contour to the right over the first N poles in the manner described in §5.1 to produce the expansion

$$I_1(x) = \frac{1}{2\pi i} \int_{-c-\infty i}^{-c+\infty i} \frac{\Gamma\left(s + \frac{1}{2}\right)}{2 \sin \frac{1}{2}\pi s} x^{-s} \, ds$$

$$= \frac{1}{\pi} \sum_{k=0}^{N-1} (-)^{k-1} \Gamma\left(2k + \frac{1}{2}\right) x^{-2k} + O(x^{-2N}).$$

The second integral in (5.6.3) can be evaluated exactly by application of (3.3.6) to find

$$I_2(x) = \frac{1}{2\pi i} \int_{-c-\infty i}^{-c+\infty i} \Gamma\left(s + \frac{1}{2}\right) \sin \frac{1}{2}\pi s \, x^{-s} \, ds = x^{\frac{1}{2}} \sin\left(x - \frac{1}{4}\pi\right).$$

Hence we obtain the asymptotic expansion of $I(x)$ in the form

$$I(x) \sim \frac{1}{\pi} \sum_{k=0}^{\infty} (-)^{k-1} \Gamma\left(2k + \frac{1}{2}\right) x^{-2k} - x^{\frac{1}{2}} \sin\left(x - \frac{1}{4}\pi\right) \qquad (5.6.4)$$

for $x \to +\infty$. We remark that a cavalier displacement of the contour in (5.6.1) over the poles in $\text{Re}(s) \geq 0$, without due regard to the growth of the integrand as $t \to \pm\infty$, would have correctly generated the algebraic expansion in (5.6.4), but would have completely failed to detect the oscillatory contribution. In Table 5.5 we show the results of numerical calculations to demonstrate the accuracy of the expansion (5.6.4). For different values of x, the second column shows the value of the difference $D(x)$ between $I(x)$ (computed from the convergent expansion (5.6.2)) and the optimally truncated algebraic expansion on the right-hand side of (5.6.4) compared with the oscillatory contribution $T(x) = -x^{\frac{1}{2}} \sin(x - \frac{1}{4}\pi)$.

Table 5.5. *Values of $D(x)$ and $T(x)$ for the expansion (5.6.4)*

x	$D(x)$	$T(x)$
5	+1.96715295262	+1.96470170184
10	−0.65973480449	−0.65975279186
15	−3.86137774526	−3.86137762731
20	−1.59651758456	−1.59651758374
25	+3.97236524898	+3.97236524898

This example can be extended by including in the denominator of (5.6.1) the factor $s - a$ for example, where (for convenience) we shall restrict a such that $0 \leq \mathrm{Re}(a) < 1$. Then we have the integral

$$J(x) = \frac{1}{2\pi i} \int_{-c-\infty i}^{-c+\infty i} \frac{\pi \operatorname{cosec} \frac{1}{2}\pi s}{2(s-a)\Gamma\left(\frac{1}{2} - s\right)} x^{-s} ds, \tag{5.6.5}$$

where $x > 0$ and we again take $0 < c < \frac{1}{2}$. The asymptotic behaviour of the integrand is now controlled by $O(|t|^{\sigma-1})$ as $t \to \pm\infty$, so that the contour cannot be displaced to the right of the line $\mathrm{Re}(s) = 1$. It is readily shown that $J(x)$ has the convergent expansion

$$J(x) = \sum_{k=1}^{\infty} \frac{(-)^{k-1} x^{2k}}{(2k+a)\Gamma\left(2k + \frac{1}{2}\right)} \qquad (0 < x < \infty).$$

The procedure for determining the asymptotic expansion of $J(x)$ is essentially the same as that for (5.6.1). We first displace the contour over the poles at $s = 0$ and $s = a$† and decompose the resulting displaced integral as above to find

$$J(x) = h(x) + \frac{1}{2\pi i} \int_{c'-\infty i}^{c'+\infty i} \frac{\Gamma\left(s + \frac{1}{2}\right)}{s - a} \left\{ \frac{1}{2\sin\frac{1}{2}\pi s} - \sin\frac{1}{2}\pi s \right\} x^{-s} ds,$$

where $\mathrm{Re}(a) < c' < 1$ and

$$h(x) = \begin{cases} \dfrac{1}{a\sqrt{\pi}} - \dfrac{x^{-a}\cos\pi a\,\Gamma\left(a + \frac{1}{2}\right)}{2\sin\frac{1}{2}\pi a} & (a \neq 0) \\[3mm] \dfrac{1}{\sqrt{\pi}}(\log 4x + \gamma) & (a = 0) \end{cases}$$

with γ denoting Euler's constant. The integrands of the first and second parts of the above integral have the controlling behaviour $O(|t|^{\sigma-1}e^{-\pi|t|})$ and $O(|t|^{\sigma-1})$, respectively, as $t \to \pm\infty$. In addition, we observe that while the integrand of the first part has retained the pole structure of the integrand of $J(x)$ in $\mathrm{Re}(s) > c'$, the integrand of the second part *has no poles on the right of the contour*. Accordingly,

---

† When $a = 0$ the pole at $s = 0$ is double with residue $-\pi^{-\frac{1}{2}}(\log 4x + \gamma)$.

we can displace the contour in the first part over the poles situated at $s = 2, 4, \ldots$
in the usual manner, while in the second part we can replace the rectilinear path
$\mathrm{Re}(s) = c'$ by an expanded loop C with endpoints at infinity in $\mathrm{Re}(s) < 0$ (cf.
Fig. 3.1) and on which the minimum value of $|s|$ can be made as large as we please.
This is permissible since the decay of the integrand of this part of the displaced
integral is controlled by $|s^s| = \exp\{\mathrm{Re}(s) \log |s|\}$ as $|s| \to \infty$. Then we obtain

$$J(x) = h(x) + \frac{1}{\pi} \sum_{k=1}^{N-1} (-)^{k-1} \frac{\Gamma(2k + \frac{1}{2})}{2k - a} x^{-2k} + O(x^{-2N})$$

$$- \frac{1}{2\pi i} \int_C \frac{\Gamma(s + \frac{1}{2}) \sin \frac{1}{2}\pi s}{s - a} x^{-s} ds.$$

To deal with the loop integral (which we call I_C) we now write

$$\frac{\Gamma(s + \frac{1}{2})}{s - a} = \frac{\Gamma(s + \frac{1}{2})\Gamma(s - a)}{\Gamma(s - a + 1)} \tag{5.6.6}$$

and, since $|s|$ is everywhere large on C, we can employ the inverse factorial
expansion in (2.2.38) given by

$$\frac{\Gamma(s + \frac{1}{2})\Gamma(s - a)}{\Gamma(s - a + 1)} = \sum_{j=0}^{M-1} (-)^j A_j \Gamma(s + \vartheta - j) + \rho_M(s)\Gamma(s + \vartheta - M)$$

$$\tag{5.6.7}$$

for $M = 1, 2, \ldots$, where $\rho_M(s) = O(1)$ as $|s| \to \infty$ in $|\arg s| \leq \pi - \epsilon, \epsilon > 0$.
From (2.2.40) the coefficients A_j satisfy $A_j = (\frac{1}{2} - a)_j$ and $\vartheta = -\frac{1}{2}$. Then we
obtain

$$I_C = \sum_{j=0}^{M-1} (-)^j A_j \frac{1}{2\pi i} \int_C \Gamma(s - \frac{1}{2} - j) \sin \frac{1}{2}\pi s \, x^{-s} ds + R_M(x)$$

$$= \sum_{j=0}^{M-1} (-)^j A_j x^{-\frac{1}{2}-j} \sin(x + \frac{1}{4}\pi + \frac{1}{2}\pi j) + R_M(x)$$

by (3.3.5), where the remainder integral

$$R_M(x) = \frac{1}{2\pi i} \int_C \rho_M(s)\Gamma(s - \frac{1}{2} - M) \sin \frac{1}{2}\pi s \, x^{-s} ds = O(x^{-M-\frac{1}{2}})$$

as $x \to +\infty$, upon expressing $\sin \frac{1}{2}\pi s$ in terms of exponentials and using Lemma
2.8. Hence we obtain the asymptotic expansion

$$J(x) \sim h(x) + \frac{1}{\pi} \sum_{k=1}^{\infty} (-)^{k-1} \frac{\Gamma(2k + \frac{1}{2})}{2k - a} x^{-2k}$$

$$+ \sum_{j=0}^{\infty} (-)^{j-1} A_j x^{-\frac{1}{2}-j} \sin(x + \frac{1}{4}\pi + \frac{1}{2}\pi j)$$

valid as $x \to +\infty$. The additional asymptotic contribution to the integral now corresponds to an oscillatory expansion.

An identical procedure can be applied when the cosec $\frac{1}{2}\pi s$ in the integrands in (5.6.1) and (5.6.5) is replaced by sec $\frac{1}{2}\pi s$. Furthermore, we point out that integrals of this type with additional factors in the denominator (or numerator) can be dealt with in the same manner. An example with the factor $s - a$ replaced by $s(s - 1)$ in the denominator of (5.6.5) has arisen in a problem in the hydrodynamics of a uniformly accelerating plate and has been considered using a different approach by King & Needham (1994) and Forehand & Olde Daalhuis (2000). For this integral the associated quotient of gamma functions in (5.6.6) is replaced by

$$\frac{\Gamma\left(s + \frac{1}{2}\right)}{s(s - 1)} = \frac{\Gamma\left(s + \frac{1}{2}\right)\Gamma(s - 1)}{\Gamma(s + 1)},$$

with its inverse factorial expansion given by (5.6.7) with $\vartheta = -\frac{3}{2}$ and the coefficients $A_j = (j + 1)(\frac{1}{2})_j$. Consequently, upon displacement of the contour over the poles at $s = 0$ and $s = 1$, we find that the integral

$$\frac{1}{2\pi i} \int_{-c-\infty i}^{-c+\infty i} \frac{\pi \operatorname{cosec} \frac{1}{2}\pi s}{2s(s - 1)\Gamma\left(\frac{1}{2} - s\right)} x^{-s} ds$$

when $0 < c < \frac{1}{2}$, has the asymptotic expansion for $x \to +\infty$ given by

$$\pi^{-\frac{1}{2}}(1 - \gamma - \log 4x) + \frac{\pi^{\frac{1}{2}}}{4x} + \frac{1}{2\pi} \sum_{k=1}^{\infty} (-)^{k-1} \frac{\Gamma\left(2k + \frac{1}{2}\right)}{k(2k - 1)} x^{-2k}$$

$$+ \sum_{j=0}^{\infty} (-)^{j-1} A_j x^{-\frac{3}{2}-j} \sin\left(x + \frac{3}{4}\pi + \frac{1}{2}\pi j\right).$$

We note that if the last integral contained a denominatorial factor of the form $s(s - 1)(s - 2)\cdots$, the resulting quotient in (5.6.6) would still involve a quotient of three gamma functions, with consequently a simple expression for the coefficients A_j. However, a factor of the form $s(s - a)$, for example, would involve a quotient of five gamma functions in (5.6.6), with the result that the coefficients A_j in its associated inverse factorial expansion would then have to be determined by the procedure discussed in §2.2.4. This is the situation encountered in the more difficult examples of the next two subsections.

5.6.2 An Integral Involving a Bessel Function

The expansion of the integral (a one-sided Hilbert transform) as $x \to +\infty$

$$I(x) = \int_0^\infty \frac{J_0^2(xt)}{1 - t} dt, \tag{5.6.8}$$

where J_0 denotes the Bessel function of zero order and the bar indicates that the integral is a Cauchy principal value, has been discussed in a problem on water waves in Ursell (1983). The closed-form evaluation of integrals of this type arising in a telecommunication problem has also been considered by Boersma & de Doelder (1979).

By means of the Parseval formula in (3.1.11) it is easily established that

$$I(x) = \frac{1}{2\pi i} \int_{c-\infty i}^{c+\infty i} x^{-s} M[J_0^2; s] \pi \cot \pi (1-s) \, ds \qquad (0 < c < 1), \qquad (5.6.9)$$

where, from the Appendix, the Mellin transform

$$M[J_0^2; s] = 2^{s-1} \frac{\Gamma(1-s)\Gamma\left(\frac{1}{2}s\right)}{\Gamma^3\left(1-\frac{1}{2}s\right)} \qquad (0 < \mathrm{Re}(s) < 1).$$

With $s = \sigma + it$, Stirling's formula (5.3.2) shows that the modulus of the integrand as $t \to \pm\infty$ has the controlling behaviour $O(|t|^{\sigma-\frac{3}{2}})$. The integration path $\mathrm{Re}(s) = c$ cannot therefore be shifted in the usual way to the right of $\mathrm{Re}(s) = \frac{3}{2}$, thereby denying access to all but the first of the poles of the integrand situated to the right of the path of integration. We shall show below how this difficulty can be surmounted by suitable decomposition of the integral (5.6.9), combined with the use of inverse factorial expansions for quotients of gamma functions. We remark that displacement of the contour to the left yields the (convergent) expansion [Ursell (1983)]

$$I(x) = \tfrac{1}{2}\pi J_0(x) Y_0(x) + \tfrac{1}{2}\pi^{\frac{1}{2}} \sum_{n=0}^{\infty} (-)^n \frac{n!}{\Gamma^3\left(n+\frac{3}{2}\right)} x^{2n+1}$$

which is suitable for computation when $x > 0$ is small.

Observing that

$$M[J_0^2; s] \pi \cot \pi (1-s) = -\frac{2^s}{4\pi} \frac{\Gamma^4\left(\frac{1}{2}s\right)}{\Gamma(s)} \frac{\sin^2 \frac{1}{2}\pi s}{\cos \frac{1}{2}\pi s} \cot \pi s$$

$$= \frac{2^s}{8\pi} \frac{\Gamma^4\left(\frac{1}{2}s\right)}{\Gamma(s)} \left(\sec^2 \tfrac{1}{2}\pi s - 2\right) \sin \tfrac{1}{2}\pi s,$$

we write

$$I(x) = \frac{1}{8\pi} \{I_1(x) - I_2(x)\}, \qquad (5.6.10)$$

where

$$I_1(x) = \frac{1}{2\pi i} \int_{c-\infty i}^{c+\infty i} \frac{\Gamma^4\left(\frac{1}{2}s\right)}{\Gamma(s)} \frac{\sin \frac{1}{2}\pi s}{\cos^2 \frac{1}{2}\pi s} \left(\tfrac{1}{2}x\right)^{-s} ds$$

and

$$I_2(x) = \frac{1}{2\pi i} \int_{c-\infty i}^{c+\infty i} \frac{\Gamma^4\left(\frac{1}{2}s\right)}{\Gamma(s)} 2 \sin \tfrac{1}{2}\pi s \left(\tfrac{1}{2}x\right)^{-s} ds.$$

For the integral $I_1(x)$, the modulus of the integrand is exponentially small with the controlling behaviour $O(|t|^{\sigma-\frac{3}{2}} e^{-\pi|t|})$ as $t \to \pm\infty$, while that of $I_2(x)$, like that of $I(x)$, is controlled by $O(|t|^{\sigma-\frac{3}{2}})$. As a consequence of this decomposition, the path of integration in $I_1(x)$ can now be displaced to the right over the poles of the integrand in the usual manner and the integrand in $I_2(x)$ has no poles in $\mathrm{Re}(s) > 0$.

Let us deal with the integral $I_1(x)$ first. This is a straightforward application of the process described in §5.1, where the integrand has a sequence of double poles at $s = 1, 3, 5, \ldots$. The residue at the double poles can be evaluated by observing that in the neighbourhood of $s = 2k + 1$ $(k = 0, 1, 2, \ldots)$ we have

$$\left(\tfrac{1}{2}x\right)^{-s}\frac{\Gamma^4\left(\tfrac{1}{2}s\right)}{\Gamma(s)}\frac{\sin\tfrac{1}{2}\pi s}{\cos^2\tfrac{1}{2}\pi s} = \left(\tfrac{1}{2}x\right)^{-2k-1}\frac{(-)^k}{\left(\tfrac{1}{2}\pi\delta\right)^2}\left(1 - \delta\log\tfrac{1}{2}x + \cdots\right)$$

$$\times\,\frac{\Gamma^4\left(k+\tfrac{1}{2}\right)}{(2k)!}\frac{\left(1+\tfrac{1}{2}\delta\psi\left(k+\tfrac{1}{2}\right)+\cdots\right)^4}{\left(1+\delta\psi(2k+1)+\cdots\right)}$$

$$= \frac{4}{\pi^2}(-)^k\frac{\Gamma^4\left(k+\tfrac{1}{2}\right)}{(2k)!}\left(\tfrac{1}{2}x\right)^{-2k-1}\left\{\frac{1}{\delta^2} + \frac{1}{\delta}\left[-\log\tfrac{1}{2}x\right.\right.$$

$$\left.\left. +\,2\psi\left(k+\tfrac{1}{2}\right) - \psi(2k+1)\right] + \cdots\right\},$$

where we have put $\delta = s - 2k - 1$ and ψ denotes the logarithmic derivative of the gamma function. Then routine displacement of the integration path over the first N poles yields the asymptotic expansion

$$I_1(x) = \frac{4}{\pi^2}\sum_{k=0}^{N-1}(-)^k\frac{\Gamma^4\left(k+\tfrac{1}{2}\right)}{(2k)!}\left(\tfrac{1}{2}x\right)^{-2k-1}$$

$$\times\,\left\{\log\tfrac{1}{2}x - 2\psi\left(k+\tfrac{1}{2}\right) + \psi(2k+1)\right\} + O(x^{-2N}) \qquad (5.6.11)$$

as $x \to +\infty$.

Now we turn to consideration of the integral $I_2(x)$. Since there are no poles of the integrand in $\mathrm{Re}(s) > 0$ we can deform the path of integration into a loop C, on which $|s|$ is everywhere large, with endpoints† at infinity in $\mathrm{Re}(s) < 0$; cf. §5.6.1. From Lemma 2.2 we obtain, for $M = 1, 2, \ldots$, the inverse factorial expansion

$$\frac{\Gamma^4\left(\tfrac{1}{2}s\right)}{\Gamma(s)} = 2^{2-2s}\pi^2\left\{\sum_{j=0}^{M-1}(-)^j A_j\Gamma(s-1-j) + \rho_M(s)\Gamma(s-1-M)\right\},$$

$$(5.6.12)$$

where $\rho_M(s) = O(1)$ as $|s| \to \infty$ in $|\arg s| \le \pi - \epsilon$, $\epsilon > 0$ and $A_0 = 2/\pi$. Application of the method described in §2.2.4 (or in §2.2.2, after use of the

---

† Alternatively, we could take C to be an expanded contour with endpoints still at infinity in the strip $0 < \mathrm{Re}(s) < \tfrac{3}{2}$.

duplication formula for the gamma function) shows that the normalised coefficients $c_j = A_j/A_0$ are given by

$$c_0 = 1, \quad c_1 = \tfrac{1}{2}, \quad c_2 = \tfrac{5}{8}, \quad c_3 = \tfrac{21}{16}, \quad c_4 = \tfrac{507}{128}, \quad c_5 = \tfrac{4035}{256},$$

$$c_6 = \tfrac{80145}{1024}, \quad c_7 = \tfrac{956025}{2048}, \quad c_8 = \tfrac{106588755}{32768}, \dots .$$

Substitution of (5.6.12) into the integrand of $I_2(x)$, where we suppose that C is sufficiently expanded to enclose the point $s = M + 1$, then yields

$$I_2(x) = 16\pi \sum_{j=0}^{M-1} (-)^j c_j \frac{1}{2\pi i} \int_C \Gamma(s - 1 - j)(2x)^{-s} \sin \tfrac{1}{2}\pi s \, ds + R_M(x)$$

$$= 16\pi \sum_{j=0}^{M-1} \frac{(-)^j c_j}{(2x)^{j+1}} \cos \left(2x + \tfrac{1}{2}\pi j\right) + R_M(x),$$

where we have used the result (3.3.5). The remainder $R_M(x)$ is given by

$$R_M(x) = \frac{8\pi^2}{2\pi i} \int_C \rho_M(s)\Gamma(s - 1 - M) \sin \tfrac{1}{2}\pi s \, (2x)^{-s} ds.$$

By Lemma 2.8, we readily find that $|R_M(x)| = O(x^{-M-1})$ as $x \to +\infty$, so that

$$I_2(x) \sim \frac{8\pi}{x} \left\{ \cos 2x \sum_{j=0}^{\infty} \frac{(-)^j c_{2j}}{(2x)^{2j}} + \sin 2x \sum_{j=0}^{\infty} \frac{(-)^j c_{2j+1}}{(2x)^{2j+1}} \right\}. \tag{5.6.13}$$

Combining the expansions (5.6.11) and (5.6.13) in (5.6.10), we then obtain the desired asymptotic expansion of $I(x)$ in the form

$$I(x) \sim \frac{1}{2\pi^3} \sum_{k=0}^{\infty} (-)^k \frac{\Gamma^4(k + \tfrac{1}{2})}{(2k)!} \left(\tfrac{1}{2}x\right)^{-2k-1}$$

$$\times \left\{ \log \tfrac{1}{2}x - 2\psi\left(k + \tfrac{1}{2}\right) + \psi(2k + 1) \right\}$$

$$+ \frac{\cos 2x}{x} \sum_{j=0}^{\infty} \frac{(-)^{j-1} c_{2j}}{(2x)^{2j}} + \frac{\sin 2x}{x} \sum_{j=0}^{\infty} \frac{(-)^{j-1} c_{2j+1}}{(2x)^{2j+1}} \tag{5.6.14}$$

valid as $x \to +\infty$.

Use of the standard properties of the ψ function [Abramowitz & Stegun (1965, p. 258)] shows that the leading terms of this expansion are given by

$$I(x) \sim \frac{1}{\pi x}(\log x + \gamma + 3\log 2) - \frac{1}{8\pi x^3}(\log x + \gamma + 3\log 2 - \tfrac{5}{2})$$
$$- \frac{\cos 2x}{x} - \frac{\sin 2x}{4x^2} + \frac{5\cos 2x}{32x^3} + \cdots,$$

where γ is Euler's constant. This result has been obtained by use of Hilbert transform theory by Ursell (1983) and Wong (1989, p. 321).

5.6.3 Example in §4.2.3 Revisited

In §4.2.3 we investigated the asymptotic expansion for the finite sum

$$S(x) = \sum_{n=1}^{N} \frac{\left(1 - n^{2/3}x^{2/3}\right)^{1/2}}{n^{2/3}}$$

as $x \to 0+$, where N denotes the largest integer such that $xN \leq 1$. From (4.2.19) together with use of the reflection formula for the gamma function, we find that

$$S(x) = -\frac{3\pi^{\frac{1}{2}}}{8\pi i} \int_{c-\infty i}^{c+\infty i} \frac{\Gamma\left(\frac{1}{2} - \frac{3}{2}s\right)}{\Gamma\left(2 - \frac{3}{2}s\right)} \zeta(s) x^{\frac{2}{3}-s} \cot \tfrac{3}{2}\pi s \, ds \qquad (c > 1), \quad (5.6.15)$$

where the integrand possesses poles at $s = 1$ and at $s = \frac{2}{3} - \frac{2}{3}k$ for nonnegative integer values of k, with the values $k = 4, 7, 10, \ldots$ deleted. Due to the algebraic growth of the modulus of the integrand given by $O(t^{-\frac{3}{2}+\mu(\sigma)} \log t)$ as $t \to \infty$, where $s = \sigma \pm it$ and $\mu(\sigma)$ is defined at (4.1.15), the path of integration cannot be shifted to the left beyond the vertical line $\mathrm{Re}(s) = -1$. Thus we are again confronted with the problem of a Mellin-Barnes integral representation for which the usual method of displacement of the contour over an arbitrary number of poles of the infinite sequence is inapplicable.

To deal with this case, we first displace the contour over the poles at $s = 1, \frac{2}{3}$ and 0 to obtain

$$S(x) = \frac{3\pi}{4x^{1/3}} - \frac{1}{2\sqrt{\pi}} \sum_{k=0}^{1} \frac{\Gamma(k - \frac{1}{2})}{k!} \zeta(\tfrac{2}{3} - \tfrac{2}{3}k)x^{\frac{2}{3}k} + \frac{3\sqrt{\pi}}{4} J(x), \qquad (5.6.16)$$

where

$$J(x) = -\frac{1}{2\pi i} \int_{c-\infty i}^{c+\infty i} \frac{\Gamma\left(\frac{1}{2} - \frac{3}{2}s\right)}{\Gamma\left(2 - \frac{3}{2}s\right)} \zeta(s) x^{\frac{2}{3}-s} \cot \tfrac{3}{2}\pi s \, ds \qquad \left(-\tfrac{2}{3} < c < 0\right).$$

We now wish to decompose the integral $J(x)$ into two parts with the following properties: (i) one part which has a sufficiently rapidly decaying integrand as $t \to \infty$ to permit displacement of the contour of integration over the sequence of poles in $\mathrm{Re}(s) < 0$, and (ii) another part which has an integrand holomorphic in $\mathrm{Re}(s) < 0$ (but for which path displacement to the left of $\mathrm{Re}(s) = -1$ is not possible).

We achieve this by employing the identity

$$\cot \tfrac{3}{2}\pi s = \cot \tfrac{1}{2}\pi s \left(1 - \frac{2\sin \tfrac{1}{2}\pi s}{\sin \tfrac{3}{2}\pi s}\right).$$

This yields the decomposition

$$J(x) = I_1(x) - I_2(x), \tag{5.6.17}$$

where

$$I_1(x) = \frac{1}{2\pi i} \int_{c-\infty i}^{c+\infty i} \frac{\Gamma\left(\tfrac{1}{2} - \tfrac{3}{2}s\right)}{\Gamma\left(2 - \tfrac{3}{2}s\right)} \zeta(s)\, x^{\tfrac{2}{3}-s}\, \frac{2\cos\tfrac{1}{2}\pi s}{\sin\tfrac{3}{2}\pi s}\, ds,$$

$$I_2(x) = \frac{1}{2\pi i} \int_{c-\infty i}^{c+\infty i} \frac{\Gamma\left(\tfrac{1}{2} - \tfrac{3}{2}s\right)}{\Gamma\left(2 - \tfrac{3}{2}s\right)} \zeta(s)\, x^{\tfrac{2}{3}-s}\, \cot\tfrac{1}{2}\pi s\, ds,$$

with $-\tfrac{2}{3} < c < 0$ in $I_1(x)$ and $-1 < c < 0$ in $I_2(x)$. Due to the presence of the ratio of trigonometric functions, the modulus of the integrand in $I_1(x)$ in $\mathrm{Re}(s) < 0$ is now seen to be $O(|t|^{-\tfrac{3}{2}+\mu(\sigma)} e^{-\pi|t|})$ as $t \to \pm\infty$, while that in $I_2(x)$ is still $O(|t|^{-\tfrac{3}{2}+\mu(\sigma)})$. However, it is most important to observe that the integrand in $I_2(x)$ has no poles in $\mathrm{Re}(s) < 0$, since the poles of $\cot\tfrac{1}{2}\pi s$ at $s = -2, -4, \ldots$ are cancelled by the 'trivial zeros' of $\zeta(s)$.

Let us consider the integral $I_1(x)$. The integrand has poles in $\mathrm{Re}(s) < 0$ at $s = \tfrac{2}{3} - \tfrac{2}{3}k$ ($k = 2, 3, \ldots$), with the values $k = 4, 7, 10, \ldots$ deleted. Because of the exponential decay of the integrand as $t \to \pm\infty$, we are free to displace the path of integration to the left over these poles to find the asymptotic expansion†
as $x \to 0+$

$$I_1(x) \sim -\frac{2}{3\pi} \sum_{k=2}^{\infty} \frac{\Gamma\left(k - \tfrac{1}{2}\right)}{k!} \zeta\left(\tfrac{2}{3} - \tfrac{2}{3}k\right) x^{\tfrac{2}{3}k}. \tag{5.6.18}$$

We now consider the integral $I_2(x)$. Upon use of the functional relation for $\zeta(s)$ in (4.1.4), we have

$$I_2(x) = \frac{\left(x^{\tfrac{2}{3}}/\pi\right)}{2\pi i} \int_{c-\infty i}^{c+\infty i} \left(\frac{x}{2\pi}\right)^s \zeta(1+s)\Upsilon(s)\cos\tfrac{1}{2}\pi s\, ds \qquad (0 < c < 1),$$

where we have defined

$$\Upsilon(s) = \frac{\Gamma(s+1)\Gamma\left(\tfrac{3}{2}s + \tfrac{1}{2}\right)}{\Gamma\left(\tfrac{3}{2}s + 2\right)}.$$

---

† We note that, when $k = 4, 7, 10, \ldots$, the sum in (5.6.18) receives no contribution since these values correspond to the 'trivial zeros' of $\zeta(s)$.

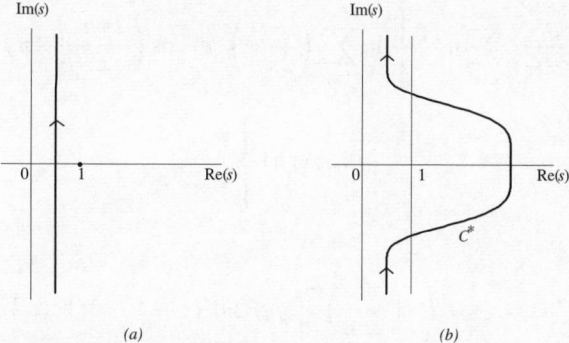

(a) *(b)*

Fig. 5.8. *(a)* The path of integration in $I_2(x)$ with $0 < c < 1$ and *(b)* the contour C^* with endpoints at infinity in the strip $0 < \text{Re}(s) < 1$.

and, for convenience, we have replaced the variable s by $-s$. Since the integrand is holomorphic in $\text{Re}(s) > 0$, we can expand the path of integration to the right, so that $|s|$ is everywhere large, but with endpoints‡ still at infinity in the strip $0 < \text{Re}(s) < 1$. This expanded contour will be denoted by C^* and is illustrated in Fig. 5.8. Lemma 2.2 then shows that $\Upsilon(s)$ has the inverse factorial expansion

$$\Upsilon(s) = \sum_{j=0}^{M-1} (-)^j A_j \Gamma\left(s - \tfrac{1}{2} - j\right) + \rho_M(s) \Gamma\left(s - \tfrac{1}{2} - M\right) \quad (M = 1, 2, \ldots),$$

where $\rho_M(s) = O(1)$ as $|s| \to \infty$ in $|\arg s| \le \pi - \epsilon$, $\epsilon > 0$ and $A_0 = (2/3)^{\frac{3}{2}}$. From §2.2.4, the normalised coefficients $c_j = A_j/A_0$ are found to have the values

$$c_0 = 1, \quad c_1 = \tfrac{9}{8}, \quad c_2 = \tfrac{2785}{1152}, \quad c_3 = \tfrac{70315}{9216},$$

$$c_4 = \tfrac{28075705}{884736}, \quad c_5 = \tfrac{1159199965}{7077888}, \cdots.$$

We now substitute the above expansion for $\Upsilon(s)$ into $I_2(x)$, where it is supposed that C^* encloses the point $s = M + \tfrac{1}{2}$, and expand the zeta function (since $\text{Re}(s) > 0$ on C^*). This yields the result

$$I_2(x) = \frac{x^{\frac{2}{3}}}{\pi} \sum_{n=1}^{\infty} n^{-1} \left\{ A_0 \sum_{j=0}^{M-1} (-)^j c_j \frac{1}{2\pi i} \int_{C^*} \left(\frac{x}{2\pi n}\right)^s \Gamma\left(s - \tfrac{1}{2} - j\right) \right.$$

$$\left. \times \cos \tfrac{1}{2}\pi s \, ds + \left(\frac{x}{2\pi n}\right)^{\frac{1}{2}} R_M(x/n) \right\} \tag{5.6.19}$$

‡ Note that it is not possible in this case to deform C^* into a loop with endpoints at infinity in $\text{Re}(s) < 0$, since, from Stirling's formula in (2.1.8), the modulus of the integrand is $O(1)$ as $|s| \to \infty$ in $\text{Re}(s) < 0$.

$$= \frac{x^{\frac{7}{6}}}{2^{\frac{1}{2}}\pi^{\frac{3}{2}}} \sum_{n=1}^{\infty} n^{-\frac{3}{2}} \left\{ A_0 \sum_{j=0}^{M-1} \left(\frac{-x}{2\pi n} \right)^j c_j \cos \left(\frac{2\pi n}{x} + \tfrac{1}{2}\pi j + \tfrac{1}{4}\pi \right) \right.$$

$$\left. + R_M(x/n) \right\},$$

where

$$R_M(x/n) = \frac{1}{2\pi i} \int_{C^*} \left(\frac{x}{2\pi n} \right)^{s-\frac{1}{2}} \rho_M(s) \Gamma\left(s - \tfrac{1}{2} - M \right) \cos \tfrac{1}{2}\pi s \, ds.$$

The integrals appearing in the sum over j in (5.6.19) have been evaluated by means of (3.3.6) (when $\arg x = 0$) to find

$$\frac{1}{2\pi i} \int_{C^*} \left(\frac{x}{2\pi n} \right)^{s-\alpha} \Gamma(s - \alpha) \cos \tfrac{1}{2}\pi s \, ds = \cos \left(\frac{2\pi n}{x} + \tfrac{1}{2}\pi \alpha \right)$$

where $\alpha = j + \frac{1}{2}$, since the rectilinear path in (3.3.6) can be reconciled with the contour C^* with endpoints at infinity in the strip $0 < \mathrm{Re}(s) < 1$. By writing $\cos \tfrac{1}{2}\pi s$ as the sum of two exponentials and bending back the contour C^* into the loop in Lemma 2.8, we have by Lemma 2.8 that $|R_M(x/n)| = O((x/n)^M)$ as $x \to 0+$ for $n = 1, 2, \ldots$. We therefore obtain the expansion

$$I_2(x) \sim \frac{2x^{\frac{7}{6}}}{(3\pi)^{\frac{3}{2}}} \sum_{j=0}^{\infty} \left(-\frac{x}{2\pi} \right)^j c_j \sum_{n=1}^{\infty} n^{-j-\frac{3}{2}} \cos \left(\frac{2\pi n}{x} + \tfrac{1}{2}\pi j + \tfrac{1}{4}\pi \right). \quad (5.6.20)$$

Collecting together the results (5.6.18) and (5.6.20) in (5.6.17), we finally obtain from (5.6.16) the asymptotic expansion

$$S(x) \sim \frac{3\pi}{4x^{1/3}} - \frac{1}{2\sqrt{\pi}} \sum_{k=0}^{\infty} \frac{\Gamma\left(k - \tfrac{1}{2} \right)}{k!} \zeta\left(\tfrac{2}{3} - \tfrac{2}{3}k \right) x^{\frac{2}{3}k}$$

$$- \frac{x^{\frac{7}{6}}}{2\pi\sqrt{3}} \sum_{j=0}^{\infty} \left(-\frac{x}{2\pi} \right)^j c_j \sum_{n=1}^{\infty} n^{-j-\frac{3}{2}} \cos \left(\frac{2\pi n}{x} + \tfrac{1}{2}\pi j + \tfrac{1}{4}\pi \right)$$

$$(5.6.21)$$

valid as $x \to 0+$. Comparison of this result with that obtained in (4.2.20) shows that the additional contribution to the expansion of $S(x)$, represented by the double sum on the right-hand side of (5.6.21), consists of an infinite number of oscillatory terms. Since the sum over n is absolutely convergent and bounded by $\zeta(j + \tfrac{3}{2})$, it follows that this contribution is $O(x^{\frac{7}{6}})$ as $x \to 0+$, as predicted in §4.2.3. The presence of this additional contribution was not accounted for in Macfarlane (1949) and Davies (1978, p. 215).

To verify the accuracy of the expansion (5.6.21) we let $x^{-1} = N + \tau$, where $0 \le \tau < 1$ and N is the integer appearing in the definition of $S(x)$. We truncate the expansion $I_1(x)$ in (5.6.18) after M terms and define the quantities

$$\mathcal{R}_M(x) = S(x) - \frac{3\pi}{4x^{\frac{1}{3}}} + \frac{1}{2\sqrt{\pi}} \sum_{k=0}^{M-1} \frac{\Gamma\left(k-\frac{1}{2}\right)}{k!} \zeta\left(\frac{2}{3}-\frac{2}{3}k\right)x^{\frac{2}{3}k},$$

$$T_r(x) = -\frac{x^{\frac{7}{6}}}{2\pi\sqrt{3}} \sum_{j=0}^{r} \left(-\frac{x}{2\pi}\right)^j c_j \sum_{n=1}^{\infty} n^{-j-\frac{3}{2}} \cos\left(2\pi n\tau + \tfrac{1}{2}\pi j + \tfrac{1}{4}\pi\right).$$

The results of numerical calculations for the case $N = 10^3$ and $M = 10$ (a sub-optimal value of the truncation index) are presented in Tables 5.6 and 5.7. The values of $\mathcal{R}_M(x)$ and $T_r(x)$ are shown for different r when $x = 10^{-3}$ and $x = (10^3 + \tau)^{-1}$ $(0 \le \tau \le 1)$, respectively. We note that the expression for $T_r(x)$ simplifies when x^{-1} is an integer (so that $\tau = 0$), since the sum over n can then be evaluated in terms of the zeta function to yield

$$T_r(x) = -\frac{x^{\frac{7}{6}}}{2\pi\sqrt{3}} \sum_{j=0}^{r} \left(-\frac{x}{2\pi}\right)^j c_j\, \zeta\left(j+\tfrac{3}{2}\right) \cos\left(\tfrac{1}{2}\pi j + \tfrac{1}{4}\pi\right) \quad (xN = 1).$$

It can be seen that the fine structure contained in $\mathcal{R}_M(x)$ as x^{-1} increases between two consecutive integer values is very accurately described by the expansion (5.6.21).

Table 5.6. *Values of $T_r(x)$ and $\mathcal{R}_{10}(x)$*
when $x = 10^{-3}$ $(\tau = 0)$

r	$T_r(x)$
0	$-5.36759\ 92482\ 8815 \times 10^{-5}$
1	$-5.36809\ 27671\ 0062 \times 10^{-5}$
2	$-5.36809\ 26253\ 3242 \times 10^{-5}$
3	$-5.36809\ 26252\ 6576 \times 10^{-5}$
$\mathcal{R}_{10}(x)$	$-5.36809\ 26252\ 6580 \times 10^{-5}$

Table 5.7. *The variation of $\mathcal{R}_{10}(x)$ and $T_r(x)$ for $r = 0, 1$*
when $x^{-1} = 10^3 + \tau$, $0 \le \tau \le 1$

τ	$\mathcal{R}_{10}(x)$	$T_0(x)$	$T_1(x)$
0	-5.36809×10^{-5}	-5.36760×10^{-5}	-5.36809×10^{-5}
0.02	-2.09705×10^{-5}	-2.09650×10^{-5}	-2.09706×10^{-5}
0.05	-0.55814×10^{-5}	-0.55755×10^{-5}	-0.55816×10^{-5}
0.10	$+0.82911\times10^{-5}$	$+0.82969\times10^{-5}$	$+0.82910\times10^{-5}$
0.20	$+2.08634\times10^{-5}$	$+2.08675\times10^{-5}$	$+2.08633\times10^{-5}$
0.40	$+2.18653\times10^{-5}$	$+2.18643\times10^{-5}$	$+2.18653\times10^{-5}$
0.60	$+0.64609\times10^{-5}$	$+0.64565\times10^{-5}$	$+0.64609\times10^{-5}$
0.80	-1.95151×10^{-5}	-1.95183×10^{-5}	-1.95151×10^{-5}
1.00	-5.36184×10^{-5}	-5.36134×10^{-5}	-5.36184×10^{-5}

6

The Stokes Phenomenon and

Hyperasymptotics

6.1 The Stokes Phenomenon

The Stokes phenomenon plays a central role in the asymptotic description of the special functions of analysis and in the important class of asymptotic phenomena in mathematical physics known as 'discontinuities'. Such discontinuities, along with those of a different origin arising in boundary layer theory and shocks in gas dynamics, are not true discontinuities in the mathematical sense but are found to depend on the scale on which the phenomenon in question is considered. When viewed under an appropriate magnification these discontinuities appear as a smooth albeit rapid transition from one form of description to another. An interesting discussion of such asymptotic phenomena is given in Friedrichs (1955).

A typical example related to the Stokes phenomenon is the boundary of the shadow which appears when a light wave passes an object. As a first approximation the shadow boundary is a sharp discontinuity, but on a smaller scale there is a transition from light to darkness which takes place in a narrow region along the shadow boundary. This corresponds to the asymptotic expansion of the solution of the wave equation taking on different forms across this boundary. Another example is the change in asymptotic form of a wave function at a caustic surface where the eikonal in the WKBJ description of the wave equation develops a singularity. Indeed, the first account of this phenomenon by Stokes in 1857 arose in this last connection in the theory of the rainbow and the formation of supernumerary arcs.

6.1.1 A Qualitative Description

The Stokes phenomenon concerns the abrupt change across certain rays in the complex plane, known as Stokes lines, exhibited by the coefficients multiplying exponentially subdominant terms in compound asymptotic expansions. The pervasive nature of this phenomenon suggests that its occurrence should have a fundamental origin. The root cause appears to be a consequence of *asymptotically approximating a given function, possessing a certain multivalued structure, in terms of approximants of a different multivalued structure.* The simplest illustration of this for solutions of linear differential equations is Airy's equation (which was also one of the examples considered by Stokes)

$$\frac{d^2y}{dz^2} = zy, \tag{6.1.1}$$

for which the solutions $y(z)$ are entire functions of z. For large $|z|$, the asymptotic solutions to leading order are given by

$$u_\pm(z) = z^{-\frac{1}{4}} \exp\left(\pm \tfrac{2}{3} z^{\frac{3}{2}}\right),$$

which are multivalued functions with a branch point at $z = 0$. Consequently, any linear combination $Cu_+(z) + Du_-(z)$ of these approximants, where C and D are arbitrary constants, cannot approximate a solution of (6.1.1) *uniformly* as one describes a complete circuit about $z = 0$, since $y(ze^{2\pi i}) = y(z)$ but $u_\pm(ze^{2\pi i}) \neq u_\pm(z)$. There must therefore be a change in the values of the constants C and D as arg z is increased by 2π.

In his review paper on the Stokes phenomenon, Meyer (1989) gives a forcible argument, particularly in the context of wave problems, for the use of the *natural metric* $\tau = \tfrac{2}{3} z^{3/2}$ in the discussion of equations of type (6.1.1). If we make the transformation $y(z) = \tau^{-1/6} W(\tau)$, the differential equation (6.1.1) takes the form

$$\frac{d^2W}{d\tau^2} - \left\{1 - \frac{\mu}{\tau^2}\right\} W = 0 \tag{6.1.2}$$

with $\mu = \tfrac{5}{36}$. The asymptotic approximants to leading order are now given by $e^{\pm\tau}$ while the solution $W(\tau)$ is associated with indicial exponents $\tfrac{1}{6}$ and $\tfrac{5}{6}$ at $\tau = 0$. In this case the situation is reversed: the approximants are entire functions of τ but the solution $W(\tau)$ possesses a branch point at $\tau = 0$, thereby again giving rise to a Stokes phenomenon.†

A similar argument applies to the modified Bessel equation

$$z^2 \frac{d^2y}{dz^2} + z\frac{dy}{dz} - (z^2 + v^2)y = 0; \tag{6.1.3}$$

---

† This argument can be extended to the equation $y'' = z^p y$ ($p > 0$), which is often cited as an example contradicting the above explanation when $p = 4, 8, \ldots$, since the solutions $y(z)$ are entire and the approximants are to leading order given by

$$u_\pm(z) = z^{-p/4} \exp\left\{\pm \frac{2}{p+2} z^{(p+2)/2}\right\}.$$

In terms of the metric $\tau = 2z^{(p+2)/2}/(p + 2)$ and the transformation $y(z) = \tau^{-p/(2p+4)} W(\tau)$, this equation reduces to the form (6.1.2) with $\mu = p(p+4)/\{4(p+2)^2\}$. The *leading-order* approximants are again $e^{\pm\tau}$ but the solution $W(\tau)$ is associated with the indicial exponents $p/(2p + 4)$ and $(p + 4)/(2p + 4)$, giving rise to a branch-point structure at $\tau = 0$ for all $p > 0$.

the details in this section are taken from the review paper by Paris & Wood (1995). For non-integer v, two linearly independent solutions of this equation are given by the modified Bessel functions $I_{\pm v}(z)$, which are associated with the indicial exponents $\pm v$ at the branch point $z = 0$. The (formal) asymptotic solutions of (6.1.3) for large $|z|$ and fixed real or complex values of v are given by

$$u_{\pm}(z) = \left(\frac{\pi}{2z}\right)^{\frac{1}{2}} e^{\pm z} \sum_{r=0}^{\infty} A_r(v)(\mp 2z)^{-r}, \qquad (6.1.4)$$

where $A_0(v) = 1$ and

$$A_r(v) = \frac{(4v^2 - 1^2)(4v^2 - 3^2) \cdots (4v^2 - (2r-1)^2)}{r! \, 2^{2r}}$$

$$= (-)^r \frac{\cos \pi v}{\pi} \frac{\Gamma\left(r + v + \frac{1}{2}\right)\Gamma\left(r - v + \frac{1}{2}\right)}{r!} \quad (r \geq 1). \qquad (6.1.5)$$

Again (provided $2v$ is not an odd integer), a Stokes phenomenon has to occur whenever a solution of (6.1.3) in the form $y(z) = AI_v(z) + BI_{-v}(z)$, where A and B are arbitrary constants, is expressed in terms of the linear combination $Cu_+(z) + Du_-(z)$ for large $|z|$, since $y(z)$ and its asymptotic approximants $u_{\pm}(z)$ have a different multivalued structure. This qualitative argument that the asymptotic approximation must be *domain-dependent* was first given by Stokes (1857);[†] see also Meyer (1989). Stokes went on to argue that the change in the constants C and D must take place where the terms of the associated expansions $u_{\pm}(z)$ come to be regularly positive: this is equivalent to the rays[‡] where $u_{\pm}(z)$ assumes maximal dominance over $u_{\mp}(z)$ for large $|z|$. Thus the values of $\theta = \arg z$ across which C and D are able to change are given by $\theta = (2n + 1)\pi$ and $\theta = 2n\pi$ ($n = 0, \pm 1, \pm 2, \ldots$), respectively.

6.1.2 The Modified Bessel Function $K_v(z)$

To be more specific we now focus our attention on the modified Bessel function $K_v(z)$. This is defined to be the combination of $I_{\pm v}(z)$ (corresponding to $-A = B = \frac{1}{2}\pi \operatorname{cosec} \pi v$) that yields $C = 0$, $D = 1$ in the sector $-\pi < \theta < \pi$, that is

$$K_v(z) \sim u_-(z) \qquad (|z| \to \infty, \quad -\pi < \theta < \pi). \qquad (6.1.6)$$

[†] It is of historical interest to note that this phenomenon (in the context of Bessel functions) had also been noticed independently by Hankel (1868) who gave essentially the same explanation for its origin.

[‡] As pointed out by Olver (1974, p. 518) there is confusion in the naming of these lines and the so-called anti-Stokes lines. We shall define the rays where maximal subdominance and rapid transition in coefficients occur as *Stokes lines*, and the rays where expansions are of the same order of magnitude as *anti-Stokes lines*. This convention is the converse of that adopted by, for example, Bender & Orzsag (1978, p. 115).

The Stokes lines for $K_\nu(z)$ are consequently situated at $\theta = \pm\pi, \pm 2\pi, \ldots$. In what follows we shall consider only the Stokes line $\theta = \pi$ across which C must change; the treatment of the other rays is similar. In the adjacent sector $\pi < \theta < 2\pi$ the value of C can be computed from the connection formula [Abramowitz & Stegun (1965, p. 376)]

$$K_\nu(ze^{m\pi i}) = e^{-m\pi\nu i} K_\nu(z) - \pi i \frac{\sin m\pi\nu}{\sin\pi\nu} I_\nu(z) \qquad (6.1.7)$$

for integer values of m. Use of this relation with $m = -1$ and $m = -2$ then shows that

$$K_\nu(z) = 2\cos\pi\nu\, K_\nu(ze^{-\pi i}) - K_\nu(ze^{-2\pi i}).$$

On replacing z in (6.1.6) by $ze^{-\pi i}$ and $ze^{-2\pi i}$ in turn and noting from (6.1.4) that $u_-(ze^{-\pi i}) = i u_+(z)$, $u_-(ze^{-2\pi i}) = -u_-(z)$, we obtain the compound asymptotic expansion

$$K_\nu(z) \sim u_-(z) + 2i\cos\pi\nu\, u_+(z) \qquad (|z| \to \infty, \quad \pi < \theta < 2\pi). \qquad (6.1.8)$$

Thus, on crossing $\theta = \pi$ in the positive sense with $|z|$ fixed, the constant C multiplying the subdominant asymptotic solution $u_+(z)$ changes from zero to the value $2i\cos\pi\nu$.

In the literature the above asymptotic expansions for $K_\nu(z)$ are normally stated in the less stringent sense of Poincaré, namely

$$K_\nu(z) \sim \begin{cases} u_-(z) & (|\theta| \le \tfrac{3}{2}\pi - \delta) \\ u_-(z) + 2i\cos\pi\nu\, u_+(z) & (\tfrac{1}{2}\pi + \delta \le \theta \le \tfrac{5}{2}\pi - \delta), \end{cases} \qquad (6.1.9)$$

where δ denotes an arbitrarily small positive constant. These expansions hold uniformly as $|z| \to \infty$ with respect to $\theta = \arg z$ in their respective sectors of validity. When 2ν is not an odd integer, the expansions (6.1.9) have the common sector of validity, $\tfrac{1}{2}\pi + \delta \le \theta \le \tfrac{3}{2}\pi - \delta$, where they differ by the inclusion of the series $u_+(z)$. The resulting contradiction, however, is only apparent, since $u_+(z)$ is uniformly exponentially small for large $|z|$ in this common sector and so is asymptotically smaller than any term in the dominant series $u_-(z)$. In the Poincaré sense, therefore, the additional term in (6.1.9b) is negligible.

Since the discovery of this phenomenon, the conventional view had been that of a discontinuous change in the constants (called *Stokes multipliers*) associated with subdominant asymptotic expansions. The fact that an analytic function could possess such a discontinuous representation in the neighbourhood of an irregular singular point had always been problematic. This 'hide-and-seek' nature of the multipliers, combined with the inherent vagueness associated with the precise location of these jumps in the complex plane, had resulted in the Stokes phenomenon being enshrouded in a certain element of mystery. Indeed, in a review paper dealing with well-constructed error bounds for asymptotic expansions, Olver (1980) re-examined the question of which expansion in (6.1.9) to adopt for $K_\nu(z)$ in the common sector of validity. By detailed analysis of the associated error bounds, he concluded that apart from a region of uncertainty R surrounding the negative real z axis, the expansion (6.1.9a) is more accurate when $\theta < \pi$ while the expansion (6.1.9b) is more accurate when $\theta > \pi$. Although his error bounds were insufficiently precise to resolve the details of the transition from one expansion to the other inside the region R, it was possible to deduce that the Stokes line should be $\theta = \pi$, since this was the only ray completely contained in R.

This picture changed dramatically when Berry (1989) argued that the coefficient of the subdominant expansion should be regarded not as a discontinuous constant but, for fixed $|z|$, as a *continuous* function of θ. Berry's theory, which was an extension of the earlier interpretative theory of asymptotic expansions by Dingle (1973), demonstrated that, when viewed on an appropriately magnified scale, the change in the subdominant multiplier near a Stokes line is in fact continuous. For a wide class of functions, the functional form of this rapid but smooth transition is found to possess a universal structure represented to leading order by an error function, whose argument is an appropriate variable describing the transition across a Stokes line.

To detect the 'birth' of such exponentially small terms it is necessary to compute the dominant series to at least a comparable accuracy. Berry achieved this by optimal truncation of the dominant asymptotic expansion after N terms† (that is, truncation just before the least term of the expansion – see §1.1.2) followed by an estimate of the associated remainder term. For the function $K_\nu(z)$, we find from (6.1.5) that optimal truncation corresponds to $N \simeq [2|z|]$, square brackets denoting the integer part. The Stokes multiplier S_N is then defined by the equation

$$K_\nu(z) = \left(\frac{\pi}{2z}\right)^{\frac{1}{2}} e^{-z} \sum_{r=0}^{N-1} A_r(\nu)(2z)^{-r} + 2i \cos \pi \nu \left(\frac{\pi}{2z}\right)^{\frac{1}{2}} e^z S_N \qquad (6.1.10)$$

which, for fixed $|z|$ (and hence fixed N) is a continuous function of θ. Since the term containing S_N is proportional to the leading term of the expansion of $u_+(z)$, we see from (6.1.6) and (6.1.8) that as we traverse the Stokes line $\theta = \pi$ (with fixed large $|z|$) the value of S_N must change from 0 to 1.

Berry was able to show that the second term on the right-hand side of (6.1.10) originates from a correct interpretation by Borel summation of the tail of the divergent expansion for $u_-(z)$ in (6.1.6). The details of his arguments are summarised in Paris & Wood (1995); see also Olver (1990). An alternative discussion of the Stokes phenomenon based on matched asymptotic expansions (a boundary-layer approach) is given in Olde Daalhuis *et al.* (1995) and an illuminating discussion from a physical viewpoint can be found in Berry & Howls (1993a) and Boyd (1999).

---

† Such optimally truncated asymptotic series are known as 'superasymptotic' [Berry (1991c)].

It is found that in the neighbourhood of the Stokes line $\theta = \pi$, when $|z|$ is large, the approximate functional form of the Stokes multiplier for $K_\nu(z)$ is described by

$$S_N \simeq \tfrac{1}{2} + \tfrac{1}{2}\mathrm{erf}\,\hat{\sigma}, \qquad \hat{\sigma} = \left(\frac{|z|}{\cos(\theta - \pi)}\right)^{\frac{1}{2}} \sin(\theta - \pi), \qquad (6.1.11)$$

where erf denotes the error function defined by

$$\mathrm{erf}\,x = \frac{2}{\sqrt{\pi}} \int_0^x e^{-u^2}\,du.$$

The change is expressed in terms of the variable $\hat{\sigma}$ which provides a *local* measure of the angular transition across $\theta = \pi$. We note that the error function for real argument lies between the values ± 1 and these limiting values are approached rapidly: for example, $|\mathrm{erf}(\pm 2)| > 0.995$. As a consequence, S_N increases smoothly but rapidly from the value 0 when $\hat{\sigma}$ is moderately large and negative (i.e., when θ is somewhat less than π) to the value 1 when $\hat{\sigma}$ is moderately large and positive (i.e., when θ is somewhat greater than π). On the Stokes line we find $S_N \simeq \tfrac{1}{2}$ to leading order.

Berry developed his theory for a wide class of functions for which the coefficients in the tail of the optimally truncated expansion possess a 'factorial divided by a power of z' dependence. The estimation of S_N in this case is analogous to that for $K_\nu(z)$ and is expressed in terms of the *singulant* $F(z)$. This quantity was defined by Dingle (1973) as the difference between the exponents of the dominant and subdominant exponentials – in the case of $K_\nu(z)$ we have e^{-z} and e^z, so that $F(z) = -2z$. The behaviour of S_N near a Stokes line for this more general class of function is similarly found to possess to leading order the error function dependence in (6.1.11), but with the variable $\hat{\sigma}$ now given locally by

$$\hat{\sigma} = \frac{\mathrm{Im}(F)}{(2\mathrm{Re}(F))^{\frac{1}{2}}}. \qquad (6.1.12)$$

A remark is in order at this point concerning the fundamental difference in philosophy between adopting an exact definition of a function $f(z)$ and working directly with the tail of a divergent asymptotic series for $f(z)$. The mathematician adopts the former viewpoint: by starting from an integral representation, or from the differential equation satisfied by $f(z)$, it is possible to develop the whole theory of the Stokes phenomenon in a rigorous fashion. This is the procedure adopted in this chapter. The physicist, on the other hand, generally approaches the problem form the other end and starts with a divergent Poincaré-type asymptotic representation for $f(z)$. This might be obtained, for example, by a perturbation expansion procedure and an exact definition of $f(z)$ is consequently not available. A central theme in Dingle (1973) is that the tail of a divergent asymptotic series $\sum c_r z^{-r}$ contains information which, if correctly interpreted, can be 'decoded' to yield the form of the dominant expansion in an adjacent sector. Indeed, in their review paper Berry & Howls (1993a) write: "[Dingle] regarded the divergent tail of a formal expansion not as indicating lack of precision but rather as a source of information to be decoded to reveal the remainder. Dingle traced the divergences of asymptotic series to singularities [...]. This led him to the result that whole classes of singularities arising in physics determine the form of the high-order (late) terms in a universal and simple way."

This astonishing behaviour of the high-order coefficients c_r, which is exhibited by a wide range of functions, is characterised by the 'factorial divided by a power of z' dependence given by

$$c_r \propto \frac{\Gamma(r + \beta)}{F^{r+\beta}} \qquad (r \gg 1), \qquad (6.1.13)$$

where β is a constant which depends on the nature of the origin of the dominant and subdominant exponential terms [Dingle (1973, pp. 111, 145, 299)] and $F(z)$ is the singulant. For example, for the Bessel function $K_\nu(z)$ we find from (6.1.5) and Stirling's formula in (2.1.8) that $\beta = 0$ and $F(z) = -2z$. By replacement of the above factorial by a well-known integral representation followed by Borel summation, it is then possible for a *convergent* representation to be extracted from the tail of the divergent series, from which the structure of the Stokes multiplier in the vicinity of a Stokes line can be obtained. Heuristic arguments for the universality of the error-function smoothing of the Stokes phenomenon for functions satisfying (6.1.13) have been given by Berry (1991b) (who also considered examples of functions where the divergence of the coefficients c_r is stronger than that in (6.1.13) – termed 'superfactorial' divergence) and by Nikishov & Ritus (1992). We point out that an example of the non-universality of the error function smoothing has been found by Chapman (1996) in the solution of an inhomogeneous delay equation (with delay parameter $1/k$) as $k \to \infty$.

6.2 Mellin-Barnes Theory

Since the appearance of Berry's seminal paper, several authors have placed this valuable theory on a rigorous foundation and have established more refined approximations to S_N. Olver (1990, 1991b) was the first to construct new uniformly exponentially-improved asymptotic expansions for functions defined by Laplace integrals. Such generalised expansions have the merit of possessing greater accuracy and a larger domain of validity than conventional Poincaré-type expansions. Independently of Olver, Jones (1990) considered the asymptotic behaviour of a certain type of Stieltjes integral and, although the question of the Stokes phenomenon was not addressed, he established that the optimal remainder has an error function dependence near the Stokes lines. Subsequently, Boyd (1990) developed an exponentially-improved asymptotic theory for functions defined by a Stieltjes transform, choosing as an example to illustrate his theory the modified Bessel function $K_\nu(z)$; for a summary of these theories, see Paris & Wood (1995).

In this section we show how the theory can be developed from a Mellin-Barnes integral description. An advantage of this approach is that integrals of this type possess a relatively wide sector of definition which generally includes a pair of Stokes lines. This obviates the need to employ analytic continuation arguments necessary in the treatment using Laplace and Stieltjes integrals (since their sectors of definition generally do not include a Stokes line). As in Olver (1990, 1991b), the theory will be developed for the confluent hypergeometric functions which include many of the commonly used special functions, such as Bessel functions, Airy functions and the parabolic cylinder functions. The theory presented here is based on Paris (1992a).

6.2.1 Exponentially-Improved Expansion

Let us consider the confluent hypergeometric function $U(a; a - b + 1; z)$ in which the parameters a and b are real or complex constants and z is a large complex variable. The asymptotic expansion for $|z| \to \infty$ is given by the Poincaré-type expansion [Abramowitz & Stegun (1965, p. 508)]

$$z^a U(a; a - b + 1; z)$$

$$\sim \begin{cases} S_1(z) & (|\arg z| \le \frac{3}{2}\pi - \delta) \\ S_1(z) \pm \dfrac{2\pi i e^{\mp \pi i \vartheta}}{\Gamma(a)\Gamma(b)} S_2(z) & (\frac{1}{2}\pi + \delta \le |\arg z| \le \frac{5}{2}\pi - \delta), \end{cases} \quad (6.2.1)$$

where the upper or lower signs are taken according as $\arg z > 0$ or $\arg z < 0$, respectively, and throughout this chapter δ denotes a small positive quantity. The formal asymptotic sums $S_1(z)$ and $S_2(z)$ are defined by

$$S_1(z) = \sum_{k=0}^{\infty} (-)^k \frac{(a)_k (b)_k}{k! \, z^k}, \qquad S_2(z) = z^\vartheta e^z \sum_{k=0}^{\infty} \frac{(1-a)_k (1-b)_k}{k! \, z^k}, \quad (6.2.2)$$

where, for convenience, we have put $\vartheta = a + b - 1$. The compound expansion in (6.2.1b) follows from use of the connection formula for $U(a; a - b + 1; z)$ given in Slater (1960, Eq. (2.2.18)); compare the procedure used to derive the compound expansion of $K_\nu(z)$ in (6.1.8). In the common sectors of validity $\frac{1}{2}\pi + \delta < |\arg z| < \frac{3}{2}\pi - \delta$, these two expansions differ because of the presence of the exponentially small series $S_2(z)$. This series is maximally subdominant on the rays $\arg z = \pm\pi$, which are consequently Stokes lines.[†]

From (3.4.7), the Mellin-Barnes integral representation for $U(a; a - b + 1; z)$ is given by

$$U(a; a - b + 1; z) = \frac{z^{-a}}{\Gamma(a)\Gamma(b)} \frac{1}{2\pi i} \int_{-\infty i}^{\infty i} \Gamma(-s)\Gamma(s+a)\Gamma(s+b) z^{-s} \, ds,$$

$$(6.2.3)$$

[†] We note that Stokes lines occur on the rays $\arg z = \pm k\pi, k = 1, 2, \ldots$.

where the integration path separates the poles of $\Gamma(-s)$ from those of $\Gamma(s+a)$ and $\Gamma(s+b)$. This separation of the poles is always possible provided a and b are not equal to zero or a negative integer (in which case $z^a U(a; a-b+1; z)$ reduces to a polynomial in z^{-1}). We note from Rule 1 in §2.4 that the sector of validity of (6.2.3) is $|\arg z| < \frac{3}{2}\pi$ which includes the first two Stokes lines on $\arg z = \pm\pi$. From Stirling's formula (2.1.8), the modulus of the integrand as $t \to \pm\infty$, when $s = \sigma + it$, is $|z|^{-\sigma} O(|t|^{\sigma+\text{Re}(a+b)-\frac{3}{2}} e^{-\Delta|t|})$, $\Delta = \frac{3}{2}\pi \mp \arg z$. Then, provided $|\arg z| < \frac{3}{2}\pi$, the contour may be displaced to the right over the first n poles of $\Gamma(-s)$ in the usual manner to find

$$z^a U(a; a-b+1; z) = \sum_{k=0}^{n-1} (-)^k \frac{(a)_k (b)_k}{k! \, z^k} + R_n(z), \qquad (6.2.4)$$

where the remainder

$$R_n(z) = -\frac{\pi}{\Gamma(a)\Gamma(b)} \frac{1}{2\pi i} \int_{-c+n-\infty i}^{-c+n+\infty i} \frac{\Gamma(s+a)\Gamma(s+b)}{\Gamma(s+1)\sin\pi s} z^{-s} \, ds \qquad (6.2.5)$$

when $0 < c < 1$. Equation (6.2.5) provides a simple representation, valid in the sector $|\arg z| < \frac{3}{2}\pi$, for the remainder $R_n(z)$ in the asymptotic expansion (6.2.1a) truncated after n terms. For the moment n is an arbitrary positive integer but will subsequently be chosen to be the optimal truncation value $n = N \simeq |z|$, when $|z|$ is large. For fixed n, it is easily seen that $|R_n(z)| = O(z^{-n+c})$ as $|z| \to \infty$ in $|\arg z| < \frac{3}{2}\pi$, so that (6.2.4) generates the asymptotic expansion (6.2.1a).

The change in structure of $R_n(z)$ across the Stokes lines $\arg z = \pm\pi$ is expressed in terms of the so-called *terminant function* $T_\nu(z)$ [cf. Dingle (1973, Chapter 23)], which is a multiple† of the incomplete gamma function $\Gamma(1-\nu, z)$ given by

$$\left. \begin{array}{ll} T_\nu(z) & = \dfrac{\Gamma(\nu)}{2\pi} \Gamma(1-\nu, z) \\[2mm] \hat{T}_\nu(z) & = -i e^{\pi i \nu} T_\nu(z). \end{array} \right\} \qquad (6.2.6)$$

The Mellin-Barnes integral representation of $T_\nu(z)$ follows immediately from that of $\Gamma(a, z)$ in (3.4.11) and is given by

$$T_\nu(z) = -\frac{z^{-\nu} e^{-z}}{4\pi i} \int_{-c-\infty i}^{-c+\infty i} \frac{\Gamma(s+\nu)}{\sin\pi s} z^{-s} \, ds, \qquad (6.2.7)$$

where $|\arg z| < \frac{3}{2}\pi$, $0 < c < 1$ and, provided $\nu \neq 0, -1, -2, \ldots$, the path of integration (suitably indented whenever $\text{Re}(\nu) \leq c$) lies to the right of all the poles of $\Gamma(s+\nu)$. In particular, we shall require the representation of the function

---

† In Olver (1991a,b) the terminant function $T_\nu(z)$ is called $F_\nu(z)$.

$T_{m+\nu}(z)$, where m is an arbitrary positive integer. This is expressed in terms of the integral (6.2.7) with a displaced contour, since (when $0 < c < 1$)

$$T_{m+\nu}(z) = -\frac{z^{-\nu}e^{-z}}{4\pi i} \int_{-c-\infty i}^{-c+\infty i} \frac{\Gamma(s+m+\nu)}{\sin \pi s} z^{-s-m} ds$$

$$= (-)^{m+1} \frac{z^{-\nu}e^{-z}}{4\pi i} \int_{-c+m-\infty i}^{-c+m+\infty i} \frac{\Gamma(s+\nu)}{\sin \pi s} z^{-s} ds \qquad (6.2.8)$$

upon replacement of the variable s by $s+m$. A discussion of the main asymptotic properties of $T_{\nu}(z)$ is deferred until §6.2.6.

To express $R_n(z)$ in (6.2.5) in terms of terminant functions, we employ the inverse factorial expansion given in (2.2.38)

$$\frac{\Gamma(s+a)\Gamma(s+b)}{\Gamma(s+1)} = \sum_{j=0}^{M-1} (-)^j A_j \Gamma(s+\vartheta-j) + \rho_M(s)\Gamma(s+\vartheta-M),$$

$$(6.2.9)$$

where M is a positive integer, ϑ is defined at (6.2.2) and, from (2.2.40), the coefficients A_j are given by

$$A_j = \frac{(1-a)_j(1-b)_j}{j!}. \qquad (6.2.10)$$

The remainder function $\rho_M(s)$ is $O(1)$ as $|s| \to \infty$ uniformly in $|\arg s| \le \pi - \delta$, $\delta > 0$. Substitution of this expansion in (6.2.5) then yields the desired result

$$R_n(z) = -\frac{\pi}{\Gamma(a)\Gamma(b)} \sum_{j=0}^{M-1} (-)^j A_j$$

$$\times \frac{1}{2\pi i} \int_{-c+n-\infty i}^{-c+n+\infty i} \frac{\Gamma(s+\vartheta-j)}{\sin \pi s} z^{-s} ds + R_{n,M}(z)$$

$$= \frac{2\pi i e^{-\pi i \vartheta}}{\Gamma(a)\Gamma(b)} z^{\vartheta} e^z \sum_{j=0}^{M-1} A_j z^{-j} \hat{T}_{n+\vartheta-j}(z) + R_{n,M}(z), \qquad (6.2.11)$$

upon identifying the integral in the finite sum in terms of (6.2.8) (with $\nu = \vartheta - j$) and use of the scaled terminant $\hat{T}_{\nu}(z)$ in (6.2.6b). The remainder term $R_{n,M}(z)$ is given by

$$R_{n,M}(z) = -\frac{\pi}{\Gamma(a)\Gamma(b)} \hat{R}_{n,M}(z),$$

where

$$\hat{R}_{n,M}(z) = \frac{1}{2\pi i} \int_{-c+n-\infty i}^{-c+n+\infty i} \rho_M(s) \frac{\Gamma(s+\vartheta-M)}{\sin \pi s} z^{-s} ds. \qquad (6.2.12)$$

The result in (6.2.11) expresses the remainder $R_n(z)$ in (6.2.4), where n is an arbitrary positive integer, as a sum of terminant functions. It has been obtained for all real or complex values of the parameters a and b (with the exception of nonpositive integer values) and for all z in the sector $|\arg z| < \frac{3}{2}\pi$. We now suppose that the series in (6.2.4) is truncated at the optimal value of n (which we denote by N); that is, when $|z|$ is large, the value of n corresponding to the term preceding the numerically smallest term. We therefore set

$$N + a + b - 1 = |z| + \alpha, \tag{6.2.13}$$

where $|\alpha|$ is bounded; this corresponds to the *superasymptotic* level of expansion. It is clear that, with N chosen in this manner, $R_N(z)$ is a piecewise continuous function of $|z|$ and a continuous function of arg z. To demonstrate that (6.2.11) constitutes an *exponentially-improved expansion* for $R_n(z)$ when $n = N$, it is necessary to derive suitable estimates for the remainder $R_{N,M}(z)$. This is carried out in the next section.

6.2.2 Estimates for $|R_{N,M}(z)|$

By Lemma 2.9, we find when $N = |z| + O(1)$ and M is finite that

$$|R_{N,M}(z)| = \begin{cases} O\left(z^{\vartheta - M - \frac{1}{2}} e^{-|z|}\right) & (|\arg z| \leq \pi - \delta) \\ O\left(z^{\vartheta - M} e^{-|z|}\right) & (|\arg z| \leq \pi) \end{cases} \tag{6.2.14}$$

as $|z| \to \infty$. An estimate for $\hat{R}_{N,M}(z)$ in the sector $\pi \leq |\arg z| \leq 2\pi - \delta$, can be obtained from its connection formula derived from the integral (6.2.12). Insertion of the identity

$$\operatorname{cosec} \pi s \equiv e^{\mp \pi i s}\left\{\frac{e^{\mp \pi i s}}{\sin \pi s} \pm 2i\right\},$$

where either the upper or lower signs are chosen, in (6.2.12) immediately yields the connection formula

$$\hat{R}_{N,M}(z) = \hat{R}_{N,M}(ze^{\pm 2\pi i}) \pm I(ze^{\pm \pi i}), \tag{6.2.15}$$

with

$$I(z) = \frac{1}{\pi} \int_{-c+N-\infty i}^{-c+N+\infty i} \rho_M(s)\Gamma(s + \vartheta - M)z^{-s}\, ds \qquad (|\arg z| < \tfrac{1}{2}\pi).$$

The integral $I(ze^{\pm \pi i})$ no longer contains a $\sin \pi s$ in the denominator and consequently, provided $|\arg(ze^{\pm \pi i})| < \frac{1}{2}\pi$, the contour of integration can be replaced by any path parallel to the imaginary s axis which is situated to the right of all the poles of the integrand. Application of Lemma 2.7 then shows that $|I(z)| = O(z^{\vartheta - M} e^{-z})$ as $|z| \to \infty$ in $|\arg z| \leq \frac{1}{2}\pi - \delta$. This estimate may be extended to the sector $|\arg z| \leq \pi - \delta$ by the device of bending back the path of integration in $I(z)$ into a loop with endpoints at infinity in $\text{Re}(s) < 0$ and use of Lemma 2.8. With the choice of the lower signs in (6.2.15), we then find

$$|\hat{R}_{N,M}(z)| \leq |\hat{R}_{N,M}(ze^{-2\pi i})| + O(z^{\vartheta-M}e^z)$$

in the sector $\pi \leq \arg z \leq 2\pi - \delta$. For this range of arg z, we have $-\pi \leq \arg(ze^{-2\pi i}) \leq -\delta$. Then, from (6.2.14b), it follows that $|\hat{R}_{N,M}(ze^{-2\pi i})| = O(z^{\vartheta-M}e^{-|z|})$ in this sector. Hence we obtain

$$|\hat{R}_{N,M}(z)| = O(z^{\vartheta-M}e^{-|z|}) + O(z^{\vartheta-M}e^z)$$

$$= O(z^{\vartheta-M}e^z) \quad \text{in} \quad \pi \leq \arg z \leq 2\pi - \delta. \tag{6.2.16}$$

A similar result using (6.2.15) with the upper signs holds for the conjugate sector, or we can appeal to symmetry.

Collecting together the results in (6.2.14) and (6.2.16) we have the asymptotic estimates for the remainder $R_{N,M}(z)$ as $|z| \to \infty$ when $N = |z| + O(1)$ given by

$$|R_{N,M}(z)| = \begin{cases} O(z^{\vartheta-M-\frac{1}{2}}e^{-|z|}) & (|\arg z| \leq \pi - \delta) \\ O(z^{\vartheta-M}e^{-|z|}) & (|\arg z| \leq \pi) \\ O(z^{\vartheta-M}e^z) & (\pi \leq |\arg z| \leq 2\pi - \delta). \end{cases} \tag{6.2.17}$$

This shows that at optimal truncation the remainder in the expansion (6.2.11) is uniformly exponentially small (controlled by $e^{-|z|}$) *throughout* the sector $|\arg z| \leq \pi$. Outside this sector, however, this exponential improvement is seen to steadily deteriorate until $|\arg z| = \frac{3}{2}\pi$, where $|e^z| = O(1)$. We thus obtain from (6.2.11) the *exponentially-improved asymptotic expansion* in the form

$$z^a U(a; a - b + 1; z) = \sum_{k=0}^{N-1} (-)^k \frac{(a)_k(b)_k}{k! z^k} + R_N(z)$$

$$R_N(z) = \frac{2\pi i e^{-\pi i\vartheta}}{\Gamma(a)\Gamma(b)} z^\vartheta e^z \sum_{j=0}^{M-1} \frac{(1-a)_j(1-b)_j}{j! z^j} \hat{T}_{N+\vartheta-j}(z) + R_{N,M}(z),$$

$$\tag{6.2.18}$$

valid for large $|z|$ in the sector $|\arg z| < 2\pi - \delta$, where N is given by (6.2.13).

This expansion agrees with that obtained by Olver (1990, 1991b) and Boyd (1990) for the particular case of the modified Bessel function $K_\nu(z)$. It is of interest to note that, by means of the connection formula for $U(a; a - b + 1; z)$, Olver (1991b, §5) showed that the estimate (6.2.17c) is valid in the wider sectors $\pi \leq |\arg z| \leq \frac{5}{2}\pi - \delta$ which are maximal. From the estimates for $T_\nu(z)$ in (6.2.59) and (6.2.6b), the expansion (6.2.18) is then seen to incorporate the three Poincaré-type expansions in (6.2.1) and, moreover, to provide a smooth transition between these expansions. However, in normal circumstances we would not need to compute the exponentially-improved expansion of $U(a; a - b + 1; z)$ outside the sector

$|\arg z| \le \pi$, since standard connection formulas are available for the construction of exponentially-improved expansions in other sectors.

6.2.3 The Stokes Multiplier

Comparison of (6.2.1) with (6.2.18) shows that the Stokes multiplier S_N is given by

$$S_N \sim \sum_{j=0}^{\infty} A_j z^{-j} \hat{T}_{N+\vartheta-j}(z); \tag{6.2.19}$$

compare (6.1.10). To examine the nature of the transition across the Stokes line $\arg z = \pi$, we truncate (6.2.19) at its first term and consider fixed large $|z|$ and vary the phase $\theta = \arg z$. Then, we find for large $|z|$

$$S_N \equiv S_N(\theta) = \hat{T}_{N+\vartheta}(z) + O(z^{-1}).$$

When $v = |z| + \alpha$, where $|\alpha|$ is bounded, the expansion of $\hat{T}_v(z)$ as $|z| \to \infty$ uniformly in the sector $-\pi + \delta \le \arg z \le 3\pi - \delta$ is, from (6.2.56),

$$\hat{T}_v(z) \sim \tfrac{1}{2} + \tfrac{1}{2}\mathrm{erf}\left\{c(\theta)\left(\tfrac{1}{2}|z|\right)^{\frac{1}{2}}\right\} - \frac{ie^{-\frac{1}{2}|z|c^2(\theta)}}{\sqrt{2\pi|z|}} \sum_{k=0}^{\infty} \left(\tfrac{1}{2}\right)_k b_{2k}(\theta,\alpha)\left(\tfrac{1}{2}|z|\right)^{-k},$$

with a conjugate expansion in the sector $-3\pi + \delta \le \arg z \le \pi - \delta$; the coefficients $b_{2k}(\theta,\alpha)$ are specified in (6.2.61) and (6.2.62). The quantity $c(\theta)$ is defined by

$$\tfrac{1}{2}c^2(\theta) = 1 + i(\theta - \pi) - e^{i(\theta-\pi)}$$

and corresponds to the branch that behaves near $\theta = \pi$ like $c(\theta) \simeq \theta - \pi$; see (6.2.52). Hence we find that

$$S_N(\theta) = \tfrac{1}{2} + \tfrac{1}{2}\mathrm{erf}\{c(\theta)\left(\tfrac{1}{2}|z|\right)^{\frac{1}{2}}\} + e^{-\frac{1}{2}|z|c^2(\theta)}O\left(|z|^{-\frac{1}{2}}\right). \tag{6.2.20}$$

In the neighbourhood of the Stokes line $\arg z = \pi$, we have $c(\theta) \simeq \theta - \pi$ and the transition of the multiplier across $\arg z = \pi$ is consequently smooth: $S_N(\theta)$ changes from approximately zero when $\theta < \pi$ to approximately 1 when $\theta > \pi$ within a zone centred on $\theta = \pi$ of angular width $O(|z|^{-\frac{1}{2}})$. A similar result is obtained for the Stokes line $\arg z = -\pi$. Berry's approximation for $S_N(\theta)$ in (6.1.11) is associated with the variable $\hat{\sigma}$ defined in (6.1.12), where for the confluent hypergeometric function $U(a; a-b+1; z)$ the singulant $F(z) = |z|e^{i(\theta-\pi)}$. Thus we have

$$\hat{\sigma} = \left(\frac{|z|}{2\cos(\theta - \pi)}\right)^{1/2} \sin(\theta - \pi),$$

and to leading order in powers of $\theta - \pi$ the approximate representation $S_N(\theta) = \tfrac{1}{2} + \tfrac{1}{2}\mathrm{erf}\,\hat{\sigma}$ agrees with the result in (6.2.20).

We draw attention to three important remarks. First, Berry's approximation in (6.1.11) is valid *only in the vicinity of a Stokes line*. The result in (6.2.18) is valid in a much wider sector, extending the standard Poincaré expansion (6.2.4), and incorporates the smooth but rapid transition of the Stokes multiplier across a Stokes line through the behaviour of the approximant $\hat{T}_\nu(z)$. Secondly, $S_N(\theta)$ has an imaginary component since $c(\theta)$ is a complex function of θ. It is of interest to note that this imaginary component becomes of comparable significance to the real component of $S_N(\theta)$ when the truncation index n is non-optimal. A discussion of the behaviour of the Stokes multiplier for non-optimal truncation is given in Berry (1989) and Middleton (1994). And thirdly, on the Stokes line $\arg z = \pi$ we have $c(\pi) = 0$, so that $S_N(\pi) = \frac{1}{2} + O(|z|^{-\frac{1}{2}})$; that is, the multiplier on the Stokes line equals $\frac{1}{2}$ to *leading order*.

The above analysis can also be applied to other solutions of the differential equation satisfied by $U(a; a - b + 1; z)$. The first of these is the function $_1F_1(a; a - b + 1; z)$, where it will again be supposed that a and b are real or complex constants and that z is a large complex variable, which may be restricted to the half-plane $|\arg z| \leq \frac{1}{2}\pi$ on account of Kummer's first transformation

$$_1F_1(a; b; z) = e^z {}_1F_1(b - a; b; -z).$$

An important difference between this function and $U(a; a - b + 1; z)$ is the fact that $_1F_1(a; a - b + 1; z)$ is an entire function of z and is consequently completely described in $-\pi < \arg z \leq \pi$.

From Slater (1960, p. 21), we have the relation expressing $_1F_1(a; c; z)$ in terms of the U hypergeometric function with argument rotated by $\pm\pi$ given by

$$_1F_1(a; c; z) = \frac{\Gamma(1 - a)\Gamma(c)e^z}{2\pi i}\{e^{\pi i c}U(c - a; c; ze^{\pi i})$$
$$- e^{-\pi i c}U(c - a; c; ze^{-\pi i})\}.$$

Application of (6.2.4) and (6.2.18) then shows that

$$_1F_1(a; a - b + 1; z) = z^{b-1}e^z\frac{\Gamma(a - b + 1)}{\Gamma(a)}\sum_{k=0}^{N-1}\frac{(1 - a)_k(1 - b)_k}{k!z^k} + R_N(z),$$

$$(6.2.21)$$

where

$$R_N(z) \sim \frac{\Gamma(a - b + 1)}{\Gamma(1 - b)}(ze^{-\pi i})^{-a}\sum_{j=0}^{\infty}(-)^j\frac{(a)_j(b)_j}{j!z^j}$$
$$\times \left\{\hat{T}_{N-\vartheta-j}(ze^{\pi i}) - e^{2\pi i b}\hat{T}_{N-\vartheta-j}(ze^{-\pi i})\right\}$$

in the sector $|\arg z| \leq \frac{1}{2}\pi$, where ϑ is defined at (6.2.2). This result is valid for all values of the parameters a and b except when b is a positive integer, in which case $_1F_1(a; a - b + 1; z)$ reduces to a polynomial in z of degree $b - 1$ multiplied by e^z, and when $a - b + 1$ is zero or a negative integer, in which case the function is not defined.

If we use the connection formula relating $\hat{T}_\nu(ze^{\pm\pi i})$ in (6.2.45), we finally obtain the result

$$R_N(z) \sim \frac{\Gamma(a - b + 1)}{\Gamma(1 - b)} z^{-a} \sum_{j=0}^{\infty} (-)^j \frac{(a)_j (b)_j}{j! z^j}$$

$$\times \left\{ e^{-\pi i a} + 2i \sin \pi a \, \hat{T}_{N-\vartheta-j}(ze^{\pi i}) \right\} \qquad (6.2.22)$$

for $|z| \to \infty$ in $|\arg z| \le \frac{1}{2}\pi$; for details, see Paris (1992a). Truncation of the expansion (6.2.22) at its first term shows that the Stokes multiplier for $_1F_1(a; a - b + 1; z)$, associated with transition across the Stokes line $\arg z = 0$, is given by

$$S_N(\theta) = e^{-\pi i a} + 2i \sin \pi a \, \hat{T}_{N-\vartheta}(ze^{\pi i}) + O(z^{-1})$$

$$= \cos \pi a + i \sin \pi a \, \mathrm{erf} \left\{ c(\theta + \pi) \left(\tfrac{1}{2} |z| \right)^{\frac{1}{2}} \right\}$$

$$+ e^{-\frac{1}{2}|z| c^2 (\theta + \pi)} O\left(|z|^{-\frac{1}{2}}\right), \qquad (6.2.23)$$

where $c(\theta)$ is given in (6.2.52). In the neighbourhood of $\arg z = 0$, we see that $c(\theta + \pi) \simeq \theta$ so that, for fixed $|z|$, the Stokes multiplier changes smoothly in a zone of angular width $O(|z|^{-\frac{1}{2}})$ from the value $e^{\pi i a}$ for $\theta > 0$ to the value $e^{-\pi i a}$ for $\theta < 0$. On the Stokes line, we have $c(\pi) = 0$ and so

$$S(0) = \cos \pi a + O\left(|z|^{-\frac{1}{2}}\right).$$

As a final remark we note a property of the coefficients in the exponentially-improved expansions for the above two hypergeometric functions. If we denote the coefficients appearing in the remainders (6.2.18) and (6.2.22) by

$$A_k = \frac{(1 - a)_k (1 - b)_k}{k!}, \qquad A'_k = \frac{(a)_k (b)_k}{k!}, \qquad (6.2.24)$$

then we see that the coefficients A'_k in the Poincaré expansion of $U(a; a - b + 1; z)$ in (6.2.18) reappear in the exponentially-improved expansion of $_1F_1(a; a - b + 1; z)$ in (6.2.22). Similarly, the coefficients A_k in the Poincaré expansion of $_1F_1(a; a - b + 1; z)$ in (6.2.21) reappear in the exponentially-improved expansion of $U(a; a - b + 1; z)$ in (6.2.18). This phenomenon is an example of *resurgence*, which is known to arise for Poincaré asymptotic expansions of integrals with saddles; see Berry & Howls (1991).

A similar situation occurs if one examines as a second solution the function $V(a; a-b+1; z)$. This function is recessive at infinity in the sector $|\arg(-z)| < \frac{1}{2}\pi$ and is defined by

$$V(a; c; z) = e^z U(c - a; c; -z);$$

see Olver (1974, p. 256). The resurgence property is again exhibited by the coefficients in the exponentially-improved expansions of $U(a; a - b + 1; z)$ and

$V(a; a - b + 1; z)$; this fact became clear from the papers of Olver(1991b, 1993, 1994) where the coefficients in the re-expansion of the remainder term for the asymptotic expansion of one solution of the confluent hypergeometric differential equation are the same as the coefficients in the other solution. In Olde Daalhuis & Olver (1994) this phenomenon was shown to extend to the solutions of the general second-order differential equation at a singularity of rank one.

6.2.4 The Stokes Multiplier for a High-Order Differential Equation

Specific examples of functions exhibiting the error-function smoothing have so far been confined to functions satisfying second-order linear differential equations. These functions possess compound expansions necessarily consisting of only one dominant and one subdominant asymptotic series. Here we briefly consider the application of the Mellin transform approach described in §6.2.1 to functions satisfying a class of differential equations of arbitrary order n. Such solutions will, in general, involve compound expansions consisting of more than one subdominant asymptotic series. The results of this section are taken from Paris (1992b).

The nth-order differential equation considered is

$$u^{(n)} - (-)^n \sum_{r=0}^{p} a_r z^r u^{(r)} = 0 \qquad (n > p \geq 0), \qquad (6.2.25)$$

where z is the independent variable and the coefficients a_r $(r = 0, 1, \ldots, p - 1)$ are arbitrary constants with $a_p = 1$ and $a_0 \neq 0$. When $n = 2$, $p = 1$ this is a transformation of Weber's equation for the parabolic cylinder function. The special feature of (6.2.25) is that the coefficients of the lower-order derivatives involve a power of z equal to the order of the derivative. Such a structure (when $n > p$) results in the coefficients of the formal power series solutions about $z = 0$ satisfying a two-term recurrence relation, with solutions consequently related to the generalised hypergeometric function. Associated with the above equation is the polynomial of degree p given by

$$a_0 + \sum_{r=1}^{p} a_r \prod_{k=0}^{r-1} (x - k) = \prod_{r=1}^{p} (x + \beta_r). \qquad (6.2.26)$$

The zeros $-\beta_r$ play a fundamental role in the asymptotic structure of solutions of (6.2.25). It will supposed, for simplicity, that none of the β_r equals a negative integer.†

Different classes of solution of (6.2.25), exhibiting different types of asymptotic behaviour in certain sectors about the irregular singular point at infinity, have been discussed in detail in Paris & Wood (1985, 1986 Chapter 3). We shall

---

† We note that $\beta_r \neq 0$ since, by hypothesis, $a_0 \neq 0$.

be concerned here only with the solution† denoted by $U_{n,p}(z)$. This solution is chosen here as it possesses a relatively simple asymptotic structure for large $|z|$. Although straightforward dominant balance arguments for large $|z|$ reveal that (6.2.25) has p algebraic-type solutions (with controlling behaviour $z^{-\beta_r}$, $r = 1, 2, \ldots, p$) together with $n - p$ exponential-type solutions, the particular solution $U_{n,p}(z)$ involves all the p algebraic-type solutions but only a *single* exponential-type solution. Other solutions of (6.2.25) which possess either an exponentially small or a single algebraic behaviour as $|z| \to \infty$ in certain sectors, and even and odd solutions (when n is even), are discussed fully in Paris & Wood (1985, 1986), where it is shown that they can be expressed as linear combinations of the fundamental system $U_{n,p}(\Omega_j z)$, $\Omega_j = \exp(2\pi i j/n)$, $j = 0, 1, \ldots, n - 1$. These solutions will therefore, in general, involve more than one exponential-type solution, each associated with its own pair of Stokes lines, and consequently possess a more complicated asymptotic structure than that of the basic solution $U_{n,p}(z)$.

From Paris & Wood (1985, 1986), the solution $U_{n,p}(z)$ has the series expansion

$$U_{n,p}(z) = \sum_{k=0}^{\infty} \frac{(-n^{p/n}z)^k}{k!} \prod_{r=1}^{p} \Gamma\left(\frac{k + \beta_r}{n}\right), \qquad (6.2.27)$$

where the β_r are defined in (6.2.26). Provided the β_r satisfy the above restriction and $n > p \geq 0$, the right-hand side of (6.2.27) defines $U_{n,p}(z)$ as a uniformly and absolutely convergent series for all finite $|z|$. The asymptotic expansion of $U_{n,p}(z)$ for large $|z|$ may be stated as follows. We define the parameters κ, ϑ by

$$\kappa = \frac{n - p}{n}, \qquad \vartheta = \frac{1}{n} \sum_{r=1}^{p} \beta_r - \tfrac{1}{2}p$$

and introduce the formal asymptotic series

$$\left.\begin{aligned} E(z) &= X^\vartheta e^X \sum_{j=0}^{\infty} A_j X^{-j}, \qquad X = \kappa z^{1/\kappa}, \\ H(z) &= n \sum_{r=1}^{p} (n^{p/n}z)^{-\beta_r} S_{n,p}(\beta_r; z), \end{aligned}\right\} \qquad (6.2.28)$$

where, provided no two of the β_r either coincide or differ by an integer multiple of n,

$$S_{n,p}(\beta_r; z) = \sum_{k=0}^{\infty} \frac{(-)^k}{k!} \Gamma(nk + \beta_r) \prod_{j=1}^{p}{}' \Gamma\left(\frac{\beta_j - \beta_r}{n} - k\right) (n^{p/n}z)^{-nk},$$

---

† We remark that in the notation of Paris & Wood (1986), this solution would be called $U_{n,p}(-z)$. The minus sign in the argument results from the factor $(-)^n$ in (6.2.25) which is introduced for convenience in the location of the Stokes lines. In addition, to ease the notation we omit the dependence on the parameters β_r.

with the prime denoting the omission of the term corresponding to $j = r$ in the product. The coefficients A_j in $E(z)$ are independent of z with

$$A_0 = (2\pi)^{\frac{1}{2}p}\kappa^{-\frac{1}{2}-\vartheta}n^{-\vartheta}.$$

The normalised coefficients $c_j = A_j/A_0$ are defined by a complicated recurrence relation given in Paris & Wood (1986, §3.4) and the first few are found to be

$$c_0 = 1, \qquad c_1 = \frac{1}{2n}\left\{a(a - n - 1) + \tfrac{1}{6}(n^2 + \tfrac{7}{2}n + 1)\right\},$$

$$c_2 = \left(\tfrac{3}{2} - a\right)c_1 + \tfrac{1}{2}c_1^2 - \frac{1}{6n^2}\left(a - \tfrac{1}{2}\right)(a - 1)\left\{\tfrac{3}{2}n^2 - (2n - 1)a\right\}, \dots.$$

$$(6.2.29)$$

Then, from §2.3, we have the asymptotic expansion† as $|z| \to \infty$ given by

$$U_{n,p}(z) \sim \begin{cases} H(z) & \text{in } |\arg z| < \tfrac{1}{2}\pi\left(1 + \dfrac{p}{n}\right) \\ H(z) + E(ze^{\mp\pi i}) & \text{in } |\arg(-z)| < \pi\left(1 - \dfrac{p}{n}\right), \end{cases} \qquad (6.2.30)$$

where the upper or lower sign is chosen according as $\arg z > 0$ or $\arg z < 0$, respectively.

The solution $U_{n,p}(z)$ is therefore exponentially large as $|z| \to \infty$ in the sector $|\arg(-z)| < \tfrac{1}{2}\pi(1 - p/n)$, described by the *single* exponential expansion $E(ze^{\mp\pi i})$, while in the sector $|\arg z| < \tfrac{1}{2}\pi(1 + p/n)$ the dominant asymptotic behaviour comprises p algebraic expansions, each with the controlling behaviour $z^{-\beta_r}$, $r = 1, 2, \dots, p$. In the common sectors of validity, $\pi p/n < |\arg z| < \tfrac{1}{2}\pi(1 + p/n)$, the expansions in (6.2.30) differ only through the presence of the exponentially small subdominant series $E(ze^{\mp\pi i})$. The rays $\arg z = \pm\pi p/n$ and $\arg z = \pi$ are Stokes lines where the exponential expansion $E(ze^{\mp\pi i})$ and the algebraic expansion $H(z)$, respectively, are maximally subdominant. The rays $\arg z = \pm\tfrac{1}{2}\pi(1 + p/n)$ are anti-Stokes lines where the exchange of dominance between the algebraic and exponential expansions takes place; see Fig. 6.1.

The Mellin-Barnes integral representation of $U_{n,p}(z)$ is given by [Paris & Wood (1986, Eq. (3.2.4))]

$$U_{n,p}(z) = \frac{n}{2\pi i}\int_{-\infty i}^{\infty i} \Gamma(ns)\prod_{r=1}^{p}\Gamma\left(-s + \frac{\beta_r}{n}\right)(n^{p/n}z)^{-ns}\,ds \qquad (6.2.31)$$

valid in the sector $|\arg z| < \tfrac{1}{2}\pi(1 + p/n)$; see Rule 1 in §2.4. Provided none of the β_r equals a negative integer, the path can always be chosen to separate the poles of $\Gamma(-s + \beta_r/n)$ ($1 \le r \le p$) from those of $\Gamma(ns)$. To illustrate the main

---

† We remark that the expansion (6.2.30b) was given in Paris & Wood (1986) only in the narrower sector $|\arg(-z)| \le \tfrac{1}{2}\pi(1 - p/n)$ where the exponential expansion $E(ze^{\mp\pi i})$ is dominant.

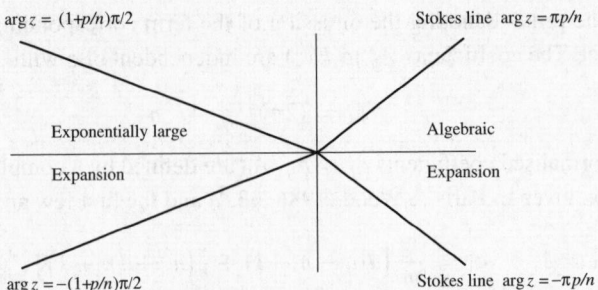

Fig. 6.1. The algebraic and exponentially large sectors for $U_{n,p}(z)$ in the complex z plane. The algebraic sector is $|\arg z| < \frac{1}{2}\pi(1 + p/n)$.

features, we present only the details for the case $p = 1$ (so that $\beta_1 = a_0 \equiv a$) and consider the Stokes line $\arg z = \pi/n$ in the algebraic sector where the subdominant exponential series 'switches on'; the treatment of the Stokes line on $\arg z = -\pi/n$ is similar. Displacement of the contour to the right in the usual manner then yields

$$U_{n,1}(z) = n \sum_{k=0}^{N-1} \frac{(-)^k}{k!} \Gamma(nk + a)(n^{1/n}z)^{-nk-a} + R_N(z), \qquad (6.2.32)$$

where

$$R_N(z) = -\frac{\pi n}{2\pi i} \int_{-c+N-\infty i}^{-c+N+\infty i} \frac{\Gamma(ns + a)}{\Gamma(s + 1) \sin \pi s} (n^{1/n}z)^{-ns-a} \, ds \qquad (6.2.33)$$

in $|\arg z| < \frac{1}{2}\pi(1 + 1/n)$ and $0 < c < 1$.

We observe that (6.2.33) provides us with a representation, valid in a sector enclosing the Stokes lines $\arg z = \pm\pi/n$ in the algebraic sector, of the remainder term $R_N(z)$ in the asymptotic expansion (6.2.30a) truncated after N terms. Although (6.2.32) holds for arbitrary positive integer N, we shall take N to correspond to the optimal truncation of the dominant algebraic series at its least term $k = N$, given by

$$(n^{1/n}|z|)^n \sim \left| \frac{\Gamma(nk + a + n)}{\Gamma(nk + a)} \right| \frac{k!}{(k + 1)!}$$

$$= \frac{(nk)^n}{k} |1 + k^{-1}(a + \tfrac{1}{2}n - \tfrac{3}{2}) + O(k^{-2})|.$$

We accordingly set

$$(n - 1)N + a + \tfrac{1}{2}n - \tfrac{3}{2} + \alpha = \frac{n - 1}{n} |z|^{n/(n-1)},$$

where $|\alpha|$ is bounded.

Proceeding along the same lines as in §6.2.1, we now use the inverse factorial expansion in Lemma 2.2

$$
\frac{\Gamma(ns+1)}{\Gamma(s+1)} = \frac{1}{2\pi} (\kappa^{n-1}/n)^{-s} \left\{ \sum_{j=0}^{M-1} (-)^j A'_j \Gamma\left(n\kappa s + a - \tfrac{1}{2} - j\right) \right.
$$

$$
\left. + \rho_M(s)\Gamma\left(n\kappa s + a - \tfrac{1}{2} - M\right) \right\},
$$

where $A'_0 = (2\pi n)^{\frac{1}{2}} (n/(n-1))^{a-1}$, M is a positive integer and the coefficients $A'_j = c_j A'_0$, with c_j given in (6.2.29). The remainder term $\rho_M(s) = O(1)$ for $|s| \to \infty$ uniformly in $|\arg s| \leq \pi - \delta$, $\delta > 0$. Since for optimal truncation $N = O(|z|^{1/\kappa})$ as $|z| \to \infty$, we may suppose N to be sufficiently large for the path of integration in (6.2.33) to lie to the right of all the poles of the integrand. Substitution of the above inverse factorial expansion into (6.2.33) followed by use of the reflection formula for the gamma function and replacement of the variable s by $s + N$ in the integrals appearing in the finite sum, then leads to the result, valid in $|\arg z| < \tfrac{1}{2}\pi(1 + 1/n)$,

$$
R_N(z) = -\frac{nA'_0(n^{1/n}z)^{-a}}{4\pi i} \int_{-c+N-\infty i}^{-c+N+\infty i} \left\{ \sum_{j=0}^{M-1} (-)^j c_j \Gamma\left(n\kappa s + a - \tfrac{1}{2} - j\right) \right.
$$

$$
\left. + \rho_M(s)\Gamma\left(n\kappa s + a - \tfrac{1}{2} - M\right) \right\} \frac{(\kappa z^{1/\kappa})^{-n\kappa s}}{\sin \pi s} ds
$$

$$
= (-)^{N+1} n\kappa (\kappa z^{1/\kappa})^{\vartheta-\mu} \sum_{j=0}^{M-1} (-)^j A_j \frac{1}{4\pi i} \int_{-c-\infty i}^{-c+\infty i} \Gamma(n\kappa s + \mu - j)
$$

$$
\times \frac{(\kappa z^{1/\kappa})^{-n\kappa s}}{\sin \pi s} ds + R_{N,M}(z),
$$

where the coefficients A_j are those in (6.2.28),

$$
\mu = n\kappa N + a - \tfrac{1}{2}, \qquad \kappa = (n-1)/n
$$

and the remainder term $R_{N,M}(z)$ is defined in an obvious manner as an integral involving $\rho_M(s)$; see (6.2.37).

The integrals appearing inside the sum over j can be expressed in terms of the *generalised terminant function* $\hat{T}_\nu^{(m)}(x)$ defined for positive integer m by

$$\hat{T}_\nu^{(m)}(x) = m(xe^{-\pi i/m})^{-\nu} \exp(xe^{-\pi i/m}) \frac{1}{4\pi} \int_{-c-\infty i}^{-c+\infty i} \frac{\Gamma(ms+\nu)}{\sin \pi s} x^{-ms} \, ds \tag{6.2.34}$$

$$(|\arg x| < \tfrac{1}{2}\pi(1 + \frac{2}{m}), \quad 0 < c < 1; \quad \nu \neq 0, -1, -2, \ldots).$$

When $m = 1$ it is seen by comparison with (6.2.7) and (6.2.6b) that $\hat{T}_\nu^{(1)}(x) \equiv \hat{T}_\nu(x)$. In Paris (1992b), it is shown that $\hat{T}_\nu^{(m)}(x)$ can be expressed as a linear combination of m 'basic' scaled terminant functions $\hat{T}_\nu(x)$ with rotated argument, given by

$$\hat{T}_\nu^{(m)}(x) = e^{-\chi} \sum_{j=0}^{m-1} \omega_j e^{\omega_j \chi} \hat{T}_\nu(\omega_j \chi), \tag{6.2.35}$$

$$\omega_j = e^{-2\pi i j/m}, \qquad \chi = xe^{\pi i(m-1)/m}.$$

The asymptotic expansion of $\hat{T}_\nu^{(m)}(x)$ can then be constructed from that of $\hat{T}_\nu(x)$ given in (6.2.56).

After a little algebraic reduction we find the final result in the form

$$R_N(z) = Z^\vartheta e^Z \sum_{j=0}^{M-1} A_j Z^{-j} \hat{T}_{\mu-j}^{(n-1)}(\kappa z^{1/\kappa}) + R_{N,M}(z), \tag{6.2.36}$$

where†

$$Z = \kappa(ze^{-\pi i})^{1/\kappa} = -\kappa(ze^{-\pi i/n})^{1/\kappa}.$$

The remainder $R_{N,M}(z)$ in the truncation of $R_N(z)$ after M terms involves the integral

$$\frac{1}{2\pi i} \int_{-c+M-\infty i}^{-c+M+\infty i} \rho_M(s) \Gamma\left(n\kappa s + a - \tfrac{1}{2} - M\right) \frac{(\kappa z^{1/\kappa})^{-n\kappa s}}{\sin \pi s} \, ds. \tag{6.2.37}$$

This integral can be treated in a similar manner to that arising in §6.2.2 to show that‡

$$|R_{N,M}(z)| = \begin{cases} O(Z^{\vartheta-M} e^{-|Z|}) & \text{in} \quad |\arg z| \leq \pi/n \\ O(Z^{\vartheta-M} e^Z) & \text{in} \quad \pi/n \leq \arg z < \pi \end{cases}$$

as $|z| \to \infty$.

---

† We note that on the Stokes line $\arg z = \pi/n$, Z is negative real.

‡ The sector of validity of the second of these estimates was given in a narrower sector in Paris (1992b, Eq. (3.14)).

From (6.2.30) and (6.2.36), the Stokes multiplier $S_N(\theta)$ for the ray arg $z = \pi/n$ is then given by (for fixed $|z|$)

$$S_N(\theta) \sim \sum_{j=0}^{\infty} c_j Z^{-j} \hat{T}_{\mu-j}^{(n-1)}(\kappa z^{1/\kappa}),$$

where the c_j are the normalised coefficients in (6.2.29). If we use only leading-order terms it can be shown that, as $|z| \to \infty$ in the neighbourhood of the Stokes line arg $z = \pi/n$, we have from (6.2.35) and (6.2.56)

$$S_N(\theta) \sim \hat{T}_{\mu}^{(n-1)}(\kappa z^{1/\kappa}) \sim \hat{T}_{\mu}\big(\kappa (ze^{-\pi i/n})^{1/\kappa} e^{\pi i}\big)$$

$$\sim \tfrac{1}{2} + \tfrac{1}{2}\mathrm{erf}\big\{c(\phi)(\tfrac{1}{2}|Z|)^{\frac{1}{2}}\big\},$$

where $\phi = \pi + \kappa^{-1}(\theta - \pi/n)$ and, from (6.2.52),

$$c(\phi) = \kappa^{-1}\left(\theta - \frac{\pi}{n}\right) + \tfrac{1}{6}i\left\{\kappa^{-1}\left(\theta - \frac{\pi}{n}\right)\right\}^2 - \tfrac{1}{36}\left\{\kappa^{-1}\left(\theta - \frac{\pi}{n}\right)\right\}^3 + \cdots.$$

The transition of the multiplier in the vicinity of arg $z = \pi/n$ is therefore smooth and described by the approximate functional form (when $n \geq 2$)

$$S_N(\theta) \sim \tfrac{1}{2} + \tfrac{1}{2}\mathrm{erf}\left\{\left(\theta - \frac{\pi}{n}\right)\left(\frac{|z|^{1/\kappa}}{2\kappa}\right)^{1/2}\right\}. \tag{6.2.38}$$

Thus when $|z|$ is large and fixed, $S_N(\theta)$ changes smoothly as θ crosses the ray arg $z = \pi/n$ from approximately 0 to 1 over a θ range of scale $|z|^{-1/2\kappa}$. A similar treatment can be given for the Stokes line on arg $z = -\pi/n$.

The situation for the solution $U_{n,p}(z)$ when $p > 1$ is dealt with in essentially the same manner. However, two complications arise which make the analysis in the neighbourhood of the Stokes lines arg $z = \pm\pi p/n$ more involved than in the case when $p = 1$. These are the necessity of optimally truncating p different algebraic expansions and the analogue of the remainder integral in (6.2.33) now involving a product of p sines in the denominator. This product can be expressed in terms of a partial fraction representation to find that the remainder integral can be decomposed into a sum of p separate integrals each containing only a single sine term in the denominator. The interested reader may find details of these calculations, together with a brief discussion of an exponentially-small solution of (6.2.25) when $n = 4$, $p = 2$, in Paris (1992b). The final result is that the Stokes multipliers in the neighbourhood of the rays arg $z = \pm\pi p/n$ possess the approximate form as $|z| \to \infty$

$$\tfrac{1}{2} \pm \tfrac{1}{2}\mathrm{erf}\left\{\left(\theta \mp \frac{\pi p}{n}\right)\left(\frac{|z|^{1/\kappa}}{2\kappa}\right)^{1/2}\right\}, \tag{6.2.39}$$

respectively, as $|z| \to \infty$ where now $\kappa = (n - p)/n$.

To conclude this section we comment on the new features introduced into the analysis of the Stokes multipliers for the nth-order differential equation (6.2.25) compared to that in §6.2.1. First, as the large $|z|$ behaviour of $U_{n,p}(z)$ consists of p algebraic expansions, each generally requires a *different* optimal truncation in order to control the magnitude of the dominant algebraic expansions to within exponential accuracy to detect the multiplier of the subdominant exponential series. Remarkably, the cumulative effect of all these different truncations is still to yield the simple error-function smoothing law given by (6.2.39). Secondly, the late terms in each algebraic expansion can be shown to correspond to divergence controlled essentially by $(k!)^{n-p}/(n^{-\kappa}z)^{nk+\beta_r}$. The validity of the error-function smoothing law for this type of 'superfactorial' divergence of the late terms is verified by considering (6.2.25) of order $n \geq 3$. And thirdly, the determination of the change of the multipliers across the Stokes lines $\arg z = \pm\pi p/n$ when $n \geq 3$ requires the introduction of a more general terminant function than $\hat{T}_\nu(x)$. This generalised terminant for the differential equation (6.2.25) involves a sum of $n - p$ 'basic' terminant functions $\hat{T}_\nu(x)$ with suitably rotated arguments.

We remark that the exponentially-improved asymptotics and the smoothing of the Stokes phenomenon for the generalised Bessel function, defined by $\sum_{k=0}^{\infty} z^k/\{k!\,\Gamma(ak+b)\}$, with $-1 < a < \infty$, b arbitrary, has been studied by Wong & Zhao (1999a,b). These authors adopted the modification of the method of steepest descent introduced by Berry & Howls (1991); see also Boyd (1993, 1994).

6.2.5 Numerical Examples

We illustrate the theory developed in the preceding sections by means of two numerical examples. The first example we consider is the modified Bessel function $K_\nu(z)$, which is expressed in terms of a confluent hypergeometric function by the relation

$$K_\nu(z) = \pi^{\frac{1}{2}} e^{-z}(2z)^\nu \, U\!\left(\nu + \tfrac{1}{2}; 2\nu + 1; 2z\right). \qquad (6.2.40)$$

From (6.1.10), the Stokes multiplier for $K_\nu(z)$ is given by (for fixed $|z|$)

$$S_N(\theta) = \left\{ K_\nu(z) - \left(\frac{\pi}{2z}\right)^{\frac{1}{2}} e^{-z} \sum_{r=0}^{N-1} A_r(\nu)(2z)^{-r} \right\} \frac{e^{-z}}{2i \cos \pi \nu} \left(\frac{\pi}{2z}\right)^{-\frac{1}{2}},$$

where the coefficients $A_r(\nu)$ are defined in (6.1.5). The parameters in the confluent hypergeometric function correspond to $a = \nu + \tfrac{1}{2}$ and $b = \tfrac{1}{2} - \nu$, and from (6.2.13) we have the optimal truncation index $N = 2|z| + \alpha$ with the parameter $\vartheta = a + b - 1 = 0$.

We take $\nu = \tfrac{1}{4}$ and $|z| = 15$ (so that $N = 30$ and $\alpha = 0$) and compute the multiplier $S_N(\theta)$ for various values of θ in the neighbourhood of the Stokes line $\arg z = \pi$. The computations were carried out using *Mathematica*, where we note that the evaluation of $K_{1/4}(z)$ for $\arg z \geq \pi$ is more easily achieved using the connection formula in (6.1.7). The results of such computations are shown in Tables 6.1 and 6.2, where the values of $S_N(\theta)$ are compared with the approximate behaviour in (6.1.11) and with the sum of the first two terms in the terminant expansion (6.2.19). In this last form the terminant functions were computed from the expansion (6.2.56) with the sum truncated at the first term (so that the expansion is valid to $O(z^{-1})$).

As a second example, we consider the Laplace integral

$$I(z) = \int_{-\infty}^{\infty} \exp\{zt - t^4\} \, dt. \tag{6.2.41}$$

Table 6.1. *Values of the Stokes multiplier $S_N(\theta)$ for $K_{1/4}(z)$ with $|z| = 15$ ($N = 30$ and $\alpha = 0$) compared with Berry's approximation (6.1.11)*

θ	$S_N(\theta)$	$\frac{1}{2} + \frac{1}{2}\mathrm{erf}\,\hat{\sigma}$
150°	$0.0021296 + 0.0008950i$	0.001626
160°	$0.0287544 - 0.0007020i$	0.026649
165°	$0.0763062 - 0.0084929i$	0.074595
170°	$0.1687651 - 0.0236490i$	0.168926
175°	$0.3141138 - 0.0405536i$	0.316225
180°	$0.4969818 - 0.0480308i$	0.500000
185°	$0.6798706 - 0.0400449i$	0.683775
190°	$0.8252815 - 0.0226347i$	0.831074
195	$0.9178438 - 0.0069793i$	0.925405
200°	$0.9655395 + 0.0013014i$	0.973351
210°	$0.9925710 + 0.0038370i$	0.998374

Table 6.2. *Values of the Stokes multiplier $S_N(\theta)$ for $K_{1/4}(z)$ with $|z| = 15$ ($N = 30$ and $\alpha = 0$) compared with the first two terms of the terminant expansion (6.2.19)*

θ	$S_N(\theta)$	Asymptotic
150°	$0.0021296 + 0.0008950i$	$0.002137 + 0.000889i$
160°	$0.0287544 - 0.0007020i$	$0.028782 - 0.000787i$
165°	$0.0763062 - 0.0084929i$	$0.076332 - 0.008687i$
170°	$0.1687651 - 0.0236490i$	$0.168767 - 0.023994i$
175°	$0.3141138 - 0.0405536i$	$0.314068 - 0.041035i$
180°	$0.4969818 - 0.0480308i$	$0.496875 - 0.048558i$
185°	$0.6798706 - 0.0400449i$	$0.679706 - 0.040490i$
190°	$0.8252815 - 0.0226347i$	$0.825078 - 0.022909i$
195°	$0.9178438 - 0.0069793i$	$0.917631 - 0.007069i$
200°	$0.9655395 + 0.0013014i$	$0.965345 + 0.001351i$
210°	$0.9925710 + 0.0038370i$	$0.992450 + 0.004014i$

This can be expressed in terms of Faxén's integral $\mathrm{Fi}(\frac{1}{4}, \frac{1}{4}; \pm z)$, defined in (5.5.1), in the form

$$I(z) = \mathrm{Fi}\left(\tfrac{1}{4}, \tfrac{1}{4}; z\right) + \mathrm{Fi}\left(\tfrac{1}{4}, \tfrac{1}{4}; -z\right).$$

Since $I(z)$ is an even function of z and $I(\bar{z}) = \overline{I(z)}$, where the bar denotes the complex conjugate, it is sufficient to confine our attention only to the first quadrant in the z plane.

The expansion of $I(z)$ can be obtained either from the expansion of $\mathrm{Fi}(a, b; z)$ in (5.5.11), where it will be seen that the algebraic expansions present in $\mathrm{Fi}(a, b; \pm z)$ cancel, or from the connection formula (5.5.12) in the form

$$I(z) = E(z) + i E(iz), \tag{6.2.42}$$

where $E(z)$ ($\equiv E_+(z)$) is defined in (5.5.9). The expansion of $E(z)$ is given in (5.5.10) as

$$E(z) \sim 4\sqrt{\frac{\pi}{3}} z^{-\frac{1}{3}} \exp\left\{\tfrac{3}{4} z^{\frac{4}{3}}\right\} \sum_{j=0}^{\infty} c_j \left(\tfrac{3}{4} z^{\frac{4}{3}}\right)^{-j}$$

as $|z| \to \infty$ in the sector $0 < \arg z < \frac{3}{2}\pi$. Since $I(z)$ is closely related to the function defined in (8.1.50), the recurrence formula for the coefficients c_j is given by the 3-term recurrence (8.1.55) in the form

$$3 j c_j = B_1 c_{j-1} + B_2 c_{j-2} + B_3 c_{j-3} \qquad (j \geq 1),$$

with $c_0 = 1$, $c_{-1} = c_{-2} = 0$ and

$$B_1 = 6j^2 - \tfrac{15}{2} j + \tfrac{31}{16},$$
$$B_2 = -4\left(j - \tfrac{7}{4}\right)^3 - \tfrac{15}{2}\left(j - \tfrac{7}{4}\right)^2 - \tfrac{23}{8}\left(j - \tfrac{7}{4}\right) + \tfrac{5}{32},$$
$$B_3 = \left(j - \tfrac{1}{2}\right)\left(j - \tfrac{5}{4}\right)(j - 2)\left(j - \tfrac{11}{4}\right).$$

Alternatively, the above recurrence formula may be obtained by application† of Lemma 2.3 to the ratio of gamma functions $\Gamma(s)/\Gamma(\frac{1}{4}s + \frac{3}{4})$ that appears in the integral representation of $E(z)$ in (5.5.9). We also note that a closed-form evaluation for the coefficients c_j is given in a footnote surrounding the discussion of (8.3.11).

The expansion of $I(z)$ in (6.2.42) thus involves two exponential series with the controlling exponential behaviours $\exp\{\frac{3}{4} z^{4/3}\}$ and $\exp\{\frac{3}{4}(iz)^{4/3}\}$. It is easily shown that in the first quadrant these two exponentials are most unequal in absolute value when $\arg z = \frac{1}{8}\pi$. On this ray the exponential factor $\exp\{\frac{3}{4} z^{4/3}\}$ is then *maximally* dominant over the other exponential factor, with the result that $\arg z = \frac{1}{8}\pi$ is a Stokes line. Examination of the saddle points of (6.2.41) and the associated

---

† The multiplication formula for the gamma function must first be applied to $\Gamma(s)$ to express it in terms of a product of gamma functions with arguments involving $\frac{1}{4}s$; compare Example 1 of §2.2.3.

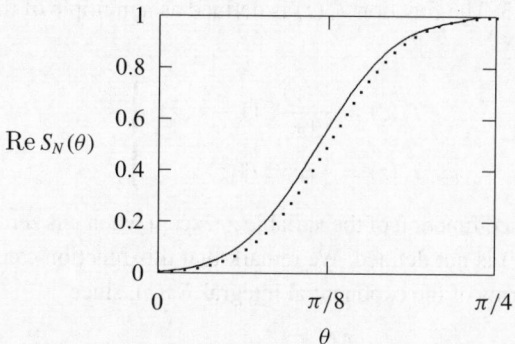

Fig. 6.2. The real part of $S_N(\theta)$ for $I(z)$ when $|z| = 10$ showing the smooth transition across the Stokes line at arg $z = \frac{1}{8}\pi$. The dotted curve represents the graph of the approximate form $\frac{1}{2} + \frac{1}{2}\operatorname{erf}\hat{\sigma}$.

paths of steepest descent connecting $-\infty$ to $+\infty$ [see Paris (1991)] readily shows that $I(z)$ consists of a single exponential series in $|\arg z| < \frac{1}{8}\pi$, whereas in $\frac{1}{2}\pi \leq \arg z < \frac{1}{8}\pi$ we have a compound expansion consisting of two exponential series; compare also the discussion surrounding the expansion of the reduced Pearcey integral in (8.3.10).

To show that (6.2.42) embodies the Stokes transition at $\theta = \arg z = \frac{1}{8}\pi$, we have calculated the variation of the Stokes multiplier $S_N(\theta)$ (for fixed $|z|$) as arg $z - \frac{1}{8}\pi$ varies from negative values to positive values. This is achieved by writing the right-hand side of (6.2.42) in the form

$$I(z) = 4\sqrt{\frac{\pi}{3}}z^{-\frac{1}{3}}\exp\left\{\tfrac{3}{4}z^{\frac{4}{3}}\right\}\sum_{j=0}^{N-1}c_j\left(\tfrac{3}{4}z^{\frac{4}{3}}\right)^{-j}$$
$$+ S_N(\theta)4\sqrt{\frac{\pi}{3}}z^{-\frac{1}{3}}\exp\left\{\tfrac{3}{4}z^{\frac{4}{3}}e^{\frac{2}{3}\pi i} + \tfrac{1}{3}\pi i\right\},$$

where N is chosen to correspond to the optimal truncation point for a given large $|z|$. The results of such computations for $|z| = 10$ are illustrated in Fig. 6.2, where it is found that the leading behaviour of $S_N(\theta)$ is real and varies smoothly from approximately zero to unity over an interval in θ of about $\frac{1}{4}\pi$. This variation agrees well with the approximate behaviour in (6.1.11) and (6.1.12) given by $S_N(\theta) \simeq \frac{1}{2} + \frac{1}{2}\operatorname{erf}\hat{\sigma}$ with $\hat{\sigma} = \operatorname{Im}(F)/(2\operatorname{Re}(F))^{\frac{1}{2}}$, where the singulant $F(z)$ is

$$F(z) = \tfrac{3}{4}z^{\frac{4}{3}} - \tfrac{3}{4}z^{\frac{4}{3}}e^{\frac{2}{3}\pi i} = \frac{3\sqrt{3}}{4}|z|^{\frac{4}{3}}\exp\left\{\tfrac{4}{3}i(\theta - \tfrac{1}{8}\pi)\right\}.$$

6.2.6 Asymptotics of the Terminant $T_\nu(z)$

In this section we give a discussion of the main asymptotic properties of the terminant function $T_\nu(z)$ that are required to determine the structure of the Stokes multiplier in §6.2.3. The function $T_\nu(z)$ is defined as a multiple of the incomplete gamma function by

$$\left.\begin{aligned} T_\nu(z) &= \frac{\Gamma(\nu)}{2\pi}\Gamma(1-\nu, z) \\[2mm] \hat{T}_\nu(z) &= -ie^{\pi i\nu}T_\nu(z) \end{aligned}\right\} \tag{6.2.43}$$

and is a multi-valued function of the variable z, except when ν is zero or a negative integer when $T_\nu(z)$ is not defined. We remark that this function can be expressed alternatively in terms of the exponential integral $E_\nu(z)$, since

$$E_\nu(z) = z^{\nu-1}\int_z^\infty \frac{e^{-t}}{t^\nu}\, dt = z^{\nu-1}\Gamma(1-\nu, z)$$

provided the integration path does not pass through $t = 0$.

Since

$$\Gamma(1-\nu, z) = e^{-z}U(\nu; \nu; z), \tag{6.2.44}$$

we can use the properties of the confluent hypergeometric function $U(a; b; z)$ to determine the main properties of $T_\nu(z)$. For example, use of the connection formula

$$e^{\pi i\nu}U(\nu; \nu; ze^{2\pi i}) = e^{-\pi i\nu}U(\nu; \nu; z) + \frac{2\pi i}{\Gamma(\nu)}e^z$$

shows that the corresponding connection formula for the scaled terminant function $\hat{T}_\nu(z)$ is given by

$$\hat{T}_\nu(ze^{\pi i}) = 1 + e^{-2\pi i\nu}\hat{T}_\nu(ze^{-\pi i}). \tag{6.2.45}$$

When dealing with the asymptotic properties, it will be found more convenient to consider the scaled terminant $\hat{T}_\nu(z)$, rather than $T_\nu(z)$. We first consider $|z| \to \infty$ and *finite* values of ν. The asymptotic expansion of $\hat{T}_\nu(z)$ in this case is found from (6.2.43), (6.2.44) and (6.2.1a) in the form

$$\hat{T}_\nu(z) \sim \frac{\Gamma(\nu)}{2\pi i}(ze^{-\pi i})^{-\nu}e^{-z}\sum_{k=0}^\infty (-)^k(\nu)_k\, z^{-k}, \tag{6.2.46}$$

valid as $|z| \to \infty$ in the sector $|\arg z| \le \frac{3}{2}\pi - \delta$, $\delta > 0$. The corresponding expansion in other phase ranges can be obtained from the above expansion and the connection formula (6.2.45). Thus, for example, we find

$$\hat{T}_\nu(z) \sim 1 + \frac{\Gamma(\nu)}{2\pi i}(ze^{-\pi i})^{-\nu}e^{-z}\sum_{k=0}^\infty (-)^k(\nu)_k\, z^{-k}, \tag{6.2.47}$$

valid as $|z| \to \infty$ in the sector $\frac{1}{2}\pi + \delta \le \arg z \le \frac{7}{2}\pi - \delta$. In the common sector $\frac{1}{2}\pi + \delta \le \arg z \le \frac{3}{2}\pi - \delta$, these expansions differ by unity; this term is exponentially small compared with the main contribution, so that in the Poincaré sense there is no discrepancy. The appearance of the term unity across the ray $\arg z = \pi$ is due to the Stokes phenomenon.

When used as an approximant in the exponentially-improved expansion developed in (6.2.18), the terminant function is associated with $\nu = |z| + O(1)$ and large $|z|$; a discussion of the situation when ν is not so restricted is given in Boyd (1990). Temme (1979) has considered the asymptotic expansion of the incomplete gamma function $\Gamma(a, z)$ for large values of a uniformly valid in the variable z. The domain of validity of this expansion, however, does not correspond to the domain of $a = 1 - \nu$ and z required here.

To determine the asymptotic expansion for this range of variables we use the Laplace integral representation for $\hat{T}_\nu(z)$. This is obtained from the Euler integral representation for $U(a; b; z)$ given in Abramowitz & Stegun (1965, p. 505, Eq. (13.2.5)) and takes the form

$$\hat{T}_\nu(z) = \frac{e^{\pi i \nu}}{2\pi i} e^{-z} \int_0^\infty e^{-zt} \frac{t^{\nu-1}}{1+t} \, dt \qquad (6.2.48)$$

when $\mathrm{Re}(\nu) > 0$ and $|\arg z| < \frac{1}{2}\pi$. Analytic continuation of this integral by rotation of the path of integration through the angle $\theta = \arg z$, followed by replacement of the variable of integration t by $\tau e^{-i\theta}$ and use of Cauchy's theorem, then shows that

$$\hat{T}_\nu(z) = \frac{e^{(\pi-\theta)i\nu}}{2\pi i} e^{-z} \int_0^\infty e^{-|z|\tau} \frac{\tau^{\nu-1}}{1+\tau e^{-i\theta}} \, d\tau \qquad (6.2.49)$$

valid when $|\theta| < \pi$. If we let $\nu = |z| + \alpha$, where $|\alpha|$ is bounded, the asymptotic behaviour of the above integral as $|z| \to \infty$ is characterised by a saddle point at $\tau = 1$ and a pole of the integrand at $\tau = -e^{i\theta}$. On the Stokes lines $\theta = \pm\pi$ these points become coincident.

For large $|z|$ with $\nu \sim |z|$, we can apply Laplace's method [see, for example, Olver (1974, p. 127)] in a straightforward manner to find that

$$\hat{T}_\nu(z) \sim -\frac{ie^{(\pi-\theta)\nu i}}{1+e^{-i\theta}} \frac{e^{-z-|z|}}{\sqrt{2\pi|z|}} \sum_{k=0}^\infty a_{2k}(\theta, \alpha)\, |z|^{-k} \qquad (6.2.50)$$

uniformly as $|z| \to \infty$ in the sector $-\pi + \delta \le \theta \le \pi - \delta$. The values of the first few coefficients $a_{2k}(\theta, \alpha)$ are given by [see Olver (1991a)]

$$a_0(\theta, \alpha) = 1, \qquad a_2(\theta, \alpha) = \tfrac{1}{12} + \Lambda_2(\theta, \alpha),$$

$$a_4(\theta, \alpha) = \tfrac{1}{288} + \tfrac{1}{12}\Lambda_2(\theta, \alpha) + 2\Lambda_3(\theta, \alpha) + 3\Lambda_4(\theta, \alpha), \dots,$$

where

$$\Lambda_r(\theta, \alpha) = \sum_{j=0}^r (-)^j \binom{\alpha}{r-j} (1 + e^{i\theta})^{-j}.$$

It is immediately apparent that this expansion breaks down as $\theta \to \pm\pi$; this is due to the pole of the integrand coinciding with the saddle point at $\theta = \pm\pi$.

A uniform expansion for $\hat{T}_\nu(z)$, which is valid in a sector enclosing the rays $\arg z = \pm\pi$, has been given by Olver (1990, 1991a). Here we present an outline of his derivation of this important expansion. We first make the standard quadratic change of variables in the integral (6.2.49)

$$\tfrac{1}{2}w^2 = \tau - \log\tau - 1,$$

so that the saddle point $\tau = 1$ corresponds to $w = 0$. The branch of w is chosen such that $w \sim \tau - 1$ as $\tau \to 1$. The pole at $\tau = e^{i(\theta-\pi)}$ then corresponds to the point $w = w_0$, where $w_0 = ic(\theta)$ and $c(\theta)$ is defined by

$$\tfrac{1}{2}c^2(\theta) = 1 + i(\theta - \pi) - e^{i(\theta-\pi)}. \tag{6.2.51}$$

With the branch for w chosen as above, this corresponds to $c(\theta) \sim \theta - \pi$ when $\theta \simeq \pi$. The Taylor series expansion of $c(\theta)$ in the neighbourhood of $\theta = \pi$ is given by

$$c(\theta) = \theta - \pi + \tfrac{1}{6}i(\theta - \pi)^2 - \tfrac{1}{36}(\theta - \pi)^3 + \cdots. \tag{6.2.52}$$

A plot of the locus of $c(\theta)$ as a function of θ in the range $-\pi \le \theta \le 3\pi$ is shown in Fig. 6.3. The significance of the quantity $c(\theta)$ is that it measures the proximity of the saddle point at $\tau = 1$ to the pole at $\tau = e^{i(\theta-\pi)}$.

The integral (6.2.49) is valid in $|\theta| < \pi$. However, an extension of the analytic continuation process used to obtain (6.2.49) shows that this integral continues to be valid in $\pi \le \theta < 2\pi$, provided the integration path is deformed to pass over the pole at $\tau = e^{i(\theta-\pi)}$. In terms of the new variable w, (6.2.49) then becomes

$$\hat{T}_\nu(z) = \frac{e^{(\pi-\theta)i\nu}}{2\pi i} e^{-z-|z|} \int_{-\infty}^{\infty} e^{-\frac{1}{2}|z|w^2} f(w)\, dw, \tag{6.2.53}$$

where

$$f(w) = \frac{\tau^{\alpha-1}}{1 + \tau e^{-i\theta}} \frac{d\tau}{dw} = \frac{w\tau^\alpha}{(\tau - 1)(1 + \tau e^{-i\theta})} \tag{6.2.54}$$

and the path is indented to pass *above* the point $w = w_0$ when $\theta \ge \pi$. The function $f(w)$ has a simple pole at $w = w_0$, and accordingly we separate off this

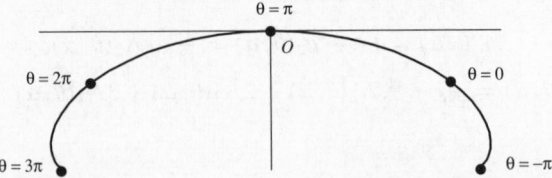

Fig. 6.3. The locus of $c(\theta)$ as a function of θ. The points $\theta = -\pi$ and $\theta = 3\pi$ correspond to $\pm 2\sqrt{\pi}e^{\mp\frac{1}{4}\pi i}$, respectively.

pole by writing†

$$f(w) = A_0 \left\{ \frac{1}{w - w_0} + g(w) \right\},$$

where $g(w)$ is analytic at the point $w = w_0$ and $A_0 = \lim_{w \to w_0} (w - w_0) f(w)$. The value of A_0 can be found by letting $\tau \to e^{i(\theta - \pi)}$ in (6.2.54) and using l'Hospital's rule to find $A_0 = -e^{(\theta - \pi)i\alpha}$.

Substituting the above expansion for $f(w)$ into (6.2.53), we obtain

$$\hat{T}_v(z) = \frac{e^{(\pi - \theta)iv}}{2\pi i} e^{-z - |z|} A_0 \left\{ \int_{-\infty}^{\infty} \frac{e^{-\frac{1}{2}|z|w^2}}{w - w_0} \, dw + \int_{-\infty}^{\infty} g(w) \, e^{-\frac{1}{2}|z|w^2} \, dw \right\}.$$

(6.2.55)

The first integral on the right-hand side of the above equation (where we recall that the path is indented to pass above the pole $w = w_0$ when $\theta \geq \pi$) can be evaluated in terms of the complementary error function

$$\int_{-\infty}^{\infty} \frac{e^{-\frac{1}{2}|z|w^2}}{w - w_0} \, dw = -\pi i e^{\frac{1}{2}|z|c^2(\theta)} \, \text{erfc} \left\{ - c(\theta)(\tfrac{1}{2}|z|)^{\frac{1}{2}} \right\}, \qquad w_0 = ic(\theta).$$

In the second integral the path may be taken to be the real axis with no indentation, since the integrand has no singularity on the path of integration. If we expand $g(w)$ as a Maclaurin series

$$g(w) = \sum_{r=0}^{\infty} b_r(\theta, \alpha) \, w^r$$

and integrate term by term, we find (formally)

$$\int_{-\infty}^{\infty} g(w) \, e^{-\frac{1}{2}|z|w^2} \, dw \sim \sum_{k=0}^{\infty} \Gamma\left(k + \tfrac{1}{2}\right) b_{2k}(\theta, \alpha) \left(\tfrac{1}{2}|z|\right)^{-k-\frac{1}{2}}$$

for large $|z|$.

Collecting together these results in (6.2.55), we finally obtain the desired expansion valid as $|z| \to \infty$ (when $v = |z| + \alpha$) uniformly in the sector $\delta \leq \arg z \leq 2\pi - \delta$ in the form

$$\hat{T}_v(z) \sim \tfrac{1}{2} + \tfrac{1}{2} \text{erf} \left\{ c(\theta)(\tfrac{1}{2}|z|)^{\frac{1}{2}} \right\} - \frac{i e^{-\frac{1}{2}|z|c^2(\theta)}}{\sqrt{2\pi |z|}} \sum_{k=0}^{\infty} \left(\tfrac{1}{2}\right)_k b_{2k}(\theta, \alpha) \left(\tfrac{1}{2}|z|\right)^{-k},$$

(6.2.56)

where, from (6.2.51), we have made use of the identity

$$\exp \left\{ - \tfrac{1}{2}|z|c^2(\theta) \right\} = \exp \left\{ -z - |z| + (\pi - \theta)i|z| \right\}$$

(6.2.57)

---

† The functions $f(w)$ and $g(w)$ are, of course, also functions of θ and α.

and $b_{2k}(\theta, \alpha)$ are coefficients independent of $|z|$. It can be shown that the expansion (6.2.56) holds uniformly in the wider sector $-\pi + \delta \leq \arg z \leq 3\pi - \delta$; for details, the reader is referred to Olver (1991a, §4). A conjugate expansion holds in the sector $-3\pi + \delta \leq \arg z \leq \pi - \delta$.

In the sector $|\arg z| \leq \pi - \delta$, the quantity $-c(\theta)$ lies in the fourth quadrant bounded away from the origin (see Fig. 6.3). The expansion of the error function in (6.2.56) can then be obtained from the result given in Abramowitz & Stegun (1965, p. 298)

$$\text{erfc } z \sim \frac{e^{-z^2}}{z\sqrt{\pi}} \sum_{k=0}^{\infty} (-)^k \left(\tfrac{1}{2}\right)_k z^{-2k} \qquad (|z| \to \infty; \; |\arg z| < \tfrac{3}{4}\pi).$$

This leads to the expansion as $|z| \to \infty$

$$\hat{T}_\nu(z) \sim -\frac{e^{-\frac{1}{2}|z|c^2(\theta)}}{\sqrt{2\pi|z|}} \sum_{k=0}^{\infty} \left(\tfrac{1}{2}\right)_k \left\{ ib_{2k}(\theta, \alpha) + \frac{(-)^k}{(c(\theta))^{2k+1}} \right\} \left(\tfrac{1}{2}|z|\right)^{-k} \qquad (6.2.58)$$

valid in $\delta \leq \arg z \leq \pi - \delta$. From this last result, together with (6.2.57) and the connection formula (6.2.45), it follows that as $|z| \to \infty$ (with $\nu = |z| + \alpha$) the (unscaled) terminant function $T_\nu(z)$ possesses the behaviour

$$T_\nu(z) = \begin{cases} e^{-z-|z|} O(|z|^{-\frac{1}{2}}) & \text{in} \quad |\arg z| \leq \pi - \delta \\ \pm i e^{\mp \pi i \nu} + e^{-z-|z|} O(|z|^{-\frac{1}{2}}) & \text{in} \quad \pi + \delta \leq \pm \arg z \leq 3\pi - \delta, \end{cases}$$
$$(6.2.59)$$

where the upper or lower signs are taken according as $\arg z > 0$ or $\arg z < 0$, respectively. In the region $\pi - \delta \leq |\arg z| \leq \pi + \delta$ enclosing the Stokes lines, the smooth transition of the leading behaviour of $T_\nu(z)$ is described by the error function.

The coefficients $b_{2k}(\theta, \alpha)$ in the above expansion can be expressed in terms of the coefficients $a_{2k}(\theta, \alpha)$ appearing in (6.2.50). This is achieved by comparison of the expansions (6.2.50) and (6.2.56) in their common sector of validity $\delta \leq \arg z \leq \pi - \delta$. The former expansion can be rearranged, with the aid of (6.2.57), in the form

$$\hat{T}_\nu(z) \sim -\frac{i e^{(\pi-\theta)i\nu}}{1 + e^{-i\theta}} \frac{e^{-\frac{1}{2}|z|c^2(\theta)}}{\sqrt{2\pi|z|}} \sum_{k=0}^{\infty} a_{2k}(\theta, \alpha) |z|^{-k} \qquad (6.2.60)$$

valid in the sector $-\pi + \delta \leq \arg z \leq \pi - \delta$. Since the uniform expansions in (6.2.58) and (6.2.60) represent the same function, they must be identical. Upon equating coefficients, we therefore find that, for $k = 0, 1, 2, \ldots$,

$$b_{2k}(\theta, \alpha) = \frac{e^{(\pi-\theta)i\alpha}}{1 + e^{-i\theta}} \frac{a_{2k}(\theta, \alpha)}{2^k \left(\tfrac{1}{2}\right)_k} + \frac{(-)^k i}{(c(\theta))^{2k+1}}. \qquad (6.2.61)$$

When $\theta = \pi$, the right-hand side is to be replaced by its limiting value. When $k = 0$, we consequently have

$$b_0(\theta, \alpha) = \frac{e^{(\pi - \theta)i\alpha}}{1 + e^{-i\theta}} + \frac{i}{c(\theta)}, \qquad (6.2.62)$$

which has the limiting value $b_0(\pi, \alpha) = \frac{2}{3} - \alpha$ when $\theta = \pi$.

6.3 Hyperasymptotics

We saw in §6.2.1 that the confluent hypergeometric function $U(a; a - b + 1; z)$ can be written as

$$z^a U(a; a - b + 1; z) = \sum_{k=0}^{n-1} (-)^k \frac{(a)_k (b)_k}{k! \, z^k} + R_n(z),$$

where n is an arbitrary positive integer and $R_n(z)$ is the remainder term defined in (6.2.5). For fixed $|z|$, the sum over k is divergent and so has to be truncated at some value of n for this representation to be of use. Either n can be chosen so that $|R_n(z)| = O(z^{-n})$ is negligibly small in $|\arg z| < \frac{3}{2}\pi$, or n can be chosen to correspond to the optimal truncation value N, thereby minimising $|R_n(z)|$ for a given fixed $|z|$. In this latter case, the remainder term $R_N(z)$ was shown to be uniformly exponentially small in the sector $|\arg z| \leq \pi$. In §6.2.1, this theory was developed from the Mellin-Barnes integral representation for $U(a; a - b + 1; z)$. Other treatments using Laplace and Stieltjes integral definitions have been given in Olver (1990, 1991a,b) and Boyd (1990). Subsequently, it was shown how to obtain such exponentially-improved asymptotic series directly from a second-order ordinary differential equation without prior knowledge of the divergent Poincaré asymptotic series or an integral representation of the solution. This was done first by Berry (1990) in a formal manner and shortly afterwards by McLeod (1992), for a special class of equation, and Olver (1993) for the confluent hypergeometric differential equation. A discussion of exponential improvement for functions defined by a series representation has been given in Jones (1994). We also note that exponential improvement of the remainder term of an integral of Laplace type, to produce a strong form of Watson's lemma, has been considered by Ursell (1990).

The idea of re-expanding the optimally-truncated remainder $R_N(z)$ into another asymptotic series, with a view to enhancing numerical accuracy, goes back† to Stieltjes (1886). This idea has been subsequently developed by several authors, notably Airey (1937) and Miller (1952), in the context of the theory of *converging factors*; for a discussion of the converging factors associated with the confluent hypergeometric functions, see Slater (1960, pp. 61–65). The use of an appropriate expansion for the converging factor leads to exponential improvement in the

---

† It should be noted that Stokes in 1857 also considered a resummation of the optimally-truncated remainder term in the expansion of the Airy function $Ai(z)$; see Paris (1996).

evaluation of an asymptotic expansion. Such a possibility was also considered formally by Dingle (1973) through the use of his terminant functions. In Olver (1974, Chapter 14) the discussion is confined to cases in which rigorous analysis can be supplied.

If the re-expansion of the remainder $R_N(z)$ is now in its turn also optimally truncated, the new remainder resulting from this re-expansion process is found to be exponentially small compared to $R_N(z)$ as $|z| \to \infty$ in a certain sector. The process can be repeated to produce a sequence of re-expanded remainder terms, each of which is exponentially smaller than its predecessor. This procedure of successive exponential improvement has been termed *hyperasymptotics* by Berry & Howls (1990, 1991), who discussed this process in the context of integrals with saddle points; see also Howls (1992) and Berry & Howls (1993b, 1994) for clusters of saddles. It is clear from the outset that, although this will yield at each stage an improved accuracy in the evaluation of a given function, the computation of the different series of terms in the re-expanded remainders will involve progressively more complicated terminant functions (called *hyperterminants*). Thus, *the price of this increased accuracy comes at the cost of considerable computational effort*.

The development of the theory of hyperasymptotic expansions has been carried out for the solutions of general second-order ordinary differential equations with an irregular singularity at infinity of rank one by Olde Daalhuis & Olver (1994, 1995a) and for arbitrary integer rank by Murphy & Wood (1997). In these references the approach used is the Stieltjes transform, initiated by Boyd (1990), which turns out to be an extremely powerful tool when dealing with the asymptotic solutions of general differential equations. Further work in this direction can be found in Olde Daalhuis & Olver (1995b) and Olde Daalhuis (1998) for nth-order differential equations. Here we shall illustrate the hyperasymptotic procedure by the Mellin-Barnes integral approach, again applied to the particular case of the confluent hypergeometric function $U(a; a-b+1; z)$; for a different treatment of this function using Laplace integrals, see Olde Daalhuis (1992) and Jones (1997, Chapter 5). Another method of hyperasymptotic evaluation based on Hadamard expansions has been described in Paris (2000b).

6.3.1 Mellin-Barnes Theory of Hyperasymptotics

We begin this section by introducing some changes in notation for convenience in the presentation of the hyperasymptotic expansion procedure. We denote the remainder term at the jth stage of the hyperasymptotic process by $R_j(z)$ and the associated truncation indices by n_j ($j = 0, 1, 2, \ldots$); we shall follow the practice used in §6.2.2 and denote an optimal truncation index by N_j. When it becomes necessary, we shall indicate the dependence of the remainder $R_j(z)$ on the *optimal* truncation indices up to the jth level by $R_j(z; N_0, N_1, \ldots, N_j)$. We define the constants $\Delta_j(a, b)$ $(j = 0, 1, 2, \ldots)$ by

$$\Delta_j(a, b) = \begin{cases} (\sin \pi a \; \sin \pi b)^{j/2} & (j \text{ even}) \\ -\dfrac{\pi (\sin \pi a \; \sin \pi b)^{(j-1)/2}}{\Gamma(1-a)\Gamma(1-b)} & (j \text{ odd}) \end{cases} \qquad (6.3.1)$$

and denote by $L(n_j)$ the path of integration

$$L(n_j) \; : \; \text{Re}(s_j) = -c_j + n_j$$

parallel to the imaginary axis of the integration variable s_j, where the constants c_j satisfy $0 < c_j < 1$.

We shall illustrate the Mellin-Barnes theory of hyperasymptotics by again using the confluent hypergeometric function $U(a; a - b + 1; z)$ as an example. From (6.2.4) and (6.2.5) we have the result

$$z^a U(a; a - b + 1; z) = \sum_{k=0}^{n_0-1} (-)^k \frac{(a)_k (b)_k}{k! z^k} + R_0(z),$$

where

$$R_0(z) = -\frac{\pi}{\Gamma(a)\Gamma(b)} \frac{1}{2\pi i} \int_{L(n_0)} \frac{\Gamma(s_0 + a)\Gamma(s_0 + b)}{\Gamma(s_0 + 1) \sin \pi s_0} z^{-s_0} \, ds_0 \qquad (6.3.2)$$

valid in the sector $|\arg z| < \frac{3}{2}\pi$. This is the *zeroth level* of the hyperasymptotic process: for finite values of the truncation index n_0, this corresponds to ordinary Poincaré asymptotics where the remainder $|R_0(z)| = O(z^{-n_0 + c_0})$ as $|z| \to \infty$ in $|\arg z| < \frac{3}{2}\pi$.

Next we use the inverse factorial expansion in (6.2.9)

$$\frac{\Gamma(s_0 + a)\Gamma(s_0 + b)}{\Gamma(s_0 + 1)} = \sum_{k=0}^{n_1-1} (-)^k A_k \Gamma(s_0 + \vartheta - k) + r_1(s_0), \qquad (6.3.3)$$

where $\vartheta = a + b - 1$ and the coefficients $A_k = (1 - a)_k (1 - b)_k / k!$. To proceed with the hyperasymptotic expansion, however, we require an exact representation of the remainder term $r_1(s_0)$. From (2.2.43) this takes the form

$$r_1(s_0) = -\frac{\pi}{\Gamma(1-a)\Gamma(1-b)} \frac{1}{2\pi i}$$

$$\times \int_{L(n_1)} \frac{\Gamma(s_1 + 1 - a)\Gamma(s_1 + 1 - b)}{\Gamma(s_1 + 1) \sin \pi s_1} \Gamma(s_0 + \vartheta - s_1) \, ds_1,$$

which is an absolutely convergent integral in $|\arg s_1| < 2\pi$. Substitution of the expansion (6.3.3) into $R_0(z)$ and identification of the integrals in terms of the terminant function† $T_\nu(z)$ in (6.2.8), then yields

$$
R_0(z) = -\frac{\pi}{\Gamma(a)\Gamma(b)} \frac{1}{2\pi i} \int_{L(n_0)} \left\{ \sum_{k=0}^{n_1-1} (-)^k A_k \Gamma(s_0 + \vartheta - k) \right.
$$

$$
\left. + r_1(s_0) \right\} \frac{z^{-s_0}}{\sin \pi s_0}\, ds_0
$$

$$
= \frac{2\pi(-)^{n_0}}{\Gamma(a)\Gamma(b)} z^\vartheta e^z \sum_{k=0}^{n_1-1} (-)^k A_k z^{-k} T_{n_0+\vartheta-k}(z) + R_1(z). \tag{6.3.4}
$$

The sum on the right-hand side of this last expression is the same as that in (6.2.11), except that we have now used the unscaled terminant. The remainder $R_1(z)$ after the first level of the hyperasymptotic expansion is given by

$$
R_1(z) = -\frac{\pi}{\Gamma(a)\Gamma(b)} \frac{1}{2\pi i} \int_{L(n_0)} r_1(s_0) \frac{z^{-s_0}}{\sin \pi s_0}\, ds_0
$$

$$
= \sin \pi a\, \sin \pi b \left(\frac{1}{2\pi i} \right)^2 \int_{L(n_1)} \frac{\Gamma(s_1 + 1 - a)\Gamma(s_1 + 1 - b)}{\Gamma(s_1 + 1)\, \sin \pi s_1}
$$

$$
\times \int_{L(n_0)} \Gamma(s_0 + \vartheta - s_1) \frac{z^{-s_0}}{\sin \pi s_0}\, ds_0\, ds_1 \tag{6.3.5}
$$

upon reversal of the order of integration, which is permissible in the sectors of absolute convergence $|\arg s_0| < \frac{3}{2}\pi$ and $|\arg s_1| < 2\pi$. The inner integral can be written in terms of a terminant function by means of (6.2.8) to find

$$
R_1(z) = \frac{2\pi(-)^{n_0}}{\Gamma(a)\Gamma(b)} z^\vartheta e^z \hat{R}_1(z), \tag{6.3.6}
$$

where

$$
\hat{R}_1(z) = \frac{\Delta_1(a, b)}{2\pi i} \int_{L(n_1)} \frac{\Gamma(s_1 + 1 - a)\Gamma(s_1 + 1 - b)}{\Gamma(s_1 + 1)\, \sin \pi s_1} z^{-s_1}\, T_{n_0+\vartheta-s_1}(z)\, ds_1. \tag{6.3.7}
$$

The above procedure is now repeated. Application of the inverse factorial expansion in (2.2.42) and (2.2.43) shows that the quotient of gamma functions appearing in the integrand of $\hat{R}_1(z)$ is

$$
\frac{\Gamma(s_1 + 1 - a)\Gamma(s_1 + 1 - b)}{\Gamma(s_1 + 1)} = \sum_{k=0}^{n_2-1} (-)^k A_k'\, \Gamma(s_1 - \vartheta - k) + r_2(s_1),
$$

---

† We shall find it more convenient in this section to use the terminant function $T_\nu(z)$, rather than the scaled terminant $\hat{T}_\nu(z)$.

where the coefficients $A'_k = (a)_k(b)_k/k!$ and

$$r_2(s_1) = -\frac{\pi}{\Gamma(a)\Gamma(b)}\frac{1}{2\pi i}\int_{L(n_2)}\frac{\Gamma(s_2+a)\Gamma(s_2+b)}{\Gamma(s_2+1)\,\sin\pi s_2}\Gamma(s_1-\vartheta-s_2)\,ds_2.$$

Substitution of this expansion into the integral for $\hat{R}_1(z)$ in (6.3.7) then leads to

$$\hat{R}_1(z) = -\frac{\Delta_1(a,b)}{2\pi i}\int_{L(n_1)}\left\{\sum_{k=0}^{n_2-1}(-)^k A'_k\,\Gamma(s_1-\vartheta-k)+r_2(s_1)\right\}$$

$$\times\frac{z^{-s_1}}{\sin\pi s_1}T_{n_0+\vartheta-s_1}(z)\,ds_1$$

$$= \Delta_1(a,b)\sum_{k=0}^{n_2-1}(-)^k A'_k\frac{1}{2\pi i}\int_{L(n_1)}\Gamma(s_1-\vartheta-k)\frac{z^{-s_1}}{\sin\pi s_1}$$

$$\times T_{n_0+\vartheta-s_1}(z)\,ds_1+\hat{R}_2(z).$$

The remainder $\hat{R}_2(z)$ is given by

$$\hat{R}_2(z) = \frac{\Delta_1(a,b)}{2\pi i}\int_{L(n_1)}r_2(s_1)\frac{z^{-s_1}}{\sin\pi s_1}T_{n_0+\vartheta-s_1}(z)\,ds_1$$

$$= \Delta_2(a,b)\left(\frac{1}{2\pi i}\right)^2\int_{L(n_2)}\frac{\Gamma(s_2+a)\Gamma(s_2+b)}{\Gamma(s_2+1)\,\sin\pi s_2}\int_{L(n_1)}\Gamma(s_1-\vartheta-s_2)$$

$$\times\frac{z^{-s_1}}{\sin\pi s_1}T_{n_0+\vartheta-s_1}(z)\,ds_1\,ds_2. \tag{6.3.8}$$

This process can clearly be continued since the quotient of gamma functions

$$\frac{\Gamma(s_2+a)\Gamma(s_2+b)}{\Gamma(s_2+1)}$$

in $\hat{R}_2(z)$ is of the same form as that in $R_0(z)$ in (6.3.2). Hence re-expansion of $\hat{R}_2(z)$ leads to the result

$$\hat{R}_2(z) = \Delta_2(a,b)\sum_{k=0}^{n_3-1}(-)^k A_k\left(\frac{1}{2\pi i}\right)^2\int_{L(n_2)}\frac{\Gamma(s_2+\vartheta-k)}{\sin\pi s_2}$$

$$\times\int_{L(n_1)}\Gamma(s_1-\vartheta-s_2)\frac{z^{-s_1}}{\sin\pi s_1}T_{n_0+\vartheta-s_1}(z)\,ds_1\,ds_2+\hat{R}_3(z),$$

where

$$\hat{R}_3(z) = \Delta_3(a,b)\left(\frac{1}{2\pi i}\right)^3\int_{L(n_3)}\frac{\Gamma(s_3+1-a)\Gamma(s_3+1-b)}{\Gamma(s_3+1)\,\sin\pi s_3}$$

$$\times\int_{L(n_2)}\frac{\Gamma(s_2+\vartheta-s_3)}{\sin\pi s_2}\int_{L(n_1)}\Gamma(s_1-\vartheta-s_2)\frac{z^{-s_1}}{\sin\pi s_1}$$

$$\times\, T_{n_0+\vartheta-s_1}(z)\,ds_1\,ds_2\,ds_3, \tag{6.3.9}$$

and so on.

These results may be expressed more compactly by introducing the sequence of integrals (the hyperterminants)

$$\mathbf{T}_0(z; x) = z^{-x} T_{n_0+\vartheta-x}(z), \tag{6.3.10}$$

$$\mathbf{T}_1(z; x) = \frac{1}{2\pi i} \int_{L(n_1)} \frac{\Gamma(s_1 - \vartheta - x)}{\sin \pi s_1} \mathbf{T}_0(z; s_1)\,ds_1,$$

$$\mathbf{T}_2(z; x) = \frac{1}{2\pi i} \int_{L(n_2)} \frac{\Gamma(s_2 + \vartheta - x)}{\sin \pi s_2} \mathbf{T}_1(z; s_2)\,ds_2$$

and, in general,

$$\mathbf{T}_j(z; x) = \frac{1}{2\pi i} \int_{L(n_j)} \frac{\Gamma(s_j + (-)^j\vartheta - x)}{\sin \pi s_j} \mathbf{T}_{j-1}(z; s_j)\,ds_j, \tag{6.3.11}$$

where $j = 1, 2, \ldots$. Thus the jth hyperterminant $\mathbf{T}_j(z; x)$ takes the form of a j-fold iterated Mellin-Barnes integral involving the terminant function $T_\nu(z)$, with $\nu = n_0 + \vartheta - x$. The above re-expansion procedure is then summarised by the following theorem, which can be established by straightforward induction.

Theorem 6.1. *For $j = 0, 1, 2, \ldots$ let the constants $\Delta_j(a, b)$ be given by (6.3.1) and the hyperterminants $\mathbf{T}_j(z; x)$ be as specified in (6.3.10) and (6.3.11). Then the expansion of the confluent hypergeometric function $U(a; a - b + 1; z)$ is given by*

$$z^a U(a; a - b + 1; z) = \sum_{k=0}^{n_0-1} (-)^k \frac{(a)_k(b)_k}{k!\, z^k} + R_0(z), \tag{6.3.12}$$

where the remainder $R_0(z)$ in the truncation of the Poincaré series after n_0 terms possesses the hyperasymptotic expansion

$$R_0(z) = \frac{2\pi(-)^{n_0}}{\Gamma(a)\Gamma(b)} z^\vartheta e^z \sum_{j=0}^{m-1} \Delta_j(a, b) \sum_{k=0}^{n_{j+1}-1} (-)^k A_k^{(j)} \mathbf{T}_j(z; k) + R_m(z) \tag{6.3.13}$$

for $m = 1, 2, \ldots$, with

$$A_k^{(j)} = \begin{cases} A_k = \dfrac{(1-a)_k(1-b)_k}{k!} & (j \text{ even}) \\[2mm] A_k' = \dfrac{(a)_k(b)_k}{k!} & (j \text{ odd}). \end{cases}$$

The remainder term $R_m(z)$ is defined by

$$R_m(z) = \frac{2\pi(-)^{n_0}}{\Gamma(a)\Gamma(b)} z^\vartheta e^z \hat{R}_m(z), \tag{6.3.14}$$

where

$$\hat{R}_m(z) = \frac{\Delta_m(a, b)}{2\pi i} \int_{L(n_m)} \frac{\Gamma(s_m + a_m)\Gamma(s_m + b_m)}{\Gamma(s_m + 1)\,\sin\pi s_m} \mathbf{T}_{m-1}(z; s_m)\,ds_m \qquad (6.3.15)$$

and

$$a_m = \begin{cases} a \\ 1-a \end{cases}, \quad b_m = \begin{cases} b & (m\ even) \\ 1-b & (m\ odd). \end{cases} \qquad (6.3.16)$$

The result in the above theorem represents the mth level of the hyperasymptotic expansion of $U(a; a - b + 1; z)$. It is apparent that the expansion in (6.3.12) involves a different asymptotic sequence $\{\mathbf{T}_r(z; k)\}$ at each level and that these sequences become progressively more complicated with increasing level. This expansion is equivalent to that obtained by Olde Daalhuis & Olver (1995a) for the hyperasymptotic expansion of solutions of general linear second-order differential equations with a singularity at infinity of rank one (which includes the confluent hypergeometric functions); in their expansion the hyperterminants were expressed as iterated Stieltjes integrals. We also note the resurgence phenomenon which takes place at each level in the expansion: the coefficients A_k and A'_k (cf. (6.2.24)) appear alternately at successive levels. Finally, we remark that the truncation indices n_j are, for the moment, arbitrary positive integers. Two different optimal truncation schemes will be discussed in the next section.

6.3.2 Optimal Truncation Schemes

We now derive estimates for the remainders $R_j(z)$, $j = 0, 1, 2, \ldots$ for two different optimal choices of truncation indices n_j. We introduce the integration variables†

$$s_j = n_j - c + it_j, \qquad (6.3.17)$$

where the t_j are real and $0 < c < 1$, and define the quantities

$$g_j(x) = \left| \frac{\Gamma(n_j - c + a_j + ix)\Gamma(n_j - c + b_j + ix)}{\Gamma(n_j + 1 - c + ix)} \right|,$$

where a_j and b_j are defined in (6.3.16). We shall also require the inequalities

$$|\mathrm{cosec}\,\pi s_j| = (\cosh^2\pi t_j - \cos^2\pi c)^{-\frac{1}{2}} \le \frac{\mathrm{cosec}\,\pi c}{\cosh\pi t_j}, \qquad (6.3.18)$$

$$\frac{e^{\theta t_j}}{\cosh\pi t_j} \le 2 \qquad (-\pi \le \theta \le \pi). \qquad (6.3.19)$$

---

† We shall suppose that each integration variable s_j is associated with the same constant c. This is not essential but is done merely to ease the presentation.

From (6.3.2) and the first of the above inequalities, we have

$$|R_0(z)| = \frac{1}{2|\Gamma(a)\Gamma(b)|} \left| \int_{-c+n_0-\infty i}^{-c+n_0+\infty i} \frac{\Gamma(s_0+a)\Gamma(s_0+b)}{\Gamma(s_0+1)\sin \pi s_0} z^{-s_0} ds_0 \right| \qquad (6.3.20)$$

$$\leq \frac{|z|^{-n_0+c}}{2|\Gamma(a)\Gamma(b)| \sin \pi c} \int_{-\infty}^{\infty} g_0(t_0) \frac{e^{\theta t_0}}{\cosh \pi t_0} dt_0,$$

where $\theta = \arg z$. For $t_0 \to \pm\infty$, the integrand is

$$O\left(|t_0|^{n_0+\vartheta-c-\frac{1}{2}} e^{-\frac{3}{2}\pi|t_0|+\theta t_0}\right)$$

by Stirling's formula. Hence, provided $|\arg z| < \frac{3}{2}\pi$, the above integral is absolutely convergent and independent of $|z|$ and so, for finite values of n_0, we obtain the estimate as $|z| \to \infty$

$$R_0(z) = O(z^{-n_0+c}) \quad \text{in} \quad |\arg z| < \frac{3}{2}\pi. \qquad (6.3.21)$$

This yields the familiar Poincaré asymptotic expansion, namely

$$z^a U(a; a-b+1; z) = \sum_{k=0}^{n_0-1} (-)^k \frac{(a)_k(b)_k}{k! \, z^k} + O(z^{-n_0+c})$$

valid for $|z| \to \infty$ in the sector $|\arg z| \leq \frac{3}{2}\pi - \delta$, $\delta > 0$, where $0 < c < 1$.

We shall initially limit the discussion of the estimates of the remainder terms to the sector $|\arg z| \leq \pi$; the extension of these estimates to wider sectors will be considered at the end of this section. We follow the method introduced in Olde Daalhuis & Olver (1995a) for the determination of the optimal truncation at each level by minimisation of the remainder terms, instead of the terms in the asymptotic series. We suppose n_0 is chosen to scale like $|z|$, so that as $|z| \to \infty$ we have $n_0 \to \infty$. Then, when $|\arg z| \leq \pi$, we obtain from (6.3.20) and Lemma 2.8 the estimate

$$R_0(z) = z^{-n_0+c} O(e^{n_0} n_0^{n_0+\vartheta-c}),$$

where we have again let $\vartheta = a+b-1$. If we now put $n_0 = \beta_0|z| + \alpha_0$, where β_0 is a positive constant to be chosen and α_0 is bounded, this estimate then becomes

$$R_0(z) = O\left(z^\vartheta (e^{-\beta_0} \beta_0^{\beta_0})^{|z|}\right).$$

The minimum value of $e^{-\beta_0} \beta_0^{\beta_0}$ is easily shown to occur for $\beta_0 = 1$, so that when[†] $n_0 = N_0 = |z| + O(1)$ we find the result

$$R_0(z) = O(z^\vartheta e^{-|z|}) \qquad (6.3.22)$$

as $|z| \to \infty$ in $|\arg z| \leq \pi$. This again shows that at optimal truncation the remainder term at the zeroth level is *uniformly* exponentially small like $e^{-|z|}$ throughout

---

[†] As in §6.2.2 we denote optimal values of n_j by N_j.

the sector $|\arg z| \leq \pi$; compare (6.2.14). As shown in §6.2.2, this exponential improvement will be found to steadily diminish outside this sector.

To deal with the remainder terms at levels corresponding to $j \geq 1$, we let

$$n_j = \beta_j |z| + \alpha_j,$$

where the β_j are to be chosen such that $\beta_j \in (0, \beta_{j-1})$ and the α_j are bounded. As $|z| \to \infty$, this implies that $n_0 \gg n_1 \gg n_2 \gg \cdots \gg 1$. When $j = 1$ we have from (6.3.5), (6.3.18) and (6.3.19)

$$|R_1(z)| = \frac{|\Delta_1(a,b)|}{4\pi |\Gamma(a)\Gamma(b)|} \left| \int_{L(n_1)} \frac{\Gamma(s_1 + 1 - a)\Gamma(s_1 + 1 - b)}{\Gamma(s_1 + 1) \sin \pi s_1} \right.$$

$$\left. \times \int_{L(n_0)} \Gamma(s_0 + \vartheta - s_1) \frac{z^{-s_0}}{\sin \pi s_0} \, ds_0 \, ds_1 \right|$$

$$\leq \frac{|z|^{-n_0+c}}{2\pi^2 \sin^2 \pi c} \int_{-\infty}^{\infty} g_1(t_1) \frac{dt_1}{\cosh \pi t_1}$$

$$\times \int_{-\infty}^{\infty} |\Gamma(n_0 - n_1 + \vartheta + iT)| \, dT, \qquad (6.3.23)$$

where we have put $T = t_0 - t_1$. From Lemmas 2.8 and 2.7, the first and second integrals are

$$O\left(e^{-n_1} n_1^{n_1 - \vartheta - c - \frac{1}{2}}\right) \quad \text{and} \quad O\left(e^{-(n_0 - n_1)}(n_0 - n_1)^{n_0 - n_1 + \vartheta}\right),$$

respectively. Hence we find

$$|R_1(z)| = |z|^{-n_0+c} O\left(e^{n_1} n_1^{n_1 - \vartheta - c - \frac{1}{2}} e^{-(n_0 - n_1)}(n_0 - n_1)^{n_0 - n_1 + \vartheta}\right)$$

$$= O\left(\left(e^{-\beta_1} \beta_1^{\beta_1}\right)^{|z|} \left(e^{\beta_1 - \beta_0}(\beta_0 - \beta_1)^{\beta_0 - \beta_1}\right)^{|z|} |z|^{-\frac{1}{2}}\right). \qquad (6.3.24)$$

This estimate is minimised when $\beta_0 - \beta_1 = \beta_1 = 1$; that is, when $\beta_0 = 2$ and $\beta_1 = 1$. Thus, with the truncation scheme at level $j = 1$ given by $n_0 = N_0 = 2|z| + O(1)$ and $n_1 = N_1 = |z| + O(1)$, we find the result

$$R_1(z) = O(z^{-\frac{1}{2}} e^{-2|z|})$$

as $|z| \to \infty$ in $|\arg z| \leq \pi$.

A similar procedure can be brought to bear on the remainder terms at levels $j \geq 2$. For example, when $j = 2$, we find from (6.3.8), (6.3.14) and (6.2.7) that when $|\arg z| \leq \pi$

$$|R_2(z)| = \frac{|\Delta_2(a,b)|}{8\pi^2 |\Gamma(a)\Gamma(b)|} \left| \int_{L(n_2)} \frac{\Gamma(s_2 + a)\Gamma(s_2 + b)}{\Gamma(s_2 + 1) \sin \pi s_2} \int_{L(n_1)} \frac{\Gamma(s_1 - \vartheta - s_2)}{\sin \pi s_1} \right.$$

$$\left. \times \int_{L(n_0)} \Gamma(s_0 + \vartheta - s_1) \frac{z^{-s_0}}{\sin \pi s_0} \, ds_0 \, ds_1 \, ds_2 \right|$$

$$\leq \frac{|\Delta_2(a,b)|}{|\Gamma(a)\Gamma(b)|} \left(\frac{|z|^{-n_0+c}}{4\pi^2\sin^3\pi c}\right) \int_{-\infty}^{\infty} \frac{g_2(t_2)}{\cosh\pi t_2}$$

$$\times \int_{-\infty}^{\infty} \frac{|\Gamma(s_1 - \vartheta - s_2)|}{\cosh\pi t_1} dt_1\, dt_2 \int_{-\infty}^{\infty} |\Gamma(n_0 - n_1 + \vartheta + iT)|\, dT$$

$$\leq \frac{|\Delta_2(a,b)|}{|\Gamma(a)\Gamma(b)|} \left(\frac{|z|^{-n_0+c}}{4\pi^2\sin^3\pi c}\right) \Gamma(n_1 - n_2 - \mathrm{Re}(\vartheta))$$

$$\times \int_{-\infty}^{\infty} g_2(t_2)\frac{dt_2}{\cosh\pi t_2} \int_{-\infty}^{\infty} |\Gamma(n_0 - n_1 + \vartheta + iT)|\, dT, \qquad (6.3.25)$$

where we have made use of the result (since $n_1 \gg n_2$)

$$\int_{-\infty}^{\infty} |\Gamma(s_1 - \vartheta - s_2)|\frac{dt_1}{\cosh\pi t_1} \leq \Gamma(n_1 - n_2 - \mathrm{Re}(\vartheta)) \int_{-\infty}^{\infty} \frac{dt_1}{\cosh\pi t_1}$$

$$= \Gamma(n_1 - n_2 - \mathrm{Re}(\vartheta)),$$

which follows from (2.5.13) and the fact that $|\Gamma(\sigma + it)| \leq \Gamma(\sigma)$ for $\sigma > 0$. The two integrals in (6.3.25) are the same type as those appearing in $|R_1(z)|$ in (6.3.23), and hence we obtain the estimate

$$|R_2(z)| = |z|^{-n_0+c} O\left(e^{-n_2} n_2^{n_2+\vartheta-c-\frac{1}{2}} e^{-(n_1-n_2)} (n_1 - n_2)^{n_1-n_2-\vartheta-\frac{1}{2}}\right.$$

$$\times \left. e^{-(n_0-n_1)}(n_0 - n_1)^{n_0-n_1+\vartheta}\right)$$

$$= O\left(\left(e^{-\beta_2}\beta_2^{\beta_2}\right)^{|z|} \left(e^{\beta_2-\beta_1}(\beta_1 - \beta_2)^{\beta_1-\beta_2}\right)^{|z|}\right.$$

$$\times \left. \left(e^{\beta_1-\beta_0}(\beta_0 - \beta_1)^{\beta_0-\beta_1}\right)^{|z|} |z|^{\vartheta-1}\right). \qquad (6.3.26)$$

This estimate is minimised by the choice $\beta_0 - \beta_1 = \beta_1 - \beta_2 = \beta_2 = 1$; that is, when $\beta_0 = 3$, $\beta_1 = 2$ and $\beta_2 = 1$. Thus we find the estimate when $N_0 = 3|z| + O(1)$, $N_1 = 2|z| + O(1)$ and $N_2 = |z| + O(1)$ given by

$$R_2(z) = O(z^{\vartheta-1} e^{-3|z|})$$

as $|z| \to \infty$ in $|\arg z| \leq \pi$.

Order estimates for the higher order remainder terms can be obtained by an induction argument to yield the final result given in the following theorem.

Theorem 6.2. *With the truncation scheme*

$$N_j = (m + 1 - j)|z| + O(1) \qquad (0 \leq j \leq m), \qquad (6.3.27)$$

the order of the remainder at level j of the hyperasymptotic expansion of the confluent hypergeometric function $U(a; a - b + 1; z)$ in Theorem 6.1 is given by

$$R_j(z) = O(z^{-\frac{1}{2}j+\delta_j\vartheta} e^{-(j+1)|z|}), \qquad (0 \leq j \leq m) \qquad (6.3.28)$$

where

$$\delta_j = \begin{cases} 1 & (j \ even) \\ 0 & (j \ odd). \end{cases}$$

This is the result obtained by Olde Daalhuis & Olver (1995a) for the more general problem of the hyperasymptotic expansion of solutions of second-order differential equations with a singularity at infinity of rank one. This is a truly remarkable result for two reasons. First, it shows that with the optimal truncation scheme in (6.3.27) the remainder at each level is exponentially smaller† than its predecessor by the factor $e^{-|z|}$ *uniformly* throughout the sector $|\arg z| \leq \pi$ and that the remainder at the mth level contains the factor $\exp\{-(m + 1)|z|\}$. And secondly, this level of accuracy is achieved by employing *post-optimal truncation* of the re-expanded series in all but the mth level. For example, if we take $m = 2$ then $N_0 = 3|z| + O(1)$, $N_1 = 2|z| + O(1)$ and $N_2 = |z| + O(1)$ so that the Poincaré asymptotic series in (6.3.12) is summed to about $[3|z|]$ terms, which is three times its optimal truncation value. Similarly the first re-expanded series in (6.3.13) is summed to about $[2|z|]$ terms with the final series then being optimally truncated after about $[|z|]$ terms. In other words, this truncation scheme shows that by extracting information contained in the divergent tails of the lower-order re-expansions it is possible to obtain theoretically an unlimited accuracy.

In numerical applications with the above truncation scheme, however, a problem can arise which may result in loss of accuracy. Once the smallest term in a re-expanded series at any level is passed, succeeding terms begin to grow. If any of these terms have an absolute value much greater than unity, there will be loss of accuracy in the evaluation of the series when working with fixed-decimal arithmetic. This difficulty has been discussed in detail by Olde Daalhuis & Olver (1995a, §8) where it is shown, by modification of the truncation scheme in (6.3.27), that a numerically stable scheme can be devised which avoids this difficulty, but at the cost of a slightly reduced overall exponential improvement.

A different truncation scheme was employed by Berry & Howls (1990, 1991). This scheme takes $N_0 = |z| + O(1)$ at the zeroth level, so that the order of the remainder $R_0(z)$ is again given by (6.3.22) in $|\arg z| \leq \pi$. Then, with N_0 *fixed at this optimal value* for the zeroth level and $N_1 = \beta_1|z| + O(1)$, the remainder at the next level is, from (6.3.24), given by

$$R_1(z) = O(z^{-\frac{1}{2}} e^{-|z|} (\beta^{\beta_1} (1 - \beta_1)^{1-\beta_1})^{|z|}).$$

This estimate is minimised when $\beta_1 \log \beta_1 + (1 - \beta_1) \log(1 - \beta_1)$ is minimum; that is, when $\beta_1 = \frac{1}{2}$. Thus, when $N_0 = |z| + O(1)$ and $N_1 = \frac{1}{2}|z| + O(1)$, we

---

† We are ignoring here the weaker algebraic scaling given by $z^{-\frac{1}{2}j+\delta_j\vartheta}$.

find

$$R_1(z) = O(z^{-\frac{1}{2}}e^{-|z|}2^{-|z|}).$$

Similarly from (6.3.26), with N_0 and N_1 fixed as above and $N_2 = \beta_2|z| + O(1)$, the remainder $R_2(z)$ now takes the form

$$R_2(z) = O\left(z^{\vartheta-1}e^{-|z|}2^{-\frac{1}{2}|z|}\left(\beta_2^{\beta_2}\left(\tfrac{1}{2} - \beta_2\right)^{\frac{1}{2}-\beta_2}\right)^{|z|}\right).$$

This is minimised when $\beta_2 = \frac{1}{4}$ so that the optimal estimate is given by

$$R_2(z) = O(z^{\vartheta-1}e^{-|z|}2^{-\frac{3}{2}|z|}).$$

Continuing in this way, we obtain the truncation scheme given by

$$N_j = 2^{-j}|z| + O(1) \tag{6.3.29}$$

associated with the remainder terms

$$R_j(z) = O(z^{-\frac{1}{2}j+\delta_j\vartheta}e^{-|z|}2^{-\lambda_j|z|}), \tag{6.3.30}$$

where δ_j is defined after (6.3.28) and

$$\lambda_j = 1 + \tfrac{1}{2} + \tfrac{1}{4} + \cdots + 2^{1-j} = 2 - 2^{1-j} \quad (j \geq 1)$$

with $\lambda_0 = 0$. The truncation index at level j is seen to be approximately half that of the preceding level, so that the Berry-Howls truncation scheme, unlike that of the Olde Daalhuis-Olver scheme in (6.3.27), comes to a halt. Since $\lambda_j < 2$, the improvement is bounded by

$$e^{-|z|}2^{-2|z|} = e^{-|z|(1+2\log 2)};$$

the maximal exponential improvement with the Berry-Howls scheme is therefore limited to a factor $e^{-\hat{\gamma}|z|}$, where $\hat{\gamma} = 1 + 2\log 2 = 2.386\ldots$

Finally, we remark that the order estimates in (6.3.28) can be extended beyond the principal sector $|\arg z| \leq \pi$ to find (not surprisingly) that the exponential improvement at a given level progressively deteriorates. However, as pointed out in §6.2.2, it would not normally be necessary to use the hyperasymptotic expansion (6.3.12) beyond $|\arg z| \leq \pi$, since standard connection formulas for $U(a; a - b + 1; z)$ are available for the construction of exponentially-improved expansions in wider sectors. By means of the device explained in §6.2.2, it can be established that the order of $R_0(z) \equiv R_0(z; N_0)$ (where we now display the dependence on the optimal truncation indices) is given by $O(z^\vartheta e^z)$ in $\pi \leq |\arg z| \leq \frac{3}{2}\pi - \delta, \delta > 0$ when $N_0 = |z| + O(1)$. This shows how the exponential improvement at the zeroth level decreases beyond $|\arg z| = \pi$.

The same procedure can be applied to $R_j(z) \equiv R_j(z; N_0, N_1, \ldots, N_j)$ or, equivalently, we can use the connection formula for the terminant function in

(6.2.45). Substitution of this latter result into (6.3.14) and (6.3.15) yields the connection formula

$$R_j(z; N_0, N_1, \ldots, N_j) = R_j(ze^{\pm 2\pi i}; N_0, N_1, \ldots, N_j)$$

$$+ \frac{2\pi i}{\Gamma(a)\Gamma(b)}(ze^{\pm \pi i})^{\vartheta} e^z R'_{j-1}(ze^{\pm \pi i}; N_1, N_2, \ldots, N_j)$$

for $j \geq 1$, where the prime on $R_{j-1}(ze^{\pm \pi i}; N_1, N_2, \ldots, N_j)$ signifies that the parameters a and b are to be replaced by $1 - a$ and $1 - b$, respectively. Use of this formula successively with $j = 1, 2, \ldots$, combined with the estimates for $R_j(z; N_0, N_1, \ldots, N_j)$ in (6.3.28), shows that when the N_j are chosen according to the optimal truncation scheme in (6.3.27) we have the general result due to Olde Daalhuis & Olver (1995a, §10)

$$R_j(z; N_0, N_1, \ldots, N_j) = \begin{cases} O(z^{\frac{1}{2}(k-j)+\delta_j \vartheta} e^{\delta_{k+1}z - (k-j-1)|z|}) \\ \qquad \text{in } k\pi \leq |\arg z| \leq (k+1)\pi \\ O(z^{\delta_j \vartheta} e^{\delta_j z}) \\ \qquad \text{in } (k+1)\pi \leq |\arg z| \leq \left(k + \frac{3}{2}\right)\pi - \delta \end{cases}$$

for nonnegative integer k as $|z| \to \infty$. Thus, for example, when $j = 2$ we find

$$R_2(z; N_0, N_1, N_2) = \begin{cases} O(z^{\vartheta - 1} e^{-3|z|}) & \text{in } \quad |\arg z| \leq \pi \\ O(z^{\vartheta - \frac{1}{2}} e^{z - 2|z|}) & \text{in } \quad \pi \leq |\arg z| \leq 2\pi \\ O(z^{\vartheta} e^{-|z|}) & \text{in } \quad 2\pi \leq |\arg z| \leq 3\pi \\ O(z^{\vartheta} e^z) & \text{in } \quad 3\pi \leq |\arg z| \leq \frac{7}{2}\pi - \delta, \end{cases}$$

which reveals how the exponential improvement progressively diminishes as $|\arg z|$ is increased beyond the principal sector $|\arg z| \leq \pi$.

6.3.3 A Numerical Example

We take as example the case $a = -\frac{1}{2}\nu, b = \frac{1}{2} - \frac{1}{2}\nu$ so that

$$U\left(-\tfrac{1}{2}\nu; \tfrac{1}{2}; z\right) = 2^{-\frac{1}{2}\nu} e^{\frac{1}{2}z} D_\nu(\sqrt{2z}),$$

where $D_\nu(x)$ denotes the Weber parabolic cylinder function; see Abramowitz & Stegun (1965, p. 510). For illustration, we shall use the values $\nu = \frac{1}{2}$ and $z = 20$; a similar example involving the modified Bessel function of the second kind of order zero (obtained by putting $a = b = \frac{1}{2}$; see (6.2.40)) has been discussed in Olde Daalhuis & Olver (1995a). The absolute value of each term in the expansions (6.3.12) and (6.3.13) up to level $j = 2$ in the hyperasymptotic expansion is shown in Fig. 6.4 as a function of its ordinal number.

At the zeroth level we set $N_0 = 20$, which terminates the Poincaré asymptotic expansion in (6.3.12) at, or near, its least term. For the first and second levels we set $N_0 = 40$, $N_1 = 20$ and $N_0 = 60$, $N_1 = 40$, $N_2 = 20$ following the truncation

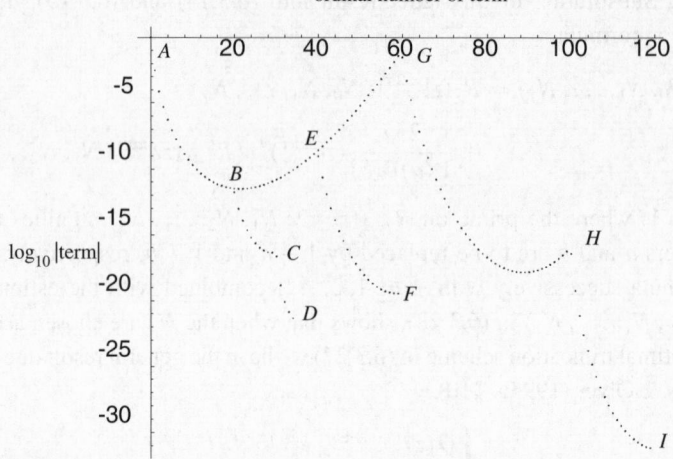

Fig. 6.4. Magnitudes of the terms in the three hyperasymptotic expansions.

scheme in (6.3.27). These different levels are represented by the sequences of points labelled AB, $ABEF$ and $ABEGHI$, respectively in Fig. 6.4. It is seen with this particular example that at the second level no term in the sequence $ABEG$ rises above unity in magnitude, thereby avoiding the numerical problem mentioned in Olde Daalhuis & Olver (1995a); this would not continue to be the case, however, at the third level (and *a fortiori* at higher levels) where the continuation of this sequence would rise well above unity in magnitude. The truncation scheme in (6.3.29) corresponds to $N_0 = 20$, $N_1 = 10$, $N_2 = 5$ and is represented by the sequence of points labelled $ABCD$. This truncation scheme yields a decreasing number of terms at successive levels and results in the smallest term of the sequence being bounded from below; compare (6.3.30).

The computations were carried out using *Mathematica*, where the terminant function $T_\nu(z)$ was evaluated as an incomplete gamma function by means of (6.2.43). The hyperterminant $\mathbf{T}_1(z; k)$, which enters at the second level, was computed by numerical evaluation of the Mellin-Barnes integral as discussed in §5.2.2. The path of integration $L(N_1)$ is chosen with $c_1 = \frac{1}{2}$ for convenience. Then, with $s_1 = N_1 - \frac{1}{2} + it$, we have from (6.3.11)

$$\mathbf{T}_1(z; k) = \frac{(-)^{N_1-1} z^{\frac{1}{2}-N_1}}{2\pi} \int_{-\infty}^{\infty} \Gamma(s_1 - \vartheta - k) \, T_{N_0+\vartheta-s_1}(z) \frac{z^{-it}}{\cosh \pi t} \, dt.$$

The above integral can be simplified when $z > 0$ since the real part of the integrand is symmetric in t, while the imaginary part is antisymmetric. We also remark that the computation of $T_1(z; k)$ for $k = 0, 1, 2, \ldots$ can be made more efficient if we denote the integrand when $k = 0$ by $f(t)$, say, so that the integrands for $k \geq 1$ are

Table 6.3. *Hyperasymptotic approximations to* $z^{-\frac{1}{4}}U(a; a - b + 1; z)$ *for* $z = 20$ *and* $a = -\frac{1}{4}$ *and* $b = \frac{1}{4}$

Level	Asymptotic	|Error|	N_0	N_1	N_2
0	1.00305609867613428596373755165508	1.7×10^{-12}	20		
1	1.00305609867443646903330402810646	2.2×10^{-17}	20	10	
2	1.00305609867443640712597748150073	1.5×10^{-22}	20	10	5
0	1.00305609867613428596373755165508	1.7×10^{-12}	20		
1	1.00305609867443640712334593608958	2.5×10^{-21}	40	20	
2	1.00305609867443640712582553811408	2.0×10^{-32}	60	40	20
Exact	1.00305609867443640712582553811406	0			

then simply given by $(-)^k f(t)/(\vartheta - s_1 + 1)_k$. The most expensive part of the evaluation of $\mathbf{T}_1(z; k)$ is the repeated computation of $f(t)$. A sequence of these values can be stored and the integrands for different k values can then be readily computed. In this way, the sequence of hyperterminants $\mathbf{T}_1(z; k)$ required for the second level can be generated in an efficient manner, so enabling the computation of the expansion of $U(a; a - b + 1; z)$ at this level to be achieved in a time which becomes competitive with the direct evaluation of the function to the same level of accuracy. The results of the calculations using different truncations are displayed in Table 6.3.

6.4 Exponentially-Improved Asymptotics for $\Gamma(z)$

Not all special functions of interest in applications satisfy a linear differential equation. A function of fundamental importance belonging to this category is the gamma function $\Gamma(z)$, which is known to be incapable of satisfying a differential equation with rational coefficients; see Whittaker & Watson (1965, p. 236). The standard asymptotic expansion of $\Gamma(z)$ for large $|z|$ is given by the well-known Stirling expansion in (2.1.9); see also (6.4.2) and (6.4.3) below. The exponentially-improved asymptotic expansion of $\Gamma(z)$, which is the subject of this section, has two striking features. The first is the fact that in the neighbourhood of the Stokes lines $\arg z = \pm\frac{1}{2}\pi$, not one but *infinitely many* subdominant exponential terms appear, each associated with its own Stokes multiplier. This peculiarity can be understood qualitatively from the fact that the gamma function satisfies a simple first-order difference equation which may be regarded as a differential equation of infinite order. The second feature is that its hyperasymptotic expansion terminates at the first level to yield an *absolutely convergent* expansion of terminant functions.

Interest in the exponentially-improved asymptotics of $\Gamma(z)$ first arose in connection with a simple model describing the leakage of energy from weakly bent optical fibre waveguides. This took the form of the eigenvalue problem defined by

$$\frac{d^2 y(x)}{dx^2} + (\lambda + \epsilon x^n)y(x) = 0 \quad \text{on} \quad (0, \infty) \quad (\epsilon \to 0+), \quad (6.4.1)$$

with

$$y'(0) + hy(0) = 0, \quad h > 0$$

together with the condition that $y(x)$ represents an outgoing wave as $x \to \infty$. This last requirement renders the problem non-self-adjoint, thereby enabling the eigenvalue to possess an imaginary part. The real part of λ can be found by straightforward regular perturbation to be $\mathrm{Re}(\lambda) = -h^2 + O(\epsilon)$, but $\mathrm{Im}(\lambda)$, which measures the rate at which energy leaks from the core of the fibre, turns out to be exponentially small in the parameter ϵ. The case $n = 1$ was studied by Paris & Wood (1989) as a model for optical tunnelling arising in the work of Kath & Kriegsmann (1989). The solution of the eigenvalue problem in this case is soluble in terms of Airy functions. An extension of this simple model to the parabolic case $n = 2$ was considered in Brazel et al. (1992). The solution of the above differential equation when $n = 2$ can again be achieved in terms of special functions (the parabolic cylinder function) to yield the exact eigenvalue relation

$$\frac{\Gamma\left(\frac{1}{4} + \frac{1}{2}a\right)}{\Gamma\left(\frac{3}{4} + \frac{1}{2}a\right)} = -\frac{2\epsilon^{\frac{1}{4}}e^{\frac{1}{4}\pi i}}{h} \frac{(1 + ie^{\pi i a})}{(1 - ie^{\pi i a})}, \qquad a = \frac{1}{2}i\lambda\epsilon^{-\frac{1}{2}}.$$

Use of the standard Poincaré asymptotics for the gamma function for large $|a|$, however, leads to an erroneous value of $\mathrm{Im}(\lambda)$. In order to obtain the correct value of this exponentially small quantity, it is necessary to take into account exponentially small contributions present in the expansion of $\Gamma(z)$ which switch on near $\arg z = \pm\frac{1}{2}\pi$. For situations with integer $n \geq 3$, there are no special functions available and the method of matched asymptotic expansions must be employed; see Liu & Wood (1991) for details. The final result is

$$\mathrm{Im}(\lambda) \sim -\frac{2h^2}{E} \exp\left\{-2h^{(n+2)/n} S(n)\epsilon^{-1/n}\right\} \quad (\epsilon \to 0+),$$

where

$$E = \begin{cases} e & (n = 1) \\ 1 & (n \geq 2) \end{cases}, \qquad S(n) = \frac{\Gamma(1 + 1/n)\Gamma\left(\frac{3}{2}\right)}{\Gamma\left(\frac{3}{2} + 1/n\right)}.$$

6.4.1 Origin of the Exponentially Small Terms

We begin by examining in detail how the infinitely many subdominant exponential terms in the expansion of $\Gamma(z)$ appear in the neighbourhood of the rays $\arg z = \pm\frac{1}{2}\pi$. From (2.1.1) we have

$$\Gamma(z) = (2\pi)^{\frac{1}{2}} z^{z-\frac{1}{2}} e^{-z+\Omega(z)}, \tag{6.4.2}$$

where as $|z| \to \infty$ in $|\arg z| \le \frac{1}{2}\pi - \delta$, $\delta > 0$, $\Omega(z)$ is given by the well-known Stirling series [see (2.1.7)]

$$\Omega(z) \sim \sum_{r=1}^{\infty} \frac{B_{2r}}{2r(2r-1)z^{2r-1}} \tag{6.4.3}$$

in terms of the Bernoulli numbers B_{2r}. We note at this point that, since successive even-order Bernoulli numbers have opposite signs, all terms in the asymptotic expansion of $\Omega(z)$ have the same phase on $\arg z = \pm\frac{1}{2}\pi$, with the consequence that the positive and negative imaginary axes are the Stokes lines for $\Gamma(z)$.

To obtain the expansion in the sectors $\frac{1}{2}\pi < |\arg z| \le \pi$, we employ the reflection formula in the form

$$\Gamma(z) = -\frac{\pi}{z \sin \pi z} \frac{1}{\Gamma(-z)}.$$

In these sectors we take $-z = ze^{\mp\pi i}$ respectively, so that $-\frac{1}{2}\pi < \arg(-z) < 0$ when z is in the second quadrant and $0 < \arg(-z) < \frac{1}{2}\pi$ when z is in the third quadrant. Then use of (6.4.2) shows that $\Omega(z)$ must satisfy the relation

$$\Omega(z) + \Omega(ze^{\mp\pi i}) = -\log(1 - e^{\pm 2\pi i z}) = \sum_{k=1}^{\infty} \frac{e^{\pm 2\pi i k z}}{k}.$$

Thus analytic continuation of $\Omega(z)$ into the upper and lower left half-plane has produced an infinite sequence of exponential terms. These terms are subdominant relative to the expansion in (6.4.3) in the upper and lower half-planes and are maximally subdominant on the Stokes lines $\arg z = \pm\frac{1}{2}\pi$, respectively. As the negative z axis is approached, however, these exponentials increase in magnitude and eventually combine to generate the poles of $\Gamma(z)$.

Based on the results in §6.2 and Berry (1989), we would expect these small exponentials to emerge from the optimally truncated divergent series in (6.4.3) in a smooth manner as one crosses the rays $\arg z = \pm\frac{1}{2}\pi$, and this is indeed found to be the case. The demonstration of the smooth transition of the leading subdominant exponential (corresponding to $k = 1$) was first given by Paris & Wood (1992); see also the account in Wood (1991). The detailed computation of the infinite sequence of smaller exponentials was carried out by Berry (1991a) who showed by a different argument, involving a sequence of increasingly delicate subtractions of optimally-truncated asymptotic series, how they all switch on smoothly across the Stokes lines according to the error-function law. The theory we present here is based on a slight modification of the Mellin-Barnes integral approach used in Paris & Wood (1992).

6.4.2 The Expansion of $\Omega(z)$

Our starting point is the representation for the logarithm of $\Gamma(z)$ in terms of the Riemann zeta function $\zeta(s)$ given in Whittaker & Watson (1965, p. 277)

$$\log \Gamma(z+1) = -\gamma z - \frac{1}{2\pi i} \int_{-c-\infty i}^{-c+\infty i} \frac{\pi z^{-s}}{s \sin \pi s} \zeta(-s) \, ds, \tag{6.4.4}$$

valid in $|\arg z| \le \pi - \delta, \delta > 0$, where $1 < c < 2$ and γ denotes Euler's constant. The integrand is exponentially decaying for large $\mathrm{Im}(s)$, so we can displace the contour to the right over the double poles at $s = -1$ and $s = 0$. Straightforward calculations using the facts that $\zeta(0) = -\frac{1}{2}$, $\zeta'(0) = -\frac{1}{2} \log 2\pi$ and $\zeta(s) \sim (s-1)^{-1} + \gamma$ near $s = 1$ yield the residues at these poles given by $z \log z - z + \gamma z$ and $\frac{1}{2} \log(2\pi z)$, respectively. It then follows that

$$\log \Gamma(z) = (z - \tfrac{1}{2}) \log z - z + \tfrac{1}{2} \log 2\pi - \frac{1}{2\pi i} \int_{c-\infty i}^{c+\infty i} \frac{\pi z^{-s}}{s \sin \pi s} \zeta(-s) \, ds,$$

where $0 < c < 1$. Comparison with (6.4.2) shows that the function $\Omega(z)$ possesses the Mellin-Barnes integral representation given by

$$\Omega(z) = -\frac{1}{2\pi i} \int_{c-\infty i}^{c+\infty i} \frac{\pi z^{-s}}{s \sin \pi s} \zeta(-s) \, ds \tag{6.4.5}$$

valid when $|\arg z| \le \pi - \delta$ and $0 < c < 1$.

We now proceed to manipulate the above integral for $\Omega(z)$ to extract the sequence of exponentially small terms. We first employ the functional relation (4.1.4) for $\zeta(s)$ in the form

$$\pi \zeta(-s) = -(2\pi)^{-s} s \Gamma(s) \zeta(1+s) \sin \tfrac{1}{2} \pi s$$

followed by use of the Dirichlet series expansion for $\zeta(1+s)$ (which is permissible since $0 < \mathrm{Re}(s) < 1$ on the path of integration), to obtain the result

$$\Omega(z) = \frac{1}{2\pi i} \int_{c-\infty i}^{c+\infty i} \frac{\Gamma(s)}{\sin \pi s} (2\pi z)^{-s} \zeta(1+s) \sin \tfrac{1}{2} \pi s \, ds$$

$$= \sum_{k=1}^{\infty} \frac{G_k(z)}{k}, \tag{6.4.6}$$

where

$$G_k(z) = \frac{1}{2\pi i} \int_{c-\infty i}^{c+\infty i} \frac{\Gamma(s)}{\sin \pi s} (2\pi k z)^{-s} \sin \tfrac{1}{2} \pi s \, ds.$$

The reversal of the order of summation and integration is justified when $|\arg z| \le \pi - \delta$ by absolute convergence. Displacement of the path to the right over the simple poles of the integrand at $s = 1, 3, \ldots, 2n_k - 1$, where n_k is (for the moment) an arbitrary positive integer, and use of Cauchy's theorem then shows that, provided $|\arg z| \le \pi - \delta$,

$$G_k(z) = \frac{1}{\pi} \sum_{r=0}^{n_k-1} \frac{(-)^r (2r)!}{(2\pi k z)^{2r+1}} + R_k(z; n_k). \tag{6.4.7}$$

With $m_k = 2n_k + 1$ and c restricted so that $0 < c < 1$, the remainder term is given by

$$R_k(z; n_k) = \frac{1}{2\pi i} \int_{-c+m_k-\infty i}^{-c+m_k+\infty i} \frac{\Gamma(s)}{\sin \pi s} (2\pi k z)^{-s} \sin \tfrac{1}{2}\pi s \, ds$$

$$= e^{2\pi i k z} \hat{T}_{m_k}(2\pi i k z) - e^{-2\pi i k z} \hat{T}_{m_k}(-2\pi i k z), \tag{6.4.8}$$

in the sector† $|\arg z| \le \pi - \delta$. The last result follows from writing $\sin \tfrac{1}{2}\pi s$ in terms of exponentials and use of the definition of the terminant function $\hat{T}_\nu(x)$ in (6.2.8) and (6.2.43). We observe that the left-hand side of (6.4.8) can also be written in terms of the generalised terminant in (6.2.35) as $e^{2\pi i k z} \hat{T}_{m_k}^{(2)}(2\pi k z)$.

From (6.4.6) and (6.4.7) we then obtain the main result of this section

$$\Omega(z) = \frac{1}{\pi} \sum_{k=1}^{\infty} \frac{1}{k} \sum_{r=0}^{n_k-1} \frac{(-)^r (2r)!}{(2\pi k z)^{2r+1}} + R(z; n_k), \tag{6.4.9}$$

where

$$R(z; n_k) = \sum_{k=1}^{\infty} \frac{R_k(z; n_k)}{k}. \tag{6.4.10}$$

The series in (6.4.10) is *absolutely convergent*, as is readily shown from the large-argument asymptotics of $\hat{T}_\nu(x)$ given in (6.2.46). The result in (6.4.9) is exact and no further expansion process is required: the hyperasymptotic process has terminated at the first level with the remainder term expressed as a convergent sum of terminant functions. The infinite sequence of exponentials $e^{\pm 2\pi i k z}$ ($k = 1, 2, \dots$) is seen to emerge from the remainder $R(z; n_k)$ on account of (6.4.8), with the terminant function $\hat{T}_{m_k}(\pm 2\pi i k z)$ respectively as coefficient.

The truncation indices n_k are arbitrary positive integers. If, for example, we choose $n_k = n$ for all k, the first sum on the right-hand side of (6.4.9) becomes

---

† We note that when $\sin \tfrac{1}{2}\pi s$ is written in terms of exponentials, the two resulting integrals appearing in $R_k(z; n_k)$ converge in the sectors $|\arg(\pm iz)| \le \tfrac{3}{2}\pi - \delta$. The combination in $R_k(z; n_k)$ has, however, $|\arg z| \le \pi - \delta$ as common sector of validity.

Stirling's series in (6.4.3), since

$$\frac{1}{\pi} \sum_{k=1}^{\infty} \frac{1}{k} \sum_{r=1}^{n} \frac{(-)^{r-1}(2r-2)!}{(2\pi kz)^{2r-1}} = \frac{1}{\pi} \sum_{r=1}^{n} \frac{(-)^{r-1}(2r-2)!}{(2\pi z)^{2r-1}} \zeta(2r)$$

$$= \sum_{r=1}^{n} \frac{B_{2r}}{2r(2r-1)z^{2r-1}},$$

upon use of the result (2.1.15) connecting $\zeta(2r)$ to the even-order Bernoulli numbers B_{2r}. With this truncation scheme we therefore obtain

$$\Omega(z) = \sum_{r=1}^{n} \frac{B_{2r}}{2r(2r-1)z^{2r-1}} + R(z; n), \qquad (6.4.11)$$

which is the result obtained in Paris & Wood (1992). Optimal truncation of the finite series in (6.4.11), given by $n = N \sim \pi |z|$, then enabled the smooth transition of the leading subdominant exponential with $k = 1$ to be established, but was not sufficiently precise to describe the transition of the smaller exponentials corresponding to $k \geq 2$.

We remark that the expansion (6.4.9) differs from that in (6.4.11) in two important ways. First, the Stirling series has been decomposed into a k-sequence of component asymptotic series with scale $2\pi kz$, each associated with its own arbitrary truncation index n_k. Secondly, the order of the terminant functions in $R(z; n_k)$ depends on n_k. It is these two features which permit *each finite series in the more refined expansion (6.4.9) to be optimally truncated* for large $|z|$ near its least term; that is, when

$$n_k = N_k \sim \pi k |z|. \qquad (6.4.12)$$

This yields the optimally-truncated expansion

$$\Omega(z) = \frac{1}{\pi} \sum_{k=1}^{\infty} \frac{1}{k} \sum_{r=0}^{N_k-1} \frac{(-)^r (2r)!}{(2\pi kz)^{2r+1}} + R(z; N_k), \qquad (6.4.13)$$

where

$$R(z; N_k) = \sum_{k=1}^{\infty} \frac{1}{k} \left\{ e^{2\pi ikz} \hat{T}_{2N_k+1}(2\pi ikz) - e^{-2\pi ikz} \hat{T}_{2N_k+1}(-2\pi ikz) \right\}. \qquad (6.4.14)$$

It is now apparent that the order and the modulus of the argument of *each terminant function* in $R(z; N_k)$ are approximately equal. When $v \simeq |x| \gg 1$, the function $\hat{T}_v(x)$ possesses the property of changing very rapidly, but smoothly, from being exponentially small to being almost a constant as arg x passes continuously through either $\pm \pi$; see §6.2.6. Use of the asymptotic properties given in (6.2.50) and (6.2.56) shows that (i) the sum $R(z; N_k)$ converges exponentially fast away from the negative real z axis and (ii) for fixed $|z|$ in the vicinity of arg $z = \frac{1}{2}\pi$, the dominant contribution to $R(z; N_k)$ arises from the terms involving $\hat{T}_{2N_k+1}(2\pi i k z)$. The coefficient (the Stokes multiplier) of each subdominant exponential $e^{2\pi i k z}$ in (6.4.14) then has the leading form given by

$$\hat{T}_{2N_k+1}(2\pi i k z) \sim \tfrac{1}{2} + \tfrac{1}{2}\mathrm{erf}\left\{c(\theta + \tfrac{1}{2}\pi)\sqrt{\pi k |z|}\right\},$$

where, from (6.2.52),

$$c(\theta + \tfrac{1}{2}\pi) = \theta - \tfrac{1}{2}\pi + \tfrac{1}{6}i(\theta - \tfrac{1}{2}\pi)^2 - \tfrac{1}{36}(\theta - \tfrac{1}{2}\pi)^3 + \cdots,$$

and $\theta = \arg z$. The treatment in the neighbourhood of $\arg z = -\tfrac{1}{2}\pi$ is similar, with the functions $\hat{T}_{2N_k+1}(-2\pi i k z)$ now controlling the dominant behaviour. The approximate functional form of the Stokes multipliers near $\arg z = \pm\tfrac{1}{2}\pi$ is then found to be

$$\tfrac{1}{2} \pm \tfrac{1}{2}\mathrm{erf}\left\{(\theta \mp \tfrac{1}{2}\pi)\sqrt{\pi k |z|}\right\} \qquad (k = 1, 2, \ldots), \tag{6.4.15}$$

respectively. This is the error-function smoothing law found in Paris & Wood (1992) for $k = 1$ and Berry (1991a) for general k. The form (6.4.15) approximately describes the birth of each subdominant exponential — on the increasingly sharp scale $(\pi k |z|)^{\frac{1}{2}}$ — in the neighbourhood of the positive and negative imaginary axes.

We have seen that a physical application of the exponentially-improved asymptotics of $\Gamma(z)$ arises in the solution of the eigenvalue problem in (6.4.1) when $n = 2$. A non-physical application has been given by Paris (1993) in an alternative derivation† of an expansion for the Riemann zeta function $\zeta(s)$. The representation chosen for this purpose was the integral

$$\zeta(s) = -\frac{e^{\pi i s}}{2\pi i} \int_{-c-\infty i}^{-c+\infty i} \left\{\frac{\Gamma'(1+z)}{\Gamma(1+z)} - \log z\right\} z^{-s}\, dz \qquad (\mathrm{Re}(s) > 0),$$

where $0 < c < 1$ and the z plane is cut along the positive real axis, which is a slightly modified form of an integral given by Kloosterman (1922); see also Titchmarsh (1986, p. 24). Use of the refined asymptotics of $\log \Gamma(z)$ in (6.4.11) shows that it is the sequence of exponential terms, and not the Stirling series component of the expansion, that contributes to the delicate structure of $\zeta(s)$. The resulting absolutely convergent expansion takes the form

$$\zeta(s) = \sum_{k=1}^{M} k^{-s} + \frac{\mu^{1-s}}{s - 1} - i\,\Gamma(1 - s) \sum_{k=1}^{\infty} (2\pi k)^{s-1}\left\{e^{\frac{1}{2}\pi i s} Q(1 - s, 2\pi i k\mu)\right.$$
$$\left. - e^{-\frac{1}{2}\pi i s} Q(1 - s, -2\pi i k\mu)\right\} \tag{6.4.16}$$

† Simpler derivations of (6.4.16) can be obtained by use of the Poisson summation formula applied to the tail of the Dirichlet series for $\zeta(s)$ [see Berry (1986) and Paris (1994b)] or by use of the exact representation (2.1.2) for $\Omega(z)$ in Kloosterman's integral.

valid throughout the s plane ($s \neq 1$), where $\mu = M + \frac{1}{2}$, M is an arbitrary non-negative integer and $Q(a, z) = \Gamma(a, z)/\Gamma(a)$ denotes the normalised incomplete gamma function. This result bears some resemblance to that given in (8.2.18) but contains some significant differences. The asymptotic consequences of (6.4.16) for the computation of $\zeta(s)$ on the critical line $\mathrm{Re}(s) = \frac{1}{2}$ have been explored in Paris (1994b).

A different type of expansion for $\Omega(z)$ in the form of an absolutely convergent Hadamard expansion, which extends a result of Hadamard (1912), has been examined in Paris (2000c). This expansion can similarly yield hyperasymptotic accuracy but suffers from the inconvenience that it requires more computational effort than that in (6.4.13).

6.4.3 A Numerical Example

We now use the optimally-truncated expansion in (6.4.13) and (6.4.14) to display numerically the smooth appearance of the first two subdominant exponentials corresponding to $k = 1, 2$. To achieve this for the pth level it is necessary to "peel off" from $\Omega(z)$ all the exponentials corresponding to $k < p$ and all larger terms of the asymptotic series in (6.4.13). As described in Berry (1991a), this latter subtraction requires a straightforward regrouping of the terms in the absolutely convergent double sum in (6.4.13). With

$$a_r = (-)^r (2r)!/(2\pi z)^{2r+1}, \qquad A_{r,k} = a_r/k^{2r+2},$$

this is given by

$$\sum_{k=1}^{\infty} \sum_{r=0}^{N_k-1} A_{r,k} = \sum_{r=0}^{N_1-1} \sum_{k=1}^{\infty} A_{r,k} + \sum_{r=N_1}^{N_2-1} \sum_{k=2}^{\infty} A_{r,k} + \sum_{r=N_2}^{N_3-1} \sum_{k=3}^{\infty} A_{r,k} + \cdots$$

$$= \sum_{m=1}^{\infty} \sum_{r=N_{m-1}}^{N_m-1} a_r b_m(r), \qquad (6.4.17)$$

where $N_0 \equiv 0$ and the coefficients $b_m(r)$ are defined by

$$b_m(r) = \sum_{k=m}^{\infty} k^{-2r-2}.$$

These latter coefficients are related to $\zeta(2r + 2)$ and we easily see that

$$b_1(r) = \zeta(2r + 2), \quad b_2(r) = b_1(r) - 1, \quad b_3(r) = b_2(r) - 2^{-2r-2}, \dots.$$

In Fig. 6.5 we show the behaviour of the terms in the sums in (6.4.17) for $1 \leq m \leq 3$ against ordinal number when $x = 5$.

The Stokes multipliers describing the appearance of the first two exponentials $e^{2\pi i p z}$, $p = 1, 2$ (at fixed $|z|$) in the vicinity of the Stokes line $\arg z = \frac{1}{2}\pi$ are then

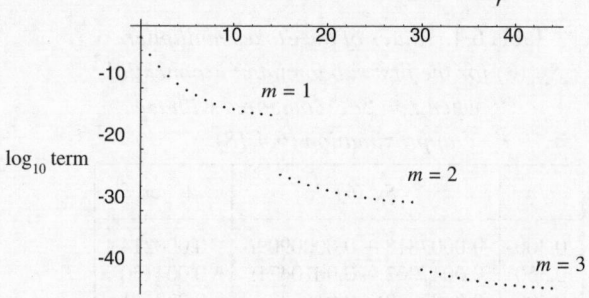

Fig. 6.5. Magnitudes of the terms in the sum (6.4.17) against ordinal number for $1 \leq m \leq 3$ when $x = 5$ and the truncation indices $N_1 = 15$, $N_2 = 30$, $N_3 = 45$.

given by

$$S_{N_1}(\theta) = e^{-2\pi i z} \left\{ \Omega(z) - \frac{1}{\pi} \sum_{r=0}^{N_1-1} a_r b_1(r) \right\}$$

and

$$S_{N_2}(\theta) = 2e^{-4\pi i z} \left\{ \Omega(z) - \frac{1}{\pi} \sum_{r=0}^{N_1-1} a_r b_1(r) - \frac{1}{\pi} \sum_{r=N_1}^{N_2-1} a_r b_2(r) - R_1(z; N_1) \right\},$$

where, as usual, $\theta = \arg z$. The procedure for the Stokes multiplier for the pth exponential $e^{2\pi i p z}$ is then obvious and is given by

$$S_{N_p}(\theta) = p e^{-2\pi i p z} \left\{ \Omega(z) - \frac{1}{\pi} \sum_{m=1}^{p} \sum_{r=N_{m-1}}^{N_m-1} a_r b_m(r) - \sum_{k=1}^{p-1} \frac{R_k(z; N_k)}{k} \right\}$$

for $p = 1, 2, \ldots$, where the optimal truncation indices N_k are specified in (6.4.12).

In Tables 6.4 and 6.5 we show the results of such computations for $z = 5e^{i\theta}$ for different values of θ in the neighbourhood of the Stokes line $\arg z = \frac{1}{2}\pi$. The values of the real part of $S_{N_p}(\theta)$ for $p = 1, 2$ are compared with the predicted approximate behaviour from (6.4.15), given by

$$\frac{1}{2} + \frac{1}{2} \operatorname{erf} \hat{\sigma}_p, \qquad \hat{\sigma}_p = \left(\theta - \frac{1}{2}\pi\right)\sqrt{\pi p |z|}. \tag{6.4.18}$$

In the computation of $S_{N_2}(\theta)$, the remainder term $R_1(z; N_1)$ has been evaluated by means of (6.4.8), where the terminant functions are computed from (6.2.43) to the same level of accuracy as the other terms. It is seen that the behaviour of the real part of the multipliers is as predicted by (6.4.18) and, moreover, that the transition of the second exponential occurs on a sharper scale than the first. We note that the above process for $p = 2$ has revealed the presence of an exponential of magnitude $e^{-20\pi} = O(10^{-28})$ hidden behind a larger exponential of magnitude $e^{-10\pi}$.

Table 6.4. *Values of the Stokes multiplier*
$S_{N_1}(\theta)$ *for the first subdominant exponential*
$e^{2\pi i z}$ *when* $z = 5e^{i\theta}$ *compared with the*
approximation (6.4.18)

θ/π	$S_{N_1}(\theta)$	$\frac{1}{2} + \frac{1}{2}\text{erf}\,\hat{\sigma}_1$
0.300	$0.0002818 + 0.0000909i$	0.000214
0.350	$0.0047002 - 0.0010471i$	0.004130
0.400	$0.0387937 - 0.0167382i$	0.039131
0.450	$0.1845516 - 0.0713116i$	0.189313
0.475	$0.3258786 - 0.1016917i$	0.329891
0.500	$0.5000000 - 0.1144236i$	0.500000
0.525	$0.6741214 - 0.1016917i$	0.670109
0.550	$0.8154484 - 0.0713116i$	0.810687
0.600	$0.9612063 - 0.0167382i$	0.960869
0.650	$0.9952998 - 0.0010471i$	0.995871
0.700	$0.9997182 + 0.0000909i$	0.999786

Table 6.5. *Values of the Stokes multiplier*
$S_{N_2}(\theta)$ *for the second subdominant*
exponential $e^{4\pi i z}$ *when* $z = 5e^{i\theta}$ *compared*
with the approximation (6.4.18)

θ/π	$S_{N_2}(\theta)$	$\frac{1}{2} + \frac{1}{2}\text{erf}\,\hat{\sigma}_2$
0.40	$0.0067412 + 0.0000860i$	0.006383
0.45	$0.1068719 - 0.0172688i$	0.106545
0.46	$0.1596065 - 0.0256381i$	0.159603
0.47	$0.2272454 - 0.0345741i$	0.227511
0.48	$0.3088606 - 0.0426487i$	0.309226
0.49	$0.4014169 - 0.0483096i$	0.401671
0.50	$0.5000000 - 0.0503487i$	0.500000
0.51	$0.5985831 - 0.0483096i$	0.598329
0.52	$0.6911395 - 0.0426487i$	0.690774
0.53	$0.7727546 - 0.0345741i$	0.772489
0.54	$0.8403935 - 0.0256381i$	0.840398
0.55	$0.8931281 - 0.0172689i$	0.893455
0.60	$0.9932588 + 0.0000860i$	0.993617

7

Multiple Mellin-Barnes Integrals

The problem of determining the asymptotic behaviour of functions defined by multiple integrals carries with it additional computational difficulties that occasionally prevent the generalisation of familiar techniques from the one-dimensional world to that of several dimensions: the method of stationary phase generalises to n-dimensional integrals, but the method of steepest descent has only recently been extended to several dimensions and is still in its infancy. Much the same can be said of the techniques used in earlier chapters for dealing with Mellin-Barnes integrals, although we shall see that, for a class of multidimensional Laplace-type integrals, the transition to higher dimensions can be accomplished with only small modifications to approaches we have taken in previous chapters.

7.1 Some Double Integrals

Before unveiling a general strategy for dealing with multiple Laplace-type integrals by means of multiple Mellin-Barnes integrals, it is worthwhile examining some special multiple integrals. It is occasionally the case that integrals that prove to be unwieldy or resistant to analysis through more standard tools such as steepest descents, will be more accommodating if a Mellin-Barnes representation is available. Perhaps surprisingly, one-dimensional integrals are sometimes more easily dealt with using a higher-dimensional representation.

Example 1. We begin with an integral from physical science [Davies (1978, p. 229)] representing a correction to the equation of state of a classical electron gas. For positive λ, put $f(x) = e^{-\lambda x}/x$. The integral in question has the form

$$F(\lambda) = \int_0^\infty \left(e^{-f(x)} - 1 + f(x) - \tfrac{1}{2}f^2(x)\right) x^2 dx, \qquad (7.1.1)$$

with the small-λ behaviour of F of particular interest. From the Mellin-Barnes integral representation of the exponential function (3.3.2), we have

$$e^{-f(x)} - 1 + f(x) - \tfrac{1}{2}f^2(x) = \frac{1}{2\pi i} \int_{c-\infty i}^{c+\infty i} \Gamma(s)f(x)^{-s}ds,$$

where $-3 < c < -2$. We may therefore recast (7.1.1) in the form

$$\begin{aligned}
F(\lambda) &= \int_0^\infty x^2 \left\{ \frac{1}{2\pi i} \int_{c-\infty i}^{c+\infty i} \Gamma(s)x^s e^{\lambda x s} ds \right\} dx \\
&= \frac{1}{2\pi i} \int_{c-\infty i}^{c+\infty i} \Gamma(s) \left\{ \int_0^\infty x^{2+s} e^{\lambda x s} dx \right\} ds \\
&= \frac{1}{2\pi i} \int_{c-\infty i}^{c+\infty i} \Gamma(s)\Gamma(s+3)(-\lambda s)^{-3-s} ds \quad (-3 < c < -2).
\end{aligned}$$

The interchange in the order of integration that occurs is legitimate as the gamma function present in the integrand grows as $|t|^{c-\frac{1}{2}} e^{-\frac{1}{2}\pi|t|}$ along the contour $s = c + it$, and the exponential function $e^{\lambda x s}$ undergoes rapid decay in view of the fact $c < 0$.

The single Mellin-Barnes integral that emerges has double poles at the integers $s = -3, -4, -5, \ldots$. The appropriate residue can be determined by setting $s = -n + \delta$ for positive integer n and small δ into $\Gamma(s)\Gamma(s+3)(-\lambda s)^{-3-s}$ and observing that

$$\begin{aligned}
&\Gamma(s)\Gamma(s+3)(-\lambda s)^{-3-s} \\
&= \frac{(\lambda n)^{n-3}\pi^2}{\sin\pi(\delta-n)\sin\pi(n-2-\delta)} \frac{(\lambda n)^{-\delta}(1-\delta/n)^{n-3-\delta}}{\Gamma(n+1-\delta)\Gamma(n-2-\delta)} \\
&= -\frac{\pi^2(\lambda n)^{n-3}}{\sin^2\pi\delta\, n!\,(n-3)!} \frac{\left(1-\delta\log(n\lambda)+\cdots\right)\left(1-(n-3-\delta)\delta/n+\cdots\right)}{\left(1-\delta\psi(n+1)+\cdots\right)\left(1-\delta\psi(n-2)+\cdots\right)} \\
&= -\frac{(\lambda n)^{n-3}}{n!\,(n-3)!}\frac{1}{\delta^2}\left\{1+\delta\left(\psi(n+1)+\psi(n-2)\right.\right. \\
&\qquad\qquad\qquad \left.\left. -\log(n\lambda)-\frac{n-3}{n}\right)+O(\delta^2)\right\};
\end{aligned}$$

here, use has been of the reflection formula (2.1.20). We conclude, therefore, that F has the small-λ expansion[†]

$$F(\lambda) = \sum_{n=3}^\infty \frac{(\lambda n)^{n-3}}{n!(n-3)!}\left\{\log(n\lambda) + \frac{n-3}{n} - \psi(n+1) - \psi(n-2)\right\}.$$

---

† Davies' computation of the small-λ expansion of F contains some computational errors. These have been corrected here.

Example 2. Consider the double integral

$$I(a) = 2^\nu a^{\mu-\nu-1} \int_0^\infty e^{-t} \int_{at}^\infty J_\nu(x) \frac{dx}{x^\mu} \, dt,$$

where $J_\nu(x)$ is the Bessel function of order ν and $a \, (> 0)$, μ, ν are real parameters; the extension to complex parameters presents no difficulty. Let us impose the temporary restrictions $\nu > 0$ and $1 < \mu - \nu < 2$; these can be subsequently relaxed by appeal to analytic continuation.

From (3.4.22) we have the representation

$$J_\nu(x) = \frac{1}{2\pi i} \int_{-\infty i}^{\infty i} \frac{\Gamma(-s)}{\Gamma(s+\nu+1)} (\tfrac{1}{2}x)^{\nu+2s} ds$$

valid for $x > 0$ and $\nu > 0$, where the path of integration is indented at $s = 0$ to lie to the left of the poles of $\Gamma(-s)$. Then we find†

$$\int_{at}^\infty J_\nu(x) \frac{dx}{x^\mu} = \frac{1}{2\pi i} \int_{-\infty i}^{\infty i} \frac{\Gamma(-s)}{\Gamma(s+\nu+1)} 2^{-\nu-2s} \left\{ \int_{at}^\infty x^{2s+\nu-\mu} dx \right\} ds$$

$$= -\frac{1}{2\pi i} \int_{-\infty i}^{\infty i} \frac{\Gamma(-s)}{\Gamma(s+\nu+1)} \frac{2^{-\nu-2s}(at)^{2s+\nu-\mu+1}}{2s+\nu-\mu+1} ds,$$

provided $\mu - \nu > 1$. Substitution of this result into the integral for $I(a)$, followed by reversal of the order of integration (when $\mu - \nu < 2$), yields the Mellin-Barnes integral representation

$$I(a) = -\frac{1}{2\pi i} \int_{-\infty i}^{\infty i} \frac{\Gamma(-s)}{\Gamma(s+\nu+1)} \frac{(\tfrac{1}{2}a)^{2s}}{2s+\nu-\mu+1} \left\{ \int_0^\infty t^{2s+\nu-\mu+1} e^{-t} \, dt \right\} ds$$

$$= -\frac{1}{2\pi i} \int_{-\infty i}^{\infty i} \frac{\Gamma(-s)\Gamma(2s+\nu-\mu+1)}{\Gamma(s+\nu+1)} (\tfrac{1}{2}a)^{2s} ds. \qquad (7.1.2)$$

Although (7.1.2) has been derived with the above restrictions, this result can be extended to values of μ and ν satisfying $-\frac{1}{2} < \mu < \nu + 2$ and $\nu \geq 0$ (at least) by analytic continuation.

The integrand in (7.1.2) has two sequences of poles at $s = 0, 1, 2, \ldots$ and $2s = \mu - \nu - 1 - k$, for nonnegative integer k. If $1 < \mu - \nu < 2$, the first pole in the second sequence lies on the right of the contour (with $s = 0$ being a double pole when $\mu - \nu = 1$), while when $\mu - \nu < 1$ all poles of the second sequence lie in $\text{Re}(s) < 0$. Thus, if we denote by C the path in (7.1.2) that is indented (when $1 < \mu - \nu < 2$) so as to *separate the sequences of poles*, we have

$$I(a) = -\frac{1}{2\pi i} \int_C \frac{\Gamma(-s)\Gamma(2s+\nu-\mu+1)}{\Gamma(s+\nu+1)} (\tfrac{1}{2}a)^{2s} ds + R_{\mu,\nu}(a),$$

---

† A particular case of this result when $\mu = 1$, $\nu = 0$ is cited in a footnote in Watson (1966, p. 434).

where

$$R_{\mu,\nu}(a) = \begin{cases} \dfrac{1}{2}\dfrac{\Gamma(\frac{1}{2}+\frac{1}{2}\nu-\frac{1}{2}\mu)}{\Gamma(\frac{1}{2}+\frac{1}{2}\mu+\frac{1}{2}\nu)}(\tfrac{1}{2}a)^{\mu-\nu-1} & (1 < \mu - \nu < 2) \\[2mm] 0 & (\mu - \nu < 1). \end{cases}$$

The second term $R_{\mu,\nu}(a)$ corresponds to the residue at the pole $s = \frac{1}{2}\mu - \frac{1}{2}\nu - \frac{1}{2}$ when this point is situated in $\mathrm{Re}(s) > 0$.

We now use the duplication formula (2.2.23) to express $\Gamma(2s + \nu - \mu + 1)$ in terms of two gamma functions with arguments involving s. This enables us to identify the integral over C as a hypergeometric function by (3.4.8) to find

$$I(a) = -\frac{\Gamma(\nu - \mu + 1)}{\Gamma(1 + \nu)}{}_2F_1\big(\tfrac{1}{2}\nu - \tfrac{1}{2}\mu + \tfrac{1}{2}, \tfrac{1}{2}\nu - \tfrac{1}{2}\mu + 1;$$

$$1 + \nu; -a^2\big) + R_{\mu,\nu}(a). \qquad (7.1.3)$$

This last result can be expressed in a more convenient form by making use of the standard property given in (4.6.15) to transform the hypergeometric function in (7.1.3) into one with argument equal to $a^2/(1 + a^2) \leq 1$, namely

$$(1 + a^2)^{\frac{1}{2}\mu - \frac{1}{2}\nu - \frac{1}{2}}{}_2F_1\big(\tfrac{1}{2}\nu - \tfrac{1}{2}\mu + \tfrac{1}{2}, \tfrac{1}{2}\mu + \tfrac{1}{2}\nu; 1 + \nu; \frac{a^2}{1 + a^2}\big).$$

In the special case $\mu - \nu = 1$, the residue of the double pole at $s = 0$ is easily shown to be $-\{\log(\frac{1}{2}a) - \frac{1}{2}\psi(1+\nu) - \frac{1}{2}\gamma\}/\Gamma(1+\nu)$, where ψ is the logarithmic derivative of the gamma function and γ denotes Euler's constant. By displacement of the contour to the right over the poles at $s = 0, 1, 2, \ldots$ we therefore find

$$I(a) = \sum_{k=1}^{\infty} \frac{(-)^{k-1}\Gamma(2k)}{k!\,\Gamma(1 + \nu + k)}(\tfrac{1}{2}a)^{2k} + R_{1+\nu,\nu}(a), \qquad (7.1.4)$$

where

$$R_{1+\nu,\nu}(a) = -\frac{1}{2\Gamma(1+\nu)}\{2\log(\tfrac{1}{2}a) - \psi(1+\nu) - \gamma\}.$$

We remark that the sum over k corresponds to the hypergeometric function in (7.1.3) with the first term deleted.

An alternative representation can be derived by displacing the contour to the left to yield the *convergent* expansion (when $a > 1$)

$$I(a) = \frac{1}{2}\sum_{k=0}^{\infty}{}' \frac{(-)^{k-1}\Gamma(\frac{1}{2}\nu - \frac{1}{2}\mu + \frac{1}{2} + \frac{1}{2}k)}{k!\,\Gamma(\frac{1}{2}\mu + \frac{1}{2}\nu + \frac{1}{2} - \frac{1}{2}k)}(\tfrac{1}{2}a)^{\mu-\nu-1-k}, \qquad (7.1.5)$$

where the prime on the summation sign indicates that when $1 \leq \mu - \nu < 2$ the first term corresponding to $k = 0$ is to be deleted. The convergence of this series is readily established by noticing that the large-k behaviour of the terms is controlled by the factor $k^{-\mu-\frac{1}{2}}a^{-k}$. This form is equivalent to that in (7.1.3) when $a > 1$ and is more suitable for computing $I(a)$ for large values of a.†

A special case of these results is given by $\mu = 1$, $\nu = 0$. From (7.1.4) and (7.1.5) we find the behaviour in the limits of small and large a

$$\int_0^\infty e^{-t} \int_{at}^\infty J_0(x)\frac{dx}{x}\,dt = \begin{cases} \log(2/a) + O(a^{-2}) & (a \to 0+) \\ \pi^{\frac{1}{2}}/a + O(a^{-2}) & (a \to \infty). \end{cases}$$

Other examples of the use of Mellin-Barnes integrals in the evaluation of definite integrals can be found in Watson (1966, p. 434 *et seq.*); for a detailed discussion on the evaluation of integrals using Mellin transforms, see Marichev (1982).

Example 3. DOUBLE MELLIN TRANSFORMS. One can extend the notion of Mellin transform, discussed in §3.1, in a straightforward fashion to accommodate functions of several variables. We illustrate this here for the case of functions of two variables.

Following Delavault (1961, Chapter 3), we define the double Mellin transform of a function $f(x, y)$ by

$$M[f; s, t] = F(s, t) = \int_0^\infty \int_0^\infty f(x, y)x^{s-1}y^{t-1}dx\,dy \qquad (7.1.6)$$

for suitable classes of functions f for which the integral makes sense. In direct analogy with the case of the Mellin transform, we also have the inverse transform given by

$$f(x, y) = \left(\frac{1}{2\pi i}\right)^2 \int_{c-\infty i}^{c+\infty i} \int_{d-\infty i}^{d+\infty i} x^{-s}y^{-t}\,F(s, t)\,ds\,dt, \qquad (7.1.7)$$

again, provided the integral exists. Convergence criteria, by no means comprehensive, can be found in Delavault (1961, pp. 26 and 29) and in Reed (1944). We will not concern ourselves overmuch with issues of convergence, but we will remark that generally, if the Mellin transform of a function $f(x, y)$ converges for a pair of values s_0, t_0, then it will converge in strips in each of the s and t planes containing s_0 and t_0, respectively.

Double Mellin transforms enjoy many of the same properties as their one-dimensional counterparts: there is a Parseval formula for double Mellin transforms; double Mellin transforms of differentiated quantities become products of polynomials and Mellin transforms of the underived quantities; premultiplication by powers of the untransformed variables x and y results in shifted arguments in the transformed function; and there is a natural choice of convolution product for the

---

† The forms in (7.1.3) and (7.1.5) are equivalent to the continuation formula for the hypergeometric function given in (3.4.9).

double Mellin transform, *viz.*, if

$$(f * g)(x, y) = \int_0^\infty \int_0^\infty f(\xi x, \eta y) g(\xi, \eta) \, d\xi \, d\eta,$$

then

$$M[f * g; s, t] = M[f; s, t] M[g; 1 - s, 1 - t].$$

The inversion of a double Mellin transform, then, presents us with a genuinely multidimensional Mellin-Barnes integral to analyse. These are in general difficult to compute exactly, although exact evaluations in some special cases are available. For example, if ϕ is some function for which

$$f(x, y) = \sum_{m=0}^\infty \sum_{n=0}^\infty (-)^{m+n} \phi(m, n) x^m y^n$$

defines an analytic function, then we note that, from Cauchy's theorem, truncated forms of the double infinite series have the representation

$$\sum_{m=0}^M \sum_{n=0}^N (-)^{m+n} \phi(m, n) x^m y^n = \left(\frac{1}{2\pi i}\right)^2 \int_{\Gamma_1} \int_{\Gamma_2} \frac{\pi^2 \phi(s, t)}{\sin \pi s \sin \pi t} x^s y^t \, dt \, ds$$

$$(7.1.8)$$

where Γ_1 is the positively oriented rectangle in the s plane with vertices at the points $-\frac{1}{2} - Mi$, $M + \frac{1}{2} - Mi$, $M + \frac{1}{2} + Mi$ and $-\frac{1}{2} + Mi$, and Γ_2 is a similarly constructed rectilinear path in the t plane with N in lieu of M. Under suitable restrictions on ϕ, principally analyticity in the half-planes $\text{Re}(s), \text{Re}(t) > -\frac{1}{2} - \epsilon$ for some small $\epsilon > 0$, and rapid evanescence as $|\text{Im}(s)|, |\text{Im}(t)| \to \infty$, the summation limits in (7.1.8) can be allowed to tend to infinity, whence we obtain the evaluation

$$M[f; s, t] = \frac{\pi^2 \phi(-s, -t)}{\sin \pi s \sin \pi t}, \qquad (7.1.9)$$

a generalisation of *Ramanujan's formula* given in (3.1.16),

$$\int_0^\infty x^{s-1} \sum_{n=0}^\infty (-)^n \phi(n) x^n \, dx = \frac{\pi}{\sin \pi s} \phi(-s).$$

The derivation of (7.1.9) together with various modifications are to be found in Reed (1944). We close this section by mentioning just one interesting evaluation found in Reed's paper: for the function

$$f(x, y) = \{(1 + x)(1 + y) - \lambda x y\}^{-1},$$

the coefficients $\phi(m, n)$ have been shown by Hardy [see Bromwich (1926, p. 193, Ex. 28)] to be given by

$$\phi(m, n) = {}_2F_1(-m, -n; 1; \lambda).$$

We then have for $0 < \mathrm{Re}(s)$, $\mathrm{Re}(t) < 1$ and $|\lambda| < 1$ the double Mellin transform

$$M\Big[\{(1+x)(1+y) - \lambda xy\}^{-1}; s, t\Big] = \frac{\pi^2 {}_2F_1(s, t; 1; \lambda)}{\sin \pi s \, \sin \pi t}.$$

7.2 Residues and Double Integrals

If we begin displacing integration contours in multiple Mellin-Barnes integrals, we find that, for each displacement, the poles of the integrand that are encountered will often depend on other integration variables, thereby complicating the type of computation we performed in earlier chapters to derive asymptotic expansions. Furthermore, each contour displacement results in unevaluated integrals, and these evaluations will not be complete until all integration contours have been displaced past poles. Nevertheless, with due care it is still possible to extract residues from multiple Mellin-Barnes integrals in a straightforward fashion and obtain asymptotic series.

We point out that the type of computation we typically undertake for multiple integrals of Mellin-Barnes type contrasts sharply with the evaluation of such integrals by other authors. In several settings† where double Mellin-Barnes integrals arise, the pole structure to be used in residue computations does not involve any linkage between poles in one integration variable and poles in the other, even though the integrand under investigation may not be a product of functions of a single variable. This is not so for the Mellin-Barnes integrals we encounter in the present case, with poles in one variable depending in an essential way on poles in the other integration variables.

The type of residue computation we encounter in the course of developing asymptotic expansions of multiple Mellin-Barnes integrals is captured well by the following example:

$$I(\lambda) = \left(\frac{1}{2\pi i}\int_{-\infty i}^{+\infty i}\right)^2 \Gamma(t_1)\Gamma(t_2)\Gamma\left(\frac{1 - m_1 t_1 - m_2 t_2}{\mu}\right)\lambda^{-(\delta_1 t_1 + \delta_2 t_2)} dt_1 dt_2$$

(7.2.1)

with μ, m_1 and m_2 real parameters satisfying the inequality $\mu > m_1 > m_2 > 0$, and where we have set

$$\delta_r = 1 - m_r/\mu \qquad (r = 1, 2);$$

(7.2.2)

evidently, $0 < \delta_1 < \delta_2 < 1$. For (7.2.1), we will not only outline the formal process of extracting its algebraic asymptotic behaviour, but will also examine the orders of the remainders in the result. Convergence of the integral is assured by virtue of Rule 1 of § 2.4.

---

† Here, we direct the reader to Example 3 of the previous section, and to the double Mellin-Barnes evaluation of Slater (1966, § 8.5) for illustrations.

The integral (7.2.1) arises in the course of the development of the asymptotic expansions of the generalised Faxén integral,

$$I(\lambda; c_1, c_2, \ldots, c_k) = \int_0^\infty \exp\left\{-\lambda(x^\mu + \sum_{r=0}^k c_r x^{m_r})\right\} dx, \qquad (7.2.3)$$

where

$$\mu > m_1 > m_2 > \cdots > m_k > 0;$$

this integral is systematically examined in Kaminski & Paris (1997). The Faxén integral itself can be expressed as

$$\mathrm{Fi}(a, b; \pm y) = \frac{y^{b/(1-a)}}{b} I(y^{b/(1-a)}; \mp 1) \qquad (y > 0, 0 \le \mathrm{Re}(a) < 1);$$

see §5.5. To keep the presentation as simple as possible, we shall restrict $k = 2$ and put the parameters c_r to unity to arrive at

$$I(\lambda; 1, 1) = \int_0^\infty \exp\{-\lambda(x^\mu + x^{m_1} + x^{m_2})\} dx. \qquad (7.2.4)$$

To each of the factors $\exp(-\lambda x^{m_r})$ appearing in the integrand of (7.2.4), apply the Mellin-Barnes integral representation of the exponential function (3.3.2) to obtain

$$I(\lambda; 1, 1) = \int_0^\infty e^{-\lambda x^\mu} \left(\frac{1}{2\pi i} \int_{-\infty i}^{\infty i}\right)^2 \Gamma(t_1)\Gamma(t_2)\lambda^{-t_1-t_2} x^{-m_1 t_1 - m_2 t_2} dt_1 dt_2 dx.$$

Because of the absolute convergence of this integral, we can interchange the order of integration and apply Euler's integral representation for the gamma function to obtain, apart from a leading factor of $\lambda^{-1/\mu}/\mu$, the integral (7.2.1), i.e., $I(\lambda) = \mu\lambda^{1/\mu} I(\lambda; 1, 1)$. We shall assume that any poles we encounter in the analysis of (7.2.1) are simple, but the presence of higher order poles will prove no obstacle to the overall strategy we detail here.

From the order estimates (2.4.4), each t_r integral in (7.2.1), $r = 1, 2$, converges absolutely in the sector

$$|\arg \lambda| < \tfrac{1}{2}\pi \frac{(1 + m_r/\mu)}{\delta_r}. \qquad (7.2.5)$$

Because $\mu > m_1 > m_2 > 0$, we see that the integral (7.2.1) converges in a sector wider than the right half-plane $\mathrm{Re}(\lambda) > 0$.

Let us translate the t_1 contour first. If we set $t_1 = \rho e^{i\theta}$, with ρ and θ real and $|\theta| < \tfrac{1}{2}\pi$, then we find that the logarithm of the modulus of the integrand has the large-ρ behaviour

$$\rho \cos\theta \log\rho \cdot \delta_1 + O(\rho)$$

which in view of $\delta_1 > 0$ must tend to ∞ as $\rho \to \infty$. Accordingly, we displace the t_1 contour to the right to obtain the large-λ asymptotic behaviour. In doing this, we encounter poles of the integrand at the sequence of t_1 values given by†

$$t_1' = t_1'(t_2) = \frac{1}{m_1}(1 + \mu r_1 - m_2 t_2), \tag{7.2.6}$$

where r_1 is a nonnegative integer. Since t_2 everywhere along its integration contour has $\mathrm{Re}(t_2) = 0$, except for an indentation to the right near the origin where $\mathrm{Re}(t_2)$ is positive but arbitrarily small, we find that every choice of nonnegative integral r_1 gives rise to a pole of the t_1 integrand.

Fix a positive integer N_1 and a positive number ϵ_1 satisfying $0 < \epsilon_1 < 1$. We shift the t_1 contour to the vertical line $\mathrm{Re}(t_1) = \{1 + \mu(N_1 + \epsilon_1)\}/m_1$ in conventional fashion by considering the integral (in t_1) taken over the rectangular contour with vertices $\pm iM$ and $\{1 + \mu(N_1 + \epsilon_1)\}/m_1 \pm iM$, $M > 0$. It is routine to show that the contributions to the value of the integral from the segments parallel to the real axis tend to zero as $M \to \infty$. The residue theorem then yields

$$I(\lambda) = \frac{\mu}{m_1} \sum_{r_1=0}^{N_1} \frac{(-)^{r_1}}{r_1!} I_1(\lambda; r_1) + R_1, \tag{7.2.7}$$

where

$$I_1(\lambda; r_1) = \frac{1}{2\pi i} \int_{-\infty i}^{\infty i} \Gamma(t_1')\Gamma(t_2)\lambda^{-(\delta_1 t_1' + \delta_2 t_2)} dt_2, \tag{7.2.8}$$

t_1' is given in (7.2.6) and the remainder term is defined by

$$R_1 = \left(\frac{1}{2\pi i}\right)^2 \int_{L_1} \int_{-\infty i}^{\infty i} \Gamma(t_1)\Gamma(t_2)\Gamma\left(\frac{1 - m_1 t_1 - m_2 t_2}{\mu}\right)\lambda^{-\delta_1 t_1 - \delta_2 t_2} dt_2 \, dt_1. \tag{7.2.9}$$

In (7.2.9), the contour L_1 is the vertical line $\mathrm{Re}(t_1) = \{1 + \mu(N_1 + \epsilon_1)\}/m_1$.

An order estimate of R_1 is easily obtained. Observe that the t_2 integral in (7.2.9) has $\mathrm{Re}(t_2) = 0$ everywhere, except for an indentation near the origin where $\mathrm{Re}(t_2)$ is positive but arbitrarily small. The t_1 variable everywhere along L_1 has $\mathrm{Re}(t_1) = \{1 + \mu(N_1 + \epsilon_1)\}/m_1$, so that

$$R_1 = O(\lambda^{-\delta_1(1 + \mu(N_1 + \epsilon_1))/m_1})$$

for $\lambda \to \infty$. Indeed, in view of the Riemann-Lebesgue lemma and the oscillatory nature of the integrand in (7.2.9), this can be strengthened to an o-estimate, if so desired.

Let us now turn to the integrals $I_1(\lambda; r_1)$ present in the finite series in (7.2.7). To determine which direction we displace the t_2 contour, we proceed as before and

---

† We write $t_1' = t_1'(t_2)$ to emphasise the functional dependence of the t_1 poles on the integration variable t_2.

set $t_2 = \rho e^{i\theta}$, $|\theta| < \frac{1}{2}\pi$, from which we see that the logarithm of the modulus of the integrand has the large-ρ behaviour

$$\rho \cos\theta \log\rho \cdot \left(\frac{m_1 - m_2}{m_1}\right) + O(\rho).$$

From $m_1 > m_2 > 0$, we see that this estimate grows to ∞ as $\rho \to \infty$ which indicates that the t_2 contour must also be displaced to the right to recover the large-λ behaviour of each $I_1(\lambda; r_1)$. In displacing the t_2 contour, we encounter poles at solutions to $t_1' = -r_2$ for r_2 a nonnegative integer; i.e., at the points

$$t_2' = \frac{1}{m_2}(1 + \mu r_1 + m_1 r_2). \tag{7.2.10}$$

Thus, we find

$$I_1(\lambda; r_1) = \frac{m_1}{m_2} \sum_{r_2=0}^{N_2(r_1)} \frac{(-)^{r_2}}{r_2!} \Gamma(t_2') \lambda^{-(\delta_1 t_1'(t_2') + \delta_2 t_2')} + R_2(r_1) \tag{7.2.11}$$

where now, since $t_1'(t_2') = -r_2$ and since

$$m_i \delta_j - m_j \delta_i = m_i - m_j \quad (i \neq j), \tag{7.2.12}$$

we may also write

$$\delta_1 t_1'(t_2') + \delta_2 t_2' = \frac{(m_1 - m_2)}{m_2} r_2 + \frac{\delta_2}{m_2}(1 + \mu r_1).$$

The remainder term $R_2(r_1)$ is given by

$$R_2(r_1) = \frac{1}{2\pi i} \int_{L_2} \Gamma(t_1')\Gamma(t_2)\lambda^{-\delta_1 t_1' - \delta_2 t_2} dt_2, \tag{7.2.13}$$

where L_2 is the vertical line $\text{Re}(t_2) = \{1 + \mu r_1 + m_1(N_2 + \epsilon_2)\}/m_2$, for some $0 < \epsilon_2 < 1$. The precise nature of the dependence of N_2 on r_1 will be detailed below.

Collecting together the approximations (7.2.7) and (7.2.11), we obtain

$$I(\lambda) = \frac{\mu}{m_2} \sum_{r_1=0}^{N_1} \left\{ \sum_{r_2=0}^{N_2(r_1)} \frac{(-)^{r_1+r_2}}{r_1! r_2!} \Gamma(t_2') \lambda^{-(\delta_1 t_1'(t_2') + \delta_2 t_2')} + R_2(r_1) \right\} + R_1,$$

with t_2' given in (7.2.10). From (7.2.13), we find that as $\lambda \to \infty$,

$$R_2(r_1) = O(\lambda^{-\text{Re}\{\delta_1 t_1'(t_2) + \delta_2 t_2\}}) = O(\lambda^{-\delta_1(1+\mu r_1)/m_1 - (m_1 - m_2)\text{Re}(t_2)/m_2})$$

$$= O(\lambda^{-\delta_2(1+\mu r_1)/m_2 - (m_1 - m_2)(N_2 + \epsilon_2)/m_2})$$

using (7.2.6), (7.2.12) and the fact that on L_2, we have $\text{Re}(t_2) = \{1 + \mu r_1 + m_1(N_1 + \epsilon_1)\}/m_2$. If the combined errors

$$\sum_{r_1=0}^{N_1} R_2(r_1)$$

are to be of the same order as the error R_1 that resulted from the initial displacement of the t_1 contour, then it suffices to have, for each r_1, the order estimate

$$R_2(r_1) = O(\lambda^{-\delta_1(1+\mu(N_1+\epsilon_1))/m_1}).$$

Then $N_2 = N_2(r_1)$ can be determined from the requirement

$$\delta_1(1 + \mu(N_1 + \epsilon_1))/m_1 = \delta_2(1 + \mu r_1)/m_2 + (m_1 - m_2)(N_2 + \epsilon_2)/m_2$$

which implies that $N_2(r_1)$ is the largest nonnegative integer strictly less than

$$N_2^*(r_1) = \frac{1}{m_1 - m_2} \left\{ \frac{\delta_1 m_2}{m_1}(1 + \mu(N_1 + \epsilon_1)) - \delta_2(1 + \mu r_1) \right\}. \qquad (7.2.14)$$

This determines the upper range of the summation in (7.2.11). One final simplification follows from the identity $1/\mu + \delta_r/m_r = 1/m_r, r = 1, 2$, whence we obtain as $\lambda \to \infty$

$$I(\lambda) = \frac{\mu \lambda^{-\delta_2/m_2}}{m_2} \sum_{r_1=0}^{N_1} \sum_{r_2=0}^{N_2(r_1)} \frac{(-)^{r_1+r_2}}{r_1! r_2!} \Gamma\left(\frac{1 + \mu r_1 + m_1 r_2}{m_2}\right) \lambda^{-\varphi(r_1,r_2)}$$

$$+ O(\lambda^{-\delta_1(1+\mu(N_1+\epsilon_1))/m_1})$$

with

$$\varphi(r_1, r_2) = \frac{\mu \delta_2}{m_2} r_1 + \frac{(m_1 - m_2)}{m_2} r_2.$$

We note that with a reordering of the powers $\varphi(r_1, r_2)$, the asymptotic series that we obtained could be put in a form in which it is clear that the expansion is of Poincaré type.

The process outlined above can readily be extended to deal with more complicated integrands and higher dimensions. This will be readily apparent in the next section, when we examine the asymptotics of Laplace-type integrals.

7.3 Laplace-Type Double Integrals

It is with Laplace-type integrals that significant benefits from a Mellin-Barnes integral approach are to be found, as we have seen in earlier chapters. Prior to considering the cases of two- and three-dimensional Laplace-type integrals, let us note some strategies for dealing with multiple Laplace-type integrals.

By a double integral of Laplace type we shall mean one of the form

$$J(\lambda) = \int\int_\Sigma e^{-\lambda f(x,y)} g(x, y) \, dx dy \qquad (7.3.1)$$

where $f(x, y)$ and $g(x, y)$, the *phase* and *amplitude,* respectively of $J(\lambda)$, satisfy some smoothness conditions on the domain Σ, usually C^∞ or analytic for the phase f. Of interest to us is the behaviour of $J(\lambda)$ as $\lambda \to \infty$.

Through the use of neutralisers (or partitions of unity) and the tools of vector analysis, it is known [Bleistein & Handelsman (1975, Chapter 8)] that the significant contributions to the asymptotic behaviour of these integrals arise from: (a) critical points of f in the interior of the domain Σ or on its boundary, $\partial\Sigma$; (b) points of $\partial\Sigma$ where the boundary fails to have a continuous tangent vector field (so-called "corner points"); and (c) points of Σ where the amplitude g, or its derivatives, is discontinuous. In this classification scheme for sources of terms in the asymptotics of Laplace-type integrals, (7.3.1) coincides with cases (a) and (b), where a critical point of the phase f is found at a corner.

Integrals of the form (7.3.1), and its oscillatory form obtained by replacing $-\lambda$ by $\pm i\lambda$, have been subjected to much study over the years, using a wide variety of tools and yielding a similarly varied range of outcomes. Laplace- and Fourier-type single integrals have a rich body of theory one can bring to bear when extracting their behaviour, but the multidimensional case is considerably more difficult, a fact reflected in the variety of approaches to the problem found in the literature, and in the more qualitative types of results obtained.

For very general phase and amplitude, work by Malgrange (1974) produced a structure theorem for the form of the large-λ asymptotic expansion of $J(\lambda)$. The work relied heavily on the use of a "resolution" of the singularity of the phase function f and was not amenable to computation. Subsequent work by a school of Russian analysts, led by Arnold et al. (1988), often for domains restricted to the positive quadrant, provided some sharper results including estimates of the order of the leading term based on properties of the Newton diagram of the phase f; the Newton diagram is described below. However, again because much of their work relied on algebraic geometric and algebraic topological methods (primarily concerned with the resolution of singularities), little is said about the problem of constructing computationally useful expansions of $J(\lambda)$. Earlier work by another member of this school, Vasil'ev (1977), produced the correct leading term and the correct order, expressed in terms of the Newton diagram of the phase, but did so in a fashion that was unsuitable for extracting later terms in the full expansion of (7.3.1).

Other efforts have been mounted in tackling this problem in the setting of information carried by the Newton diagram. One novel approach by Denef & Sargos (1989) used structures "dual" to the faces of the Newton diagram and sidestepped the use of a resolution of singularities by resorting to a dissection of the integration domain: each constituent piece of the dissected domain was associated with a particular change of variables determined by the Newton diagram.† Denef & Sargos obtained detailed information about the poles of the distribution $f_+^s \equiv \max\{f^s, 0\}$,

---

† In a sense, Denef & Sargos have introduced the resolution of singularities favoured by the theoretical work of the Russian and French schools, but in a fashion more accessible than had been done previously.

whose action on test functions is given by

$$\phi \mapsto \int_{R^n} f_+^s(x_1, \dots, x_n)\phi(x_1, \dots, x_n)\, dx_1 \cdots dx_n.$$

The style of argument employed by Denef & Sargos leads the reader to expect that each face in the Newton diagram ought to play a role in the asymptotics of f_+^s. Their result can be rendered in the form of a Laplace- or Fourier-type integral by use of a Mellin transform, after the fashion discussed by Wong & McClure (1981, p. 518).

A more elementary approach for integrals taken over the positive quadrant, was developed by Dostal & Gaveau (1987, 1989) where an argument employing a rescaling of a polynomial phase f, with a different rescaling for each face of the Newton diagram, resulted in an asymptotic approximation for $J(\lambda)$. The resulting approximation, however, was incomplete in the sense that the coefficients in the resulting terms comprised infinite integrals of exponential functions.

Sharper results are available if one is willing to discard the link between the geometry of the Newton diagram and the phase function under examination. One efficient approach relies on tools from vector analysis, and is well described in a number of standard reference works [Bleistein & Handelsman (1975); Wong (1989)]. In short, the method proceeds by integration by parts using Green's theorem in the plane for examining asymptotic contributions from boundary points of the domain, and "splits off" the contribution from interior critical points in the domain of integration for subsequent analysis by other means.

One efficient approach for dealing with these interior critical points (which may also lie on the boundary of the integration domain) involves representing the integral as the Laplace (or Fourier) transform of a lower-dimensional integral, as is done by Jones & Kline (1958) or Wong & McClure (1981), deduce the asymptotics of the function defined by this lower-dimensional integral and subsequently apply a Watson's-lemma style of argument. This approach has much in its favour, but it becomes unwieldy when the phase function has something other than a non-degenerate Hessian matrix at the critical point of interest.†

The approach we describe in this chapter has much to commend it: we are able to retain links to geometric information carried by the Newton diagram; we are able to develop full asymptotic expansions; the nature of the critical point permitted to exist at the origin is of little direct consequence, as its singular character is captured by the Newton diagram; and the computational cost associated with developing the asymptotic expansions, although moderately high, is entirely bound up in relatively straightforward residue evaluations of Mellin-Barnes integrals.

---

† We do note that Jones & Kline did analyse a few degenerate cases in their seminal paper of 1958. However, the computational costs associated with increasingly singular critical points quickly diminish the effectiveness of their approach. Additionally, the process of determining suitable coordinate transformations to invoke becomes more obscure as the dimensionality of the problem increases.

However, we do pay a cost for these benefits: the phases to which the method applies are limited to a class of polynomial-like functions, the phase must be free of critical points in the positive quadrant, apart from an isolated critical point at the origin, and the integration domain restricted to the positive quadrant.

The restricted class of double Laplace-type integrals we shall examine are of the form

$$I(\lambda) = \int_0^\infty \int_0^\infty e^{-\lambda f(x,y)} g(x, y) \, dx dy, \qquad (7.3.2)$$

where f is a 'polynomial'

$$f(x, y) = x^\mu + \sum_{r=1}^k c_r x^{m_r} y^{n_r} + y^\nu$$

where all exponents are positive and all constants c_r are nonnegative real numbers (or complex with positive real part). We observe that such a phase function may possess a highly singular critical point at the origin, and the hypothesis of nonnegative c_r ensures that no critical points exist in the positive quadrant of the real xy plane.[†]

The function $g(x, y)$ is assumed to be analytic and free of singularity in the positive quadrant, with a large x, y growth rate that is eventually damped out by the decay of the 'phase' term $e^{-\lambda f(x,y)}$ for all sufficiently large λ. For the time being, we will simplify matters by taking $g(x, y)$ to be unity; more general choices for $g(x, y)$ will be discussed later. For a more complete discussion of two-dimensional Laplace-type integrals with general amplitude functions $g(x, y)$, see Paris & Liakhovetski (2000b). The integral under examination is therefore

$$I(\lambda) = \int_0^\infty \int_0^\infty \exp\left\{-\lambda(x^\mu + \sum_{r=1}^k c_r x^{m_r} y^{n_r} + y^\nu)\right\} dx dy \qquad (7.3.3)$$

with λ large and positive; we will see later that it is often possible to relax this restriction and allow λ to be complex for a certain range of values for its argument.

We also mention that an elaborate treatment of the hyperasymptotic expansion of multidimensional Laplace-type integrals with doubly infinite paths of integration has recently been considered in Howls (1997), for the case when the critical points of the phase function $f(x, y)$ lie in the domain of integration, and in Delabaere & Howls (1999), for the case when the boundary itself gives rise to non-degenerate critical points.

[†] This is easily seen: the partial derivatives f_x and f_y everywhere have positive real part in the positive quadrant $x > 0$, $y > 0$, so f can have no critical point there. The fact that terms x^μ and y^ν are present in f guarantees that any critical point at the origin must be isolated.

7.3.1 Mellin-Barnes Integral Representation

Let us define for each $r = 1, 2, \ldots, k$

$$\delta_r = 1 - \frac{m_r}{\mu} - \frac{n_r}{\nu}, \tag{7.3.4}$$

and set

$$\vec{m} = (m_1, m_2, \ldots, m_k), \quad \vec{t} = (t_1, t_2, \ldots, t_k),$$
$$\vec{n} = (n_1, n_2, \ldots, n_k), \quad \vec{\delta} = (\delta_1, \delta_2, \ldots, \delta_k). \tag{7.3.5}$$

Apply the Mellin-Barnes integral representation of the exponential function, given in (3.3.2) to each factor $\exp(-\lambda c_r x^{m_r} y^{n_r})$ in the integrand of (7.3.3) so that we can recast the integral in (7.3.3), after interchanging the order of integration and applying the Euler integral representation of the gamma function as we did in §7.2 for (7.2.4), as

$$I(\lambda) = \frac{\lambda^{-1/\mu-1/\nu}}{\mu\nu} \left(\frac{1}{2\pi i} \int_{-\infty i}^{\infty i} \right)^k \Gamma(\vec{t}) \Gamma\left(\frac{1 - \vec{m} \cdot \vec{t}}{\mu} \right) \Gamma\left(\frac{1 - \vec{n} \cdot \vec{t}}{\nu} \right) \vec{c}^{-\vec{t}} \lambda^{-\vec{\delta} \cdot \vec{t}} d\vec{t}, \tag{7.3.6}$$

where we have set

$$\Gamma(\vec{t}) = \Gamma(t_1)\Gamma(t_2) \cdots \Gamma(t_k), \vec{c}^{-\vec{t}} = c_1^{-t_1} c_2^{-t_2} \cdots c_k^{-t_k}, \text{ and } d\vec{t} = dt_1 \, dt_2 \cdots dt_k.$$

The 'dot' appearing between two vector quantities is just the usual Euclidean dot product. The integration contours are indented to the right away from the origin, to avoid the pole of the integrand present there.

By considering each integral in (7.3.6) separately, for example

$$J_r = \frac{1}{2\pi i} \int_{-\infty i}^{\infty i} \Gamma(t_r) \Gamma\left(\frac{1 - \vec{m} \cdot \vec{t}}{\mu} \right) \Gamma\left(\frac{1 - \vec{n} \cdot \vec{t}}{\nu} \right) c_r^{-t_r} \lambda^{-\delta_r t_r} dt_r,$$

we can determine, reasoning as in §2.4, that part of the complex λ plane in which the integral J_r converges. By (2.4.4), we have for $t_r = \rho e^{i\theta}$, $\rho \to \infty$,

$$\log \left| \Gamma(t_r) \Gamma\left(\frac{1 - \vec{m} \cdot \vec{t}}{\mu} \right) \Gamma\left(\frac{1 - \vec{n} \cdot \vec{t}}{\nu} \right) c_r^{-t_r} \lambda^{-\delta_r t_r} \right|$$
$$\sim \delta_r \rho \cos\theta \log\rho + A\rho + B \log\rho, \tag{7.3.7}$$

where

$$A = -\delta_r(\theta \sin\theta + \cos\theta) - \cos\theta \left(\delta_r \log|\lambda| + \log|c_r| \right) + \sin\theta \left(\delta_r \arg\lambda + \arg c_r \right)$$

$$- \left(\frac{m_r}{\mu} + \frac{n_r}{\nu} \right) \pi |\sin\theta| - \cos\theta \cdot \log \left\{ \left(\frac{m_r}{\mu} \right)^{m_r/\mu} \left(\frac{n_r}{\nu} \right)^{n_r/\nu} \right\},$$

$$B = \text{Re}\left\{ \frac{1}{\mu}(1 - \vec{m}^* \cdot \vec{t}^*) + \frac{1}{\nu}(1 - \vec{n}^* \cdot \vec{t}^*) \right\} - \frac{3}{2};$$

here, we temporarily allow c_r to be complex, and write $\vec{m}^*$, $\vec{n}^*$ and $\vec{t}^*$ for those quantities in (7.3.5) with the rth entry deleted. Since the integral defining J_r is taken over the imaginary axis (except near the origin), we set $\theta = \pm\frac{1}{2}\pi$ to find the estimate for the logarithm of the dominant real part of the integrand given by

$$-\tfrac{1}{2}\pi\delta_r\rho - \left(\frac{m_r}{\mu} + \frac{n_r}{\nu}\right)\pi\rho \pm \rho(\delta_r \arg\lambda + \arg c_r)$$

$$= -\tfrac{1}{2}\pi\rho\left(1 + \frac{m_r}{\mu} + \frac{n_r}{\nu}\right) \pm \rho(\delta_r \arg\lambda + \arg c_r)$$

for large ρ. The integrand of J_r decays exponentially if this estimate tends to $-\infty$ as $\rho \to \infty$ and hence, treating the cases of positive and negative δ_r separately, shows that absolute convergence is assured if

$$\left| \arg\lambda + \frac{1}{\delta_r}\arg c_r \right| < \tfrac{1}{2}\pi\frac{(1 + m_r/\mu + n_r/\nu)}{|\delta_r|}; \qquad (7.3.8)$$

the convergence of the integral J_r in the case of $\delta_r = 0$ (when J_r is independent of λ) is handled similarly. Evidently, the factor following $\frac{1}{2}\pi$ on the right-hand side of (7.3.8) is greater than unity, so that each integral J_r defines an analytic function of λ in a sector including the imaginary λ axis when c_r is positive.

7.3.2 The Newton Diagram

For an analytic function f with Maclaurin series

$$f(x, y) = \sum_{m,n \geq 0} a_{mn}x^m y^n,$$

define the *carrier* of f to be the set of ordered pairs of nonnegative integers $\{(m, n) : a_{mn} \neq 0\}$; for a polynomial f, this is just the set of powers of monomials comprising f.

For each point P in the carrier of f, consider the positive quadrant $\mathbf{R}_+^2 = \{(x, y) : x > 0, y > 0\}$ translated by P so that the origin is sited at P. Call this translate of the positive quadrant $P + \mathbf{R}_+^2$. Form the union of these translated quadrants, and then its convex hull. The boundary of the convex hull will be composed of two rays parallel to the coordinate axes, as well as a polygonal path composed of finitely many line segments. This polygonal path is termed the *Newton diagram* or *Newton polygon* of f. Figure 7.1 displays the two-dimensional Newton diagram for the polynomial $2x^5 + x^3y^2 + xy^2 + 3y^5$; in the notation of (7.3.2), we have $\mu = \nu = 5$.

This definition of the Newton diagram, from Brieskorn & Knörrer (1986), can easily be extended to functions f with arbitrary nonnegative real powers for the monomials comprising f, an extension which we adopt throughout this chapter.

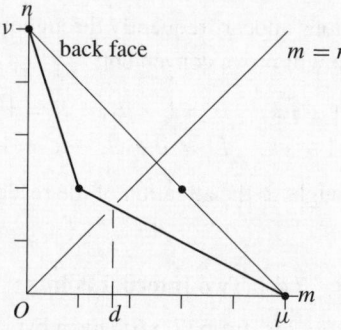

Fig. 7.1. Newton diagram for $2x^5 + x^3y^2 + xy^2 + 3y^5$ showing the line $m = n$ and the distance d to the Newton diagram. The back face has also been drawn, connecting the vertices $(\mu, 0) = (5, 0)$ and $(0, \nu) = (0, 5)$.

Extension to functions of three or more variables is also straightforward: one replaces the use of the positive quadrant in the construction by the use of the positive orthant of appropriate dimension.

Given a Newton diagram for a function $f(x, y)$, we form the ray issuing from the origin with direction vector $\vec{e} = (1, 1)$. Because the Newton diagram meets both coordinate axes, there is a number, d say, for which $d\vec{e}$ is a point of the Newton diagram. The number d is the *distance* to the Newton diagram, and the quantity $-1/d$ is termed the *remoteness* of the Newton diagram. The points (m, n) used to construct the Newton diagram we term *internal points* if they do not lie on coordinate axes. If these points lie on the Newton diagram, they are termed *vertices*.

For a function of two variables $f(x, y)$, let μ and ν be the lowest positive powers of x and y respectively for which monomials x^m and y^n appear in the Maclaurin series for f. The *back face* will be the line segment joining the points $(\mu, 0)$ and $(0, \nu)$. For functions of more than two variables, the back face can similarly be defined as the hyperplane containing the points of the carrier of f of lowest degree found on the coordinate axes.

7.4 Asymptotics of $I(\lambda)$

We will illustrate the process by which the Mellin-Barnes representation (7.3.6) for (7.3.3) can be used to extract the asymptotic behaviour of $I(\lambda)$. To keep the discussion to a moderate length, we shall only examine the case $k = 2$ in detail, and provide a short overview of how the analysis proceeds in the case $k = 3$. The simpler case $k = 1$ has been discussed in §2.3.1. The reader interested in a more complete account is directed to Kaminski & Paris (1998a).†

---

† A treatment of the case $k = 1$ is included in Example 1 of §7.7.

A number of expressions appear frequently throughout our analysis for which a more compact notation will prove convenient:

$$K = 1 + \mu k, \quad L = 1 + \mu l, \quad R = 1 + \mu r,$$
$$K' = 1 + \nu k, \quad L' = 1 + \nu l, \quad R' = 1 + \nu r. \tag{7.4.1}$$

These definitions are brought to the attention of the reader when first used.

7.4.1 Two Internal Points

The integral $I(\lambda)$ with $k = 2$ is, from (7.3.6), given by

$$I(\lambda) = \frac{\lambda^{-1/\mu - 1/\nu}}{\mu\nu} \left(\frac{1}{2\pi i} \int_{-\infty i}^{+\infty i} \right)^2 \Gamma(t_1)\Gamma(t_2)$$
$$\times \Gamma\left(\frac{1 - \vec{m} \cdot \vec{t}}{\mu} \right) \Gamma\left(\frac{1 - \vec{n} \cdot \vec{t}}{\nu} \right) \lambda^{-\vec{\delta} \cdot \vec{t}} dt_1 dt_2, \tag{7.4.2}$$

with the vector representations in (7.3.5) in use. The constants c_1 and c_2 in (7.3.3) have been set, for ease of discussion, to unity and it is a simple matter to restore these constants. We assume throughout this section that all poles of the integrand are simple.

With two internal points present, say $P_1 = (m_1, n_1)$ and $P_2 = (m_2, n_2)$, we see that the Newton diagram can assume a number of forms: we can have (a) both internal points acting as vertices of the Newton diagram, so that the diagram is formed from three non-collinear line segments, as depicted in Fig. 7.2(a); (b) both internal points on the Newton diagram, but with two of the three line segments collinear, depicted in Fig. 7.2(b); (c) one point only on the Newton diagram, with the other internal point behind any of the line segment "faces" of the Newton diagram, as in Fig. 7.2(c); or finally, (d) both internal points could lie on or behind the back face joining $(\mu, 0)$ to $(0, \nu)$, so that the Newton diagram is just the back face, as illustrated in Fig. 7.2(d). We shall keep the exposition limited to only the first instance, the *convex case*. The interested reader will find the other cases discussed in Kaminski & Paris (1998a).

Let us impose some structure on our points P_1 and P_2: we shall assume that $\mu > m_1 > m_2, n_1 < n_2 < \nu$, so that the quantities $M \equiv m_1 - m_2$ and $N \equiv n_2 - n_1$ are positive. We shall also find it convenient to introduce the quantity

$$\Delta \equiv m_1 n_2 - m_2 n_1. \tag{7.4.3}$$

Observe that Δ is the signed area of the parallelogram generated by the vectors $\vec{P}_1$, $\vec{P}_2$, in that order, where $\vec{P}$ is the position vector defined by P. Since the ordering imposed on P_1 and P_2 gives $\vec{P}_1$ and $\vec{P}_2$ a positive orientation, the quantity Δ must be positive.

Additionally, elementary analytic geometry reveals that the line through P_1 and P_2 cuts the m and n axes at Δ/N and Δ/M, respectively. By considering the

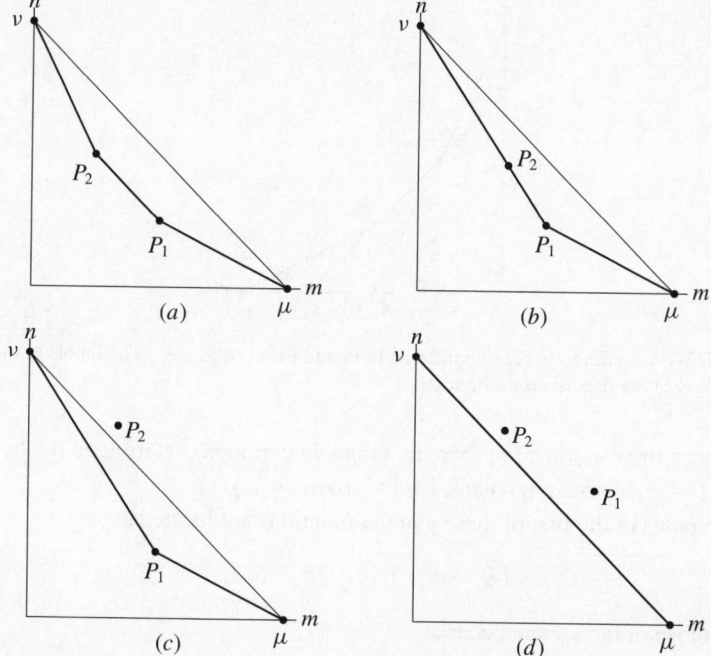

Fig. 7.2. Possible arrangements for two internal points and their Newton diagrams (drawn with heavier lines). In illustration (d), the Newton diagram and the back face coincide.

intersection points of lines through $(\mu, 0)$ and P_1, and P_2 and $(0, \nu)$, with the coordinate axes, we also find

$$\frac{\nu m_2 N}{\nu - n_2} < \Delta < \mu N \quad \text{and} \quad \frac{\mu n_1 M}{\mu - m_1} < \Delta < \nu M; \tag{7.4.4}$$

the intersection of the line through $(0, \nu)$ and P_2 with the m axis is indicated as (1_m) in Fig. 7.3; that of the line through P_1 and P_2 with the m axis is labelled (2_m), and so on. The labelled intersection points (1_m), (2_m), (1_n) and (2_n) correspond to $m = m_2\nu/(\nu - n_2)$, $m = \Delta/N$, $n = \mu n_1/(\mu - m_1)$ and $n = \Delta/M$, respectively.

Let us begin by displacing the t_2 contour in (7.4.2) first. With $t_2 = \rho e^{i\theta}$, $|\theta| < \frac{1}{2}\pi$, we see that the logarithm of the modulus of the integrand of (7.4.2) exhibits the large-ρ behaviour

$$\delta_2 \rho \cos\theta \log\rho + O(\rho).$$

Since P_2 is in front of the back face

$$\frac{m}{\mu} + \frac{n}{\nu} - 1 = 0, \tag{7.4.5}$$

the quantity δ_2 must be positive from which it follows that the order estimate above tends to $+\infty$ as $\rho \to \infty$. Thus, we displace the t_2 contour to the right to pick up contributions to the asymptotics of $I(\lambda)$. Candidate poles arise in two

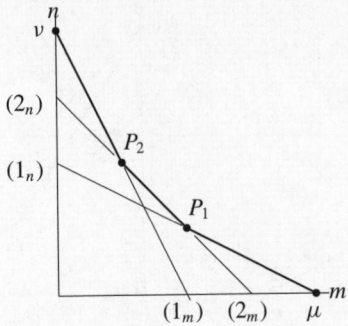

Fig. 7.3. Newton diagram for two internal points in the convex case. The labels (1_m), (2_m), (1_n) and (2_n) are described in the text.

sequences (here assumed to share no points in common) determined by the poles of $\Gamma((1 - m_1 t_1 - m_2 t_2)/\mu)$ and $\Gamma((1 - n_1 t_1 - n_2 t_2)/\nu)$.

The poles of the first of these gamma functions are located at

$$t_2^{(1)} = (K - m_1 t_1)/m_2 \quad (k = 0, 1, 2, \ldots), \tag{7.4.6}$$

while those of the second occur at

$$t_2^{(2)} = (K' - n_1 t_1)/n_2 \quad (k = 0, 1, 2, \ldots). \tag{7.4.7}$$

Poles from the $t_2^{(1)}$ sequence give rise to the formal sum

$$I_1 = \frac{\lambda^{-1/\mu - 1/\nu}}{m_2 \nu} \sum_k \frac{(-)^k}{k!} \frac{1}{2\pi i} \int_{-\infty i}^{\infty i} \Gamma(t_1) \Gamma(t_2^{(1)}) \Gamma\left(\frac{1 - n_1 t_1 - n_2 t_2^{(1)}}{\nu}\right) \lambda^{-\vec{\delta} \cdot \vec{t}} dt_1, \tag{7.4.8}$$

with $\vec{\delta} \cdot \vec{t} = \delta_1 t_1 + \delta_2 t_2^{(1)}$. Poles from the $t_2^{(2)}$ sequence similarly give rise to a formal series

$$I_2 = \frac{\lambda^{-1/\mu - 1/\nu}}{\mu n_2} \sum_k \frac{(-)^k}{k!} \frac{1}{2\pi i} \int_{-\infty i}^{\infty i} \Gamma(t_1) \Gamma(t_2^{(2)}) \Gamma\left(\frac{1 - m_1 t_1 - m_2 t_2^{(2)}}{\mu}\right) \lambda^{-\vec{\delta} \cdot \vec{t}} dt_1, \tag{7.4.9}$$

with $\vec{\delta} \cdot \vec{t} = \delta_1 t_1 + \delta_2 t_2^{(2)}$.

The integrals in the I_1 series have integrands which reduce to

$$\Gamma(t_1) \Gamma\left(\frac{K - m_1 t_1}{m_2}\right) \Gamma\left(\frac{m_2 - n_2 K + \Delta t_1}{m_2 \nu}\right) \lambda^{-\vec{\delta} \cdot \vec{t}},$$

for which, with $t_1 = \rho e^{i\theta}$, $|\theta| < \frac{1}{2}\pi$, the logarithm of the modulus behaves for large ρ as

$$\rho \cos \theta \log \rho \cdot (\Delta - \nu M)/m_2 \nu + O(\rho). \tag{7.4.10}$$

From (7.4.4), we see that the factor following $\log \rho$ is negative, whence the logarithm of the modulus of the integrand tends to $-\infty$ as $\rho \to \infty$. We must therefore displace the t_1 contour in each integral in I_1 to the left. We find that there are three sequences of candidate poles arising: one each from the gamma functions appearing in the integrands in I_1.

The factor $\Gamma(t_1)$ has its poles at $t_1^{(1)} = -l, l = 0, 1, 2, \ldots$, whereas the second factor, $\Gamma(t_2^{(1)}) = \Gamma((K - m_1 t_1)/m_2)$ would have its poles at points $(K + m_2 l)/m_1$, l a nonnegative integer. However, it is clear that these points are positive, for all nonnegative integer values of k and l, so no poles of this factor are encountered in displacing the t_1 contour to the left. The final sequence of candidate poles arises from poles of $\Gamma((1 - n_1 t_1 - n_2 t_2^{(1)})/\nu) = \Gamma((m_2 - n_2 K + \Delta t_1)/m_2 \nu)$, given by

$$
t_1^{(2)} = \frac{n_2}{\Delta} K - \frac{m_2}{\Delta}(1 + \nu l) = \frac{n_2 K - m_2 L'}{\Delta}, \qquad (7.4.11)
$$

for nonnegative integer k and l satisfying $t_1^{(2)} \leq 0$. Here, we have set $1 + \nu l \equiv L'$; recall (7.4.1).

The $t_1^{(1)}$ sequence of poles in I_1 gives rise to the formal asymptotic series

$$
I_{11} = \frac{\lambda^{-1/\mu - 1/\nu}}{m_2 \nu} \sum_{k,l} \frac{(-)^{k+l}}{k! \, l!} \Gamma(t_2^{(1)}(t_1^{(1)})) \Gamma\left(\frac{1 - n_1 t_1^{(1)} - n_2 t_2^{(1)}(t_1^{(1)})}{\nu}\right) \lambda^{-\vec{\delta} \cdot \vec{t}}
$$

$$
= \frac{\lambda^{-1/\mu - 1/\nu}}{m_2 \nu} \sum_{k,l} f_{11}(k, l), \qquad (7.4.12)
$$

where

$$
f_{11}(k, l) \equiv \frac{(-)^{k+l}}{k! \, l!} \Gamma\left(\frac{K + m_1 l}{m_2}\right) \Gamma\left(\frac{m_2 - n_2 K - \Delta l}{m_2 \nu}\right) \lambda^{-\vec{\delta} \cdot \vec{t}}, \qquad (7.4.13)
$$

with

$$
-\vec{\delta} \cdot \vec{t} = -\delta_1 t_1^{(1)} - \delta_2 t_2^{(1)}(t_1^{(1)}) = -\frac{\delta_2}{m_2} K + \frac{\Delta - \nu M}{m_2 \nu} l. \qquad (7.4.14)
$$

The $t_1^{(2)}$ sequence of poles in I_1 generates the formal asymptotic series

$$
I_{12} = \frac{\lambda^{-1/\mu - 1/\nu}}{\Delta} \sum_{k,l'} \frac{(-)^{k+l}}{k! \, l!} \Gamma(t_1^{(2)}) \Gamma(t_2^{(1)}(t_1^{(2)})) \lambda^{-\vec{\delta} \cdot \vec{t}}
$$

$$
= \frac{\lambda^{-1/\mu - 1/\nu}}{\Delta} \sum_{k,l'} f_{12}(k, l), \qquad (7.4.15)
$$

where

$$
f_{12}(k, l) \equiv \frac{(-)^{k+l}}{k! \, l!} \Gamma\left(\frac{n_2 K - m_2 L'}{\Delta}\right) \Gamma\left(\frac{m_1 L' - n_1 K}{\Delta}\right) \lambda^{-\vec{\delta} \cdot \vec{t}}, \qquad (7.4.16)
$$

with

$$-\vec{\delta} \cdot \vec{t} = -\delta_1 t_1^{(2)} - \delta_2 t_2^{(1)}(t_1^{(2)}) = \frac{\Delta - \mu N}{\mu \Delta} K + \frac{\Delta - \nu M}{\nu \Delta} L'. \qquad (7.4.17)$$

The prime attached to the summation index l appearing in (7.4.15) indicates that the index is subject to the restriction imposed earlier on (7.4.11), namely that $t_1^{(2)} \le 0$.

We repeat the analysis for the integrals appearing in the series I_2. The integrals in (7.4.9) have integrands which reduce to

$$\Gamma(t_1) \Gamma\left(\frac{K' - n_1 t_1}{n_2}\right) \Gamma\left(\frac{n_2 - m_2 K' - \Delta t_1}{\mu n_2}\right) \lambda^{-\vec{\delta} \cdot \vec{t}},$$

which has, with $t_1 = \rho e^{i\theta}$, $|\theta| < \pi/2$, the logarithm of the modulus behaving for large ρ as

$$\rho \cos \theta \log \rho \cdot (\mu N - \Delta)/\mu n_2 + O(\rho).$$

From (7.4.4), we see that the factor following $\log \rho$ is positive whence the logarithm of the modulus of the integrand tends to $+\infty$ as $\rho \to \infty$. Accordingly, we must displace the contour to the right to pick up poles contributing to the asymptotics. We find that there are two sequences of relevant poles: the first, from poles of $\Gamma((K' - n_1 t_1)/n_2)$, is given by

$$t_1^{(1)} = (K' + n_2 l)/n_1; \qquad (7.4.18)$$

and the second, from poles of $\Gamma((n_2 - m_2 K' - \Delta t_1)/\mu n_2)$, is given by

$$t_1^{(2)} = \frac{n_2}{\Delta}(1 + \mu l) - \frac{m_2 K'}{\Delta} = \frac{n_2 L - m_2 K'}{\Delta}, \qquad (7.4.19)$$

where we have set $1 + \mu l \equiv L$; see (7.4.1). In (7.4.18) and (7.4.19), the parameter l is a nonnegative integer, but in the case of $t_1^{(2)}$, l is subject to the additional constraint $t_1^{(2)} > 0$.

The $t_1^{(1)}$ sequence of poles in (7.4.18) gives rise to the formal asymptotic series

$$I_{21} = \frac{\lambda^{-1/\mu - 1/\nu}}{\mu n_1} \sum_{k,l} \frac{(-)^{k+l}}{k! \, l!} \Gamma(t_2^{(2)}(t_1^{(1)})) \Gamma\left(\frac{1 - m_1 t_1^{(1)} - m_2 t_2^{(2)}(t_1^{(1)})}{\mu}\right) \lambda^{-\vec{\delta} \cdot \vec{t}}$$

$$= \frac{\lambda^{-1/\mu - 1/\nu}}{\mu n_1} \sum_{k,l} f_{21}(k, l), \qquad (7.4.20)$$

where

$$f_{21}(k, l) \equiv \frac{(-)^{k+l}}{k! \, l!} \Gamma\left(\frac{K' + n_2 l}{n_1}\right) \Gamma\left(\frac{n_1 - m_1 K' - \Delta l}{\mu n_1}\right) \lambda^{-\vec{\delta} \cdot \vec{t}}, \qquad (7.4.21)$$

with

$$-\vec{\delta} \cdot \vec{t} = -\delta_1 t_1^{(1)} - \delta_2 t_2^{(2)}(t_1^{(1)}) = -\frac{\delta_1}{n_1} K' + \frac{\Delta - \mu N}{\mu n_1} l. \qquad (7.4.22)$$

The $t_1^{(2)}$ sequence gives rise to

$$I_{22} = \frac{\lambda^{-1/\mu-1/\nu}}{\Delta} \sum_{k,l'} \frac{(-)^{k+l}}{k!\,l!} \Gamma(t_1^{(2)}) \Gamma(t_2^{(2)}(t_1^{(2)})) \lambda^{-\vec{\delta}\cdot\vec{t}}$$

$$= \frac{\lambda^{-1/\mu-1/\nu}}{\Delta} \sum_{k,l'} f_{12}(l,k), \qquad (7.4.23)$$

where the prime attached to l indicates that it is subject to the restriction $t_1^{(2)} > 0$. Collecting together the series I_{11}, I_{12}, I_{21} and I_{22}, we find that

$$I(\lambda) \sim \lambda^{-1/\mu-1/\nu}(I_{11} + I_{12} + I_{21} + I_{22})$$

for large λ. However, we note that, if we relabel the summation indices in I_{22} by putting $k \to l$ and $l \to k$, then the linear restriction $t_1^{(2)} = n_2 L/\Delta - m_2 K'/\Delta > 0$ becomes the constraint $n_2 K/\Delta - m_2 L'/\Delta > 0$, exactly the complement of the constraint $t_1^{(2)} = n_2 K/\Delta - m_2/L'/\Delta \leq 0$ which governs I_{12}. Thus, the two series I_{12} and I_{22} may be fused into a single series without constraint (save that of k and l being nonnegative integers), to give the final asymptotic form

$$I(\lambda) \sim \lambda^{-1/\mu-1/\nu} \left\{ \frac{1}{m_2\nu} \sum_{k,l} f_{11}(k,l) + \frac{1}{\Delta} \sum_{k,l} f_{12}(k,l) + \frac{1}{\mu n_1} \sum_{k,l} f_{21}(k,l) \right\}$$

$$(7.4.24)$$

for $\lambda \to \infty$. Observe that the first series associates in a natural way with the face of the Newton diagram joining the vertices $(0, \nu)$ and P_2, the second with the face joining P_1 and P_2, and the third with that joining $(\mu, 0)$ with P_1.

An outline of the computations leading to (7.4.24) is presented as Fig. 7.4.

7.4.2 Three and More Internal Points

The approach of the previous section can be applied to deal with circumstances where more than two internal points are present to define the Newton diagram. If $k = 3$ in (7.3.3), then (7.3.6) becomes a more imposing treble integral. However, the number of gamma functions present in the integrand remains unchanged, and the arguments of those gamma functions remain affine functions of the integration variables.

Rather than detail the residue computations, as we have done in the previous section and in §7.2, we shall merely provide an overview of the main steps encountered in such an analysis. What differs in the case of three or more internal points to that of the previous section is the high degree of cancellation or augmentation of various formal asymptotic series, a feature that allows us to reduce the complexity of the final form of the asymptotic expansion of $I(\lambda)$ for $k \geq 3$. We will discuss this in some detail, but for a fuller account of other computations, we direct the interested reader to Kaminski & Paris (1998a).

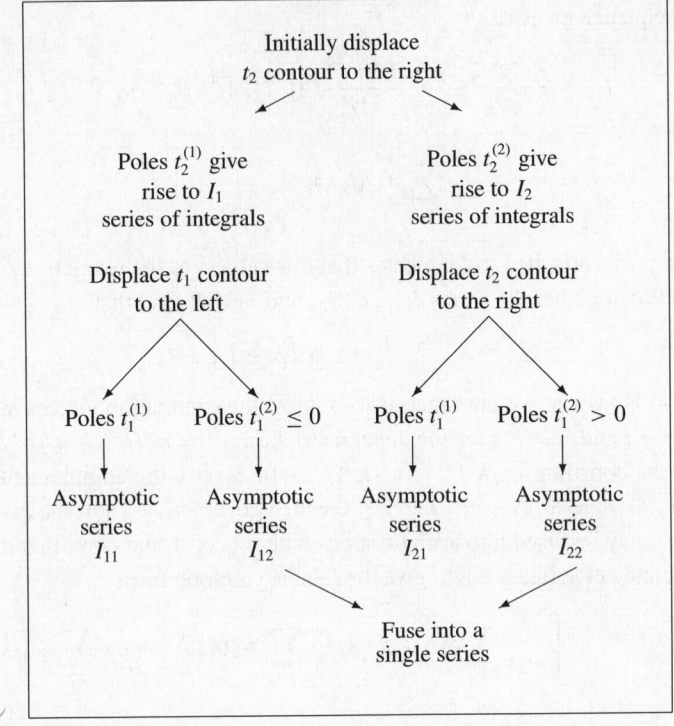

Fig. 7.4. Outline of the computations for the convex case with two internal points

Now with $k = 3$ in (7.3.3), suppose that the internal points $P_i = (m_i, n_i)$, $i = 1, 2, 3$, are labelled so that $\mu > m_1 > m_2 > m_3$ and $n_1 < n_2 < n_3 < \nu$. The points P_1, P_2 and P_3 then appear as shown in Fig. 7.5. We shall find it convenient to define analogues of M, N and Δ from the section dealing with two internal points, and supply a small set of identities relating these quantities. For the quantities

$$M_{ij} = m_i - m_j, \; N_{ij} = n_j - n_i, \; \Delta_{ij} = m_i n_j - n_i m_j, \qquad (7.4.25)$$

where $i, j = 1, 2, 3$, $i < j$, it is easily seen that

$$\begin{aligned} m_i \delta_j - m_j \delta_i &= M_{ij} - \Delta_{ij}/\nu, \quad n_j \delta_i - n_i \delta_j = N_{ij} - \Delta_{ij}/\mu, \\ m_3 \Delta_{12} + m_1 \Delta_{23} &= m_2 \Delta_{13}, \qquad n_3 \Delta_{12} + n_1 \Delta_{23} = n_2 \Delta_{13} \end{aligned} \qquad (7.4.26)$$

and

$$\delta_1 \Delta_{23} - \delta_2 \Delta_{12} + \delta_3 \Delta_{12} = \Delta_{12} - \Delta_{13} + \Delta_{23}.$$

With the ordering imposed on the points P_1, P_2 and P_3, we have all the M_{ij}, N_{ij} positive, and the quantities Δ_{ij} can be seen to be the areas of parallelograms generated by pairs of position vectors $\vec{P}_i$, $\vec{P}_j$ which form positively ordered bases for the mn plane. Accordingly, each Δ_{ij} is positive.

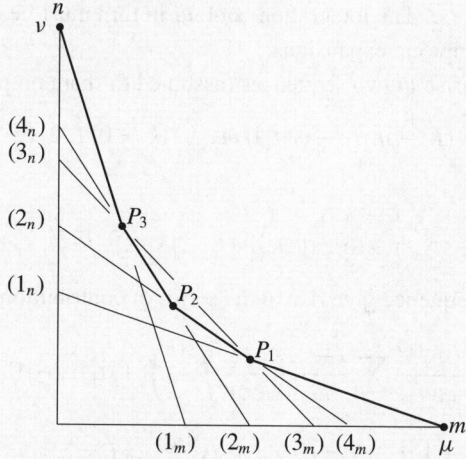

Fig. 7.5. The Newton diagram for three internal points in the convex case. The values of the labels (1_m), (2_m), ... are discussed in the text.

The lines constituting the faces in the Newton diagram meet the coordinate axes in a regular fashion, giving rise to a number of useful inequalities. The intersection points on the m axis, indicated in Fig. 7.5 by labels (1_m), (2_m), ... yield the chain of inequalities

$$\frac{m_3 \nu}{\nu - n_3} < \frac{\Delta_{23}}{N_{23}} < \frac{\Delta_{13}}{N_{13}} < \frac{\Delta_{12}}{N_{12}} < \mu,$$

while those intersection points on the n axis, indicated in the figure by labels (1_n), (2_n), ... give

$$\frac{\mu n_1}{\mu - m_1} < \frac{\Delta_{12}}{M_{12}} < \frac{\Delta_{13}}{M_{13}} < \frac{\Delta_{23}}{M_{23}} < \nu. \tag{7.4.27}$$

Under the simplifying assumption $c_1 = c_2 = c_3 = 1$, (7.3.3) for $k = 3$ becomes

$$I(\lambda) = \frac{\lambda^{-1/\mu - 1/\nu}}{\mu \nu} \left(\frac{1}{2\pi i} \int_{-\infty i}^{\infty i} \right)^3 \Gamma(t_1) \Gamma(t_2) \Gamma(t_3) \tag{7.4.28}$$

$$\times \Gamma \left(\frac{1 - \vec{m} \cdot \vec{t}}{\mu} \right) \Gamma \left(\frac{1 - \vec{n} \cdot \vec{t}}{\nu} \right) \lambda^{-\vec{\delta} \cdot \vec{t}} dt_1 \, dt_2 \, dt_3,$$

where the vector representations in (7.3.5) are being used. If we begin our investigation by displacing the t_3 contour first, setting $t_3 = \rho e^{i\theta}$ with $|\theta| < \frac{1}{2}\pi$, then we find that the logarithm of the modulus of the integrand is dominated by

$$\delta_3 \rho \cos \theta \log \rho + O(\rho)$$

for large ρ. Since all the points P_i in the Newton diagram lie in front of the back face, the quantity δ_3 is positive, from which it follows the above estimate must tend to $+\infty$ as $\rho \to \infty$. The integration contour in turn must be displaced to the right to develop asymptotic expansions.

Candidate poles arise in two sequences (assumed to share no points):

$$t_3^{(1)} = (K - m_1t_1 - m_2t_2)/m_3 \qquad (k = 0, 1, 2, \dots)$$

and

$$t_3^{(2)} = (K' - n_1t_1 - n_2t_2)/n_3 \qquad (k = 0, 1, 2, \dots).$$

Poles from the $t_3^{(1)}$ sequence give rise to the series of contributions

$$I_1 = \frac{\lambda^{-1/\mu - 1/\nu}}{m_3\nu} \sum_k \frac{(-)^k}{k!} \left(\frac{1}{2\pi i} \int_{-\infty i}^{\infty i}\right)^2 \Gamma(t_1)\Gamma(t_2)\Gamma(t_3^{(1)})$$

$$\times \Gamma\left(\frac{1 - n_1t_1 - n_2t_2 - n_3t_3^{(1)}}{\nu}\right) \lambda^{-\vec{\delta}\cdot\vec{t}} dt_1\, dt_2 \qquad (7.4.29)$$

with $\vec{\delta} \cdot \vec{t} = \delta_1t_1 + \delta_2t_2 + \delta_3t_3^{(1)}$. Similarly, the $t_3^{(2)}$ sequence of poles yields the series

$$I_2 = \frac{\lambda^{-1/\mu - 1/\nu}}{\mu n_3} \sum_k \frac{(-)^k}{k!} \left(\frac{1}{2\pi i} \int_{-\infty i}^{\infty i}\right)^2 \Gamma(t_1)\Gamma(t_2)\Gamma(t_3^{(2)})$$

$$\times \Gamma\left(\frac{1 - m_1t_1 - m_2t_2 - m_3t_3^{(2)}}{\mu}\right) \lambda^{-\vec{\delta}\cdot\vec{t}} dt_1\, dt_2 \qquad (7.4.30)$$

with $\vec{\delta} \cdot \vec{t} = \delta_1t_1 + \delta_2t_2 + \delta_3t_3^{(2)}$.

The integrals in the I_1 series (7.4.29) have integrands reducing to

$$\Gamma(t_1)\Gamma(t_2)\Gamma\left(\frac{K - m_1t_1 - m_2t_2}{m_3}\right)\Gamma\left(\frac{m_3 - n_3K + \Delta_{13}t_1 + \Delta_{23}t_2}{m_3\nu}\right)\lambda^{-\vec{\delta}\cdot\vec{t}}$$

from which it follows, upon setting $t_2 = \rho e^{i\theta}$ with $|\theta| < \frac{1}{2}\pi$, the logarithm of the modulus of the integrand behaves as

$$\rho \cos\theta \log\rho \cdot (\Delta_{23} - M_{23}\nu)/m_3\nu + O(\rho)$$

for large ρ. From (7.4.27), we have $\Delta_{23} < M_{23}\nu$ whence it follows that the above estimate tends to $-\infty$ as $\rho \to \infty$. The next step in determining the asymptotics arising from (7.4.29) involves displacing the t_2 contour to the left. Because of the presence of an additional gamma function in the integrand, there are now three sequences of poles that must be considered: poles of $\Gamma(t_2)$, poles of $\Gamma((K - m_1t_1 - m_2t_2)/m_3)$ and poles of $\Gamma((m_3 - n_3K + \Delta_{12}t_1 + \Delta_{23}t_2)/m_3\nu)$. However, poles of the second gamma function listed, of the form $(K + m_3l - m_1t_1)/m_2$, l a nonnegative integer, fail to satisfy the requirement that the real part be negative (recall that along the integration contour in the t_1 plane, $\text{Re}\, t_1 = 0$, except for an indentation to the right near the origin where $\text{Re}(t_1)$ is positive, but arbitrarily small). As a result there are only two sequences of poles that can contribute to the asymptotic series arising from (7.4.29):

$$t_2^{(1)} = -l,$$

$$t_2^{(2)} = -\frac{m_3}{\Delta_{23}}L' + \frac{n_3}{\Delta_{23}}K - \frac{\Delta_{13}}{\Delta_{23}}t_1 \leq 0, \tag{7.4.31}$$

where l is a nonnegative integer, and the poles listed in (7.4.31) are subject to the indicated restrictions. These sequences of poles give rise to formal series I_{11} and I_{12}, in a fashion similar to what we saw in the case of two internal points, each of which contains Mellin-Barnes integrals of the form

$$\frac{1}{2\pi i}\int_{-\infty i}^{\infty i}\Gamma(t_1)\Gamma(t_3^{(1)}(-l))\Gamma\left(\frac{1 - n_1 t_1 + n_2 l - n_3 t_3^{(1)}(-l)}{v}\right)\lambda^{-\vec{\delta}\cdot\vec{t}}\,dt_1$$

and

$$\frac{1}{2\pi i}\int_{-\infty i}^{\infty i}\Gamma(t_1)\Gamma(t_2^{(3)})\Gamma(t_3^{(1)}(t_2^{(3)}))\lambda^{-\vec{\delta}\cdot\vec{t}}\,dt_1,$$

respectively, which in turn have integrands that reduce to

$$\Gamma(t_1)\Gamma\left(\frac{K - m_1 t_1 + m_2 l}{m_3}\right)\Gamma\left(\frac{m_3 - n_3 K - \Delta_{23}l + \Delta_{13}t_1}{m_3 v}\right)\lambda^{-\vec{\delta}\cdot\vec{t}}$$

and

$$\Gamma(t_1)\Gamma\left(\frac{n_3 K - m_3 L' - \Delta_{13}t_1}{\Delta_{23}}\right)\Gamma\left(\frac{-n_2 K + m_2 L' + \Delta_{12}t_1}{\Delta_{23}}\right)\lambda^{-\vec{\delta}\cdot\vec{t}},$$

respectively. Use of the inequalities (7.4.27) allows us to determine straightaway that the t_1 contour in the integrals appearing in the series I_{11} must be displaced to the left. For the integrals in the series I_{12}, however, we find we must determine the sign of the quantity $\Delta_{23} - \Delta_{13} + \Delta_{12}$ which arises in the course of determining the asymptotic behaviour of these integrals for large $|t_1|$ taken along the semicircular arc $t_1 = \rho e^{i\theta}$, $|\theta| < \frac{1}{2}\pi$.

The quantity $\Delta_{23} - \Delta_{13} + \Delta_{12}$ is, in fact, a measure of *relative convexity*. To see this, consider the line through P_1 and P_3,

$$\frac{m}{M_{13}} + \frac{n}{N_{13}} - \frac{m_1}{M_{13}} - \frac{n_1}{N_{13}} = 0. \tag{7.4.32}$$

If P_2 lies below this line (i.e., on the same side of the line as the origin), then the left-hand side of (7.4.32) must be negative, whereas if P_2 lies above the line, the left-hand side must be positive. (If P_2 lies on (7.4.32), the Newton diagram has a pair of collinear faces, a case we have excluded from the present discussion.) Consequently, the case of three internal points in the convex case requires that

$$\frac{m_2}{M_{13}} + \frac{n_2}{N_{13}} - \frac{m_1}{M_{13}} - \frac{n_1}{N_{13}} = \frac{-\Delta_{13} + \Delta_{23} + \Delta_{12}}{M_{13}N_{13}} < 0,$$

from which we conclude that the t_1 contour appearing in the integrals for the I_{12} series must be translated to the left, as is the case for the integrals in the I_{11} series.

For the I_{11} series, we find there are two sequences of poles to consider,

$$
t_1^{(1)} = -r,
$$

$$
t_1^{(2)} = \frac{n_3}{\Delta_{13}} K + \frac{\Delta_{23}}{\Delta_{13}} l - \frac{m_3}{\Delta_{13}} (1 + vr)
$$

$$
= \frac{n_3 K + \Delta_{23} l - m_3 R'}{\Delta_{13}} \leq 0,
$$

giving rise to the expansions

$$
I_{111} = \frac{\lambda^{-1/\mu - 1/\nu}}{m_3 \nu} \sum_{k,l,r} f_{111}(k, l, r), \tag{7.4.33}
$$

$$
I_{112} = \frac{\lambda^{-1/\mu - 1/\nu}}{\Delta_{13}} \sum_{k,l,r'} f_{112}(k, l, r), \tag{7.4.34}
$$

where

$$
f_{111}(k, l, r) = \frac{(-)^{k+l+r}}{k! \, l! \, r!} \Gamma\left(\frac{K + m_1 r + m_2 l}{m_3}\right)
$$

$$
\times \Gamma\left(\frac{m_3 - n_3 K - \Delta_{13} r - \Delta_{23} l}{m_3 \nu}\right) \lambda^{-\vec{\delta} \cdot \vec{t}}, \tag{7.4.35}
$$

$$
-\vec{\delta} \cdot \vec{t} = -\frac{\delta_3}{m_3} K + \frac{\Delta_{23} - M_{23} \nu}{m_3 \nu} l + \frac{\Delta_{13} - M_{13} \nu}{m_3 \nu} r,
$$

and

$$
f_{112}(k, l, r) = \frac{(-)^{k+l+r}}{k! \, l! \, r!} \Gamma\left(\frac{n_3 K + \Delta_{23} l - m_3 R'}{\Delta_{13}}\right)
$$

$$
\times \Gamma\left(\frac{-n_1 K + \Delta_{12} l + m_1 R'}{\Delta_{13}}\right) \lambda^{-\vec{\delta} \cdot \vec{t}}, \tag{7.4.36}
$$

$$
-\vec{\delta} \cdot \vec{t} = \frac{\Delta_{13} - \mu N_{13}}{\mu \Delta_{13}} K + \frac{\Delta_{13} - \Delta_{12} - \Delta_{23}}{\Delta_{13}} l + \frac{\Delta_{13} - M_{13} \nu}{\nu \Delta_{13}} R'.
$$

In these expressions, we have nonnegative integer indices k, l and r, and restrictions apply in those sums with primes on the summation indices. We have also set $R' \equiv 1 + vr$; see (7.4.1).

For the integrals that arise in the I_{12} series, we find there are three sequences of poles to consider,

$$t_1^{(1)} = -r,$$

$$t_1^{(2)} = \frac{n_3}{\Delta_{13}}K - \frac{m_3}{\Delta_{13}}L' + \frac{\Delta_{23}}{\Delta_{13}}r \leq 0, \tag{7.4.37}$$

$$t_1^{(3)} = \frac{n_2}{\Delta_{12}}K - \frac{m_2}{\Delta_{12}}L' - \frac{\Delta_{23}}{\Delta_{12}}r \leq 0, \tag{7.4.38}$$

giving rise to the expansions

$$I_{121} = \frac{\lambda^{-1/\mu-1/\nu}}{\Delta_{23}} \sum_{k,l',r} f_{121}(k,l,r),$$

$$I_{122} = -\frac{\lambda^{-1/\mu-1/\nu}}{\Delta_{13}} \sum_{k,l',r'} f_{112}(k,r,l), \tag{7.4.39}$$

$$I_{123} = \frac{\lambda^{-1/\mu-1/\nu}}{\Delta_{12}} \sum_{k,l',r'} f_{123}(k,l,r),$$

with

$$f_{121}(k,l,r) = \frac{(-)^{k+l+r}}{k!\,l!\,r!}\Gamma\left(\frac{n_3K - m_3L' + \Delta_{13}r}{\Delta_{23}}\right)$$

$$\times \Gamma\left(\frac{-n_2K + m_2L' - \Delta_{12}r}{\Delta_{23}}\right)\lambda^{-\vec{\delta}\cdot\vec{t}}, \tag{7.4.40}$$

$$-\vec{\delta}\cdot\vec{t} = \frac{\Delta_{23} - \mu N_{23}}{\mu\Delta_{23}}K + \frac{\Delta_{23} - \nu M_{23}}{\nu\Delta_{23}}L' + \frac{\Delta_{12} - \Delta_{13} + \Delta_{23}}{\Delta_{23}}r,$$

and

$$f_{123}(k,l,r) = \frac{(-)^{k+l+r}}{k!\,l!\,r!}\Gamma\left(\frac{n_2K - m_2L' - \Delta_{23}r}{\Delta_{12}}\right)$$

$$\times \Gamma\left(\frac{-n_1K + m_1L' + \Delta_{13}r}{\Delta_{12}}\right)\lambda^{-\vec{\delta}\cdot\vec{t}}, \tag{7.4.41}$$

$$-\vec{\delta}\cdot\vec{t} = \frac{\Delta_{12} - \mu N_{12}}{\mu\Delta_{12}}K + \frac{\Delta_{12} - M_{12}\nu}{\nu\Delta_{12}}L' + \frac{\Delta_{12} - \Delta_{13} + \Delta_{23}}{\Delta_{12}}r.$$

The chain of residue evaluations leading to this point are presented in Fig. 7.6, and show the order in which we have encountered the various sequences of poles and the resulting formal asymptotic series thereby generated.

If we repeat our analysis for (7.4.30), then we find that displacement of the t_2 integrals in I_2 must occur to the right, and the poles encountered come in two

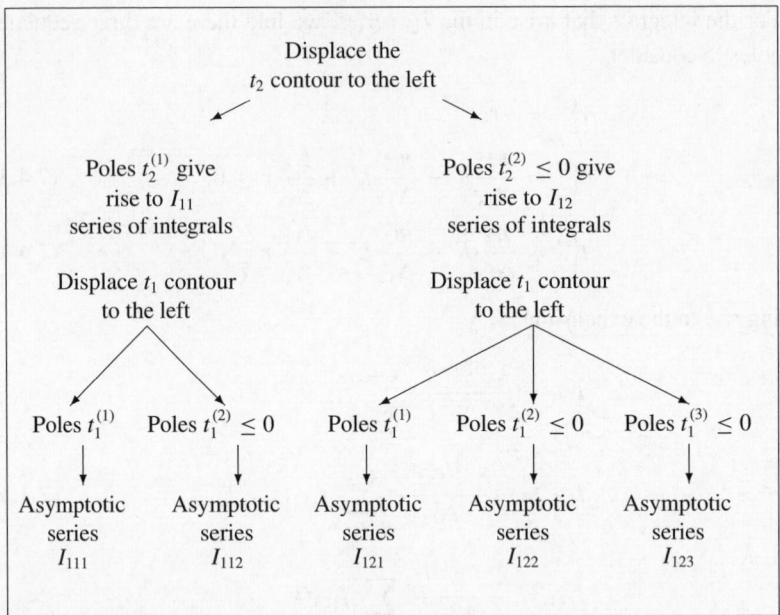

Fig. 7.6. Order of the computations for the determination of the asymptotic series resulting from the sequence of poles $t_3^{(1)}$ (the I_1 formal asymptotic series)

series,

$$t_2^{(1)} = (K' + n_3 l - n_1 t_1)/n_2 > 0, \tag{7.4.42}$$

$$t_2^{(2)} = -\frac{m_3}{\Delta_{23}} K' + \frac{n_3}{\Delta_{23}} L - \frac{\Delta_{13}}{\Delta_{23}} t_1 > 0. \tag{7.4.43}$$

The $t_2^{(1)}$ sequence (7.4.42) gives rise to a series of contributions, I_{21} by analogy with the convex case with two internal points, the integrals in which must be displaced to the right. In doing so, two further sequences of poles,

$$t_1^{(1)} = (K' + n_2 l + n_3 r)/n_1,$$

$$t_1^{(2)} = \frac{-m_2}{\Delta_{12}} K' - \frac{\Delta_{23}}{\Delta_{12}} l + \frac{n_2}{\Delta_{12}} (1 + \mu r)$$

$$= \frac{-m_2 K' - \Delta_{23} l + n_2 R}{\Delta_{12}} > 0, \tag{7.4.44}$$

give rise to the asymptotic series

$$I_{211} = \frac{1}{\mu n_1} \sum_{k,l,r} f_{211}(k, l, r),$$

$$I_{212} = \frac{1}{\Delta_{12}} \sum_{k,l,r'} f_{123}(r, k, l),$$

where

$$f_{211}(k,l,r) = \frac{(-)^{k+l+r}}{k!\,l!\,r!}\Gamma\left(\frac{K'+n_3l+n_2r}{n_1}\right)$$

$$\times \Gamma\left(\frac{n_1 - m_1K' - \Delta_{13}l - \Delta_{12}r}{\mu n_1}\right)\lambda^{-\vec{\delta}\cdot\vec{t}}, \qquad (7.4.45)$$

$$-\vec{\delta}\cdot\vec{t} = -\frac{\delta_1}{n_1}K' + \frac{\Delta_{13}-\mu N_{13}}{\mu n_1}l + \frac{\Delta_{12}-\mu N_{12}}{\mu n_1}r,$$

and where f_{123} is given in (7.4.41). Here, we have set $R = 1 + \mu r$; recall (7.4.1).

The $t_2^{(2)}$ sequence (7.4.43) similarly gives rise to a series of contributions, say I_{22}, the integrals in which must be displaced to the left. Poles that are encountered in the displacement give rise to three sequences:

$$t_1^{(1)} = -r,$$

$$t_1^{(2)} = \frac{-m_3}{\Delta_{13}}K' + \frac{n_3}{\Delta_{13}}L + \frac{\Delta_{23}}{\Delta_{13}}r \leq 0, \qquad (7.4.46)$$

$$t_1^{(3)} = \frac{-m_2}{\Delta_{12}}K' + \frac{n_2}{\Delta_{12}}L - \frac{\Delta_{23}}{\Delta_{12}}r \leq 0. \qquad (7.4.47)$$

These, in turn, generate the asymptotic series

$$I_{221} = \frac{1}{\Delta_{23}}\sum_{k,l',r'} f_{121}(l,k,r),$$

$$I_{222} = -\frac{1}{\Delta_{13}}\sum_{k,l',r'} f_{112}(l,r,k), \qquad (7.4.48)$$

$$I_{223} = \frac{1}{\Delta_{12}}\sum_{k,l',r'} f_{123}(l,r,k),$$

where f_{121}, f_{112} and f_{123} are given, respectively, in (7.4.40), (7.4.36) and (7.4.41). This collection of residue evaluations is summarised in the accompanying Fig. 7.7, and displays the families of poles in the order in which they were encountered in our computations.

The triply-subscripted asymptotic series that we have obtained can be collected together in a more attractive form. In Table 7.1, we gather together the restrictions attached to each of the formal asymptotic sums $I_{111}, I_{112}, \ldots, I_{223}$; for ease of discussion, we have also displayed the denominator that appears in the factor before the Σ in each of the definitions of these sums. The origins of the three-term inequalities present in the table arise from the restrictions in place on the final sequences of poles resulting from displacement, in order, of t_3, t_2 and t_1 contours and recorded in (7.4.37), (7.4.38), (7.4.44), (7.4.46) and (7.4.47). The two-term inequalities arise at intermediate stages (after the second displacement of t_2) from (7.4.31) and (7.4.43); the t_1 dependence there does not appear in the table since the

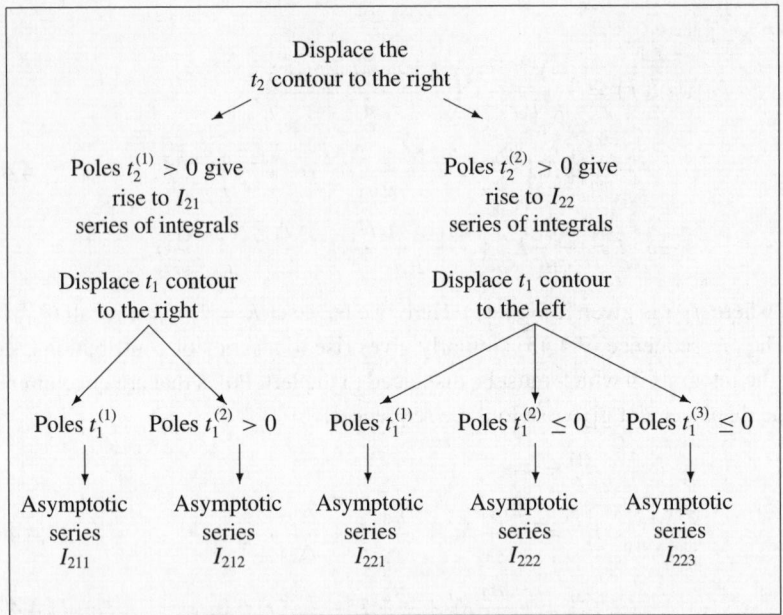

Fig. 7.7. Order of the computations for the determination of the asymptotic series resulting from the sequence of poles $t_3^{(2)}$ (the I_2 formal asymptotic series)

Table 7.1. *Linear restrictions imposed on the asymptotic series. The quantities K, K', L, L', R and R' are defined in (7.4.1)*

Asymptotic series	Denominator	Inequalities
I_{111}	$m_3\nu$	none
I_{112}	Δ_{13}	$(n_3 K + \Delta_{23}l - m_3 R')/\Delta_{13} \leq 0$
I_{121}	Δ_{23}	$(n_3 K - m_3 L')/\Delta_{23} \leq 0$
I_{122}	$-\Delta_{13}$	$(n_3 K - m_3 L')/\Delta_{23} \leq 0$
		$(n_3 K - m_3 L' + \Delta_{23}r)/\Delta_{13} \leq 0$
I_{123}	Δ_{12}	$(n_3 K - m_3 L')/\Delta_{23} \leq 0$
		$(n_2 K - m_2 L' - \Delta_{23}r)/\Delta_{12} \leq 0$
I_{211}	μn_1	none
I_{212}	Δ_{12}	$(-m_2 K' - \Delta_{23}l + n_2 R)/\Delta_{12} > 0$
I_{221}	Δ_{23}	$(-m_3 K' + n_3 L)/\Delta_{23} > 0$
I_{222}	$-\Delta_{13}$	$(-m_3 K' + n_3 L)/\Delta_{23} > 0$
		$(-m_3 K' + n_3 L + \Delta_{23}r)/\Delta_{13} \leq 0$
I_{223}	Δ_{12}	$(-m_3 K' + n_3 L)/\Delta_{23} > 0$
		$(-m_2 K' + n_2 L - \Delta_{23}r)/\Delta_{12} \leq 0$

inequality must still hold prior to the (final) displacement of the t_1 contour, when $\mathrm{Re}(t_1) = 0$ everywhere except near the origin in the t_1 plane where it is positive but arbitrarily small. We also mention that series in the table sharing a common denominator for the leading factor (modulo a minus sign) also share a common summand, albeit with different inequalities applying to the summand's arguments.

It is this last feature that permits substantial simplification, which we illustrate in detail only for one case.

Let us examine the series with denominator $\pm\Delta_{13}$. This corresponds to the entries for I_{112}, I_{122} and I_{222}. To effect a comparison, let us make the changes of summation indices $r \to l$, $l \to r$ in I_{122} and $r \to l$, $l \to k$, $k \to r$ in I_{222}, so that all summands in the series appear as $f_{112}(k, l, r)$; recall (7.4.34), (7.4.39), (7.4.48) and (7.4.36). These changes of variables also effect changes to the inequalities governing I_{122} and I_{222}, rendering those for I_{122} as

$$\frac{n_3}{\Delta_{23}}K - \frac{m_3}{\Delta_{23}}R' \leq 0,$$

$$\frac{n_3}{\Delta_{13}}K + \frac{\Delta_{23}}{\Delta_{13}}l - \frac{m_3}{\Delta_{13}}R' \leq 0,$$

and those for I_{222} as

$$\frac{n_3}{\Delta_{23}}K - \frac{m_3}{\Delta_{23}}R' > 0,$$

$$\frac{n_3}{\Delta_{13}}K + \frac{\Delta_{23}}{\Delta_{13}}l - \frac{m_3}{\Delta_{13}}R' \leq 0.$$

The second of each of these pairs of inequalities are identical, and the first of each of these pairs are complementary. As a result, we can combine the series I_{122} and I_{222} into a single sum, which we continue to denote by I_{122}, subject to the single linear inequality

$$\frac{n_3}{\Delta_{13}}K + \frac{\Delta_{23}}{\Delta_{13}}l - \frac{m_3}{\Delta_{13}}R' \leq 0.$$

This inequality, however, is precisely that governing I_{112}, and since I_{112} and our 'fused' I_{122} are negatives, we see that all the series in Table 7.1 associated with the factor Δ_{13} annihilate.

In similar fashion, we can deduce that the I_{123}, I_{212} and I_{223} series sum into a single series, which we continue to denote by I_{123}, subject to no linear constraints, and that the I_{121} and I_{221} series unite in another series, still denoted by I_{121}, also free of linear constraint (except, of course, that all indices be nonnegative integers).

Thus, we arrive at the large-λ asymptotic expansion of (7.4.28)

$$I(\lambda) \sim \lambda^{-1/\mu-1/\nu}(I_{111} + I_{121} + I_{123} + I_{211}),$$

where the sums have no linear constraint, or more explicitly,

$$I(\lambda) \sim \lambda^{-1/\mu-1/\nu}\left\{\frac{1}{m_3\nu}\sum_{k,l,r}f_{111}(k,l,r) + \frac{1}{\Delta_{23}}\sum_{k,l,r}f_{121}(k,l,r)\right.$$

$$\left. + \frac{1}{\Delta_{12}}\sum_{k,l,r}f_{123}(k,l,r) + \frac{1}{\mu n_1}\sum_{k,l,r}f_{211}(k,l,r)\right\} \quad (7.4.49)$$

as $\lambda \to \infty$. The summands in (7.4.49) are given in (7.4.35), (7.4.40), (7.4.41) and (7.4.45), and the indices range over nonnegative integers k, l and r. The principle of one asymptotic series per face of the Newton diagram continues to hold, with the first series corresponding to the face joining $(0, \nu)$ to P_3, the next corresponding to the face joining P_3 to P_2 and so on.

7.4.3 Other Double Integrals

The detailed analyses of the previous two sections can, of course, be brought to bear on other Laplace-type integrals, and we shall outline here some straightforward extensions.

First, if the internal points in the Newton diagram are all vertices of the diagram, then the setting is precisely that of the previous two sections: the convex case. For such a setting, one can carry forward the analysis in precisely the fashion we have here, with little modification in the organisation of the computation but, of course, with increasing complexity as the number of vertices grows. The annihilation and fusion of constituent asymptotic series that we saw for two and three internal points will continue to operate, allowing one to reduce the number of constituent series in the compound asymptotic series approximating (7.3.3) to a setting with one component series per face of the Newton diagram.

Such a rule may appear to be unjustified following merely two examples, but we point out that it is possible to generate the compound asymptotic expansion of $I(\lambda)$ in a more general framework using some elementary linear algebra, without having to step through the sequence of residue computations as we have done in the previous two sections. The ansatz for determining this compound expansion is available in Kaminski & Paris (1998a, p. 612), and the interested reader is directed there for details.

If the integral (7.3.3) contains terms in the phase function that correspond to internal points that are *not* on the Newton diagram, then the method described above can still be employed, but without any linear-algebraic computational scheme to bypass the explicit residue computations we have undertaken here. We still find that the resulting compound expansion has one constituent series per face of the Newton diagram, but the presence of non-vertex internal points is not felt until the later terms in each constituent series in the asymptotic expansion. That is, the leading terms in the compound expansion are still determined solely by the faces of the Newton diagram, and the contribution of other vertices in the carrier of the phase do not appear until further along in the asymptotic series for each face.

If, instead of beginning with an integral (7.3.3), we have an integral of the form

$$\tilde{I}(\lambda) = \int_0^\infty \int_0^\infty \exp\left\{-\lambda\left(x^\mu + \sum_{p=1}^k c_p x^{m_p} y^{n_p} + y^\nu\right)\right\} x^\alpha y^\beta dx\, dy,$$

where $\alpha > -1$, $\beta > -1$, then the Mellin-Barnes representation used in this section still applies. To see this, we observe that $\tilde{I}(\lambda)$ may be transformed into an integral of the form (7.3.3) upon application of the simple change of variables $x = X^p, y = Y^q$. The differential $x^\alpha y^\beta dx\, dy$ then becomes $pq X^{p-1+\alpha p} Y^{q-1+\beta q} dX\, dY$, whence the choice $p = 1/(1+\alpha)$, $q = 1/(1+\beta)$ removes the powers of X and Y from the differential. The result is an integral of the type (7.3.3), albeit with different powers appearing in the phase. The ensuing compound asymptotic expansion is then derived as we have shown.

For more general analytic 'phases' $f(x, y)$ and 'amplitudes' $g(x, y)$ in (7.3.2), we can proceed in the following fashion. For a phase $f(x, y)$ with no saddle points in the positive xy quadrant, we write $f(x, y) = f_0(x, y) + f_1(x, y)$, where $f_0(x, y)$ is the polynomial formed from the terms of the Maclaurin expansion of $f(x, y)$ that correspond to vertices of the Newton diagram of $f(x, y)$. The remainder of the phase is then incorporated into the amplitude function along with $g(x, y)$, so that our integral has the form

$$I(\lambda) = \int_0^\infty \int_0^\infty e^{-\lambda f_0(x,y)} \left\{ e^{-\lambda f_1(x,y)} g(x, y) \right\} dx dy.$$

We next develop our 'new' amplitude $e^{-\lambda f_1(x,y)} g(x, y)$ into its Maclaurin series, say $\sum_{r,s} (-\lambda)^{v(r,s)} p_{r,s}(\lambda) x^r y^s$ and termwise integrate, to produce a formal series expansion of the type

$$I(\lambda) \sim \sum_{r,s} (-\lambda)^{v(r,s)} p_{r,s}(\lambda) I_{r,s}(\lambda)$$

where each $I_{r,s}(\lambda)$,

$$I_{r,s}(\lambda) \equiv \int_0^\infty \int_0^\infty e^{-\lambda f_0(x,y)} x^r y^s\, dx dy,$$

is amenable to analysis through its Mellin-Barnes integral representation after the fashion employed here. However, easily used formulas, after the fashion of Kaminski & Paris (1998a, §6(b)), are not available to us in this case, principally due to difficulties associated with constructing the Maclaurin series expansion of the amplitude function $e^{-\lambda f_1(x,y)} g(x, y)$. For a discussion and examples of the case of general amplitude functions $g(x, y)$, see Paris & Liakhovetski (2000b).

7.5 Geometric Content

One novel consequence of the treatment in the previous section is the robust connection between the geometry of the Newton diagram of the phase function and the asymptotics of the Laplace-type integral associated with such a phase. There are (at least) two interesting directions to pursue: one relates the order of the leading term of the asymptotic expansion of an integral (7.3.3) with the remoteness of the Newton diagram (recall §7.3.1), and the other describes the powers of the leading terms in the expansion of (7.3.3) in terms of a triangular decomposition of the region bounded by the Newton diagram and its back face.

7.5.1 Remoteness

Let us return to the convex case with two internal points, beginning at (7.4.2). The leading terms in the three-component series in the asymptotic expansion of (7.4.2) are, from (7.4.24), (7.4.13), (7.4.16) and (7.4.21),

$$I(\lambda) \sim \lambda^{-1/\mu - 1/\nu} \left\{ \frac{1}{m_2\nu} \Gamma\left(\frac{1}{m_2}\right) \Gamma\left(\frac{m_2 - n_2}{m_2\nu}\right) \lambda^{-\delta_2/m_2} \right.$$

$$+ \frac{1}{\Delta} \Gamma\left(\frac{n_2 - m_2}{\Delta}\right) \Gamma\left(\frac{m_1 - n_1}{\Delta}\right) \lambda^{(\Delta - \mu N)/\mu\Delta + (\Delta - M\nu)/\nu\Delta}$$

$$\left. + \frac{1}{\mu n_1} \Gamma\left(\frac{1}{n_1}\right) \Gamma\left(\frac{n_1 - m_1}{\mu n_1}\right) \lambda^{-\delta_1/n_1} \right\}. \tag{7.5.1}$$

There are only three possible cases to address: both internal points P_1 and P_2 lie below the $m = n$ line; the points "straddle" the $m = n$ line (i.e., one lies above the line and one lies below); and both lie above the $m = n$ line. By symmetry considerations, we can reduce this further by discarding one of the cases where both points lie on the same side of the diagonal. Let m_1^*, m_2^* and m_3^* be the abscissas of the points of intersection of the line $m = n$ with each of the lines L_1, L_2 and L_3, defined respectively as the line generated by $(0, \nu)$ and P_2, P_2 and P_1, and finally P_1 and $(\mu, 0)$; see Fig. 7.8. Elementary analytical geometry gives us

$$m_1^* = \frac{m_2\nu}{m_2 + \nu - n_2}, \quad m_2^* = \frac{\Delta}{M + N}, \quad m_3^* = \frac{\mu n_1}{\mu - m_1 + n_1}.$$

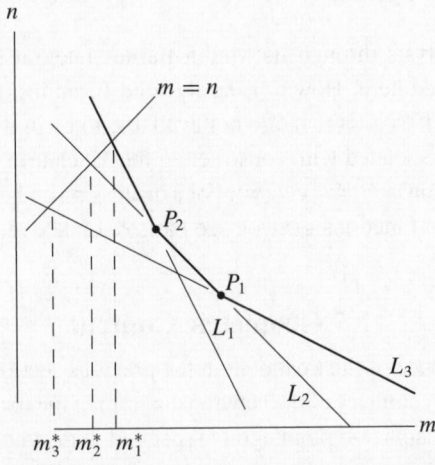

Fig. 7.8. Portion of the Newton diagram for two internal points in the case where P_1 and P_2 both lie below the diagonal $m = n$.

After distributing the factor $\lambda^{-1/\mu-1/\nu}$ across all three terms in (7.5.1), we see that the orders of the leading terms have a particularly elegant expression in terms of the quantities m_1^*, m_2^*, m_3^*: the first term has order $-1/m_1^*$, the second has order $-1/m_2^*$ and the third has order $-1/m_3^*$.

Let us suppose, then, that both points P_1 and P_2 lie below the $m = n$ line. Evidently, we have the ordering $m_3^* < m_2^* < m_1^*$, whence $-1/m_3^* < -1/m_2^* < -1/m_1^*$. The dominant leading term is therefore the one of order $-1/m_1^*$, corresponding to the remoteness of the Newton diagram. For this case, $I(\lambda)$ has the dominant asymptotic behaviour

$$I(\lambda) \sim \frac{\lambda^{-1/m_1^*}}{m_2\nu} \Gamma\left(\frac{1}{m_2}\right) \Gamma\left(\frac{m_2 - n_2}{m_2\nu}\right)$$

for $\lambda \to \infty$. Observe that all arguments of the gamma functions in this leading term are positive in view of the fact that P_2 below the $m = n$ line implies $m_2 > n_2$.

When the points P_1 and P_2 straddle the diagonal $m = n$, as depicted in Fig. 7.9, we arrive at a different ordering of the abscissas m_i^*. In this case, we find $m_1^* < m_3^* < m_2^*$ which in turn yields $-1/m_1^* < -1/m_3^* < -1/m_2^*$. From this, it is apparent that the dominant leading term in (7.5.1) is the second one corresponding to $-1/m_2^*$, namely

$$I(\lambda) \sim \frac{\lambda^{-1/m_2^*}}{\Delta} \Gamma\left(\frac{n_2 - m_2}{\Delta}\right) \Gamma\left(\frac{m_1 - n_1}{\Delta}\right).$$

The quantity $-1/m_2^*$ is the remoteness of the Newton diagram in this instance. Since P_2 lies above the diagonal and P_1 lies below the diagonal, we must also

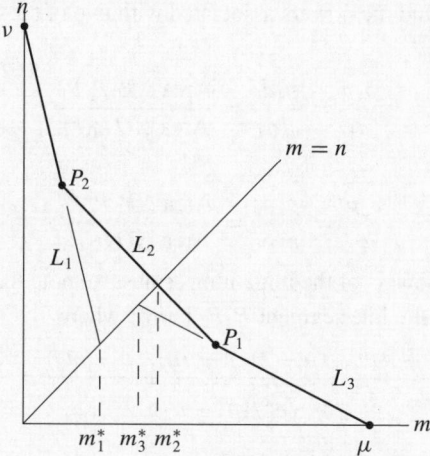

Fig. 7.9. The Newton diagram for two internal points in the case where P_1 and P_2 "straddle" the diagonal $m = n$.

have $n_2 > m_2$ and $m_1 > n_1$ so that the gamma functions in the leading term have positive arguments.

This type of analysis can be conducted for more general settings, and the result is the same, in every case: the order of the leading term in the asymptotic expansion of $I(\lambda)$ given by (7.3.3) is always obtainable through the Newton diagram for the phase. This is entirely in accord with results reported by both the French and Russian schools of analysts, described in the preamble to §7.3.

Extensions to higher dimensional Laplace-type integrals are formed in a similar manner.

7.5.2 Asymptotic Scales

The various asymptotic scales of λ (displayed as (7.4.14), (7.4.17) and (7.4.22)) found in the asymptotic sums over k and l in (7.4.24), can be given a simple geometric interpretation based on the Newton diagram. From the result for the perpendicular distance p from a point (x', y') to a given line $Ax + By = C$ given by

$$p = \frac{C - Ax' - By'}{\sqrt{A^2 + B^2}} \quad (C > 0),$$

we easily see that $\delta_i = p_i/p_0$ $(i = 1, 2)$, where

$$p_i = \frac{1 - m_i/\mu - n_i/\nu}{\sqrt{1/\mu^2 + 1/\nu^2}}, \qquad p_0 = \frac{1}{\sqrt{1/\mu^2 + 1/\nu^2}},$$

are the perpendicular distances of the vertex $P_i = (m_i, n_i)$ and the origin, respectively, to the back face — presented in Fig. 7.10 — as $\overline{P_0 P_3}$, with $P_0 \equiv (\mu, 0)$ and $P_3 \equiv (0, \nu)$. If we write $p_0 = \mu\nu/d_0$, where $d_0 = P_0 P_3$ is the length of the back face, then we find the powers associated with two of these asymptotic scales given by

$$\frac{\delta_1 \nu}{n_1} = \frac{p_1 d_0}{\mu n_1} = \frac{\text{Area} \triangle P_0 P_3 P_1}{\text{Area} \triangle O P_0 P_1}$$

and

$$\frac{\delta_2 \mu}{m_2} = \frac{p_2 d_0}{m_2 \nu} = \frac{\text{Area} \triangle P_0 P_3 P_2}{\text{Area} \triangle O P_2 P_3}.$$

To interpret the powers of the remaining scales, we note that the perpendicular distance from P_1 to the line segment $\overline{P_0 P_2}$ is p_{11}, where

$$p_{11} = \frac{n_1 + n_2 m_1/(\mu - m_2) - \mu n_2/(\mu - m_2)}{\sqrt{1 + n_2^2/(\mu - m_2)^2}} = \frac{\Delta - \mu N}{d_1}$$

with $d_1 = P_0 P_2$, the length of $\overline{P_0 P_2}$. Then

$$\frac{\Delta - \mu N}{\mu n_1} = \frac{p_{11} d_1}{\mu n_1} = \frac{\text{Area} \triangle P_0 P_2 P_1}{\text{Area} \triangle O P_0 P_1}.$$

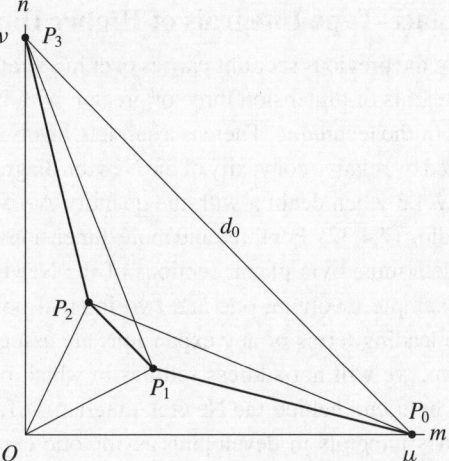

Fig. 7.10. Triangular regions in the Newton diagram for the convex case with two internal points.

and, in a similar fashion,

$$\frac{\Delta - \nu M}{m_2 \nu} = \frac{p_{22} d_2}{m_2 \nu} = \frac{\text{Area} \triangle P_3 P_1 P_2}{\text{Area} \triangle O P_2 P_3},$$

where p_{22} denotes the perpendicular distance from P_2 to the line segment $\overline{P_1 P_3}$ and $d_2 = P_1 P_3$. Finally, since the area of triangle $O P_1 P_2$ is $\frac{1}{2} \Delta$, we have

$$\frac{\Delta - \mu N}{\Delta} = \frac{p_{11} d_1}{\Delta} = \frac{\text{Area} \triangle P_0 P_1 P_2}{\text{Area} \triangle O P_1 P_2}$$

and

$$\frac{\Delta - \nu M}{\Delta} = \frac{\text{Area} \triangle P_3 P_1 P_2}{\text{Area} \triangle O P_1 P_2}.$$

Hence the powers associated with the different asymptotic scales of λ appearing in the expansion (7.4.24) can be interpreted as ratios of triangular areas in the Newton diagram.

For double integrals (7.3.3) with more than two internal points, a similar triangular decomposition of the region bounded by the Newton diagram and the back face is possible, but the number of triangular sections possible grows dramatically with the number of vertices of the Newton diagram. We add, too, that for higher dimensional analogues of (7.3.3), the region bounded by the Newton "diagram" and the back face will be a polyhedral solid of appropriate dimension, and instead of using triangles in constructing the decomposition, one would use simplices of appropriate dimension.

7.6 Laplace-Type Integrals of Higher Dimension

The methodology of the previous sections carries over in a straightforward fashion
to Laplace-type integrals of dimension three or greater, with little increase in the
computational cost of the technique. There is a subtlety involving the increasingly
important role played by relative convexity of the Newton diagram,† met in simpler
circumstances in §7.4.3 when dealing with the quantity $\Delta_{23} - \Delta_{13} + \Delta_{12}$ in the
discussion surrounding (7.4.32). For three and more dimensions, matters of relative
convexity can be determined via planar sections of the Newton diagram, and is
illustrated in the example involving one and two internal points in the Newton
diagram. Since the leading terms of any expansions are associated with faces of
the Newton diagram, we will not address settings in which points of the carrier
of the phase are to be found behind the Newton diagram. A fuller account of the
use of Mellin-Barnes integrals in developing asymptotic expansions for higher
dimensional integrals is to be found in Kaminski & Paris (1998b).

7.6.1 Representation of Treble Integrals

Let us define, in a fashion analogous to that carried out in §7.3.1 and (7.4.1), the
quantities

$$\delta_r = 1 - \frac{m_r}{\mu} - \frac{n_r}{\nu} - \frac{p_r}{\eta} \qquad (r = 1, \ldots, k)$$

and set

$$\boldsymbol{m} = (m_1, m_2, \ldots, m_k), \boldsymbol{n} = (n_1, n_2, \ldots, n_k), \boldsymbol{p} = (p_1, p_2, \ldots, p_k),$$

$$\boldsymbol{t} = (t_1, t_2, \ldots, t_k), \boldsymbol{\delta} = (\delta_1, \delta_2, \ldots, \delta_k)$$

with the conventions of §7.3.1 holding. As we have done before in §§7.2, 7.3, let
us apply the Mellin-Barnes integral representation of the exponential function in
(3.3.2) to each of the exponential factors $\exp(-\lambda c_r x^{m_r} y^{n_r} z^{p_r})$ in the integrand of

$$I(\lambda) = \int_0^\infty \int_0^\infty \int_0^\infty \exp\left\{-\lambda\left(x^\mu + y^\nu + z^\eta + \sum_{r=1}^k c_r x^{m_r} y^{n_r} z^{p_r}\right)\right\} dx \, dy \, dz$$

$$(7.6.1)$$

to arrive, after interchanging the order of integration, at the representation

$$I(\lambda) = \frac{\lambda^{-1/\mu - 1/\nu - 1/\eta}}{\mu\nu\eta} \left(\frac{1}{2\pi i} \int_{-\infty i}^{\infty i}\right)^k \Gamma(\boldsymbol{t})$$

$$\times \Gamma\left(\frac{1 - \boldsymbol{m} \cdot \boldsymbol{t}}{\mu}\right) \Gamma\left(\frac{1 - \boldsymbol{n} \cdot \boldsymbol{t}}{\nu}\right) \Gamma\left(\frac{1 - \boldsymbol{p} \cdot \boldsymbol{t}}{\eta}\right) \boldsymbol{c}^{-\boldsymbol{t}} \lambda^{-\boldsymbol{\delta} \cdot \boldsymbol{t}} d\boldsymbol{t}. \quad (7.6.2)$$

---

† We continue to use the term *Newton diagram* in this setting, even though it would more properly
be termed the *Newton polyhedron*.

The Laplace-type treble integral (7.6.1) is clearly an extension to three variables of the type of integrals under examination in §§7.3, 7.4, with a phase function that is of the same form as that examined in those earlier sections. The factor $\lambda^{-1/\mu-1/\nu-1/\eta}$ appearing in (7.6.2) recurs throughout this section, so we will rescale $I(\lambda)$ accordingly and put

$$I(\lambda) \equiv \lambda^{-1/\mu-1/\nu-1/\eta} \tilde{I}(\lambda) \tag{7.6.3}$$

for notational convenience.

Observe that, as in §7.3, the number of independent variables $t_1, \ldots, t_k$ in (7.6.2) is governed by the number of terms in the sum in the integrand of (7.6.1), but unlike the situation in §7.3, the higher dimensionality of (7.6.1) is reflected in the integrand of (7.6.2) by the presence of an additional gamma function, $\Gamma((1 - \vec{p} \cdot \vec{t})/\nu)$, corresponding to the additional spatial variable z. Because of this, we can expect the asymptotic expansions we develop to contain asymptotic series with an additional gamma function appearing in the summand of each series.

The representation (7.6.2) can be readily extended to deal with higher-dimensional Laplace-type integrals. Each additional dimension of such a Laplace integral manifests itself in the Mellin-Barnes integral (7.6.2) by the appearance of another factor such as $\Gamma((1 - \vec{m} \cdot \vec{t})/\mu)$, while the dimensionality of (7.6.2) is governed by the number of terms in the summation present in the phase, as presented in (7.6.1). The dimensionality of the associated Newton diagram will also increase and with it, the accompanying difficulty of geometric analysis in four or more dimensions.

7.6.2 Asymptotics with One Internal Point

By setting $k = 1$, we can recast (7.6.1) and (7.6.2) (in rescaled form (7.6.3)) as

$$\tilde{I}(\lambda) = \lambda^{1/\mu+1/\nu+1/\eta} \int_0^\infty \int_0^\infty \int_0^\infty \exp\big\{ -\lambda(x^\mu + y^\nu$$
$$+ z^\eta + x^{m_1} y^{n_1} z^{p_1}) \big\} \, dx \, dy \, dz$$
$$= \frac{1}{\mu\nu\eta} \frac{1}{2\pi i} \int_{-\infty i}^{\infty i} \Gamma(t) \Gamma\left(\frac{1 - m_1 t}{\mu}\right) \Gamma\left(\frac{1 - n_1 t}{\nu}\right) \Gamma\left(\frac{1 - p_1 t}{\eta}\right) \lambda^{-\delta_1 t} dt; \tag{7.6.4}$$

where, for ease of presentation, the value of the constant c_1 in (7.6.1) has been set to unity. This is a routine one-variable problem in residue computation, and we restrict attention to just the convex case, where the point $P_1 = (m_1, n_1, p_1)$ lies between the origin and the back face.

The back face of the Newton diagram for our class of phases is simply the plane generated by $(\mu, 0, 0)$, $(0, \nu, 0)$ and $(0, 0, \eta)$ and given by

$$\frac{m}{\mu} + \frac{n}{\nu} + \frac{p}{\eta} - 1 = 0. \tag{7.6.5}$$

If P_1 lies on the same side of the back face as the origin (i.e., in front of the back face), then P_1 must satisfy $m_1/\mu + n_1/\nu + p_1/\eta - 1 < 0$ or equivalently, $\delta_1 > 0$. Otherwise, the reverse inequalities hold if P_1 lies behind the back face (if P_1 lies on the back face, then (7.6.5) must be satisfied, in which case $\delta_1 = 0$).

In the convex case, δ_1 is positive and the Newton diagram comprises three triangular faces, each formed by the vertex P_1 and any two of $(\mu, 0, 0)$, $(0, \nu, 0)$ and $(0, 0, \eta)$; see Fig. 7.11. Together with the back face, these triangular faces generate a tetrahedron. The planes generated by each of the faces of the Newton diagram are easy to obtain. If we denote by $\Pi(Q_1, Q_2, Q_3)$ the plane through the points Q_1, Q_2 and Q_3, then we find that $\Pi_{\mu\nu p_1} \equiv \Pi((\mu, 0, 0), (0, \nu, 0), P_1)$ has equation

$$\frac{m}{\mu} + \frac{n}{\nu} + \frac{\mu\nu - \mu n_1 - m_1\nu}{\mu\nu p_1}p = 1,$$

$\Pi_{\mu n_1 \eta} \equiv \Pi((\mu, 0, 0), P_1, (0, 0, \eta))$ has equation

$$\frac{m}{\mu} + \frac{\mu\eta - m_1\eta - \mu p_1}{\mu n_1 \eta}n + \frac{p}{\eta} = 1$$

and $\Pi_{m_1 \nu \eta} \equiv \Pi(P_1, (0, \nu, 0), (0, 0, \eta))$ has equation

$$\frac{\nu\eta - \nu p_1 - n_1\eta}{m_1\nu\eta}m + \frac{n}{\nu} + \frac{p}{\eta} = 1.$$

The asymptotics of (7.6.4) are easily obtained. As in §7.3, we find the dominant behaviour of the logarithm of the modulus of the integrand, putting $t = \rho e^{i\theta}$, $|\theta| < \frac{1}{2}\pi$, to be

$$\delta_1\rho\cos\theta\log\rho + O(\rho)$$

which, since P_1 is in front of the back face, tends to $+\infty$ as $\rho \to \infty$. Accordingly, the asymptotics of (7.6.4) can be obtained by displacing the integration contour to the right. As we do so, we find that poles in three sequences must be considered. From poles of the factor $\Gamma((1 - m_1 t)/\mu)$ we obtain the formal series

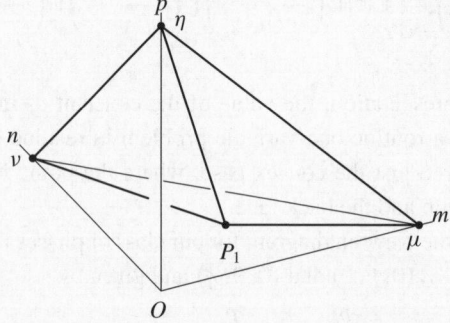

Fig. 7.11. Newton diagram for the convex case with one internal point.

of contributions

$$I_{m_1 v \eta} = \frac{1}{m_1 v \eta} \sum_k \frac{(-)^k}{k!} \Gamma\left(\frac{K}{m_1}\right) \Gamma\left(\frac{m_1 - n_1 K}{m_1 v}\right) \Gamma\left(\frac{m_1 - p_1 K}{m_1 \eta}\right) \lambda^{-\delta_1 K/m_1},$$

(7.6.6)

while those of $\Gamma((1 - n_1 t)/v)$ and $\Gamma((1 - p_1 t)/\eta)$ similarly give rise to

$$I_{\mu n_1 \eta} = \frac{1}{\mu n_1 \eta} \sum_k \frac{(-)^k}{k!} \Gamma\left(\frac{n_1 - m_1 K'}{\mu n_1}\right) \Gamma\left(\frac{K'}{n_1}\right) \Gamma\left(\frac{n_1 - p_1 K'}{n_1 \eta}\right) \lambda^{-\delta_1 K'/n_1},$$

(7.6.7)

$$I_{\mu v p_1} = \frac{1}{\mu v p_1} \sum_k \frac{(-)^k}{k!} \Gamma\left(\frac{p_1 - m_1 K''}{\mu p_1}\right) \Gamma\left(\frac{p_1 - n_1 K''}{v p_1}\right) \Gamma\left(\frac{K''}{p_1}\right) \lambda^{-\delta_1 K''/p_1},$$

(7.6.8)

respectively, provided none of the three sequences of poles shares a common point, and where we have written

$$K = 1 + \mu k, \ K' = 1 + v k, \ K'' = 1 + \eta k, \tag{7.6.9}$$

for notational convenience. Assembling these together and recalling (7.6.3), we arrive at the asymptotic expansion

$$I(\lambda) \sim \lambda^{-1/\mu - 1/v - 1/\eta} (I_{m_1 v \eta} + I_{\mu n_1 \eta} + I_{\mu v p_1}) \tag{7.6.10}$$

as $\lambda \to \infty$. The correspondence we observed in earlier sections between the number of series constituting the compound asymptotic expansion of $I(\lambda)$ and the number of faces of the Newton diagram continues to hold here, with the series $I_{m_1 v \eta}$ corresponding to the face with vertices P_1, $(0, v, 0)$ and $(0, 0, \eta)$, the series $I_{\mu n_1 \eta}$ corresponding to the face with vertices P_1, $(\mu, 0, 0)$ and $(0, 0, \eta)$, and finally $I_{\mu v p_1}$ corresponding to the face generated by P_1, $(\mu, 0, 0)$ and $(0, v, 0)$.

Additionally, in the case where the diagonal punctures a face, it is possible to have double or treble poles in the higher order terms. This will happen whenever the arguments of the gamma functions in (7.6.6)–(7.6.8) (i.e., expressions such as $(m_1 - n_1 K)/m_1 v$, $(n_1 - p_1 K')/n_1 \eta$, $(p_1 - m_1 K'')/\mu p_1$ and so on) is a nonpositive integer. Only by choosing special values for μ, v, η, m_1, n_1 and p_1 can this be avoided.

If the diagonal meets the Newton diagram in an edge, then two of m_1, n_1 and p_1 must be the same, say $m_1 = n_1 > p_1$. In this event, the gamma functions appearing in $I_{m_1 v \eta}$ and $I_{\mu n_1 \eta}$ will have $k = 0$ corresponding to a double pole, so that the leading term in the asymptotic expansion of $I(\lambda)$ will contain a logarithmic factor. Furthermore, if μ/v is rational, other double poles will appear and if $\mu = v$, all terms in the series will have double poles. Higher order terms may even have some treble poles, but these will not be present in the leading behaviour.

Finally, if the diagonal meets the Newton diagram at the vertex P_1, then $m_1 = n_1 = p_1$ and the $k = 0$ terms in each of the expansions will arise from a treble pole. The possibility of double or treble poles in other terms in the expansion depends on the rationality of ratios of pairs of μ, ν and η. The completely symmetrical case of $\mu = \nu = \eta$ produces poles that are all of order three and our three series (7.6.6)–(7.6.8) collapse into a single series of evaluations of treble poles.

The relationship between symmetry and the relative frequency of higher order poles appearing in the asymptotics is explored in some detail in §7.7.

7.6.3 Asymptotics with Two Internal Points

Let us suppose that there are now two internal points in front of the back face, $P_1 = (m_1, n_1, p_1)$ and $P_2 = (m_2, n_2, p_2)$. Because they lie on the same side of the back face (7.6.5), we must have δ_1 and δ_2 both positive. With $t_2 = \rho e^{i\theta}$ and $|\theta| < \frac{1}{2}\pi$ in the integrand of (7.6.2) with $k = 2$, we find the logarithm of the modulus of the integrand to exhibit the dominant behaviour

$$\delta_2 \rho \cos\theta \log\rho + O(\rho)$$

for large ρ. Since $\delta_2 > 0$, we see that this tends to $+\infty$ as $\rho \to \infty$ leading us to displace the t_2 integration contour to the right to obtain the asymptotic behaviour of $\tilde{I}(\lambda)$. Three sequences of poles are encountered in shifting the contour to the right: one each for the gamma functions present in the integrand, except $\Gamma(t_1)$ and $\Gamma(t_2)$.

The functions $\Gamma((1 - \vec{m} \cdot \vec{t})/\mu)$, $\Gamma((1 - \vec{n} \cdot \vec{t})/\nu)$ and $\Gamma((1 - \vec{p} \cdot \vec{t})/\eta)$ have their poles occurring at the points

$$t_2^{(1)} = (K - m_1 t_1)/m_2, \qquad t_2^{(2)} = (K' - n_1 t_1)/n_2, \qquad t_2^{(3)} = (K'' - p_1 t_1)/p_2,$$
$$\text{(7.6.11)}$$

respectively, where K, K' and K'' are given in (7.6.9). We assume that these sequences share no points. In turn, these sequences give rise to formal asymptotic series

$$I_1 = \frac{1}{m_2\nu\eta} \sum_k \frac{(-)^k}{k!} \frac{1}{2\pi i} \int_{-i\infty}^{i\infty} \Gamma(t_1)\Gamma\left(t_2^{(1)}\right)\Gamma\left(\frac{1 - n_1 t_1 - n_2 t_2^{(1)}}{\nu}\right)$$

$$\times \Gamma\left(\frac{1 - p_1 t_1 - p_2 t_2^{(1)}}{\eta}\right)\lambda^{-\vec{\delta}\cdot\vec{t}} dt_1, \qquad \text{(7.6.12)}$$

$$I_2 = \frac{1}{\mu n_2\eta} \sum_k \frac{(-)^k}{k!} \frac{1}{2\pi i} \int_{-i\infty}^{i\infty} \Gamma(t_1)\Gamma\left(t_2^{(2)}\right)\Gamma\left(\frac{1 - m_1 t_1 - m_2 t_2^{(2)}}{\mu}\right)$$

$$\times \Gamma\left(\frac{1 - p_1 t_1 - p_2 t_2^{(2)}}{\eta}\right)\lambda^{-\vec{\delta}\cdot\vec{t}} dt_1, \qquad \text{(7.6.13)}$$

$$I_3 = \frac{1}{\mu\nu p_2} \sum_k \frac{(-)^k}{k!} \frac{1}{2\pi i} \int_{-i\infty}^{i\infty} \Gamma(t_1)\Gamma(t_2^{(3)})\Gamma\left(\frac{1 - m_1 t_1 - m_2 t_2^{(3)}}{\mu}\right)$$

$$\times \Gamma\left(\frac{1 - n_1 t_1 - n_2 t_2^{(3)}}{\nu}\right) \lambda^{-\vec{\delta}\cdot\vec{t}} dt_1, \tag{7.6.14}$$

with $\vec{\delta}\cdot\vec{t}$ equal to $\delta_1 t_1 + \delta_2 t_2^{(i)}$, $i = 1, 2, 3$, respectively. If we set $t_1 = \rho e^{i\theta}$, $|\theta| < \frac{1}{2}\pi$, in each of the integrands in the series I_1, I_2 and I_3, then we find they have dominant large-ρ behaviour

$$\rho\cos\theta\log\rho \cdot (\Delta_{12}/\nu + \Delta_{12}''/\eta - M_{12})/m_2 + O(\rho),$$

$$\rho\cos\theta\log\rho \cdot (\Delta_{12}'/\eta - \Delta_{12}/\mu - N_{12})/n_2 + O(\rho), \tag{7.6.15}$$

$$\rho\cos\theta\log\rho \cdot (-\Delta_{12}'/\nu - \Delta_{12}''/\mu - P_{12})/p_2 + O(\rho),$$

respectively; the various subscripted items appearing in these expressions are defined, for $i, j = 1, 2, 3$ $(i \neq j)$, by

$$\Delta_{12} = m_1 n_2 - n_1 m_2, \quad \Delta_{13} = n_1 m_3 - m_1 n_3, \quad \Delta_{23} = n_2 m_3 - m_2 n_3,$$
$$\Delta_{12}' = n_1 p_2 - n_2 p_1, \quad \Delta_{13}' = n_1 p_3 - n_3 p_1, \quad \Delta_{23}' = n_2 p_3 - n_3 p_2,$$
$$\Delta_{12}'' = m_1 p_2 - m_2 p_1, \quad \Delta_{13}'' = m_1 p_3 - m_3 p_1, \quad \Delta_{23}'' = m_3 p_2 - m_2 p_3,$$
$$\tag{7.6.16}$$

$$M_{ij} = m_i - m_j, N_{ij} = n_i - n_j, P_{ij} = p_i - p_j. \tag{7.6.17}$$

To determine which direction to shift the integration contours in (7.6.12), (7.6.13) and (7.6.14), we must determine the sign of each of the parenthesised quantities following the factor $\rho\cos\theta\log\rho$. We shall see presently that the sign of each of these parenthesised quantities in (7.6.15) has a geometric interpretation.

Let $\overline{P_1 P_2}$ denote the line generated by P_1 and P_2. Elementary geometry reveals that if $\overline{P_1 P_2}$ meets the mn, mp and np planes, then it does so in points with co-ordinates $(-\Delta_{12}''/P_{12}, -\Delta_{12}'/P_{12}, 0)$, $(-\Delta_{12}/N_{12}, 0, \Delta_{12}'/N_{12})$ and $(0, \Delta_{12}/M_{12},$ $\Delta_{12}''/M_{12})$, respectively. (The possibility of $\overline{P_1 P_2}$ being parallel to a coordinate plane is discussed later.) If the parenthesised expressions in (7.6.15) are set to zero and rescaled by dividing through, respectively, by M_{12}, N_{12} and P_{12} then the following equations result:

$$\frac{\Delta_{12}/M_{12}}{\nu} + \frac{\Delta_{12}''/M_{12}}{\eta} - 1 = 0,$$

$$\frac{-\Delta_{12}/N_{12}}{\mu} + \frac{\Delta_{12}'/N_{12}}{\eta} - 1 = 0, \tag{7.6.18}$$

$$\frac{-\Delta_{12}''/P_{12}}{\mu} + \frac{-\Delta_{12}'/P_{12}}{\nu} - 1 = 0.$$

The first equation is equivalent to the statement that the point of intersection of $\overline{P_1 P_2}$ with the np plane lies on the line of intersection of the back face of the

Newton diagram with the np plane. The second and third equations, similarly, assert the presence on the line of intersection of the back face with either of the mp or mn planes of the point of intersection of $\overline{P_1 P_2}$ with each of these coordinate planes, respectively.

The determination of the sign of the parenthesised quantities in (7.6.15) is therefore equivalent to deciding on which side of the line of intersection of the back face with a coordinate plane a point of intersection of $\overline{P_1 P_2}$ happens to lie. The difficulty in determining this stems from the fact that the distribution in space of P_1 and P_2 can lead to M_{12}, N_{12} and P_{12} being of either sign. Let us therefore fix the sign of these quantities so as to permit further analysis: let us suppose that $\overline{P_1 P_2}$ punctures the mn and np planes in the positive octant, and that $M_{12} > 0$, $P_{12} < 0$. This setting is depicted in Fig. 7.12, with a view assuming the observer is behind the back face, looking towards the origin.

Because $M_{12} > 0$, $P_{12} < 0$ and $\overline{P_1 P_2}$ meets the mn and np planes in the positive octant, we may deduce that $\Delta_{12} > 0$ and $\Delta_{12}'' > 0$. Consequently, the statement that the point of intersection of $\overline{P_1 P_2}$ with the np plane lie below the line of intersection of the back face with that coordinate plane is equivalent to the inequality

$$\frac{\Delta_{12}/M_{12}}{\nu} + \frac{\Delta_{12}''/M_{12}}{\eta} - 1 < 0,$$

and the equivalent statement involving the point of intersection of $\overline{P_1 P_2}$ and the mn plane results in

$$\frac{-\Delta_{12}''/P_{12}}{\mu} + \frac{-\Delta_{12}'/P_{12}}{\nu} - 1 < 0.$$

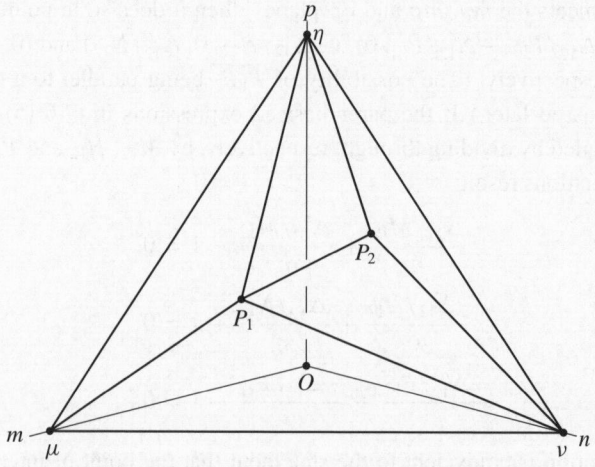

Fig. 7.12. Newton diagram with two internal points.

Both of these inequalities follow from the assumption of a convex configuration for the internal points P_1 and P_2: one needs only to form the plane $\Pi(O, P_1, P_2)$ and intersect it with the Newton diagram to see this.

The second equation in (7.6.18) on first glance appears to be redundant, but in fact captures additional convexity information regarding the Newton diagram. This second equation is a statement regarding the location on the mp plane of the point $(-\Delta_{12}/N_{12}, 0, \Delta'_{12}/N_{12})$, the point of intersection of the mp plane with the line $\overline{P_1 P_2}$. If $N_{12} > 0$, then this intersection point lies behind ($m < 0$) the np plane and above ($p > 0$) the mn plane. If $N_{12} < 0$, then the intersection point lies below ($p < 0$) the mn plane, but in front ($m > 0$) of the np plane. Where it lies relative to the back face (7.6.5) is determined by the sign of the left-hand side of the second equation of (7.6.18).

Consider the plane $\Pi(O, P_1, P_2)$ generated by the origin and the internal points P_1 and P_2. If $(-\Delta_{12}/N_{12}, 0, \Delta'_{12}/N_{12})$ does not satisfy $m/\mu + p/\eta - 1 < 0$, then $(-\Delta_{12}/N_{12}, 0, \Delta'_{12}/N_{12})$ lies on the back face or in the half-space determined by the back face not containing the origin. Suppose, for the purpose of illustration, that $(-\Delta_{12}/N_{12})/\mu + (\Delta'_{12}/N_{12})/\eta - 1 > 0$. Then, in the plane $\Pi(O, P_1, P_2)$, we see that the intersection point $(-\Delta_{12}/N_{12}, 0, \Delta'_{12}/N_{12})$ is positioned in such a way that P_1 cannot be a vertex of the Newton diagram. This situation is depicted in Fig. 7.13, with the line of intersection of $\Pi(O, P_1, P_2)$ and the mp plane corresponding to $\overline{OA}$, the intersection of $\overline{P_1 P_2}$ and the mp plane labelled as I, the line of intersection of the back face with the plane $\Pi(O, P_1, P_2)$ appearing as $\overline{AB}$ and the line $\overline{OB}$ corresponding to the line of intersection of $\Pi(O, P_1, P_2)$ with one of the other coordinate planes (either the mn or np plane).

Because the convex hull used in the construction of the Newton diagram is (tautologically) convex (recall §7.3.2) and the half-space defined by the back face

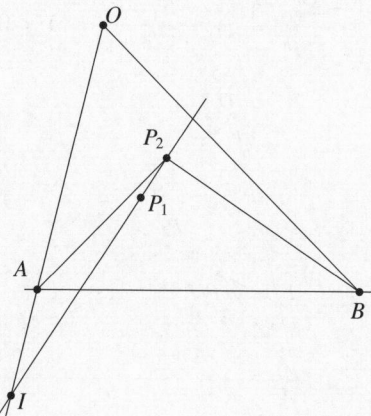

Fig. 7.13. Planar section $\Pi(O, P_1, P_2)$ of the Newton diagram in the case where the intersection point $I = (-\Delta_{12}/N_{12}, 0, \Delta'_{12}/N_{12})$ lies on the side of the back face not containing the origin.

containing the origin is convex, their intersection must be a bounded convex solid. By intersecting this with the plane $\Pi(O, P_1, P_2)$, a convex planar region must result. This region is $\triangle ABP_2$, and since we have assumed that the intersection point I lies on the side of the back face not containing the origin, it follows that P_1 lies in the interior of $\triangle ABP_2$, a violation of the hypothesis that P_1 and P_2 be vertices of the Newton diagram.

If instead we assume that the intersection point I lies on the back face, i.e., $m/\mu + p/\eta - 1 = 0$ is satisfied by I, then a planar section through P_1 and P_2 will reveal that P_1 lies on the line segment $\overline{AP_2}$. In this circumstance, the hypothesis that P_1 and P_2 both be vertices fails: P_1 will be a point of the Newton diagram, but not an extreme point.

If the intersection point I lies on the same side of the back face as the origin, i.e., $m/\mu + p/\eta - 1 < 0$ is satisfied by I, then the geometry of the planar section of the Newton diagram changes. The result, depicted in Fig. 7.14, reveals the planar section of the bounded convex region formed from the convex hull, used to construct the Newton diagram, intersected with the half-space defined by the back face containing the origin; this polygonal region has vertices ABP_2P_1, in the notation of the previous paragraphs.

We see, therefore, that regardless of the sign of N_{12}, we must have the intersection point I and the origin in the same half-space determined by the back face. Convexity then permits us to assert the following inequalities:

$$\frac{\Delta_{12}/M_{12}}{\nu} + \frac{\Delta_{12}''/M_{12}}{\eta} - 1 < 0,$$

$$\frac{-\Delta_{12}/N_{12}}{\mu} + \frac{\Delta_{12}'/N_{12}}{\eta} - 1 < 0, \qquad (7.6.19)$$

$$\frac{-\Delta_{12}''/P_{12}}{\mu} + \frac{-\Delta_{12}'/P_{12}}{\nu} - 1 < 0.$$

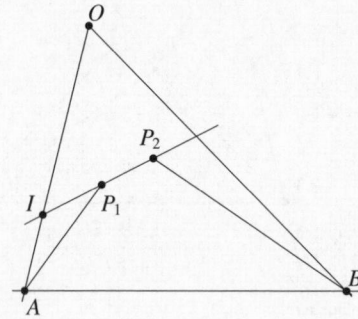

Fig. 7.14. Planar section $\Pi(O, P_1, P_2)$ of the Newton diagram in the case where the intersection point $I = (-\Delta_{12}/N_{12}, 0, \Delta_{12}'/N_{12})$ lies on the side of the back face containing the origin.

With these inequalities at hand, we can determine the large-ρ behaviour of each of the estimates in (7.6.15), and so determine in which direction the t_1 integration contours in (7.6.12)–(7.6.14) must be shifted in order to extract the asymptotic behaviour of each of the integrals in these formal series. Let us assume that $N_{12} < 0$, so that our assumptions on the distribution of P_1 and P_2 in space become

$$M_{12} > 0, N_{12} < 0, P_{12} < 0. \qquad (7.6.20)$$

Returning to the order estimates (7.6.15), we can now assert that the first estimate tends to $-\infty$ and the second and third to $+\infty$ as $\rho \to \infty$. Accordingly, the t_1 contours in the integrals in I_1, I_2 and I_3 of (7.6.12)–(7.6.14) must be shifted left, right and right, respectively, to determine their asymptotic behaviour.

In displacing the t_1 contour in (7.6.12) to the left, we encounter poles of the integrand at $t_1^{(1)} = -l$ from the factor $\Gamma(t_1)$ in the integrand of (7.6.12) and at

$$t_1^{(2)} = (n_2 K - m_2 L')/\Delta_{12} \leq 0, \quad t_1^{(3)} = (p_2 K - m_2 L'')/\Delta_{12}'' \leq 0, \qquad (7.6.21)$$

from the factors $\Gamma((1 - n_1 t_1 - n_2 t_2^{(1)})/\nu)$ and $\Gamma((1 - p_1 t_1 - p_2 t_2^{(1)})/\eta)$, respectively. The parameter l is a nonnegative integer, and we have set

$$L = 1 + \mu l, L' = 1 + \nu l, L'' = 1 + \eta l; \qquad (7.6.22)$$

recall (7.6.9). The factor $\Gamma(t_2^{(1)})$ encounters no poles as the t_1 contour is shifted to the left.

As we displace the t_1 contour to the left past the poles $t_1^{(1)} = -l$ and those given in (7.6.21), we generate, for each sequence of poles, formal asymptotic series which we will denote by I_{11}, I_{12} and I_{13}, coinciding with the ordering $t_1^{(1)}$, $t_1^{(2)}$ and $t_1^{(3)}$. Because of the restrictions present on the poles in (7.6.21), the corresponding series I_{12} and I_{13} have linear inequalities present governing the ranges of the summation indices.

For I_2 in (7.6.13) we must displace the integration contour to the right, and in so doing, encounter poles of the integrands at

$$t_1^{(1)} = (K' + n_2 l)/n_1, \quad t_1^{(2)} = \frac{(-m_2 K' + n_2 L)}{\Delta_{12}} > 0, \quad t_1^{(3)} = \frac{(p_2 K' - n_2 L'')}{\Delta_{12}'} > 0.$$
$$(7.6.23)$$

Displacement of the integrals in the I_2 series to the right past these sequences of poles gives rise to formal asymptotic series in a manner similar to the situation arising for the series I_1. Let us call these series, corresponding to the sequences of poles, I_{21}, I_{22} and I_{23}, and observe that, since the sequences $t_1^{(2)}$ and $t_1^{(3)}$ in (7.6.23) are subject to restrictions, so too are the double sums appearing in I_{22} and I_{23}.

For I_3 in (7.6.14) we shift the integration contour to the right as for I_2, and encounter poles of the integrands at

$$t_1^{(1)} = (K'' + p_2 l)/p_1, \quad t_1^{(2)} = \frac{(-m_2 K'' + p_2 L)}{\Delta_{12}''} > 0,$$

$$t_1^{(3)} = \frac{(-n_2 K'' + p_2 L')}{\Delta_{12}'} > 0. \tag{7.6.24}$$

These in turn spawn formal asymptotic series I_{31}, I_{32}, I_{33} as the t_1 contours in the integrals in I_3 are shifted to the right. Since the sequences $t_1^{(2)}$ and $t_1^{(3)}$ are subject to linear inequalities, so too are the series — I_{32} and I_{33}, respectively — that arise from these sequences.

We can now assemble the expansions I_{11}, I_{12}, I_{13}, I_{21}, I_{22}, I_{23}, I_{31}, I_{32} and I_{33} into an expansion of $\tilde{I}(\lambda)$ with $k = 2$ and $c_1 = c_2 = 1$. Before doing so, we observe that some simplification is possible. The reader will find that the series I_{12} and I_{22} are of similar form, and through performing a simple change of summation indices, will further notice that the series can be rendered in precisely the same form, albeit with different restrictions imposed on the summation indices of each series. For I_{22}, the restriction governing its indices was $t_1^{(2)} > 0$, or

$$(-m_2 K' + n_2 L)/\Delta_{12} > 0;$$

recall (7.6.23). After the change of variables $k \to l$, $l \to k$, K' will become L' and L will become K and the restriction assumes the form

$$(n_2 K - m_2 L')/\Delta_{12} > 0.$$

However, from (7.6.21), we see that the inequality governing I_{12} is

$$(n_2 K - m_2 L')/\Delta_{12} \leq 0,$$

the complement of the inequality governing our transformed I_{22}. Therefore, I_{12} and I_{22} may be fused into a single series, $I_{\eta \Delta_{12}}$ say, with no restrictions placed on the nonnegative summation indices k and l.

In the same way, the reader will find that, apart from a minus sign attached to I_{23}, the series I_{23} and I_{33} are of similar form, and with a simple change of variables (effectively just a relabelling of summation indices), the two series can be seen to be additive inverses of each other, governed by a common inequality. Thus, the series I_{23} and I_{33} annihilate each other.

Finally, the two series I_{13} and I_{32} governed by inequalities that remain fuse into a single series with no restriction on the summation indices involved (apart from the fact that the indices must be nonnegative integers).

Collecting together the series that remain after this exercise is complete produces the large-λ expansion of $I(\lambda)$ [cf. (7.6.3)]

$$I(\lambda) \sim \lambda^{-1/\mu - 1/\nu - 1/\eta}(I_{m_2 \nu \eta} + I_{\mu n_1 \eta} + I_{\mu \nu p_1} + I_{\eta \Delta_{12}} + I_{\nu \Delta_{12}''}),$$

where

$$I_{m_2 v\eta} = \frac{1}{m_2 v\eta} \sum f_{11}(k, l), \quad I_{\mu n_1 \eta} = \frac{1}{\mu n_1 \eta} \sum f_{21}(k, l),$$

$$I_{\mu v p_1} = \frac{1}{\mu v p_1} \sum f_{31}(k, l), \quad I_{\eta \Delta_{12}} = \frac{1}{\eta \Delta_{12}} \sum f_{12}(k, l),$$

$$I_{v\Delta_{12}''} = \frac{1}{v\Delta_{12}''} \sum f_{13}(k, l).$$

The functions in the formal sums are, in turn, of the form

$$f_{ij}(k, l) = \frac{(-1)^{k+l}}{k! \, l!} \Gamma(A_{ij}(k, l)) \Gamma(B_{ij}(k, l)) \Gamma(C_{ij}(k, l)) \lambda^{-D_{ij}(k, l)},$$

with A_{ij}, B_{ij}, C_{ij} and D_{ij} linear functions of k and l, and in particular, D_{ij} is positive for $k, l \geq 0$. Additional details regarding the functions A_{ij}, B_{ij}, C_{ij} and D_{ij} are to be found in Kaminski & Paris (1998b, §3).

The correspondence between the series in this compound expansion and the faces of the Newton diagram is easily seen (as indicated by the subscripts used in the series on the right-hand side): $I_{m_2 v\eta}$ corresponds to the face with vertices P_2, $(0, v, 0)$ and $(0, 0, \eta)$; $I_{\mu n_1 \eta}$ corresponds to the face with vertices $(\mu, 0, 0)$, P_1 and $(0, 0, \eta)$; $I_{\mu v p_1}$ corresponds to the face with vertices $(\mu, 0, 0)$, $(0, v, 0)$ and P_1; $I_{\eta \Delta_{12}}$ corresponds to the face with vertices P_1, P_2 and $(0, 0, \eta)$; and finally, $I_{v\Delta_{12}''}$ corresponds to the face with vertices P_1, P_2 and $(0, v, 0)$. The reader may want to review the arrangement of faces by referring to Fig. 7.12.

7.6.4 Other Considerations for Treble Integrals

Much of the commentary supplied in §7.4.3 applies to treble and higher dimensional integrals and so will not be discussed here. We do point out, though, some novelties with treble and higher dimensional integrals not present in the case of double integrals.

First, the "atomic unit" in building up faces in the Newton diagram is, for double integrals, just a line segment, while that for treble integrals is a triangle. Carrying this forward to higher dimensional integrals, one realises that for an n-dimensional integral of Laplace type, the basic building block of the Newton diagram that emerges from our method is that of an $(n - 1)$-simplex. Thus, double integrals have Newton diagrams built from 1-simplices (line segments), treble integrals have Newton diagrams built from 2-simplices (triangles) and n-fold Laplace-type integrals will use $(n - 1)$-dimensional analogues of tetrahedra.

However, for integrals of three or more dimensions, the Newton diagrams can have faces that are not formed exclusively from simplicial solids. For example, a treble integral with two internal points can have a quadrilateral face in its Newton diagram [Kaminski & Paris (1998b, §6)], and with more vertices on the Newton diagram, more possibilities arise.

These circumstances are still amenable to analysis by the techniques of this chapter, and the reader will find that our methods effectively decompose non-triangular (or non-simplicial, for higher dimensional integrals) faces into unions of triangular (resp. simplicial) faces. The asymptotic scales for the series corresponding to each face in the simplicial decomposition will coincide with the scales used in the series for another face in the decomposition of the non-simplicial face into simplicial components, so such asymptotic series can ultimately be fused into a series with a common asymptotic scale. The principle of one series per face of the Newton diagram thus survives (as long as the poles encountered in the development of the expansion remain distinct).

The link between remoteness of the Newton diagram and the order of the leading term of the asymptotic expansion of a multidimensional integral of Laplace type continues to hold, although a formal study of this relationship is fraught with greater complexity, owing to the greater variety of possible configurations for the Newton diagram in higher-dimensional cases. An account of remoteness for treble integrals with a single internal point serving as a vertex of the Newton diagram is supplied in Kaminski & Paris (1998b, §2).

A geometric interpretation of the asymptotic scales arising from each face of the Newton diagram, by analogy with §7.5.2, can also be constructed, with suitable reformulations of the decomposition of the Newton diagram and the convex solid formed with the back face. In short, instead of using ratios of areas of various triangles, one uses volumes of simplicial solids of appropriate dimension. However, the effort one must expend in the visualisation process in three or more dimensions, and the greater variety of possible shapes for the Newton diagram available to consider, make this extension an exercise of more limited utility.

A linear-algebraic formulation of the computations used to develop the asymptotic expansion of treble and higher-dimensional integrals is also available, paralleling the situation for double integrals of Laplace type. For details, the reader is directed to Kaminski & Paris (1998b, §5).

In any event, if an asymptotic expansion of a Laplace-type integral with the phase of the sort we have examined here is required, with no need to capture the geometry of the Newton diagram, then the application of the method is a much simpler matter.

7.7 Numerical Examples

We gather together here some computational illustrations of the expansions that we have obtained in the course of the investigations of this chapter. These examples showcase the expansions in a variety of settings, including their use in wider sectors of convergence with the large parameter $\lambda \to \infty$ in something other than the direction of the positive real axis, in illustrating how one deals with cases when the sequences of poles share points and in revealing how increasing symmetry of the

Newton diagram spawns a higher frequency of logarithmic terms present in the expansions.

Example 1. We consider the integral with a single internal point, i.e., $k = 1$ in (7.3.3), given by

$$I(\lambda) = \int_0^\infty \int_0^\infty \exp\{-\lambda(x^\mu + x^{m_1} y^{n_1} + y^\nu)\}\, dx\, dy, \qquad (7.7.1)$$

where we assume that $\delta_1 = 1 - m_1/\mu - n_1/\nu > 0$, i.e., the vertex $P_1 = (m_1, n_1)$ lies in front of the back face of the Newton diagram joining the points $(\mu, 0)$ and $(0, \nu)$. Without loss of generality we have put the constant c_1 multiplying the term $x^{m_1} y^{n_1}$ equal to unity, since a simple scaling of the variables x, y and the parameter λ enables the general case to be expressed in the above form. From (7.3.6) this has the Mellin-Barnes integral representation

$$I(\lambda) = \frac{\lambda^{-1/\mu - 1/\nu}}{\mu\nu} \frac{1}{2\pi i} \int_{-\infty i}^{\infty i} \Gamma(t)\Gamma\left(\frac{1 - m_1 t}{\mu}\right)\Gamma\left(\frac{1 - n_1 t}{\nu}\right)\lambda^{-\delta_1 t}\, dt \qquad (7.7.2)$$

valid in the sector $|\arg\lambda| < \frac{1}{2}\pi(1 + m_1/\mu + n_1/\nu)/\delta_1$; see (7.3.8).

When all the poles of the integrand in (7.7.2) are simple, the asymptotic expansion of $I(\lambda)$ is readily developed. In the integrand of (7.7.2) let us set $t = \rho e^{i\theta}$, $|\theta| < \frac{1}{2}\pi$. By employing the estimates (2.4.4), we find that the dominant real part of the logarithm of the integrand of (7.7.2) behaves as

$$\rho\cos\theta\log\rho \cdot \left(1 - \frac{m_1}{\mu} - \frac{n_1}{\nu}\right) = \delta_1\rho\cos\theta\log\rho$$

for large ρ. This tends to $+\infty$ as $\rho \to \infty$, from which we conclude that contributions to the asymptotic behaviour of (7.7.2) result from poles obtained by displacing the integration contour to the right. Candidate poles arise in two sequences: one from poles of $\Gamma((1 - m_1 t)/\mu)$, and one from poles of $\Gamma((1 - n_1 t)/\nu)$. For the present, we will assume that these two sequences share no points, so that all such poles are simple.

The poles of $\Gamma((1 - m_1 t)/\mu)$ are easily seen to occur at the points

$$t^{(1)} = (1 + \mu k)/m_1 \qquad (k = 0, 1, 2, \ldots), \qquad (7.7.3)$$

while poles of $\Gamma((1 - n_1 t)/\nu)$ occur at the points

$$t^{(2)} = (1 + \nu k)/n_1 \qquad (k = 0, 1, 2, \ldots). \qquad (7.7.4)$$

Poles from the $t^{(1)}$ sequence give rise to the formal asymptotic series

$$I_1 = \frac{\lambda^{-1/\mu - 1/\nu}}{m_1 \nu} \sum_k \frac{(-)^k}{k!}\Gamma\left(\frac{1 + \mu k}{m_1}\right)\Gamma\left(\frac{m_1 - n_1(1 + \mu k)}{m_1 \nu}\right)\lambda^{-\delta_1(1 + \mu k)/m_1},$$

$$(7.7.5)$$

whereas poles from the $t^{(2)}$ sequence give rise to the asymptotic series

$$I_2 = \frac{\lambda^{-1/\mu - 1/\nu}}{\mu n_1} \sum_k \frac{(-)^k}{k!} \Gamma\left(\frac{1 + \nu k}{n_1}\right) \Gamma\left(\frac{n_1 - m_1(1 + \nu k)}{\mu n_1}\right) \lambda^{-\delta_1 (1 + \nu k)/n_1};$$

(7.7.6)

in each of the formal asymptotic sums constituting I_1 and I_2, the index k ranges over all nonnegative integers.

Upon assembling these two asymptotic series, we obtain the asymptotic expansion, for $\lambda \to \infty$,

$$I(\lambda) \sim \lambda^{-1/\mu - 1/\nu} (I_1 + I_2), \tag{7.7.7}$$

where $I(\lambda)$, I_1 and I_2 are given in (7.7.2), (7.7.5) and (7.7.6), respectively. In the case $\mu = \nu = 3$ and $m_1 = \frac{3}{2}$, $n_1 = 1$, (7.7.7) yields the expansion

$$I(\lambda) \sim \tfrac{2}{9} \lambda^{-7/9} \sum_{k=0}^{\infty} \frac{(-)^k}{k!} \Gamma\left(\tfrac{2}{3} + 2k\right) \Gamma\left(\tfrac{1}{9} - \tfrac{2}{3}k\right) \lambda^{-k/3}$$

$$+ \tfrac{1}{3} \lambda^{-5/6} \sum_{k=0}^{\infty} \frac{(-)^k}{k!} \Gamma(1 + 3k) \Gamma\left(-\tfrac{1}{6} - \tfrac{3}{2}k\right) \lambda^{-k/2}$$

associated with the scales $\lambda^{-1/3}$ and $\lambda^{-1/2}$, respectively.

In Table 7.2 we present the results of numerical calculations for different values of the parameters μ, ν, m_1 and n_1. The second column shows the value of $I(\lambda)$ calculated either from (7.7.1) by standard numerical quadrature or from its equivalent representation as a generalised hypergeometric function.† The third column shows the asymptotic value of $I(\lambda)$ obtained by optimal truncation of each asymptotic series, that is, truncation just before the numerically smallest term.

An example of an integral of type (7.7.1) involving double poles is given by $\mu = \nu = 5$ and $m_1 = n_1 = 2$ for which the expansion (7.7.7) becomes nugatory. In this case the integrand in (7.7.2) possesses a sequence of double poles at $t = \frac{1}{2} + \frac{5}{2}k$, $k = 0, 1, 2, \ldots$. The residues at these poles are given by the coefficient of x^{-1} in the Maclaurin expansion of

$$\Gamma\left(\tfrac{1}{2} + \tfrac{5}{2}k + x\right) \Gamma^2\left(-k - \tfrac{2}{5}x\right) \lambda^{-k/2 - 1/10 - x/5}$$

$$= \frac{\lambda^{-k/2 - 1/10} \Gamma\left(\tfrac{1}{2} + \tfrac{5}{2}k\right) \left\{1 + \psi\left(\tfrac{1}{2} + \tfrac{5}{2}k\right)x + \cdots\right\}\left\{1 - \tfrac{x}{5}\log\lambda + \cdots\right\}}{(2x/5)^2 (k!)^2 \left\{1 + \tfrac{2}{5}\psi(1 + k)x + \cdots\right\}^2}$$

---

† Such a representation follows from (7.7.2) by displacing the integration contour to the *left*, so as to obtain the small-λ expansion given by

$$I(\lambda) = \frac{\lambda^{-1/\mu - 1/\nu}}{\mu \nu} \sum_{k=0}^{\infty} \frac{(-)^k}{k!} \Gamma\left(\frac{1 + m_1 k}{\mu}\right) \Gamma\left(\frac{1 + n_1 k}{\nu}\right) \lambda^{\delta_1 k} \qquad (\lambda \neq 0).$$

Table 7.2. *Comparison of the asymptotic*
values of $I(\lambda)$ with one internal point

$\mu = v = 3$ $\quad m_1 = \frac{3}{2}$, $n_1 = 1$ $\quad \delta_1 = \frac{1}{6}$		
λ	$I(\lambda)$	Asymptotic value
1.0×10^2	0.02643 82192	0.01952 49523
1.0×10^3	0.00511 42751	0.00507 50847
5.0×10^3	0.00160 24135	0.00160 24277
1.0×10^4	0.00096 91553	0.00096 91550
5.0×10^4	0.00029 95087	0.00029 95086

$\mu = v = 4$ $\quad m_1 = \frac{1}{2}$, $n_1 = \frac{4}{5}$ $\quad \delta_1 = \frac{27}{40}$		
λ	$I(\lambda)$	Asymptotic value
1.0×10^1	0.07981 13640	0.05396 33427
2.0×10^1	0.03703 09129	0.03707 77826
5.0×10^1	0.01235 59580	0.01235 59563
8.0×10^1	0.00686 99475	0.00686 99466
1.0×10^2	0.00517 85557	0.00517 85557

$\mu = v = 5$ $\quad m_1 = n_1 = 2$ $\quad \delta_1 = \frac{1}{5}$		
λ	$I(\lambda)$	Asymptotic value
1.0×10^2	0.11047 54348	0.10919 62332
5.0×10^2	0.05523 15228	0.05520 87617
1.0×10^3	0.04087 48889	0.04087 42206
1.0×10^4	0.01488 32370	0.01488 32334
5.0×10^4	0.00728 03604	0.00728 03598

$$
= \tfrac{25}{4} \lambda^{-k/2 - 1/10} \frac{\Gamma(\frac{1}{2} + \frac{5}{2}k)}{x^2 (k!)^2} \Big\{ 1 + x \big[\psi(\tfrac{1}{2} + \tfrac{5}{2}k)
$$

$$
- \tfrac{1}{5} \log \lambda - \tfrac{4}{5} \psi(1 + k) \big] + \cdots \Big\},
$$

where ψ denotes the logarithmic derivative of the gamma function. The asymptotic expansion of $I(\lambda)$ in this case therefore consists of the single expansion

$$
I(\lambda) \sim \frac{\lambda^{-1/2}}{20} \sum_{k=0}^{\infty} \frac{\Gamma(\frac{1}{2} + \frac{5}{2}k)}{(k!)^2} \big\{ \log \lambda + 4\psi(1 + k) - 5\psi(\tfrac{1}{2} + \tfrac{5}{2}k) \big\} \lambda^{-k/2}
$$

as $\lambda \to \infty$.

It can be observed in the examples chosen that there is a considerable variation in the asymptotic scales according to the proximity of the vertex P_1 to the back face. This results in a rather wide range of λ-values necessary to achieve an adequate description of the behaviour of $I(\lambda)$ by the corresponding asymptotic formula. In the symmetrical case with $\mu = v$ and $m_1 = n_1$, all the poles in the integrand in (7.7.2) are double and each term in the expansion involves a term in log λ. If we relax the degree of symmetry of the Newton diagram by taking $\mu \neq \nu$ (but with m_1, n_1 still situated on the 45° line), it is then easily seen (when μ/ν is rational) that both single and double poles can arise, with the first pole of the sequence always being double. This means that, although the leading term again contains log λ, the other logarithmic contributions now appear as higher order terms distributed amongst the algebraic terms in the expansion. When there is no longer any symmetry in the Newton diagram (i.e., when $\mu \neq \nu$ and $m_1 \neq n_1$), the poles can either be all simple (as in the first two examples in Table 7.2) or there can be double poles embedded within the sequence of single poles. In either case, however, the leading term of the expansion in the unsymmetric case is always algebraic.

Finally, as mentioned in §7.3.1, the expansion (7.7.7) (and the corresponding expansion involving log λ in the case of double poles) is valid for complex λ in a sector enclosing the rays $\arg \lambda = \pm\frac{1}{2}\pi$. To illustrate this fact we show in Table 7.3 the values of $I(\lambda)$ when $\mu = \nu = 5$ and $m_1 = 1$, $n_1 = \frac{1}{2}$ computed from its generalised hypergeometric representation (see the footnote on p. 342) for varying phase of λ with fixed $|\lambda|$. In this case the sector of validity of the integral (7.7.2), and hence that of the expansion (7.7.7) (in the sense of Poincaré), is $|\arg \lambda| < 13\pi/14$. We remark that $\arg \lambda = \pm\frac{1}{2}\pi$ corresponds to a Fourier integral with a single critical point at the origin.

Example 2. As an example of an integral with two internal points, we consider

$$I(\lambda) = \int_0^\infty \int_0^\infty \exp\{-\lambda(x^\mu + c_1 x^{m_1} y^{n_1} + c_2 x^{m_2} y^{n_2} + y^\nu)\}\, dx\, dy,$$

where the parameters will be chosen to correspond to a convex Newton diagram. From (7.3.6) with $k = 2$, $I(\lambda)$ has the Mellin-Barnes integral representation

$$\frac{\lambda^{-1/\mu - 1/\nu}}{\mu\nu} \left(\frac{1}{2\pi i} \int_{-\infty i}^{\infty i}\right)^2 \Gamma(t_1)\Gamma(t_2)\Gamma\left(\frac{1 - m_1 t_1 - m_2 t_2}{\mu}\right)\Gamma\left(\frac{1 - n_1 t_1 - n_2 t_2}{\nu}\right)$$

$$\times c_1^{-t_1} c_2^{-t_2} \lambda^{-\delta_1 t_1 - \delta_2 t_2}\, dt_1\, dt_2 \qquad\qquad (7.7.8)$$

Table 7.3. *Comparison of the asymptotic values of $I(\lambda)$ with a single internal point for complex λ when $\mu = \nu = 5$, $m_1 = 1$, $n_1 = \frac{1}{2}$ $(\delta_1 = \frac{7}{10})$ and $\lambda = 50e^{i\theta}$*

θ/π	$I(\lambda)$	Asymptotic value
0.0	$0.0157804781 + 0.0000000000i$	$0.0157804780 + 0.0000000000i$
0.1	$0.0148190028 - 0.0055301069i$	$0.0148190025 - 0.0055301067i$
0.2	$0.0120210294 - 0.0104467200i$	$0.0120210301 - 0.0144672144i$
0.3	$0.0076478949 - 0.0141701485i$	$0.0076478957 - 0.0141701492i$
0.4	$0.0021395833 - 0.0161966016i$	$0.0021395841 - 0.0161965996i$
0.5	$-0.0038877161 - 0.0161567121i$	$-0.0038877151 - 0.0161567144i$

defined in the sector (see (7.3.8))

$$|\arg \lambda| < \min_{i=1,2} \left\{ \tfrac{1}{2}\pi(1 + m_i/\mu + n_i/\nu)/\delta_i \right\},$$

when c_1, c_2 are assumed to be positive.

When all the poles of the integrand in (7.7.8) are simple, the asymptotic expansion of $I(\lambda)$ consists of three series given by (7.4.24) in the case $c_1 = c_2 = 1$. An illustration of such a situation is $\mu = 5$, $\nu = 4$ with $(m_1, n_1) = (\tfrac{5}{2}, 1)$ and $(m_2, n_2) = (1, 2)$. In this case we have $\delta_1 = \tfrac{1}{4}$, $\delta_2 = \tfrac{3}{10}$, $\Delta = 4$ and $\Delta - \mu N = -1$, $\Delta - \nu M = -2$ so that, from (7.4.4), the associated Newton diagram is convex. The expansion of $I(\lambda)$ then takes the form

$$I(\lambda) \sim \lambda^{-9/20} \left\{ \tfrac{1}{4} \sum_{k,l} f_{11}(k, l) + \tfrac{1}{4} \sum_{k,l} f_{12}(k, l) + \tfrac{1}{5} \sum_{k,l} f_{21}(k, l) \right\} \qquad (7.7.9)$$

as $\lambda \to \infty$, where $k, l = 0, 1, 2, \ldots$ and, from (7.4.13), (7.4.14), (7.4.16), (7.4.17), (7.4.21) and (7.4.22),

$$f_{11}(k, l) = \frac{(-)^{k+l}}{k! \, l!} \Gamma(1 + 5k + \tfrac{5}{2}l) \Gamma(-\tfrac{1}{4} - \tfrac{5}{2}k - l) \lambda^{-3/10 - 3k/2 - l/2},$$

$$f_{12}(k, l) = \frac{(-)^{k+l}}{k! \, l!} \Gamma(\tfrac{1}{4} + \tfrac{5}{2}k - l) \Gamma(\tfrac{3}{8} - \tfrac{5}{4}k + \tfrac{5}{2}l) \lambda^{-7/40 - k/4 - l/2},$$

$$f_{21}(k, l) = \frac{(-)^{k+l}}{k! \, l!} \Gamma(1 + 4k + 2l) \Gamma(-\tfrac{3}{10} - 2k - \tfrac{4}{5}l) \lambda^{-1/4 - k - l/5}.$$

When c_1, $c_2 \neq 1$, additional factors involving powers of c_1 and c_2 appear in each of the expansions in (7.7.9). We can restore the appropriate powers of c_1 and c_2 in these terms by observing that, from (7.7.9), each term must be accompanied by factors $c_1^{-t_1} c_2^{-t_2}$. Through the use of (7.4.6), (7.4.7), (7.4.11) and the fact that $t_1^{(1)} = -l, l = 0, 1, \ldots$ (cf. §7.4.1), the summands in (7.7.9) are then replaced by

$$f_{11}(k, l) c_1^l c_2^{-(1+5k+5l/2)},$$

$$f_{12}(k, l) c_1^{-(1/4+5k/2-l)} c_2^{-(3/8-5k/4+5l/2)},$$

$$f_{21}(k, l) c_1^{-(1+4k+2l)} c_2^l,$$

respectively. In Table 7.4 we present the results of numerical calculations of $I(\lambda)$ when $c_1 = c_2 = 1$ and $c_1 = 2$, $c_2 = 3$.

The second case we examine has $\mu = \nu = 4$, $c_1 = c_2 = 1$ with $(m_1, n_1) = (2, 1)$ and $(m_2, n_2) = (1, 2)$, so that the internal points are again in front of the back face but now straddle the 45° line in a symmetrical manner. Not unexpectedly, it will be found that the high degree of symmetry in the Newton diagram in this case will involve double poles which generate terms in $\log \lambda$. We follow the procedure

Table 7.4. *Comparison of the asymptotic values of*
$I(\lambda)$ with two internal points in the two-dimensional
case

$\mu = 5, \nu = 4$ $\quad (m_1, n_1) = (\frac{5}{2}, 1), \quad (m_2, n_2) = (1, 2)$ $c_1 = 1, c_2 = 1$		
λ	$I(\lambda)$	Asymptotic value
1.0×10^3	0.01649 86855	0.01530 34083
5.0×10^3	0.00660 18949	0.00662 55882
1.0×10^4	0.00442 75336	0.00442 22631
1.0×10^5	0.00115 31511	0.00115 33535
5.0×10^5	0.00044 41628	0.00044 40313
1.0×10^6	0.00029 36705	0.00029 36729

$\mu = 5, \nu = 4$ $\quad (m_1, n_1) = (\frac{5}{2}, 1), \quad (m_2, n_2) = (1, 2)$ $c_1 = 2, c_2 = 3$		
λ	$I(\lambda)$	Asymptotic value
1.0×10^3	0.01078 82133	0.01228 82165
5.0×10^3	0.00418 68491	0.00421 90510
1.0×10^4	0.00277 62231	0.00278 25077
1.0×10^5	0.00070 10746	0.00070 10886
5.0×10^5	0.00026 56345	0.00026 56349
1.0×10^6	0.00017 45786	0.00017 45787

$\mu = \nu = 4$ $\quad (m_1, n_1) = (2, 1), \quad (m_2, n_2) = (1, 2)$ $c_1 = c_2 = 1$		
λ	$I(\lambda)$	Asymptotic value
1.0×10^3	0.01185 73873	0.01173 42624
5.0×10^3	0.00447 05490	0.00447 33081
1.0×10^4	0.00292 42695	0.00292 60266
1.0×10^5	0.00070 23219	0.00070 23687
5.0×10^5	0.00025 58167	0.00025 58160
1.0×10^6	0.00016 51223	0.00016 51222

outlined in §7.4.1 where the t_2 contour in (7.7.8) is displaced to the right to yield
the result

$$I(\lambda) = \lambda^{-1/2}(I_1 + I_2), \qquad (7.7.10)$$

where, from (7.4.8) and (7.4.9),

$$I_1 = \frac{1}{4} \sum_k \frac{(-)^k}{k!} \lambda^{-k-1/4} \frac{1}{2\pi i} \int_{-\infty i}^{\infty i} \Gamma(t_1)\Gamma(1 + 4k - 2t_1)$$
$$\times \Gamma\left(-\tfrac{1}{4} - 2k + \tfrac{3}{4}t_1\right)\lambda^{t_1/4}dt_1,$$

$$I_2 = \frac{1}{8} \sum_k \frac{(-)^k}{k!} \lambda^{-k/2-1/8} \frac{1}{2\pi i} \int_{-\infty i}^{\infty i} \Gamma(t_1) \Gamma\left(\frac{1}{2} + 2k - \frac{1}{2}t_1\right)$$

$$\times \Gamma\left(\frac{1}{8} - \frac{1}{2}k - \frac{3}{8}t_1\right) \lambda^{-t_1/8} dt_1.$$

As shown in §7.4.1, the t_1 contour in each integral in I_1 must be displaced to the left. We thus encounter two sequences of poles at $t_1^{(1)} = 0, -1, -2, \ldots$ and $t_1^{(2)} = \frac{1}{3} + \frac{8}{3}k - \frac{4}{3}l$, $l = 0, 1, 2, \ldots$, where the latter sequence is subject to the constraint $4l \geq 8k + 1$ to ensure that $t_1^{(2)} \leq 0$. When $l = 1 + 2k + 3n$, $n = 0, 1, 2, \ldots$, the corresponding poles in the $t_1^{(2)}$ sequence coincide with poles of the $t_1^{(1)}$ sequence to form double poles at $t = -1 - 4n$ with residue

$$\frac{4}{3}(-)^n \frac{\Gamma(3 + 4k + 8n)}{\Gamma(2 + 4n)\Gamma(2 + 2k + 3n)} \lambda^{-k-n-1/2}$$

$$\times \left\{\frac{1}{4}\log\lambda - 2\psi(3 + 4k + 8n) + \psi(2 + 4n) + \frac{3}{4}\psi(2 + 2k + 3n)\right\}.$$

$$(7.7.11)$$

Evaluation of the contribution from the simple poles is as described in §7.4.1 and we consequently find that for large λ

$$I_1 \sim \frac{1}{4}\lambda^{-1/4} \sum_{\substack{k=0 \\ l \neq 1+4n}}^{\infty} \sum_{l=0}^{\infty} f_{11}(k, l) + \frac{1}{3}\lambda^{-1/6} \sum_{\substack{k=0 \\ l \neq 1+2k+3n \\ 4l \geq 8k+1}}^{\infty} \sum_{l=0}^{\infty} f_{12}(k, l)$$

$$+ \frac{1}{3}\lambda^{-1/2} \sum_{k=0}^{\infty} \sum_{l=0}^{\infty} g(k, l),$$

where n is a nonnegative integer and, from (7.4.13), (7.4.14), (7.4.16) and (7.4.17)

$$f_{11}(k, l) = \frac{(-)^{k+l}}{k! \, l!} \Gamma(1 + 4k + 2l)\Gamma(-\tfrac{1}{4} - 2k - \tfrac{3}{4}l)\lambda^{-k-l/4},$$

$$f_{12}(k, l) = \frac{(-)^{k+l}}{k! \, l!} \Gamma(\tfrac{1}{3} + \tfrac{8}{3}k - \tfrac{4}{3}l)\Gamma(\tfrac{1}{3} - \tfrac{4}{3}k + \tfrac{8}{3}l)\lambda^{-k/3-l/3}$$

and

$$g(k, l) = \frac{(-)^{k+l}}{k!} \frac{\Gamma(3 + 4k + 8l)}{\Gamma(2 + 4l)\Gamma(2 + 2k + 3l)} \lambda^{-k-l}$$

$$\times \left\{\frac{1}{4}\log\lambda - 2\psi(3 + 4k + 8l) + \psi(2 + 4l) + \frac{3}{4}\psi(2 + 2k + 3l)\right\}.$$

Proceeding in a similar fashion for I_2 upon displacement of the t_1 contour to the right, we encounter two sequences of poles at $t_1^{(1)} = 1 + 4k + 2l$ and $t_1^{(2)} = \frac{1}{3} - \frac{4}{3}k + \frac{8}{3}l$, $l = 0, 1, 2, \ldots$, where the latter sequence is subject to the constraint $4k < 8l + 1$ to ensure that $t_1^{(2)} > 0$. Double poles arise at $t = 3 + 4k + 8n$, $n = 0, 1, 2, \ldots$, corresponding to the values of l in $t_1^{(2)}$ given by $l = 1 + 2k + 3n$, with residues equal to a factor of -2 times the value in (7.7.11). Hence, as $\lambda \to \infty$, we find

$$I_2 \sim \tfrac{1}{4}\lambda^{-1/4} \sum_{\substack{k=0 \\ l\neq 1+4n}}^{\infty} \sum_{l=0}^{\infty} f_{11}(k,l) + \tfrac{1}{3}\lambda^{-1/6} \sum_{\substack{k=0 \\ l\neq 1+2k+3n \\ 4k<8l+1}}^{\infty} \sum_{l=0}^{\infty} f_{12}(l,k)$$

$$+ \tfrac{1}{3}\lambda^{-1/2} \sum_{k=0}^{\infty} \sum_{l=0}^{\infty} g(k,l).$$

Collecting together I_1 and I_2 according to (7.7.10), we find that the series involving $\lambda^{-1/4}$ and $\lambda^{-1/2}$ are equal while those involving $\lambda^{-1/6}$ yield

$$\tfrac{1}{3}\lambda^{-1/6}\left\{ \sum_{\substack{k=0 \\ l\neq 1+2k+3n \\ 4l\geq 8k+1}}^{\infty} \sum_{l=0}^{\infty} f_{12}(k,l) + \sum_{\substack{k=0 \\ l\neq 1+2k+3n \\ 4k<8l+1}}^{\infty} \sum_{l=0}^{\infty} f_{12}(l,k) \right\}.$$

If we relabel the summation indices by putting $k \to l$ and $l \to k$ in the second expansion, we see that the constraint $4l \geq 8k+1$ on the first expansion is exactly the complement of the constraint $4l < 8k + 1$ on the second expansion. Consequently, the two expansions can be expressed as a single expansion subject only to the constraints $k \neq 1 + 2l + 3n$ and $l \neq 1 + 2k + 3n$.

Hence, we finally obtain the expansion of $I(\lambda)$ in the form (for $n = 0, 1, 2, \ldots$)

$$I(\lambda) \sim \tfrac{1}{2}\lambda^{-3/4} \sum_{\substack{k=0 \\ l\neq 1+4n}}^{\infty} \sum_{l=0}^{\infty} f_{11}(k,l) + \tfrac{1}{3}\lambda^{-2/3} \sum_{\substack{k=0 \\ k\neq 1+2l+3n \\ l\neq 1+2k+3n}}^{\infty} \sum_{l=0}^{\infty} f_{12}(k,l)$$

$$+ \tfrac{2}{3}\lambda^{-1} \sum_{k=0}^{\infty} \sum_{l=0}^{\infty} g(k,l) \tag{7.7.12}$$

as $\lambda \to \infty$. We note that the restrictions appearing on the first two expansions in (7.7.12) simply correspond to the deletion from these sums of the singular terms resulting from the double poles. The numerical values of $I(\lambda)$ in this case are shown in the last entry in Table 7.4 and are compared with the asymptotic values computed from (7.7.12).

Example 3. We consider the integral with a single internal point given by

$$I(\lambda) = \int_0^\infty \int_0^\infty \int_0^\infty \exp\{-\lambda(x^\mu + y^\nu + z^\eta + x^{m_1}y^{n_1}z^{p_1})\}\,dx\,dy\,dz, \tag{7.7.13}$$

where we assume that $\delta_1 = 1 - m_1/\mu - n_1/\nu - p_1/\eta > 0$, so that the vertex $P_1 = (m_1, n_1, p_1)$ lies in front of the back face of the Newton diagram, and without loss of generality we have put the constant $c_1 = 1$. From (7.6.4), this has the Mellin-Barnes integral representation

$$
I(\lambda) = \frac{\lambda^{-1/\mu-1/\nu-1/\eta}}{\mu\nu\eta} \frac{1}{2\pi i} \int_{-\infty i}^{\infty i} \Gamma(t)\Gamma\left(\frac{1-m_1 t}{\mu}\right)
$$

$$
\times \Gamma\left(\frac{1-n_1 t}{\nu}\right)\Gamma\left(\frac{1-p_1 t}{\eta}\right)\lambda^{-\delta_1 t}dt, \quad (7.7.14)
$$

which provides the analytic continuation of $I(\lambda)$ from the half-plane $\mathrm{Re}(\lambda) > 0$ to a sector including the Fourier case with $\arg \lambda = \pm\frac{1}{2}\pi$. Although we are primarily concerned here with positive real λ, we remark that, like the integral (7.7.2), the asymptotic expansions developed for (7.7.14) are valid (in the Poincaré sense) for $\lambda \to \infty$ in the sector $|\arg \lambda| < \frac{1}{2}\pi(1 + m_1/\mu + n_1/\nu + p_1/\eta)/\delta_1$; see (7.3.8).

When all the poles of the integrand in (7.7.14) are simple, the expansion of $I(\lambda)$ is given by (7.6.10). As an example of this simple-pole structure, we show in Table 7.5 the results of numerical computations for the particular case $\mu = \nu = \eta = 4$, with $m_1 = \frac{2}{5}$, $n_1 = \frac{1}{2}$ and $p_1 = \frac{3}{5}$. The second column shows the value of $I(\lambda)$ determined by numerical evaluation of the integral (7.7.13) or from the representation of $I(\lambda)$ as a generalised hypergeometric function; see the footnote on p. 342. The third column shows the asymptotic value of $I(\lambda)$ obtained by optimally truncating each constituent asymptotic series in the expansion.

As mentioned in §7.6.2, the incidence of higher order poles depends in part on the degree of symmetry of the Newton diagram. To illustrate this fact, let us first consider the most symmetrical situation with $\mu = \nu = \eta$ and $m_1 = n_1 = p_1$, so that the back face is an equilateral triangle and the vertex P_1 lies on the $(1,1,1)$ line. In this case the Mellin-Barnes integral in (7.7.14) becomes

$$
I(\lambda) = \frac{\lambda^{-3/\mu}}{\mu^3} \frac{1}{2\pi i} \int_{-\infty i}^{\infty i} \Gamma(t)\Gamma^3\left(\frac{1-mt}{\mu}\right)\lambda^{-\delta t}dt, \quad \delta = 1 - 3m/\mu,
$$

where, for convenience, we have momentarily set $m_1 = m$ and $\delta_1 = \delta$. The three sequences of poles in the right half-plane in the general case have now collapsed into a single sequence of treble poles situated at $t = (1+\mu k)/m$, $k = 0, 1, 2, \ldots$. The residues are given by the coefficient of x^{-1} in the Maclaurin expansion of

$$
\Gamma\left(\frac{1+\mu k}{m} + x\right)\Gamma^3\left(-k - \frac{mx}{\mu}\right)\lambda^{-\delta x - \delta(1+\mu k)/m}
$$

$$
= \pi^3(-)^{k-1}\lambda^{-\delta(1+\mu k)/m} \times \frac{\lambda^{-\delta x}\Gamma(x + (1+\mu k)/m)}{\sin^3(\pi mx/\mu)\,\Gamma^3(k+1+mx/\mu)};
$$

that is, by

$$
(-)^{k-1}\frac{(\mu/m)^3}{2(k!)^3}\Gamma\left(\frac{1+\mu k}{m}\right)\lambda^{-\delta(1+\mu k)/m}C_k(\lambda),
$$

Table 7.5. *Comparison of the asymptotic values of $I(\lambda)$ with one internal point in the three-dimensional case*

$\mu = \nu = \eta = 4$	$m_1 = \frac{2}{5},\ n_1 = \frac{1}{2},\ p_1 = \frac{3}{5}$	$\delta_1 = \frac{5}{8}$
λ	$I(\lambda)$	Asymptotic value
2.0×10^1	0.02234 75076	0.02155 91346
4.0×10^1	0.00876 25593	0.00876 06317
5.0×10^1	0.00639 39397	0.00639 38264
6.0×10^1	0.00492 03593	0.00492 02589
8.0×10^1	0.00322 97878	0.00322 97876

$\mu = \nu = \eta = 4$	$m_1 = 1,\ n_1 = 1,\ p_1 = 1$	$\delta_1 = \frac{1}{4}$
λ	$I(\lambda)$	Asymptotic value
1.0×10^2	0.01775 29863	0.01729 72404
2.0×10^2	0.01013 84495	0.01005 67565
5.0×10^2	0.00480 37523	0.00480 91387
1.0×10^3	0.00271 67652	0.00271 69815
5.0×10^3	0.00071 15152	0.00071 15151

$\mu = \nu = \eta = 3$	$m_1 = \frac{1}{2},\ n_1 = \frac{1}{2},\ p_1 = \frac{1}{2}$	$\delta_1 = \frac{1}{2}$
λ	$I(\lambda)$	Asymptotic value
5.0×10^1	0.00371 27773	0.00370 54488
8.0×10^1	0.00185 42982	0.00185 45509
1.0×10^2	0.00132 44252	0.00132 44556
1.5×10^2	0.00071 09237	0.00071 09234
2.0×10^2	0.00045 36285	0.00045 36285

$\mu = \nu = \eta = 5$	$m_1 = n_1 = 1,\ p_1 = \frac{1}{2}$	$\delta_1 = \frac{1}{2}$
λ	$I(\lambda)$	Asymptotic value
5.0×10^1	0.03756 74263	0.03755 80382
8.0×10^1	0.02537 48359	0.02537 46738
1.0×10^2	0.02099 22663	0.02099 25949
1.5×10^2	0.01479 77289	0.01479 77269
2.0×10^2	0.01150 23481	0.01150 23481

where

$$C_k(\lambda) = (\delta \log \lambda)^2 - 2\delta \log \lambda \left\{ \psi\left(\frac{1+\mu k}{m}\right) - \frac{3m}{\mu}\psi(1+k) \right\}$$

$$+ \psi'\left(\frac{1+\mu k}{m}\right) + \psi^2\left(\frac{1+\mu k}{m}\right)$$

$$- \frac{6m}{\mu}\psi(1+k)\psi\left(\frac{1+\mu k}{m}\right) + 3\left(\frac{m}{\mu}\right)^2 \{3\psi^2(1+k)$$

$$- \psi'(1+k)\} + (\pi m/\mu)^2$$

and ψ denotes the logarithmic derivative of the gamma function. When $\delta_1 > 0$, the expansion of $I(\lambda)$ in this case is therefore given by

$$I(\lambda) \sim \frac{\lambda^{-1/m}}{2m^3} \sum_{k=0}^{\infty} \frac{(-)^k}{(k!)^3} \Gamma\left(\frac{1+\mu k}{m}\right) C_k(\lambda) \, \lambda^{-\{(\mu/m)-3\}k},$$

as $\lambda \to \infty$.

Examples of this highly symmetrical case with $\mu = 4$, $m = 1$ ($\delta_1 = \frac{1}{4}$) and $\mu = 3$, $m = \frac{1}{2}$ ($\delta_1 = \frac{1}{2}$) are also given in Table 7.5. It will be observed that the two cases differ mainly in the proximity of the vertex P_1 to the back face, as can be seen from the corresponding values of δ_1. This results in the expansions being associated with different asymptotic scales, and hence different ranges of λ values.

The final example we give shows the type of expansion which arises when the $(1,1,1)$ line intersects an edge of the Newton diagram. The situation considered again has a symmetrical back face ($\mu = \nu = \eta$), but now with $m_1 = n_1 > p_1$, so that the edge connecting the vertex P_1 with $(0, 0, \eta)$ meets the $(1,1,1)$ line. The integral (7.7.14) now possesses two sequences of poles in the right half-plane: a sequence of double poles at $t = (1 + \mu k)/m$ and a sequence of simple poles at $t = (1 + \mu k)/p$, $k = 0, 1, 2, \ldots$. Provided these two sequences have no point in common, the expansion of $I(\lambda)$ when $\delta_1 > 0$ is then given by

$$I(\lambda) \sim \frac{\lambda^{-3/\mu}}{\mu^2 p_1} \sum_{k=0}^{\infty} \frac{(-)^k}{k!} \Gamma\left(\frac{1+\mu k}{p_1}\right) \Gamma^2\left(\frac{p_1 - m_1 - m_1 \mu k}{\mu p_1}\right) \lambda^{-\delta_1(1+\mu k)/p_1}$$

$$+ \frac{\lambda^{-3/\mu}}{\mu m_1^2} \sum_{k=0}^{\infty} \frac{1}{(k!)^2} \Gamma\left(\frac{1+\mu k}{m}\right) \Gamma\left(\frac{m_1 - p_1 k}{\mu m_1}\right) \lambda^{-\delta_1(1+\mu k)/m_1}$$

$$\times \left\{ \delta_1 \log \lambda - \psi\left(\frac{1+\mu k}{m_1}\right) \right.$$

$$\left. + \frac{p_1}{\mu} \psi\left(\frac{m_1 - p_1 - p_1 \mu k}{\mu m_1}\right) + \frac{2m_1}{\mu} \psi(1+k) \right\} \qquad (7.7.15)$$

as $\lambda \to \infty$. An example of the expansion (7.7.15) is shown in Table 7.5 where $\mu = \nu = \eta = 5$ with $m_1 = n_1 = 1$ and $p_1 = \frac{1}{2}$ (so that $\delta_1 = \frac{1}{2}$). This separation of the poles will certainly exist for integer values of m_1 and p_1, but not necessarily for noninteger values. In this case the expansion (7.7.15) would be modified by the formation of a sequence of treble poles which will generate additional terms involving $(\log \lambda)^2$ in the expansion.

As the symmetry of the Newton diagram is reduced the higher order poles become progressively more sparsely distributed. Even when the $(1,1,1)$ line punctures a face of the Newton diagram (cf. the first example in Table 7.5 where all poles are simple), it is still possible to select values of the parameters which generate either double or treble poles, although the leading term of each expansion in these cases is always algebraic in λ.

8

Application to Some Special Functions

This final chapter is concerned with the application of the theory of the Mellin-Barnes integral developed in the earlier chapters. We draw examples from a selection of asymptotic problems that illustrate the power and usefulness of this technique.

The first example concerns the asymptotic expansion of the generalised Euler-Jacobi series given by

$$S_p(a) = \sum_{n=0}^{\infty} \exp(-an^p) \quad (p > 0),$$

as the parameter $a \to 0+$. We shall be particularly interested in the appearance of exponentially small expansions in this limit as p increases beyond the classical value $p = 2$. For this problem we present two approaches: the first uses the method of steepest descent applied to a certain integral arising in the analogue of the classical Poisson-Jacobi transformation when $p = 2$, and the second adopts a Mellin-Barnes integral description.

The second example concerns a new asymptotic formula for the Riemann zeta function $\zeta(s)$ high up on the so-called critical line $\mathrm{Re}(s) = \frac{1}{2}$. The classical expansions, known as the Gram and Riemann-Siegel formulas, have been known for the better part of a century, but recent work has shown that other, more accurate (although computationally more expensive) expansions exist. One such expansion, which we establish by means of a Mellin-Barnes integral, involves the terms in the original Dirichlet series (valid in $\mathrm{Re}(s) > 1$) 'smoothed' by the incomplete gamma function associated with a free parameter. Suitable choice of this parameter then enables us to establish expansions of different character.

The last example deals with a novel approach to the asymptotics of the well-known Pearcey integral arising in diffraction theory. This integral is a Fourier integral involving three coalescing stationary points and two variables. The standpoint of our discussion is based on a Mellin-Barnes integral representation of this integral which involves the Weber parabolic cylinder function. We show how the large-variable asymptotics of the Pearcey integral can be obtained without recourse to the stationary points.

8.1 Asymptotics of the Euler-Jacobi Series

8.1.1 Introduction

The generalised Euler-Jacobi series, defined by

$$S_p(a) = \sum_{n=0}^{\infty} e^{-an^p} \qquad (\operatorname{Re}(a) > 0, \quad p > 0), \qquad (8.1.1)$$

has applications in statistical mechanics and, when $p = 2$, in the theory of Jacobian theta functions and the Riemann zeta function. We mention that a related problem, when the parameter a is pure imaginary and $p = 2$ (with the upper limit in the sum in (8.1.1) replaced by the finite limit N), gives rise to the study of "curlicues" and finds application in optical diffraction when $N \gg 1$; see Dekking & Mendès-France (1981), Berry & Goldberg (1988).[†] A detailed discussion of the function $S_p(a)$ in the limit $a \to 0+$ has recently been given in the monograph by Kowalenko et al. (1995) for the case when p is a rational fraction. Here we shall consider the sum in (8.1.1) for arbitrary[‡] $p > 0$. The classical case with $p = 2$ yields the remarkable and well-known result

$$S_2(a) = \tfrac{1}{2}\sqrt{\frac{\pi}{a}} + \tfrac{1}{2} + \sqrt{\frac{\pi}{a}} \sum_{n=1}^{\infty} e^{-\pi^2 n^2/a} \qquad (\operatorname{Re}(a) > 0) \qquad (8.1.2)$$

called the Poisson-Jacobi transformation [Whittaker & Watson (1965), p. 124; see also §4.1.2]. This transformation enables the asymptotic behaviour of $S_p(a)$ to be determined in the limit $a \to 0$ in $\operatorname{Re}(a) > 0$, since the terms in the infinite sum then decay very rapidly with increasing n.

The analogous transformation of $S_p(a)$ for $p > 1$ can be obtained from Poisson's formula [Titchmarsh (1975), p. 61]. This states that if a function $f(t)$ is of bounded variation in $(0, \infty)$, is continuous§ and belongs to $L(0, \infty)$, then

$$\sqrt{\alpha}\left\{\tfrac{1}{2}f(0+) + \sum_{n=1}^{\infty} f(n\alpha)\right\} = \sqrt{\beta}\left\{\tfrac{1}{2}\mathcal{F}_c(0) + \sum_{n=1}^{\infty} \mathcal{F}_c(n\beta)\right\},$$

[†] A detailed study of this finite sum in a number theoretic context is given in a memoir by Hardy & Littlewood (1914).

[‡] When $p = 1$ we have the trivial result $S_1(a) = 1/(1 - e^{-a})$.

§ The formula can also be adapted to deal with functions $f(t)$ which possess finite discontinuities in $[0, \infty)$; see Titchmarsh [1975, Eq. (2.8.3)].

where $\alpha\beta = 2\pi$, $\alpha > 0$ and $\mathcal{F}_c(w)$ denotes the Fourier cosine transform given by

$$\mathcal{F}_c(w) = \sqrt{\frac{2}{\pi}} \int_0^\infty f(t) \cos wt \, dt.$$

We apply this result with $f(t) = \exp(-t^p)$, $\alpha = a^{1/p}$ and perform a straightforward evaluation of the associated Fourier cosine transform. This can be achieved by expansion of the cosine term followed by reversal of the order of summation and integration, and use of the duplication formula for the gamma function, to obtain

$$\sqrt{\frac{2}{\pi}} \int_0^\infty e^{-t^p} \cos wt \, dt = \frac{1}{p}\sqrt{\frac{2}{\pi}} \sum_{k=0}^\infty \frac{(-w^2)^k}{(2k)!} \int_0^\infty e^{-u} u^{(2k+1)/p-1} \, du$$

$$= \frac{1}{p}\sqrt{\frac{2}{\pi}} \sum_{k=0}^\infty \frac{(-w^2)^k}{(2k)!} \Gamma\left(\frac{2k+1}{p}\right) = \frac{2^{\frac{1}{2}}}{p} F_p(w),$$

$$(8.1.3)$$

where we have defined

$$F_p(z) = \sum_{k=0}^\infty \frac{(-)^k (\frac{1}{2}z)^{2k}}{k!\Gamma(k+\frac{1}{2})} \Gamma\left(\frac{2k+1}{p}\right) \qquad (p > 1; \, |z| < \infty). \qquad (8.1.4)$$

The function $F_p(z)$ is a generalised hypergeometric function and is uniformly and absolutely convergent for all finite $|z|$ when $p > 1$ [Braaksma (1963); see also §2.3].

Poisson's formula then yields the desired transformation in the form

$$S_p(a) = \frac{1}{2} + \frac{\Gamma(1/p)}{pa^{1/p}} + \frac{2\pi^{\frac{1}{2}}}{pa^{1/p}} \sum_{n=1}^\infty F_p(2\pi na^{-1/p}) \quad (p > 1). \qquad (8.1.5)$$

This relation has been given previously in Paris (1994a) in connection with an asymptotic expansion for the Riemann zeta function $\zeta(s)$. When $p = 2$, we have $F_2(z) = \exp(-\frac{1}{4}z^2)$ and (8.1.5) reduces to the classical result (8.1.2). The transformation (8.1.5) is equivalent to that obtained in (2.14) and (2.16) in Kowalenko et al. (1995), where the function $F_p(z)$ has effectively been expressed as a finite sum of P generalised hypergeometric functions (or Meijer G functions in the case of (2.16)) of the type $_Q F_{P-1}$, where $p = P/Q$ with P, Q denoting relatively prime integers. Thus, the equivalent transformation formula developed in Kowalenko et al. (1995) can only deal with values of p that are rational fractions, whereas the form (8.1.5) is much more compact and is valid for all $p > 1$. The case $0 < p < 1$ is dealt with in §8.1.4.

The dominant asymptotics of $S_p(a)$ as $a \to 0$ in $\mathrm{Re}(a) > 0$ can be deduced from (8.1.5) by employing the asymptotic expansion of $F_p(z)$ as $|z| \to \infty$ given in (2.3.10); see also Paris & Wood (1986, pp. 36–51). This is obtained from (2.3.6) (when we replace z in (2.3.10) by $(\frac{1}{2}z)^2 e^{\pm \pi i}$), followed by use of the reflection and duplication formulas for the gamma function, in the form

$$F_p(z) \sim \pi^{-\frac{1}{2}} p \sum_{k=1}^{\infty} \frac{(-)^{k-1}}{k!} \Gamma(kp+1) \sin(\tfrac{1}{2}\pi kp) z^{-kp-1} \qquad (8.1.6)$$

valid as $|z| \to \infty$ in the sector $|\arg z| < \pi/2p$. We remark that the expansion of $F_p(z)$ is exponentially large in the complementary sector, although we do not require its precise form here. The expansion (8.1.6) is valid in the sense of Poincaré only: exponentially subdominant terms present in $F_p(z)$ are not included in (8.1.6); see Kaminski & Paris (1996) and also the discussion surrounding (5.5.13).

Thus the desired asymptotic behaviour of $F_p(2\pi n a^{-1/p})$ appearing in the infinite sum in (8.1.5) is described by the above algebraic expansion with $z = 2\pi n a^{-1/p}$, since $|\arg z| = (1/p)|\arg a| < \pi/2p$. Upon evaluation of the resulting sum over n in terms of the zeta function, we finally obtain from (8.1.5)

$$S_p(a) - \frac{\Gamma(1/p)}{pa^{1/p}} - \tfrac{1}{2} \sim H_p(a) \qquad (p > 1), \qquad (8.1.7)$$

where $H_p(a)$ denotes the formal expansion

$$H_p(a) = \frac{1}{\pi} \sum_{k=1}^{\infty} \frac{(-)^{k-1}}{k!} \Gamma(kp+1) \sin(\tfrac{1}{2}\pi kp) \zeta(kp+1)(a/(2\pi)^p)^k. \qquad (8.1.8)$$

This result yields the dominant algebraic expansion of $S_p(a)$ as $a \to 0$ in $\mathrm{Re}(a) > 0$ and agrees with the Berndt-Ramanujan result [Berndt (1985)]. We remark however, as pointed out in Kowalenko et al. (1995), that when p is an even integer, $H_p(a)$ vanishes on account of the term $\sin(\tfrac{1}{2}\pi kp)$. This loss of asymptotic information when p is an even integer is the signal that there exist exponentially small terms which have not been accounted for when using the Poincaré asymptotics of $F_p(z)$ in (8.1.6). Such transcendentally small terms, which, until recently, have largely been neglected in asymptotics, are of vital importance for an accurate description of the function $S_p(a)$, particularly for small but finite values of a.

The method adopted in Kowalenko et al. (1995), which considers $a > 0$, consists of use of the asymptotic properties of the hypergeometric function ${}_P F_Q$ given in Luke (1969). Elaborate algebraic manipulation then removes the exponentially large terms present in these functions and enables recursion relations for the coefficients in the exponentially small expansions to be found. In our presentation here, we shall also restrict our attention to $a > 0$, although the methods used can be adapted to deal with complex a. We shall show how the asymptotics of $S_p(a)$ in the limit $a \to 0+$ can be derived by two different methods. The first uses the transformation formula (8.1.5) combined with a straightforward application of the method of steepest descent. This approach is included here as it is more familiar and proves to be instructive in shedding light on certain features of the asymptotic structure of $S_p(a)$: namely, the appearance of exponentially small expansions (manifested by a Stokes phenomenon) as the parameter p increases through certain values, and the disappearance of the algebraic asymptotic expansion for even integer values of p. The second approach, in keeping with the main theme of this book, shows how the asymptotic expansion can be derived more directly from suitable manipulation of a Mellin-Barnes integral representation involving the Riemann zeta function. It will be found that the role of the transformation formula (8.1.5) is now played by the well-known functional relation for $\zeta(s)$.

8.1.2 The Saddle Point Approach

We shall suppose throughout that $a > 0$ and introduce the notation

$$X_n = \kappa (hx)^{1/\kappa}, \quad x = 2\pi n a^{-1/p}, \quad A_0 = (2\pi)^{\frac{1}{2}} \kappa^{-\frac{1}{2}-\vartheta} p^{-\vartheta},$$

$$\kappa = (p-1)/p, \quad h = (1/p)^{1/p}, \quad \vartheta = (2-p)/2p. \tag{8.1.9}$$

We start with the transformation formula (8.1.5) that generalises the classical Poisson-Jacobi transformation in (8.1.2). Using (8.1.3), we write the function $F_p(x)$ as the integral†

$$F_p(x) = \pi^{-\frac{1}{2}} p \, \mathrm{Re}\{I_p(x)\}, \tag{8.1.10}$$

where

$$I_p(x) = \int_0^\infty e^{-t^p + ixt} dt = (hx^{1/p})^{1/\kappa} \int_0^\infty e^{-X_n f(\tau)} d\tau, \tag{8.1.11}$$

with $f(\tau) = (\tau^p/p - i\tau)/\kappa$. For $n = 1, 2, \ldots$, we see that $x \to +\infty$ as $a \to 0+$. As a consequence, the parameter X_n appearing in the exponential in (8.1.11) is large so that the asymptotics of the integral $I_p(x)$ can be obtained by the method of steepest descent.

Saddle points of $f(\tau)$ (corresponding to points where $f'(\tau) = 0$) are given by

$$\tau_r = \exp\left\{(2r + \tfrac{1}{2})\pi i/(p-1)\right\} \quad (r = 0, \pm 1, \pm 2, \ldots). \tag{8.1.12}$$

For integer values of p, there are exactly $p - 1$ saddles situated on the unit circle in the τ plane. When p is noninteger, however, the τ plane must be cut along the negative real axis: in this case there will, in general, be saddles situated on the principal sheet $-\pi < \arg \tau \le \pi$, with other saddles situated on adjacent Riemann sheets. When $1 < p < \frac{3}{2}$, there are no saddles situated on the principal sheet. As p increases from $\frac{3}{2}$, the saddle τ_0 crosses the branch cut $\arg \tau = \pi$ from the upper adjacent sheet onto the principal sheet and steadily moves round the unit circle, passing through the point $\tau = i$ when $p = 2$. As p continues to increase, a second saddle τ_{-1} moves off the lower adjacent sheet, crossing the branch cut $\arg \tau = -\pi$

---

† If a were complex then we would need to write $F_p(x) = \frac{1}{2}\pi^{-\frac{1}{2}} p\{I_p(x) + I_p(-x)\}$ and treat the integrals $I_p(\pm x)$ separately.

when $p = \frac{5}{2}$, and moves round the unit circle in the opposite sense. This pattern of oppositely-directed rotation of the saddles towards the point $\tau = 1$ continues as p increases further, with a new saddle crossing over (alternately from the upper and lower adjacent sheets) onto the principal sheet each time p equals a half-integer value.

The distribution of the saddles and the paths of steepest descent for $I_p(x)$ are illustrated in Fig. 8.1 for $1 < p < 3$. The valleys V_k (resp. ridges R_k) of $f(\tau)$, corresponding to the directions along which $f(\tau) \to +\infty$ (resp. $-\infty$) as $|\tau| \to \infty$, are given by the rays $\arg \tau = 2\pi k/p$ (resp. $\arg \tau = (2k+1)\pi/p$), where $k = 0, \pm 1, \pm 2, \ldots$. For $1 < p < 2$, the path of steepest descent from the origin on the principal sheet does not pass over a saddle, with the result that the path of integration in (8.1.11) can be reconciled with this contour. As $p \to 2-$, the

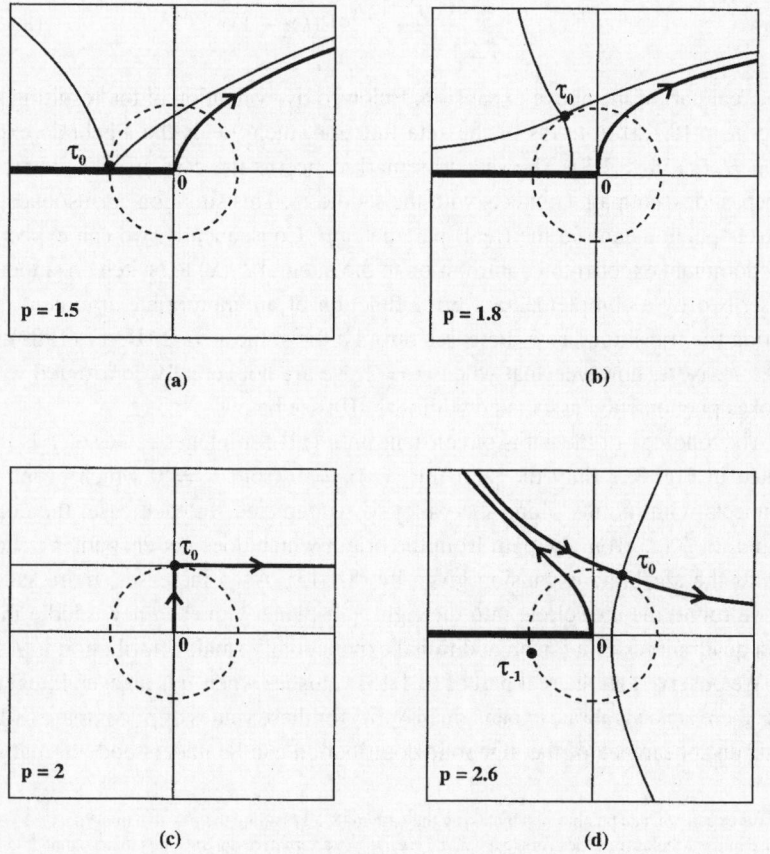

Fig. 8.1. The paths of steepest descent when $1 < p < 3$. The branch cut along the negative real axis is indicated by the heavy line. Arrows indicate the integration path.

path $\text{Im}(f(\tau)) = 0$ becomes progressively deformed in the neighbourhood of the saddle τ_0 until, when $p = 2$, it connects with this saddle. For $p > 2$, the path of steepest descent from the origin now passes to infinity down the valley V_1. In this case, the path of integration in (8.1.11) can be deformed to coincide with this path together with the path that connects V_1 to the adjacent valley V_0 passing over the saddle τ_0. In the usual interpretation of the different contributions resulting from such composite paths, the contribution to $I_p(x)$ from the path from the origin (not passing over a saddle) yields an algebraic expansion, while that from the path passing over the saddle τ_0 yields an exponential expansion. Most importantly, this latter contribution is exponentially small since the contributory saddle satisfies $\text{Re}(f(\tau_0)) = -\cos(\pi/2\kappa) > 0$.

From this discussion we can make two remarks. The first concerns the case $1 < p < 2$ where the steepest descent path does not pass over a saddle. In this case, $I_p(x)$ will consist of only an algebraic expansion given by†

$$I_p(x) \sim i \sum_{k=0}^{\infty} \frac{(-)^k}{k!} e^{\frac{1}{2}\pi i k p} \Gamma(kp+1) x^{-kp-1}. \tag{8.1.13}$$

The real part of the above expansion, followed by evaluation of the resulting sum over n in (8.1.5) in terms of the zeta function, then yields the algebraic expansion $H_p(a)$ in (8.1.8). The second remark concerns the case $p = 2$, where the steepest descent path connects with the saddle τ_0. This situation corresponds to a Stokes phenomenon in the (real) parameter p. Consequently, we can expect the subdominant exponential contribution to the integral $I_p(x)$ to switch on smoothly, described by a complementary error function of an appropriate argument measuring the transition, as p increases through the value $p = 2$ [Berry (1989); see §8.1.7]. Note, however, that when $p = 2$, we are not actually confronted with a Stokes phenomenon associated with (8.1.10); see below.

The topology of the paths of constant $\text{Im}(f(\tau))$ for integer values of p is illustrated in Fig. 8.2; only the path $\text{Im}(f(\tau)) = 0$ from $\tau = 0$ which eventually connects with infinity along the valley V_0 is depicted. In each case, the contribution to $I_p(x)$ from the path from the origin which does not encounter a saddle yields the algebraic expansion given by (8.1.13). As p increases, more saddles move round the unit circle into the right half-plane with each new saddle in the first quadrant resulting in an additional exponentially small contribution to $I_p(x)$.

We observe that the real part of (8.1.13) vanishes when p is an even integer, so that there is no algebraic expansion in $S_p(a)$ for these values of p; compare (8.1.8). This disappearance of the algebraic contribution can be understood alternatively

† This expansion can be obtained by taking the path in (8.1.11) along the positive imaginary τ axis and use of the Maclaurin series for $\exp\{-X_n \tau^p / (\kappa p)\}$. An alternative derivation is to note that $I_p(x)$ can also be expressed [see (8.1.49)] in terms of the function $U_{p,1}(ihx)$ defined in Paris & Wood [1986, Eq. (3.10.12)], followed by use of the asymptotic expansion of this latter function in Eq. (3.5.6) of this reference.

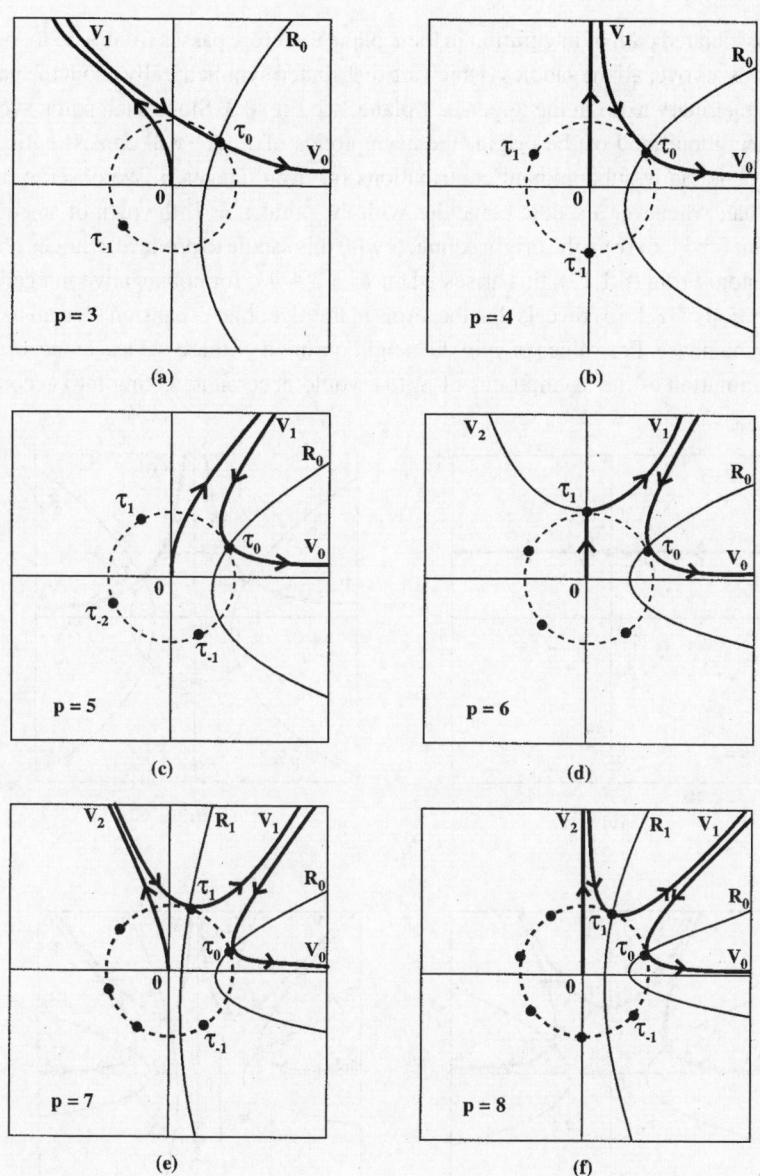

Fig. 8.2. The paths of steepest descent for integer values of p.

from the fact that, when $p = 2m$, we can write

$$F_{2m}(x) = \pi^{-\frac{1}{2}} m \int_{-\infty}^{\infty} e^{-t^{2m} + ixt} dt \qquad (m = 1, 2, \dots). \qquad (8.1.14)$$

The associated path of integration in the τ plane therefore passes from $-\infty$ to $+\infty$ and crosses over all the saddles (which are distributed symmetrically about the positive imaginary axis) in the upper-half plane; see Fig. 8.3. Since such paths avoid the neighbourhood of the origin, the asymptotics of $F_p(x)$ will consist entirely of exponentially subdominant contributions for even integer p. We observe further that, whenever a saddle coincides with the point $\tau = i$, the path of steepest descent for $I_p(x)$ from the origin connects with this saddle to produce a Stokes phenomenon. From (8.1.12), this arises when $p = 2 + 4k$, for nonnegative integer k; inspection of (8.1.13) reveals that the terms in the algebraic expansion then all have the same phase. For values of p in the neighbourhood of these values, an accurate determination of the asymptotics of $S_p(a)$ would necessitate taking into account

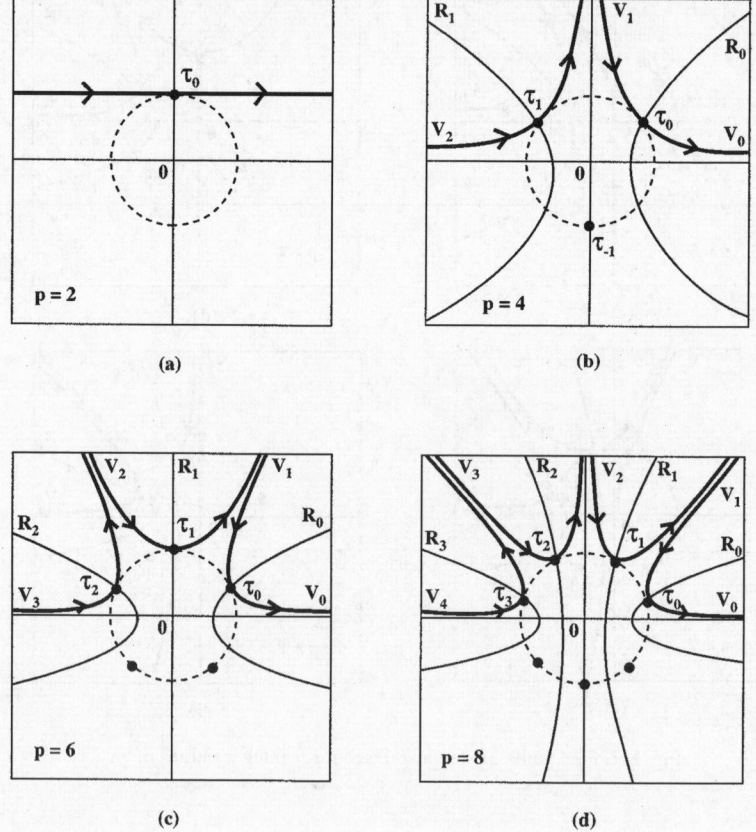

Fig. 8.3. The paths of steepest descent when p is an even integer.

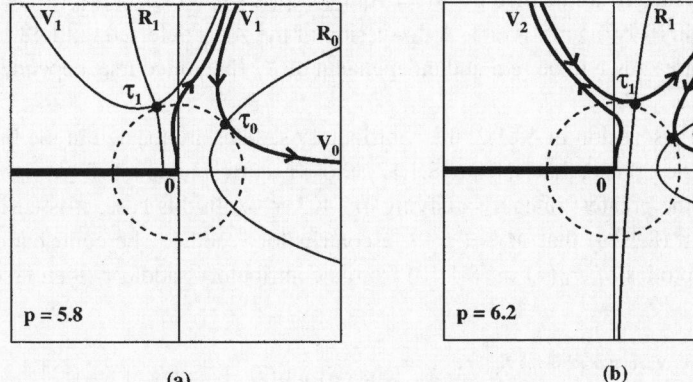

(a)   (b)

Fig. 8.4. The paths of steepest descent in the neighbourhood of $p = 6$.

the smooth 'switching on' of the additional exponentially small contribution. The situation arising in the neighbourhood of $p = 6$ is shown in Fig. 8.4. However, we note from (8.1.14) that, when $p = 2 + 4k$, the expansion of $F_p(x)$ consists only of exponentially small terms, so that the asymptotics of $S_p(a)$ for these special values of p can be obtained without reference to the Stokes phenomenon.

8.1.3 The Asymptotics of $S_p(a)$

To determine the contribution to the asymptotics of $F_p(x)$ from a saddle τ_r, we put

$$\arg f''(\tau_r) = (2r + \tfrac{1}{2})\pi(p-2)/(p-1) = 2\omega + 2\pi M_r \quad (|\omega| < \pi), \quad (8.1.15)$$

where M_r denotes a nonnegative integer. Then, in the neighbourhood of τ_r we have $f(\tau) = f(\tau_r) + \tfrac{1}{2}(\tau - \tau_r)^2|f''(\tau_r)|e^{2i\omega} + \cdots$, so that the directions of the paths of steepest descent from the saddle τ_r are given by $\arg(\tau - \tau_r) = -\omega$ and $\pi - \omega$. Straightforward application of the method of steepest descent [Olver (1974, p. 127); Wong (1989, p. 90)] then shows that as $X_n \to +\infty$ the contribution to the asymptotic expansion of $I_p(x)$ associated with the saddle τ_r is given by

$$(hx^{1/p})^{1/\kappa} \left(\frac{2\pi}{|f''(\tau_r)|} \right)^{\frac{1}{2}} e^{-i\omega - X_n f(\tau_r)} \sum_{k=0}^{\infty} C_{2k} X_n^{-k-\frac{1}{2}} \frac{\Gamma(k + \frac{1}{2})}{\Gamma(\frac{1}{2})} \quad (8.1.16)$$

$$= \frac{(-)^{M_r}}{p} X_{nr}^{\vartheta} e^{X_{nr}} \sum_{k=0}^{\infty} A_k X_{nr}^{-k}, \quad X_{nr} = X_n e^{(2r+\frac{1}{2})\pi i/\kappa}, \quad (8.1.17)$$

since routine algebra shows that

$$f(\tau_r) = -e^{(2r+\frac{1}{2})\pi i/\kappa}, \quad (hx^{1/p})^{1/\kappa} \left(\frac{2\pi}{X_n f''(\tau_r)} \right)^{\frac{1}{2}} = A_0 X_{nr}^{\vartheta}/p.$$

The coefficients A_k are defined by $A_k = A_0 C_{2k} \exp\{(2r + \frac{1}{2})\pi i k/\kappa\}(\frac{1}{2})_k$, where A_0 is given in (8.1.9) and $C_0 = 1$. A discussion of the A_k is deferred until §8.1.6 where they are shown to be *real* and independent of X_n (but, of course, dependent on p).

From the discussion in §8.1.2, the contributory saddles are located in the first quadrant in the τ plane; that is, from (8.1.12), those τ_r corresponding to $0 \leq r \leq r'$, where r' is the greatest integer satisfying $4r' + 2 \leq p$. In this case, it is easily verified from (8.1.15) that $M_r = r$ for a contributory saddle. The contribution to the asymptotics of $F_p(x)$ in (8.1.10) from a contributory saddle τ_r then takes the form

$$\frac{(-)^r}{\pi^{\frac{1}{2}}} X_n^\vartheta e^{X_n \cos\{(2r+\frac{1}{2})\pi/\kappa\}} \sum_{k=0}^{\infty} A_k X_n^{-k} \cos\left\{X_n \sin(2r + \tfrac{1}{2})\pi/\kappa + \Theta_k(r)\right\},$$

$$(8.1.18)$$

as $a \to 0+$, where

$$\Theta_k(r) = (2r + \tfrac{1}{2})(\vartheta - k)\pi/\kappa$$

and we recall that X_n is related to the summation variable n by (8.1.9). Each contribution in (8.1.18) is seen to be exponentially small, since $\cos[(2r+\frac{1}{2})\pi/\kappa] < 0$ for $0 \leq r \leq r'$. The transformation formula (8.1.5) then shows that, as $a \to 0+$, a contributory saddle τ_r gives rise to the exponentially small expansion to $S_p(a)$ given by

$$e_r(a; p) = \frac{2(-)^r}{pa^{1/p}} \sum_{n=1}^{\infty} X_n^\vartheta e^{X_n \cos\{(2r+\frac{1}{2})\pi/\kappa\}} \sum_{k=0}^{\infty} A_k X_n^{-k}$$

$$\times \cos\left\{X_n \sin(2r + \tfrac{1}{2})\pi/\kappa + \Theta_k(r)\right\}. \qquad (8.1.19)$$

Each expansion $e_r(a; p)$ $(0 \leq r \leq r')$ represents an infinite sum of exponentially small terms, and shows how the classical Poisson-Jacobi result in (8.1.2) is modified when $p \neq 2$.

We can now give the asymptotic expansion of $S_p(a)$ as $a \to 0+$ for different values of p in the form

$$S_p(a) - \frac{\Gamma(1/p)}{pa^{1/p}} - \tfrac{1}{2} \sim E_p(a) + H_p(a), \qquad (8.1.20)$$

where $E_p(a)$ and $H_p(a)$ denote the formal exponentially small and algebraic asymptotic expansions respectively, with $H_p(a)$ defined in (8.1.8). When $1 < p < 2$, the expansion of $S_p(a)$ is algebraic with $E_p(a) \equiv 0$ and

$$S_p(a) - \frac{\Gamma(1/p)}{pa^{1/p}} - \tfrac{1}{2} \sim H_p(a) \qquad (1 < p < 2). \qquad (8.1.21)$$

We note that as $p \to 2-$, an accurate determination of $S_p(a)$ would need to take into account the Stokes phenomenon to obtain the 'smooth' appearance of the exponentially small contribution resulting from the saddle τ_0 as it crosses the imaginary τ axis. When $p = 2$, the algebraic expansion in (8.1.8) vanishes and the contribution to $F_p(x)$ from the saddle $\tau_0 = i$ is given by half the value in (8.1.18), since in this case we only have the path of steepest descent from the saddle† in one direction. From (8.1.9), we find when $p = 2$ the values $\kappa = \frac{1}{2}$, $\vartheta = 0$, $X_n = (\pi n)^2/a$ and the coefficients $A_k = 0$ for $k \geq 1$; see (8.1.57). From (8.1.19), the exponentially small contribution to $S_2(a)$ is then given by $\frac{1}{2}e_0(a; 2)$, which becomes the (convergent) expansion

$$E_2(a) = \sqrt{\frac{\pi}{a}} \sum_{n=1}^{\infty} e^{-(\pi n)^2/a}$$

in accordance with the Poisson-Jacobi result in (8.1.2).

For $2 < p < 6$, there is just one contributory saddle τ_0 and the expansion of $S_p(a)$ for this range of p is therefore given by (8.1.20), where $E_p(a) = e_0(a; p)$; that is,

$$E_p(a) = \frac{2}{pa^{1/p}} \sum_{n=1}^{\infty} X_n^{\vartheta} e^{X_n \cos(\pi/2\kappa)} \sum_{k=0}^{\infty} A_k X_n^{-k} \cos\{X_n \sin\frac{\pi}{2\kappa} + \frac{\pi}{2\kappa}(\vartheta - k)\},$$

$$(2 < p < 6). \quad (8.1.22)$$

For example, when $p = 3$, we have $\kappa = \frac{2}{3}$, $\vartheta = -\frac{1}{6}$. Then, from (8.1.8), (8.1.9) and (8.1.22), we find

$$H_3(a) = -\frac{1}{\pi} \sum_{k=1}^{\infty} \frac{1}{k!} \Gamma(3k + 1) \sin(\tfrac{1}{2}\pi k)\zeta(3k + 1)(a/(2\pi)^3)^k, \quad (8.1.23)$$

and

$$E_3(a) = \frac{2}{3a^{1/3}} \sum_{n=1}^{\infty} X_n^{-\frac{1}{6}} e^{-X_n/\sqrt{2}} \sum_{k=0}^{\infty} A_k X_n^{-k} \cos\left\{X_n/\sqrt{2} - \tfrac{3}{4}\pi k - \tfrac{1}{8}\pi\right\},$$

$$(8.1.24)$$

where $X_n = 2(\frac{2}{3}\pi n)^{\frac{3}{2}} a^{-\frac{1}{2}}$; we remark that only odd values of k contribute to $H_3(a)$. Similarly, when $p = 4$, so that $\kappa = \frac{3}{4}$ and $\vartheta = -\frac{1}{4}$, the expansion $H_4(a)$ vanishes and we have‡

$$E_4(a) = \frac{1}{2a^{1/4}} \sum_{n=1}^{\infty} X_n^{-\frac{1}{4}} e^{-\frac{1}{2}X_n} \sum_{k=0}^{\infty} A_k X_n^{-k} \cos\left\{\tfrac{\sqrt{3}}{2}X_n - \tfrac{2}{3}\pi k - \tfrac{1}{6}\pi\right\}, \quad (8.1.25)$$

---

† This results from the fact that, although the contribution to the asymptotic sum in (8.1.16) is now given by $\frac{1}{2}\sum_{k=0}^{\infty} C_k X_n^{-k/2-1/2} \Gamma(\tfrac{1}{2}k + \tfrac{1}{2})/\Gamma(\tfrac{1}{2})$, the contribution from the odd coefficients C_{2k+1} vanishes when the real part is taken in (8.1.10), since C_{2k+1} are purely imaginary. This can most readily be seen from (8.1.14), where the path of integration in $F_p(x)$ can be extended from $-\infty$ to $+\infty$ for even values of p; see Fig. 8.3.

‡ There would appear to be a misprint in the equivalent expression given in Kowalenko et al. (1995, Eq. (6.30)): instead of $-\frac{2}{3}\pi k$ in the cosine term in the sum over k, they have $-\frac{1}{3}\pi k$.

where $X_n = 3(\frac{1}{2}\pi n)^{\frac{4}{3}} a^{-\frac{1}{3}}$. We note that in the neighbourhood of $p = 2$ and $p = 6$, an accurate determination of the asymptotics of $S_p(a)$ would require the Stokes phenomenon to be taken into account.

When $p = 6$, a second saddle τ_1 crosses the positive imaginary τ axis with the consequence that, when $6 < p < 10$, there are now two exponentially small contributions which result from these saddles. In this case, we have $E_p(a) = e_0(a; p) + \Lambda_p e_1(a; p)$, so that

$$
E_p(a) = \frac{2}{pa^{1/p}} \sum_{n=1}^{\infty} X_n^{\vartheta} \sum_{k=0}^{\infty} A_k X_n^{-k} \left\{ e^{X_n \cos(\pi/2\kappa)} \cos\left\{ X_n \sin\frac{\pi}{2\kappa} + \frac{\pi}{2\kappa}(\vartheta - k) \right\} \right.
$$

$$
\left. - \Lambda_p e^{X_n \cos(5\pi/2\kappa)} \cos\left\{ X_n \sin\frac{5\pi}{2\kappa} + \frac{5\pi}{2\kappa}(\vartheta - k) \right\} \right\} \quad (6 \le p < 10),
$$

(8.1.26)

where $\Lambda_p = 1 - \frac{1}{2}\delta_{6p}$, with δ_{6p} denoting the Kronecker delta symbol. As with the case $p = 2$, the presence of the factor Λ_p results from the saddle τ_1 making only half the contribution to $F_p(x)$ when $p = 6$. Thus, when $p = 6$ (so that $\kappa = \frac{5}{6}$ and $\vartheta = -\frac{1}{3}$), we find

$$
E_6(a) = \frac{1}{3a^{1/6}} \sum_{n=1}^{\infty} X_n^{-\frac{1}{3}} \sum_{k=0}^{\infty} A_k X_n^{-k} \left\{ e^{X_n \cos\frac{3}{5}\pi} \cos\{ X_n \sin\frac{3}{5}\pi - \frac{3}{5}\pi k - \frac{1}{5}\pi \} \right.
$$

$$
\left. + \frac{1}{2}(-)^k e^{-X_n} \right\},
$$

(8.1.27)

where $X_n = 5(\frac{1}{3}\pi n)^{\frac{6}{5}} a^{-\frac{1}{5}}$. We remark that the second exponential e^{-X_n} is subdominant with respect to the first exponential $\exp\{X_n \cos\frac{3}{5}\pi\}$.

The expansion given in (8.1.20), when the coefficients $A_k = A_0 c_k$ are evaluated according to (8.1.57), agrees with the results given in Kowalenko et al. (1995) for integer values of p. In particular, the expansions in (8.1.23)–(8.1.25) and (8.1.27) correspond to (6.82), (6.30) and (6.48) of this reference. The cases $p = 5$ and $p = 7$ given in (6.105) and (6.129), respectively agree with our general expansions in (8.1.20), (8.1.22) and (8.1.26). The pattern which emerges in the exponentially small contribution $E_p(a)$ continues to hold for higher values of p. Thus, each time p increases by 4, a new saddle in (8.1.12) crosses over into the first quadrant of the τ plane and results in an additional exponentially small contribution to the asymptotics of $S_p(a)$. The formulation of a general statement of the expansion of $S_p(a)$ is given at the end of §8.1.5.

The numerical interpretation of the expansion (8.1.20) in the case $p = 3$ has been examined in considerable detail in Kowalenko et al. (1995, §7), where an interesting discussion is given concerning the meaning of the use of such composite expansions involving algebraic and exponentially small terms. By employing optimal truncation of both the algebraic expansion $H_3(a)$ and the exponentially small expansion $E_3(a)$ (for which the coefficients A_k in this case can be given explicitly – see (8.1.53)), these authors studied the accuracy of the resulting expansions. In addition, Borel resummation of the divergent tails of both expansions was then considered to verify, for a wide range of values of the parameter a, the validity of (8.1.20). We do not repeat these calculations here.

8.1.4 Mellin-Barnes Integral Approach

In §8.1.3 the asymptotics of $S_p(a)$ when $p > 1$ were obtained using the analogue of the Poisson-Jacobi transformation in (8.1.5) combined with an application of the method of steepest descent. In this section we show how these results can be obtained using a Mellin-Barnes integral approach,† which does not directly make use of (8.1.5) and which also supplies the expansion when $0 < p < 1$. In fact, it will be seen during the course of the analysis that this relation is effectively embodied in the well-known functional relation for the Riemann zeta function.

Our starting point is the identity, with $c > 0$ so that the contour passes to the right of the origin,‡

$$e^{-z} = \frac{1}{2\pi i} \int_{c-\infty i}^{c+\infty i} \Gamma(s) z^{-s} \, ds \qquad (|\arg z| < \tfrac{1}{2}\pi; \quad z \neq 0); \qquad (8.1.28)$$

see §3.3.1. Use of this result enables us to recast the sum of exponentials in $S_p(a)$ in the form

$$S_p(a) - 1 = \sum_{n=1}^{\infty} e^{-an^p} = \sum_{n=1}^{\infty} \frac{1}{2\pi i} \int_{c-\infty i}^{c+\infty i} \Gamma(s)(an^p)^{-s} \, ds$$

$$= \frac{1}{2\pi i} \int_{c-\infty i}^{c+\infty i} \Gamma(s)\zeta(ps)a^{-s} \, ds \qquad (c > 1/p), \qquad (8.1.29)$$

upon reversal of the order of integration and summation, which is justified when $c > 1/p$, and evaluation of the inner sum in terms of the Riemann zeta function. The path of integration in the Mellin-Barnes integral representation (8.1.29) is shown in Fig. 8.5(a).

The domain of convergence of the integral (8.1.29) can be determined by examining the behaviour of the integrand as $|s| \to \infty$ in the usual manner. Since $c > 1/p$, we have on the path of integration $|\zeta(ps)| < \sum_{n=1}^{\infty} n^{-pc} = O(1)$ so that the convergence is controlled by the factor $\Gamma(s)$. It therefore follows, by comparison with (8.1.28), that (8.1.29) defines $S_p(a)$ in the sector $|\arg a| < \tfrac{1}{2}\pi$. We shall see that it is not possible when $p > 1$ to bend the contour of integration in (8.1.29) to pass to infinity along a path with $|\arg s| \neq \tfrac{1}{2}\pi$ in either $\mathrm{Re}(s) > 0$ or

---

† It has been brought to the authors' attention that a similar treatment using Mellin-Barnes integrals had been developed by Boersma and Braaksma in two unpublished notes written in 1992 and 1994.

‡ The constant c is used as a generic parameter. Its value can change as the associated Mellin-Barnes integrals are manipulated. We shall indicate the restriction on c at each occurrence.

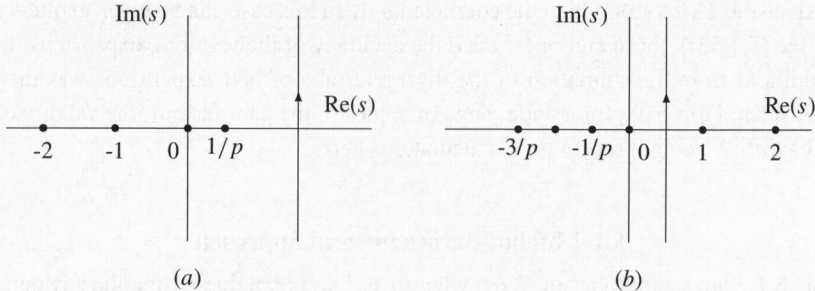

Fig. 8.5. Paths of integration in the s plane (a) for the integral in (8.1.30) and (b) for the integral in (8.1.34). The heavy points denote poles.

$\mathrm{Re}(s) < 0$. When $s = Re^{i\theta}$, $R \to \infty$ the logarithm of the dominant real part of the integrand from Stirling's formula in (2.1.8) is controlled by

$$R \cos\theta \log R + \log |\zeta(ps)| + O(R).$$

Since $|\zeta(ps)| = O(1)$ to the right of $\mathrm{Re}(s) = c$, this expression tends to $+\infty$ when $\cos\theta > 0$. Use of the functional relation for the zeta function in (4.1.4) shows that the behaviour of $\zeta(ps)$ in the left-hand half-plane is now dominated by $\Gamma(1 - ps)$. In this case, the logarithm of the dominant real part of the integrand is controlled by

$$R \cos\theta \log R - pR \cos\theta \log R + O(R) = (1 - p)R \cos\theta \log R + O(R),$$

which, when $\cos\theta < 0$, tends to $\pm\infty$ according as $p > 1$ or $p < 1$, respectively. Thus, deformation of the contour in (8.1.29) into the right or left-hand half-planes (when $p > 1$) is prohibited by the growth of $\Gamma(s)$ and $\zeta(ps)$, respectively, in these half-planes. However, when $0 < p < 1$, the contour in (8.1.29) can be bent back to enclose the poles of the integrand on the negative real axis to yield the result

$$S_p(a) - \frac{\Gamma(1/p)}{pa^{1/p}} - \tfrac{1}{2} = H_p(a) \qquad (0 < p < 1),$$

where we recall that $H_p(a)$ is defined in (8.1.8). We remark that the sum $H_p(a)$ is *convergent* when $0 < p < 1$ and that consequently the above result is an equality.

To deal with the case $p > 1$, we shall now suppose, as in §8.1.2, that $a > 0$. The integral (8.1.29) has simple poles at $s = -k$ (where k is a nonnegative integer) together with a simple pole at $s = 1/p$ resulting from $\zeta(ps)$. We now displace the path over the poles at $s = 1/p$ and $s = 0$, where the residues are given by $\Gamma(1/p)/(pa^{1/p})$ and $-\tfrac{1}{2}$, respectively. This yields

$$S_p(a) - \frac{\Gamma(1/p)}{pa^{1/p}} - \tfrac{1}{2} = \frac{1}{2\pi i} \int_{c-\infty i}^{c+\infty i} \Gamma(-s)\zeta(-ps)a^s \, ds$$

$$= \frac{1}{2\pi i} \int_{c-\infty i}^{c+\infty i} \zeta(1+ps) \frac{\Gamma(1+ps)}{\Gamma(1+s)}$$

$$\times \left(\frac{\sin \frac{1}{2}\pi ps}{\sin \pi s} \right) \left(\frac{2\pi}{a^{1/p}} \right)^{-ps} ds, \qquad (8.1.30)$$

where $0 < c < 1$ and, for convenience, we have made the change of variable $s \to -s$ followed by use of the functional relation for $\zeta(s)$ in (4.1.4) in the form

$$\zeta(-s) = -2^{-s} \pi^{-s-1} \zeta(1+s) \Gamma(1+s) \sin \tfrac{1}{2}\pi s. \qquad (8.1.31)$$

Since $\mathrm{Re}(1+ps) > 1$ on the path of integration in (8.1.30), we can expand the zeta function in the integrand to obtain, after interchanging the order of summation and integration, the result

$$S_p(a) - \frac{\Gamma(1/p)}{pa^{1/p}} - \tfrac{1}{2} = 2\pi a^{-1/p} \sum_{n=1}^{\infty} G_n(a) \qquad (p > 1), \qquad (8.1.32)$$

where

$$G_n(a) = \frac{1}{2\pi i} \int_{c-\infty i}^{c+\infty i} \frac{\Gamma(1+ps)}{\Gamma(1+s)} \left(\frac{\sin \frac{1}{2}\pi ps}{\sin \pi s} \right) \left(\frac{2\pi n}{a^{1/p}} \right)^{-ps-1} ds \qquad (8.1.33)$$

with $0 < c < 1$. The integrals $G_n(a)$ will form the basis of our investigation into the asymptotics of $S_p(a)$ by the Mellin-Barnes integral approach. In general, there are two sequences of simple poles of the integrand: one sequence, resulting from $\Gamma(1+ps) \sin(\frac{1}{2}\pi ps)$, is given by $s = -(2k+1)/p$ for nonnegative k, while the second, resulting from $\sin \pi s$, is given by positive integer values of s; the path of integration in (8.1.33) is shown in Fig. 8.5(b). For certain values of p (namely integer values or rational fractions) some of these poles can become ordinary points. In particular, when p is an even integer, the sequence of poles from $\sin \pi s$ vanishes, due to the term $\sin \frac{1}{2}\pi ps$ in the numerator, and there are then no poles to the right of the contour. We shall see below that this corresponds to the vanishing of the algebraic contribution to $S_p(a)$.

We note that $G_n(a)$ is related to the generalised hypergeometric function $F_p(2\pi n a^{-1/p})$ appearing in the analogue of the Poisson-Jacobi transformation in (8.1.5). If we displace the contour in (8.1.33) to the left over the poles of the integrand at $s = -(2k+1)/p$, and compare (8.1.32) with (8.1.5), we find that

$$G_n(a) \equiv \frac{\pi^{-\frac{1}{2}}}{p} F_p(2\pi n a^{-1/p}).$$

8.1.5 The Asymptotic Expansion of $S_p(a)$ for $a \to 0+$

We employ the result in (5.1.20)

$$\frac{\sin \frac{1}{2}\pi p s}{\sin \pi s} = 2 \sum_{r=0}^{N-1} \cos\left(\tfrac{1}{2}p - 2r - 1\right)\pi s + \frac{\sin\left(\tfrac{1}{2}p - 2N\right)\pi s}{\sin \pi s}, \qquad (8.1.34)$$

where we choose† $N = [p/4]$. Substitution of (8.1.34) into (8.1.33) then yields

$$G_n(a) = \sum_{r=0}^{N-1}(J_r^+ + J_r^-) + K, \qquad (8.1.35)$$

where

$$J_r^{\pm} = \frac{1}{2\pi i}\int_{c-\infty i}^{c+\infty i} \frac{\Gamma(1+ps)}{\Gamma(1+s)}\left(\frac{2\pi n}{a^{1/p}}\right)^{-ps-1} e^{\pm(\frac{1}{2}p-2r-1)\pi i s}ds$$

$$= (-)^r e^{\pm(2r+1)\pi i \vartheta}\frac{1}{2\pi i p}\int_{c-\infty i}^{c+\infty i} \frac{\Gamma(u)}{\Gamma(\kappa + u/p)}$$

$$\times \left(e^{\mp(2r+1)\pi i}X_{nr}^{\pm}\right)^{-\kappa u}(h\kappa^{\kappa})^u du \qquad (8.1.36)$$

upon making the change of variable‡ $u = 1 + ps$, and

$$K = \frac{1}{2\pi i}\int_{c-\infty i}^{c+\infty i} \frac{\Gamma(1+ps)}{\Gamma(1+s)}\frac{\sin(\frac{1}{2}p - 2N)\pi s}{\sin \pi s}\left(\frac{2\pi n}{a^{1/p}}\right)^{-ps-1}ds, \qquad (8.1.37)$$

where $0 < c < 1$. Here, we have put $X_{nr}^{\pm} = X_n e^{\pm(2r+\frac{1}{2})\pi i/\kappa}$, where we recall that X_n is defined in (8.1.9); compare (8.1.17). Note that the restriction on c in (8.1.36) (but not that in (8.1.37)) can be relaxed to $c > 0$, since there are no poles of the integrand in $\mathrm{Re}(u) > 0$.

We now introduce the inverse factorial expansion, obtained from Stirling's expansion, for the ratio of gamma functions in (8.1.36). From Lemma 2.2 (for convenience, we have extracted a factor κ from the coefficients A_k and the remainder function $\rho_M(u)$), we find

$$\frac{\Gamma(u)}{\Gamma(\kappa + u/p)} = \frac{\kappa}{2\pi}(h\kappa^{\kappa})^{-u}\left\{\sum_{k=0}^{M-1}(-)^k A_k \Gamma(\kappa u + \vartheta - k)\right.$$

$$\left. + \rho_M(u)\Gamma(\kappa u + \vartheta - M)\right\}, \qquad (8.1.38)$$

valid as $|u| \to \infty$ in the sector $|\arg u| < \pi$, where the quantities h, κ, ϑ are defined in (8.1.9) and $M = 1, 2, \ldots$. The function $\rho_M(u)$ has poles at $u = 0, -1, -2, \ldots$

---

† We use [ ] in this section to denote the *nearest* integer part; hence, $[x] = N$ when x is in the interval $(N - \frac{1}{2}, N + \frac{1}{2}]$. Note that when $N = 0$ (i.e., when $1 < p \le 2$) the expansion (8.1.34) contains no information.

‡ In arriving at (8.1.36), it is helpful to note that the factor $(2r+1)/p - \frac{1}{2}$ appearing in the exponential has been written both as $2r + \frac{1}{2} - (2r + 1)\kappa$ and $(2r + 1)\vartheta + r$.

and is such that $\rho_M(u) = O(1)$ as $|u| \to \infty$ in $|\arg u| < \pi$. The coefficients A_k are real (see §8.1.6) and will turn out to be the same coefficients in the exponential expansions in (8.1.19), with A_0 given in (8.1.9). Substitution of (8.1.38) into (8.1.36) then leads to

$$
J_r^\pm = \frac{(-)^r \kappa}{2\pi p} e^{\pm(2r+1)\pi i \vartheta} \left\{ \sum_{k=0}^{M-1} (-)^k A_k \frac{1}{2\pi i} \int_{c-\infty i}^{c+\infty i} \Gamma(\kappa u + \vartheta - k) \right.
$$

$$
\times \left(e^{\mp(2r+1)\pi i} X_{nr}^\pm \right)^{-\kappa u} du + \frac{1}{2\pi i} \int_{c-\infty i}^{c+\infty i} \rho_M(u)\Gamma(\kappa u + \vartheta - M)
$$

$$
\times \left. \left(e^{\mp(2r+1)\pi i} X_{nr}^\pm \right)^{-\kappa u} du \right\}. \tag{8.1.39}
$$

Since there are no poles to the right of $u = 0$ in (8.1.36), we can choose c sufficiently large and positive so that the path in each integral appearing in the sum over k in (8.1.39) lies to the right of all the poles of its respective integrand. The resulting integrals can then be evaluated by employing (8.1.28) in the slightly modified form

$$
\frac{1}{2\pi i} \int_{c-\infty i}^{c+\infty i} \Gamma(\kappa u + \vartheta - k) z^{-\kappa u} du = \frac{1}{\kappa} z^{\vartheta - k} e^{-z} \quad (|\arg z| < \tfrac{1}{2}\pi),
$$

where the contour lies to the right of the poles of the integrand. Identifying z with $\exp\{\mp(2r+1)\pi i\} X_{nr}^\pm$, so that $\arg z = \pm(2r+1 - \tfrac{1}{2}p)\pi/(p-1)$, we note that the convergence condition $|\arg z| < \tfrac{1}{2}\pi$ is satisfied for the integrals in the sum over k in (8.1.39), since

$$
|(2r+1 - \tfrac{1}{2}p)\pi/(p-1)| < \tfrac{1}{2}\pi \quad (0 \le r \le N-1) \tag{8.1.40}
$$

when $N \ge 1$. We therefore obtain the expansion as $a \to 0+$

$$
J_r^\pm = \frac{(-)^r}{2\pi p} (X_{nr}^\pm)^\vartheta e^{X_{nr}^\pm} \left\{ \sum_{k=0}^{M-1} A_k (X_{nr}^\pm)^{-k} + R_M^\pm \right\}, \tag{8.1.41}
$$

where the remainders $R_M^\pm$ result from the last integral involving $\rho_M(u)$ in (8.1.39). The order of the remainder term in such expansions is considered in Lemma 2.7, where it is shown that $R_M^\pm = O((X_{nr}^\pm)^{-M})$; we omit these details here.

The integral K in (8.1.37) involves an integrand which, for general p, possesses simple poles situated to the right of the contour at $s = 1, 2, \ldots$. Following the standard method of asymptotic estimation of Mellin-Barnes integrals [see §5.1], we displace the path of integration to the right over these poles to obtain the algebraic expansion

$$
K \sim \frac{1}{\pi} \sum_{k=1}^{\infty} \frac{(-)^{k-1}}{k!} \Gamma(kp+1) \sin\left(\tfrac{1}{2}\pi kp\right) \left(a^{1/p}/2\pi n\right)^{kp+1}. \tag{8.1.42}
$$

This last expansion vanishes when p is an even integer. In this case, we observe that

$$\frac{\sin\left(\frac{1}{2}p - 2N\right)\pi s}{\sin \pi s} = \begin{cases} 0 & (p/2 \text{ even}) \\ 1 & (p/2 \text{ odd}) \end{cases}$$

when $N = [p/4]$. Thus, when $p/2$ is even (i.e., when $p = 4N$) the integral $K \equiv 0$, but when $p/2$ is odd (i.e., when $p = 4N + 2$), K makes an exponentially small contribution to $G_n(a)$. This is most easily seen by writing the finite Fourier expansion (8.1.34), when $p = 4N + 2$, in the form

$$\frac{\sin \frac{1}{2}\pi p s}{\sin \pi s} = 2 \sum_{r=0}^{N} \Lambda_r \cos 2(N - r)\pi s,$$

where $\Lambda_r = 1 - \frac{1}{2}\delta_{rN}$, with δ_{ij} being the Kronecker delta symbol. The analysis of the integrals $J_r^{\pm}$ can therefore be extended to deal with the term corresponding to $r = N$ in the above sum (since (8.1.40) is also satisfied when $r = N$). Consequently, we see that when $p/2$ is odd, there is an additional exponentially subdominant contribution to $G_n(a)$ given by

$$J_N^{\pm} = \frac{X_n^{\vartheta}}{4\pi p} e^{-X_n} \sum_{k=0}^{\infty} (-)^k A_k X_n^{-k}, \qquad (8.1.43)$$

since $X_{nN}^{\pm} = X_n e^{\pm(2N+1)\pi i} = -X_n$ when $p = 4N + 2$.

From (8.1.35), (8.1.41) and (8.1.43), the expansion of $G_n(a)$ for $p > 1$ is therefore given by

$$G_n(a) \sim \frac{1}{\pi} \sum_{k=1}^{\infty} \frac{(-)^{k-1}}{k!} \Gamma(kp + 1) \sin\left(\frac{1}{2}\pi kp\right) \left(a^{1/p}/2\pi n\right)^{kp+1}$$

$$+ \frac{X_n^{\vartheta}}{\pi p} \sum_{r=0}^{N-1} (-)^r e^{X_n \cos\{(2r+\frac{1}{2})\pi/\kappa\}}$$

$$\times \sum_{k=0}^{\infty} A_k X_n^{-k} \cos\left\{X_n \sin\left(2r + \frac{1}{2}\right)\pi/\kappa + \Theta_k(r)\right\}$$

$$+ \delta_{pp'} \frac{X_n^{\vartheta}}{2\pi p} e^{-X_n} \sum_{k=0}^{\infty} (-)^k A_k X_n^{-k},$$

where we have put $p' = 4N + 2$ and X_n, $\Theta_k(r)$ are defined in (8.1.9) and at (8.1.18). Evaluation of the sum over n according to (8.1.32) then produces the expansion for $S_p(a)$, which we state in the final form:

Theorem 8.1. *With the notation of §8.1.1, we have the asymptotic expansion*

$$S_p(a) - \frac{\Gamma(1/p)}{pa^{1/p}} - \frac{1}{2} \begin{cases} = H_p(a) & (0 < p < 1) \\ \sim E_p(a) + H_p(a) & (p > 1) \end{cases} \qquad (8.1.44)$$

as $a \to 0+$, where $E_p(a)$ and $H_p(a)$ are the exponentially small and algebraic contributions given by

$$E_p(a) = \sum_{r=0}^{N-1} e_r(a; p) + \tfrac{1}{2} e_N(a; p)\delta_{pp'} \quad (p' = 4N + 2) \tag{8.1.45}$$

and

$$H_p(a) = \frac{1}{\pi} \sum_{k=1}^{\infty} \frac{(-)^{k-1}}{k!} \Gamma(kp + 1) \sin(\tfrac{1}{2}\pi kp)\zeta(kp + 1)(a/(2\pi)^p)^k, \tag{8.1.46}$$

with N being the nearest integer part of $\tfrac{1}{4}p$. When $0 < p < 1$, the sum $H_p(a)$ is convergent and the first equation in (8.1.44) is an equality; when $p > 1$ this sum is divergent. The expansions $e_r(a; p)$ $(0 \leq r \leq N - 1)$ are defined by

$$e_r(a; p) = \frac{2(-)^r}{pa^{1/p}} \sum_{n=1}^{\infty} X_n^\vartheta e^{X_n \cos\{(2r+\frac{1}{2})\pi/\kappa\}} \sum_{k=0}^{\infty} A_k X_n^{-k}$$

$$\times \cos\{X_n \sin(2r + \tfrac{1}{2})\pi/\kappa + \Theta_k(r)\}, \tag{8.1.47}$$

where X_n is related to the summation variable n through (8.1.9), the angle $\Theta_k(r) = (2r + \tfrac{1}{2})(\vartheta - k)\pi/\kappa$ and

$$e_N(a; p) = \frac{2}{pa^{1/p}} \sum_{n=1}^{\infty} X_n^\vartheta e^{-X_n} \sum_{k=0}^{\infty} (-)^k A_k X_n^{-k}. \tag{8.1.48}$$

The coefficients A_k are discussed in §8.1.6 and $A_0 = (2\pi)^{\frac{1}{2}} \kappa^{-\frac{1}{2}-\vartheta} p^{-\vartheta}$.

This result agrees with that obtained in §8.1.3 by the method of steepest descent. We remark that the algebraic expansion can be expressed alternatively in the form

$$H_p(a) = \sum_{k=1}^{\infty} \frac{(-)^k}{k!} \zeta(-kp)a^k$$

by (4.1.4). The sum of exponential terms $e_r(a; p)$ corresponds to the contribution obtained from all the contributory saddles in (8.1.19), while the presence of the term involving $e_N(a; p)$ is equivalent to the appearance of an additional contributory saddle on the positive imaginary τ axis when $p = p'$. We remark that the expansion $E_p(a)$ in (8.1.45) does not hold uniformly as $p \to p'$. The treatment of this case would require taking into account the smooth appearance (a Stokes phenomenon) of the additional exponentially small expansion $e_N(a; p)$ in the neighbourhood of $p = p'$. We discuss this problem in §8.1.7.

8.1.6 The Coefficients A_k

In this section we determine the coefficients A_k appearing in the exponential expansions $e_r(a; p)$ in (8.1.47) and (8.1.48). For convenience, we shall normalise these coefficients by writing $A_k = A_0 c_k$, so that $c_0 = 1$. For integer values of p we can proceed by observing that the integral $I_p(x)$ in (8.1.11) is related to the function $U_{p,1}(z)$ defined in Paris & Wood (1986, §3.2 and Eq. (3.10.12)) by

$$I_p(x) = \frac{1}{p} \sum_{k=0}^{\infty} \frac{(ix)^k}{k!} \Gamma\left(\frac{k+1}{p}\right) = \frac{1}{p} U_{p,1}(ihx), \qquad (8.1.49)$$

where h is defined in (8.1.9). The coefficients c_k in the exponential expansion of $U_{p,1}(ihx)$ and those in $I_p(x)$ are consequently the same. The p-term recurrence relation for the c_k is given by [Paris & Wood (1986, Eqs. (3.5.4), (3.4.13))]

$$(p-1)kc_k = \sum_{j=1}^{p-1} c_{k-j} P_{p-j-1}(j-k), \qquad (8.1.50)$$

where $c_k = 0$ for $k < 0$. The coefficients $P_{p-j-1}(j-k)$ are

$$P_{p-j-1}(j-k) = \sum_{r=0}^{j} \mathfrak{S}_{p-1-r}^{(p-1-j)} \sum_{m=0}^{r} \kappa^{k-1} (\vartheta + j - k)^{r-m} \binom{p-m}{r-m} S_p^{(p-m)}, \qquad (8.1.51)$$

where S and $\mathfrak{S}$ are the Stirling numbers of the first and second kinds, respectively; see Temme (1996, §1.3). Since all parameters in (8.1.51) are real, it follows that the coefficients c_k are real.

For example, when $p = 3$, we have the 3-term recurrence relation given by

$$2kc_k = P_1(1-k)c_{k-1} + P_0(2-k)c_{k-2} \qquad (k \geq 1), \qquad (8.1.52)$$

with

$$P_1(1-k) = 3\left(k - \tfrac{5}{6}\right)^2 + \left(k - \tfrac{5}{6}\right) - \tfrac{1}{9},$$
$$P_0(2-k) = \left(\tfrac{11}{6} - k\right)\left(\tfrac{7}{6} - k\right)\left(\tfrac{1}{2} - k\right).$$

This agrees with the coefficients obtained from Kowalenko et al. (1995, Eq. (6.85)) where their coefficients $N_k = 2^{-k} c_k$. Through use of the theory of the generalised hypergeometric function, Kowalenko et al. were able to show that the recurrence relation (8.1.52) reduces to a two-term recurrence with the closed-form solution

$$c_k = \frac{\Gamma\left(k + \tfrac{1}{6}\right)\Gamma\left(k + \tfrac{5}{6}\right)}{2^k k! \Gamma\left(\tfrac{1}{6}\right)\Gamma\left(\tfrac{5}{6}\right)}; \qquad (8.1.53)$$

see Eq. (6.89) of this reference. When $p = 4$, we obtain the 4-term recurrence relation

$$3kc_k = P_2(1-k)c_{k-1} + P_1(2-k)c_{k-2} + P_0(3-k)c_{k-3} \qquad (k \geq 1), \qquad (8.1.54)$$

with

$$P_2(1 - k) = 6k^2 - \tfrac{15}{2}k + \tfrac{31}{16},$$

$$P_1(2 - k) = -4\left(k - \tfrac{7}{4}\right)^3 - \tfrac{15}{2}\left(k - \tfrac{7}{4}\right)^2 - \tfrac{23}{8}\left(k - \tfrac{7}{4}\right) + \tfrac{5}{32},$$

$$P_0(3 - k) = \left(k - \tfrac{1}{2}\right)\left(k - \tfrac{5}{4}\right)(k - 2)\left(k - \tfrac{11}{4}\right).$$

This yields the same coefficients as obtained in Kowalenko *et al.* (1995, Eq. (6.31)).

The recurrence relation (8.1.50) was obtained from the pth-order ordinary differential equation satisfied by $U_{p,1}(z)$. It is clear that, for noninteger values of p, this process will no longer be valid. To obtain the coefficients c_k for general values of p, we turn to the inverse factorial expansion in (8.1.38) and employ the approach described in §2.2.4. If we introduce the scaled gamma function $\Gamma^*(z)$ defined in (2.1.10), then we find that

$$\frac{\Gamma(u)}{\Gamma(\kappa + u/p)\Gamma(\kappa u + \vartheta)} = \frac{\kappa A_0}{2\pi}(h\kappa^\kappa)^{-u} R(u)\Upsilon(u),$$

where

$$R(u) = e^{\frac{1}{2}}\left(1 + \frac{\kappa p}{u}\right)^{\vartheta - u/p}\left(1 + \frac{\vartheta}{\kappa u}\right)^{\kappa - \kappa u} = 1 + \frac{1}{8\kappa u} + O(u^{-2}),$$

$$\Upsilon(u) = \frac{\Gamma^*(u)}{\Gamma^*(\kappa + u/p)\Gamma^*(\kappa u + \vartheta)} = 1 + \frac{\gamma_1}{\kappa u}(p - \kappa) + O(u^{-2})$$

and $\gamma_1 = -\tfrac{1}{12}$. Then (8.1.38) can be rewritten in the form

$$R(u)\Upsilon(u) = \sum_{k=0}^{M-1} \frac{c_k}{(-\kappa u - \vartheta + 1)_k} + \frac{(-)^M \rho_M(u)}{A_0(-\kappa u - \vartheta + 1)_M} \qquad (M = 1, 2, \ldots),$$

$$(8.1.55)$$

where $(\alpha)_k = \Gamma(\alpha + k)/\Gamma(\alpha)$. Equating coefficients of u^{-1} in the expansion of both sides of (8.1.55) we obtain

$$c_1 = \frac{1}{24p}(2p - 1)(p - 2) \qquad (8.1.56)$$

valid for general values of p. It is worth noting that (8.1.56) corresponds to the value of c_1 given in Paris & Wood (1986, p. 74) for integer values of p.

The higher coefficients can be obtained by continuation of this process. If we define

$$c_k = \frac{(2p - 1)(p - 2)}{k(24p)^k}\hat{c}_k \qquad (k \geq 1), \qquad (8.1.57)$$

we find, with the aid of *Mathematica*, for general values of $p \geq 2$

$$\hat{c}_1 = 1, \quad \hat{c}_2 = (2p^2 + 19p + 2),$$

$$\hat{c}_3 = -\tfrac{1}{10}(556p^4 - 1628p^3 - 9093p^2 - 1628p + 556),$$

$$\hat{c}_4 = -\tfrac{1}{30}(4568p^6 + 226668p^5 - 465702p^4 - 2013479p^3 - 465702p^2$$
$$+ 226668p + 4568),$$

$$\hat{c}_5 = \tfrac{1}{168}(2622064p^8 - 12598624p^7 - 167685080p^6 + 302008904p^5$$
$$+ 1115235367p^4 + 302008904p^3 - 167685080p^2$$
$$- 12598624p + 2622064),$$

$$\hat{c}_6 = \tfrac{1}{4200}(167898208p^{10} + 22774946512p^9 - 88280004528p^8$$
$$- 611863976472p^7 + 1041430242126p^6$$
$$+ 3446851131657p^5 + 1041430242126p^4 - 611863976472p^3$$
$$- 88280004528p^2 + 22774946512p + 167898208),$$

$$\hat{c}_7 = \tfrac{1}{3600}(34221025984p^{12} - 226022948160p^{11} - 5067505612464p^{10}$$
$$+ 18868361443936p^9 + 86215425028308p^8$$
$$- 143500920544692p^7 - 437682618704613p^6$$
$$- 143500920544692p^5 + 86215425028308p^4$$
$$+ 18868361443936p^3 - 5067505612464p^2$$
$$- 226022948160p + 34221025984),$$

$$\hat{c}_8 = \tfrac{1}{25200}(573840801152p^{14} + 156998277198784p^{13}$$
$$- 898376974770592p^{12} - 8622589006459984p^{11}$$
$$+ 32874204024803560p^{10} + 111492707520083828p^9$$
$$- 184768503480287646p^8 - 528612016938984183p^7$$
$$- 184768503480287646p^6 + 111492707520083828p^5$$
$$+ 32874204024803560p^4 - 8622589006459984p^3$$
$$- 898376974770592p^2 + 156998277198784p + 573840801152).$$

We remark that these values satisfy the recurrence relations in (8.1.52) and (8.1.54) when $p = 3$ and $p = 4$, and that $c_k = 0$ $(k \geq 1)$ when $p = 2$.

8.1.7 The Stokes Phenomenon for $p \simeq 2$

To conclude this discussion of the Euler-Jacobi series, we consider the exponential expansion $E_p(a)$ in (8.1.45) when p is in the neighbourhood of the classical value $p = 2$. As we saw in §8.1.2, such values of p correspond to the first saddle point

τ_0 of the integrand in (8.1.11) crossing the positive imaginary axis into the first quadrant (as p increases) to yield the exponentially small expansion $e_0(a; p)$. The detailed structure of the appearance of each of the infinite subdominant exponentials contained in $e_0(a; p)$ is controlled by a Stokes phenomenon, and we therefore expect this structure to exhibit (to leading order) an error function dependence with an argument measuring the transition proportional to $p - 2$; see §6.2.

To investigate the birth of the expansion† $e_0(a; p)$ when $p \simeq 2$, we consider the integral for $G_n(a)$ in (8.1.33), where we can take the contour to be the line $(-c - \infty i, -c + \infty i)$ with $0 < c < 1/p$. Such displacement of the path across the origin is permissible because the integrand is regular at $s = 0$; cf. Fig. 8.5(b). We let m_n ($n = 1, 2, \ldots$) denote arbitrary positive integers and displace the contour of integration to the right over the first $m_n - 1$ poles of the integrand on the positive real axis to find

$$G_n(a) = \frac{1}{\pi} \sum_{k=1}^{m_n-1} \frac{(-)^{k-1}}{k!} \Gamma(kp + 1) \sin\left(\tfrac{1}{2}\pi kp\right) \left(a^{1/p}/2\pi n\right)^{kp+1} + R_n(a; m_n),$$

$$(8.1.58)$$

where the remainder $R_n(a; m_n)$ is given by

$$R_n(a; m_n) = \frac{1}{2\pi i} \int_{-c+m_n-\infty i}^{-c+m_n+\infty i} \frac{\Gamma(1 + ps)}{\Gamma(1 + s)} \frac{\sin \tfrac{1}{2}\pi ps}{\sin \pi s} \left(\frac{2\pi n}{a^{1/p}}\right)^{-ps-1} ds \quad (8.1.59)$$

with $0 < c < 1/p$. The finite sum in (8.1.58) corresponds to the first $m_n - 1$ terms in the algebraic expansion of $G_n(a)$ in (8.1.42).

In order to detect the appearance of $e_0(a; p)$ we shall need to choose the m_n to correspond to the optimal truncation (see §1.1.2) of the algebraic expansion in (8.1.58) for each n. This is achieved when the terms in the expansion attain their numerically smallest value and is given by balancing the approximate magnitudes of the $(k - 1)$th and kth terms to yield

$$\frac{(kp - p)!}{(k - 1)!} \left(\frac{a^{1/p}}{2\pi n}\right)^{(k-1)p+1} \simeq \frac{(kp)!}{k!} \left(\frac{a^{1/p}}{2\pi n}\right)^{kp+1};$$

that is, on the assumption that $k \gg 1$ and use of Stirling's formula in (2.1.8) to approximate the factorials, when

$$\left(\frac{2\pi n}{a^{1/p}}\right)^p \simeq \frac{(kp)!}{(kp - p)! k} \simeq e^{-p} \frac{(kp)^p}{k} (1 + \tfrac{1}{k})^{kp} \simeq k^{p-1} p^p.$$

---

† The procedure adopted here is essentially the same as that employed in the discussion of the exponential improvement of the asymptotics of the gamma function in §6.4.2.

The optimal truncation index $k \equiv M_n$ for $G_n(a)$ is therefore given by

$$M_n \simeq \left(\frac{2\pi n}{p a^{1/p}} \right)^{1/\kappa} = \frac{X_n}{p-1}, \tag{8.1.60}$$

where we recall that κ and X_n are defined in (8.1.9).

We now choose the integers m_n to be the optimal truncation indices M_n (for $n = 1, 2, \ldots$) and proceed to manipulate the remainder integral (8.1.59) into a form from which we can extract the so-called terminant function $\hat{T}_\nu(z)$ defined in (6.2.6) and (6.2.7). From §6.2.6, it is seen that the leading asymptotic behaviour of $\hat{T}_\nu(z)$ when $\nu \sim |z| \to \infty$ and $|\arg z| \simeq \pi$ incorporates the error function smoothing that we seek. Since in the limit $a \to 0+$ the optimal truncation indices $M_n \to \infty$, it follows that the arguments in the gamma functions in the integrand of $R_n(a; M_n)$ are both large on the (displaced) path of integration. Consequently, we can express the quotient of gamma functions in (8.1.59) to leading order as

$$\frac{\Gamma(1+ps)}{\Gamma(1+s)} \sim \frac{\kappa A_0}{2\pi} (h\kappa^\kappa)^{-ps-1} \Gamma(\kappa ps + \tfrac{1}{2});$$

compare (8.1.38). Then we find approximately (when $m_n = M_n$)

$$\begin{aligned}
R_n(a; M_n) &\sim \frac{\kappa A_0}{2\pi} X_n^{\vartheta - \frac{1}{2}} \frac{1}{2\pi i} \int_{-c+M_n-\infty i}^{-c+M_n+\infty i} \Gamma\left(\kappa ps + \tfrac{1}{2}\right) \frac{\sin \frac{1}{2}\pi ps}{\sin \pi s} X_n^{-\kappa ps} \, ds \\
&= (-)^{M_n} \frac{A_0}{2\pi p} X_n^{\vartheta - \nu_n} \frac{1}{2\pi i} \int_{-c-\infty i}^{-c+\infty i} \Gamma(\xi + \nu_n) \\
&\quad \times \frac{\sin \frac{1}{2}\pi p(s + M_n)}{\sin \pi s} X_n^{-\xi} \, d\xi \\
&= (-)^{M_n} \frac{A_0}{4\pi i p} X_n^{\vartheta - \nu_n} \frac{1}{2\pi i} \int_{-c-\infty i}^{-c+\infty i} \frac{\Gamma(\xi + \nu_n)}{\sin \pi \xi} \\
&\quad \times \left\{ e^{\frac{1}{2}\pi i p M_n} (e^{-\pi i / 2\kappa} X_n)^{-\xi} \right. \\
&\quad \left. - e^{-\frac{1}{2}\pi i p M_n} (e^{\pi i / 2\kappa} X_n)^{-\xi} \right\} \frac{\sin \pi (p-1)s}{\sin \pi s} \, d\xi.
\end{aligned} \tag{8.1.61}$$

In the above integrals the constant c can be chosen to satisfy $0 < c < 1$, and we have let $s \to s + M_n$, introduced the new variable $\xi = \kappa ps = (p-1)s$ and put $\nu_n = \kappa p M_n + \tfrac{1}{2}$.

From the discussion in §5.2.2, it can be seen that the dominant contribution to the integral of type (8.1.61) containing the factor $(X_n e^{i\theta})^{-\xi}$ arises from the vertical contour in the upper half plane when $\theta > 0$, where $\sin \pi(p-1)s/$

$\sin \pi s \sim \exp\{-\pi i(p-2)s\}$, and in the lower half plane when $\theta < 0$, where $\sin \pi(p-1)s / \sin \pi s \sim \exp\{\pi i(p-2)s\}$. This observation then enables us to estimate the integral (8.1.61) in the form

$$
R_n(a; M_n) \sim (-)^{M_n} \frac{A_0}{4\pi i p} X_n^{\vartheta - \nu_n} \frac{1}{2\pi i} \int_{-c-\infty i}^{-c+\infty i} \frac{\Gamma(\xi + \nu_n)}{\sin \pi \xi} \left\{ e^{\frac{1}{2}\pi i p M_n} (X_n e^{-i\phi})^{-\xi} \right.
$$
$$
\left. - e^{-\frac{1}{2}\pi i p M_n} (X_n e^{i\phi})^{-\xi} \right\} d\xi, \tag{8.1.62}
$$

where we have put

$$
\phi = \frac{\pi}{2\kappa} + \frac{\pi(p-2)}{p-1} = \pi + \tfrac{1}{2}\pi \left(\frac{p-2}{p-1} \right).
$$

We observe that $\phi \simeq \pi$ in the transition region $p \simeq 2$.

From the definition of the terminant function $\hat{T}_\nu(z)$ given in (6.2.6) and (6.2.7) in the form of a Mellin–Barnes integral, we can immediately identify the integrals in (8.1.62) in terms of the terminant functions $\hat{T}_\nu(X_n e^{\pm i\phi})$ to find

$$
R_n(a; M_n) \sim \frac{A_0}{2\pi p} X_n^\vartheta e^{-\pi i \nu_n} \left\{ e^{X_n e^{i\phi} + i\omega_n} \hat{T}_{\nu_n}(X_n e^{i\phi}) \right.
$$
$$
\left. - e^{X_n e^{-i\phi} - i\omega_n} \hat{T}_{\nu_n}(X_n e^{-i\phi}) \right\},
$$

where we have defined $\omega_n = \phi\nu_n + (1 - \tfrac{1}{2}p)\pi M_n$. Since M_n corresponds to the optimal truncation value in (8.1.60), we require the asymptotics of $\hat{T}_{\nu_n}(X_n e^{\pm i\phi})$ for $X_n \to \infty$ when $\nu_n = \kappa p M_n + \tfrac{1}{2} \sim X_n$ and $\phi \simeq \pi$. From (6.2.56), we then have† (for $n = 1, 2, \ldots$)

$$
\hat{T}_{\nu_n}(X_n e^{i\phi}) \sim \tfrac{1}{2} + \tfrac{1}{2}\mathrm{erf}\{(\phi - \pi)(\tfrac{1}{2}X_n)^{\frac{1}{2}}\}
$$

and‡

$$
\hat{T}_{\nu_n}(X_n e^{-i\phi}) \sim e^{2\pi i \nu_n} \left\{ -\tfrac{1}{2} + \tfrac{1}{2}\mathrm{erf}\{(\pi - \phi)(\tfrac{1}{2}X_n)^{\frac{1}{2}}\} \right\}
$$

in the neighbourhood of $\phi = \pi$.

Let us introduce the variable $\epsilon = p - 2$ to measure transition through the classical Euler–Jacobi case. Use of the identities $\omega_n = \pi\nu_n - \pi\vartheta/(2\kappa)$ and $\phi = 2\pi - \pi/(2\kappa)$, together with the fact that $\mathrm{erf}(-z) = -\mathrm{erf}(z)$, then shows that we

---

† The quantity $c(\phi)$ appearing in the argument of the error function in (6.2.56) satisfies $c(\phi) \simeq \phi - \pi$ in the neighbourhood of $\phi = \pi$.

‡ This follows, for example, from the connection formula (6.2.45) and use of the above result for $\hat{T}_{\nu_n}(X_n e^{i\phi})$.

can define the Stokes multipliers $\mathcal{S}_n(\epsilon)$ by writing $R_n(a; M_n)$ in the form

$$R_n(a; M_n) = \frac{A_0}{\pi p} X_n^{\vartheta} e^{X_n \cos(\pi/2\kappa)} \cos\{X_n \sin\frac{\pi}{2\kappa} + \frac{\pi\vartheta}{2\kappa}\} \mathcal{S}_n(\epsilon).$$

Then in the neighbourhood of $p = 2$ (i.e., when $\phi \simeq \pi$) we have the leading behaviour of the Stokes multipliers given by

$$\mathcal{S}_n(\epsilon) \simeq \tfrac{1}{2} + \tfrac{1}{2}\mathrm{erf}\{(\phi - \pi)(\tfrac{1}{2}X_n)^{\frac{1}{2}}\}$$

$$\simeq \tfrac{1}{2} + \tfrac{1}{2}\mathrm{erf}\left\{\tfrac{1}{2}\pi \frac{\epsilon}{1+\epsilon}(\tfrac{1}{2}X_n)^{\frac{1}{2}}\right\} \qquad (8.1.63)$$

as $a \to 0+$, where we recall that $X_n = \kappa(2\pi h n/a^{1/p})^{1/\kappa}$.

From (8.1.58) and (8.1.32), we finally arrive at our desired result in the form

$$S_p(a) - \frac{\Gamma(1/p)}{pa^{1/p}} - \tfrac{1}{2} = \mathbf{E}_p(a) + \mathbf{H}_p(a), \qquad (8.1.64)$$

where

$$\mathbf{H}_p(a) = \frac{1}{\pi} \sum_{n=1}^{\infty} \frac{1}{n} \sum_{k=1}^{M_n-1} \frac{(-)^{k-1}}{k!} \Gamma(kp+1) \sin(\tfrac{1}{2}\pi kp)(a/(2\pi n)^p)^k \qquad (8.1.65)$$

and

$$\mathbf{E}_p(a) = \frac{2\pi}{a^{1/p}} \sum_{n=1}^{\infty} R_n(a; M_n)$$

$$= \frac{2A_0}{pa^{1/p}} \sum_{n=1}^{\infty} X_n^{\vartheta} e^{X_n \cos(\pi/2\kappa)} \cos\left\{X_n \sin\frac{\pi}{2\kappa} + \frac{\pi\vartheta}{2\kappa}\right\} \mathcal{S}_n(\epsilon). \qquad (8.1.66)$$

We thus see that *each* exponential in the infinite sum over n in (8.1.66) is associated with its own Stokes multiplier, each of which is described by (8.1.63) in the limit $a \to 0+$ as we pass through the value $p = 2$.

We can now summarise the principal finding of this section. When the algebraic expansion associated with each value of n is optimally truncated according to (8.1.60), the transition of the Stokes multipliers $\mathcal{S}_n(\epsilon)$ ($n = 1, 2, \ldots$) in the neighbourhood of the classical value $p = 2$ is smooth: $\mathcal{S}_n(\epsilon)$ changes from approximately zero when $p < 2$ ($\epsilon < 0$) to approximately 1 when $p > 2$ ($\epsilon > 0$) within a region centred on $p = 2$ of width $O((a^{1/p}/n)^{p/2(p-1)})$ as $a \to 0+$. We note that this transition region becomes sharper as n increases. Thus, outside this transition region, when $p < 2$ the expansion of $S_p(a)$ is given by the algebraic expansion $\mathbf{H}_p(a)$ only (which is equivalent to $H_p(a)$ in (8.1.46)), while when $p > 2$ the expansion in (8.1.66) agrees with the leading exponential terms in the expansion $e_0(a; p)$ in (8.1.45) (with $N = 1$). When $p = 2$ we have $\mathcal{S}_n(0) \simeq \frac{1}{2}$. However, in this case we know from §8.1.2 that $\mathcal{S}_n(0) = \frac{1}{2}$ *exactly* (and not just to leading order).

Two additional remarks on the nature of the appearance of the expansion $e_0(a; p)$ are in order. First, the transition is characterised by the (dominant) algebraic expansion $\mathbf{H}_p(a)$ actually vanishing when $p = 2$. And secondly, we remark that the variation through $p = 2$ is *asymmetric* in the transition variable ϵ due to the presence of the factor $\epsilon/(1+\epsilon)$ in the argument of the error function in (8.1.63). This will result in a skewed distribution of the Stokes multiplier as a function of p.

To verify these conclusions we have carried out numerical calculations for the transition of the leading exponential term in (8.1.66) with $n = 1$. We subtract off the corresponding algebraic expansion (with $n = 1$) from the left-hand side of (8.1.64) and compute the value of

$$
\mathcal{S}_1(\epsilon) \equiv \left\{ S_p(a) - \frac{\Gamma(1/p)}{pa^{1/p}} - \frac{1}{2} \right.
$$
$$
\left. - \frac{1}{\pi} \sum_{k=1}^{M_1-1} \frac{(-)^{k-1}}{k!} \Gamma(kp+1) \sin(\tfrac{1}{2}\pi kp)(a/(2\pi)^p)^k \right\}
$$
$$
\div \frac{2A_0}{pa^{1/p}} X_n^{\vartheta} e^{X_n \cos(\pi/2\kappa)} \cos\left\{ X_n \sin \frac{\pi}{2\kappa} + \frac{\pi\vartheta}{2\kappa} \right\}
$$

as a function of p for a fixed a, where the truncation index M_1 corresponds to the optimal truncation scheme in (8.1.60) and $X_1 = \kappa(2\pi h/a^{1/p})^{1/\kappa}$. The results are presented in Fig. 8.6 for the particular case $a = 0.2$. It can be seen that the error function approximation (8.1.63) agrees reasonably well (for this relatively

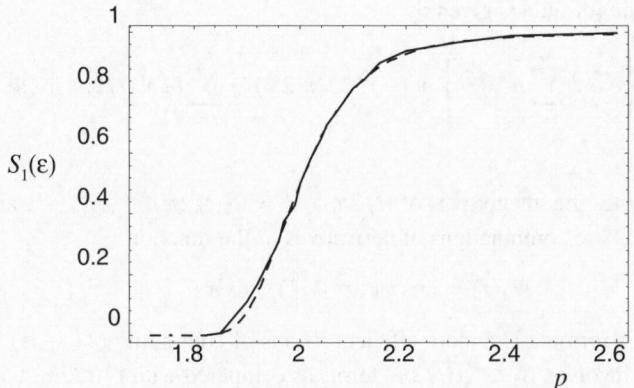

Fig. 8.6. The behaviour of the Stokes multiplier $\mathcal{S}_1(\epsilon)$ (where $\epsilon = p - 2$) in the neighbourhood of $p = 2$ when $a = 0.2$. The computed values are shown by the solid curve and the approximate dependence in (8.1.63) is shown by the dashed curve.

large value of a) with the calculated Stokes multiplier profile. In particular, the asymmetric nature of the profile is plainly evident.

8.2 An Asymptotic Formula for $\zeta\left(\frac{1}{2} + it\right)$

8.2.1 Introduction

The earliest published method of computing the Riemann zeta function $\zeta(s)$ as a function of the complex variable $s = \sigma + it$ is the so-called Gram formula [Gram (1903)]. This results from an application of the Euler-Maclaurin summation formula and is an asymptotic expansion valid for all s ($\neq 1$) given by [Haselgrove (1963); Bender & Orszag† (1978, p. 379)]

$$\zeta(s) \sim \sum_{n=1}^{N-1} n^{-s} + \frac{N^{1-s}}{s-1} + \tfrac{1}{2}N^{-s} + N^{-s} \sum_{m=1}^{\infty} \frac{B_{2m}\,(s)_{2m-1}}{(2m)!\,N^{2m-1}}, \tag{8.2.1}$$

where B_{2m} are the Bernoulli numbers and $(\alpha)_m = \Gamma(\alpha+m)/\Gamma(\alpha)$. The asymptotic parameter in this expansion is N. Inspection of the terms in the sum over m shows that for this to possess an asymptotic character (i.e., for the absolute value of the terms to initially decrease before ultimately diverging), N must be chosen such that $N > |s|/2\pi$. On the critical line $\sigma = \frac{1}{2}$, the number of terms in the finite main sum (over n) in (8.2.1) when t is large is thus of $O(|t|/2\pi)$.

A more powerful means of computing $\zeta(s)$ for large t (> 0) on the critical line is the Riemann-Siegel formula‡ [Edwards (1974); Haselgrove (1963)]. This is usually presented for the *real*, even function $Z(t) = \exp\{i\vartheta(t)\}\zeta(\frac{1}{2}+it)$, where the phase-angle

$$\vartheta(t) = \arg\Gamma\left(\tfrac{1}{4} + \tfrac{1}{2}it\right) - \tfrac{1}{2}t\log\pi.$$

This asymptotic formula is given by

$$Z(t) \sim 2\mathrm{Re}\left\{e^{i\vartheta(t)} \sum_{n=1}^{N_t} n^{-\frac{1}{2}-it}\right\} + (-)^{N_t-1}(t/2\pi)^{-\frac{1}{4}} \sum_{r=0}^{\infty}(-)^r (t/2\pi)^{-\frac{1}{2}r}\Psi_r(z),$$

$$\tag{8.2.2}$$

where N_t denotes the integer part of $(t/2\pi)^{\frac{1}{2}}$, $z = 2\{(t/2\pi)^{\frac{1}{2}} - N_t\} - 1$ and the functions $\Psi_r(z)$ are combinations of derivatives of the function

$$\Psi_0(z) = \cos\pi(\tfrac{1}{2}z^2 + \tfrac{3}{8})/\cos\pi z.$$

For $t \gg 1$, this formula is a more efficient means of computing $\zeta(\frac{1}{2} + it)$, since the main sum involves $N_t \simeq \sqrt{(t/2\pi)}$ terms as compared with $O(t/2\pi)$ terms in

---

† There is a misprint in this reference: the $n!$ in Eq. (8.1.23) should be $(2n)!$.
‡ This formula, which was discovered by Siegel in the posthumous papers of Riemann dating from the late 1850s, was finally published in 1932.

the Gram formula. For a recent investigation of the behaviour of the coefficients $\Psi_r(z)$ for large r, see Berry (1995).

Recently, some new asymptotic formulas for $Z(t)$ have been developed in which the terms in the original Dirichlet series $\sum_{n=1}^{\infty} n^{-s}$ (valid in $\sigma > 1$) are smoothed by a function which is approximately unity for $n \lesssim N_t$ and decays to zero for $n \gtrsim N_t$. The first is an expansion derived by Berry & Keating (1992), in which the leading term is given by the *convergent* sum

$$2\mathrm{Re}\, e^{i\vartheta(t)} \sum_{n=1}^{\infty} n^{-\frac{1}{2}-it} \times \tfrac{1}{2}\mathrm{erfc}\left\{\sqrt{(t/2)}\,\xi(n,t)/q(\kappa,t)\right\}, \tag{8.2.3}$$

where $\xi(n,t) = \log n - \vartheta'(t)$, $q^2(\kappa, t) = \kappa^2 - it\vartheta''(t)$ and κ is a real free parameter. For large t, $\xi(n,t) \simeq \log(n/N_t)$ and $q^2(\kappa, t) \simeq \kappa^2 - \frac{1}{2}i$, so that (8.2.3) resembles the finite main sum of the Riemann-Siegel formula, but with the sharp cut-off after N_t terms smoothed away by the complementary error function. One advantage of such a smoothed formula over the Riemann-Siegel formula is the removal of the discontinuity in the main sum as a function of t resulting from the discrete upper limit N_t. Other expansions for $Z(t)$, with the normalised incomplete gamma function

$$Q(a, z) = \frac{\Gamma(a, z)}{\Gamma(a)} = \frac{1}{\Gamma(a)} \int_z^{\infty} u^{a-1} e^{-u} \, du \quad (|\arg z| < \pi) \tag{8.2.4}$$

as the smoothing function, have been developed in Paris (1994b) and Paris & Cang (1997b). These expansions rely on the uniform asymptotics of $Q(a, z)$ (when the phase of a and z are close to $\frac{1}{2}\pi$) and also yield an expansion that involves a main sum smoothed by a complementary error function, although with a different argument to that in (8.2.3). It is found that in these expansions the smoothing function is approximately unity for $n \lesssim N_t$ and decays to zero for $n \gtrsim N_t$; the cut-off therefore occurs after roughly N_t terms so that these formulas are all of the computationally more powerful Riemann-Siegel type.

In this section we derive an expansion which involves the incomplete gamma function with a free parameter a. Suitable choice of this parameter then enables us to present expansions of different asymptotic character. When $a = O(1)$, we obtain a main sum which is smoothed either by a simple exponential factor or a complementary error function of real argument. In these cases it is found that for these terms to possess an asymptotic character, it is necessary to take the cut-off in the smoothed main sum to occur after $O(t/2\pi)$ terms, resulting in a formula of the less powerful Gram-type. On the other hand, when we choose $a = \frac{1}{2}s$, we obtain the Riemann-Siegel-type expansion developed in Paris & Cang (1997b).

8.2.2 An Expansion for $\zeta(s)$

Let p be a positive real number and a, K denote arbitrary complex parameters which will be restricted to satisfy $|\arg K| < \pi/4p$ and $\mathrm{Re}(a) \geq 0$. We consider the sum of normalised incomplete gamma functions given by

$$S = \sum_{n=1}^{\infty} n^{-s} Q(a, (n/K)^{2p}), \tag{8.2.5}$$

where $s = \sigma + it$ and $Q(a, z)$ is defined in (8.2.4). From the asymptotic behaviour [Abramowitz & Stegun (1965, p. 263)] $\Gamma(a, z) \sim z^{a-1} e^{-z}$ as $|z| \to \infty$ in $|\arg z| < \frac{3}{2}\pi$, it can be seen that the late terms in S are controlled by $n^{2p(a-1)-\sigma} \exp\{-(n/K)^{2p}\}$ (when $n/|K| \gg 1$), so that, provided $|\arg K| < \pi/4p$, the sum S converges absolutely for all a and s.

We now employ the Mellin-Barnes integral representation of the incomplete gamma function $\Gamma(a, x)$ given by (3.4.12)

$$\Gamma(a, x) = \frac{1}{2\pi i} \int_{c-\infty i}^{c+\infty i} \Gamma(a + u) x^{-u} \frac{du}{u} \qquad (|\arg x| < \tfrac{1}{2}\pi),$$

where $c > 0$. Substitution of this integral into the sum S, followed by reversal of the order of summation and integration (which is justified by absolute convergence) and introduction of the new variable $z = 2pu$, then leads to the Perron-type formula

$$S = \frac{1}{2\pi i \Gamma(a)} \int_{c-\infty i}^{c+\infty i} \Gamma\left(a + \frac{z}{2p}\right) K^z \zeta(s + z) \frac{dz}{z}, \tag{8.2.6}$$

$$|\arg K| < \pi/4p, \quad c > \max\{0, 1 - \sigma\}.$$

The integrand in (8.2.6) has poles at $z = 0$ and $z = 1 - s$, together with (in general) an infinite sequence of poles in $\mathrm{Re}(z) \leq 0$ resulting from the gamma function. As we are primarily interested in s values situated in the critical strip $0 < \sigma < 1$, we shall suppose for convenience in presentation that $\sigma \leq 1$. This ensures that the pole at $z = 1 - s$ lies in $\mathrm{Re}(z) \geq 0$, thereby avoiding the possibility of the formation of a double pole. Displacement of the path of integration and evaluation of the residues at the simple poles $z = 0$ and $z = 1 - s$ then shows that

$$\zeta(s) = S + \frac{\Gamma(a + (1 - s)/2p)}{\Gamma(a)} \frac{K^{1-s}}{s - 1} + J, \tag{8.2.7}$$

where, provided $|\arg K| < \pi/4p$,

$$J = \frac{i}{2\pi \Gamma(a)} \int_{-c-\infty i}^{-c+\infty i} \Gamma\left(a + \frac{z}{2p}\right) K^z \zeta(s + z) \frac{dz}{z}, \tag{8.2.8}$$

and $0 < c/2p < \text{Re}(a)$. We remark that the restriction on σ is easily removed to include the case $\sigma > 1$ by suitable indentation (when necessary) of the path of integration in (8.2.8) to lie to the left of the pole $z = 1 - s$ but to the right of the poles of the gamma function.†

Straightforward displacement of the path of integration in (8.2.8) to the left over the simple poles of the gamma function at $z = -2p(m + a)$ for nonnegative integer m, combined with the functional relation (8.1.31) expressed in the form

$$\zeta(s) = \chi(s)\zeta(1 - s), \tag{8.2.9}$$

where

$$\chi(s) = 2^s \pi^{s-1} \sin \tfrac{1}{2}\pi s \, \Gamma(1 - s) = \pi^{s-\frac{1}{2}} \frac{\Gamma(\frac{1}{2} - \frac{1}{2}s)}{\Gamma(\frac{1}{2}s)}, \tag{8.2.10}$$

then leads to the result

$$J = \frac{\chi(s)}{\Gamma(a)} \sum_{m=0}^{M-1} \frac{(-)^m}{m!} \frac{(2\pi K)^{-2p(m+a)}}{m + a} A_m + R_M, \tag{8.2.11}$$

where $M = 1, 2, \ldots$. The coefficients A_m are defined by

$$A_m = \frac{\sin \tfrac{1}{2}\pi(s - 2p(m + a))}{\sin \tfrac{1}{2}\pi s} \frac{\Gamma(1 - s + 2p(m + a))}{\Gamma(1 - s)}$$

$$\times \zeta(1 - s + 2p(m + a)) \tag{8.2.12}$$

and the remainder R_M, when $|\arg K| < \pi/4p$, is given by

$$R_M = \frac{i}{2\pi \Gamma(a)} \int_{-c_M - \infty i}^{-c_M + \infty i} \Gamma\left(a + \frac{z}{2p}\right) K^z \zeta(s + z) \frac{dz}{z}, \tag{8.2.13}$$

where $c_M = c + 2p(M - 1 + \text{Re}(a))$ and $0 < c < 2p$.

Combination of (8.2.7) and (8.2.11) then gives the desired expansion for $\zeta(s)$ in the form [Paris & Cang (1997a)]

$$\zeta(s) = \sum_{n=1}^{\infty} n^{-s} Q(a, (n/K)^{2p}) + \frac{\Gamma(a + (1 - s)/2p)}{\Gamma(a)} \frac{K^{1-s}}{s - 1}$$

$$+ \frac{\chi(s)}{\Gamma(a)} \sum_{m=0}^{M-1} \frac{(-)^m}{m!} \frac{(2\pi K)^{-2p(m+a)}}{m + a} A_m + R_M. \tag{8.2.14}$$

This representation is seen to involve the original Dirichlet series smoothed by the incomplete gamma function. We remark that the coefficients A_m in (8.2.12) enjoy the unusual property of involving the zeta function itself. The expansion (8.2.14)

---

† This separation of the poles will not be possible when $s = 1 + 2p(m + a)$, $m = 0, 1, 2, \ldots$, due to the presence of a double pole at $z = 1 - s$. In this case, the second term on the right-hand side of (8.2.7) would disappear. Since this situation can only arise for $\sigma > 1$ we do not consider this further.

will therefore be of computational use only if $\text{Re}(a) > \sigma/2p$, for then the zeta functions in (8.2.12) can be computed simply from the convergent Dirichlet series.

8.2.3 An Exponentially-Smoothed Gram-Type Expansion

To simplify the expansion (8.2.14) we now choose $a = 1$, since $Q(1, z) = e^{-z}$, and let K be a real positive parameter. This choice† results in the main sum (over n) being smoothed by the real exponential factor $\exp\{-(n/K)^{2p}\}$. If we further suppose p to be a positive integer, the coefficients A_m in (8.2.12) then assume the simpler form

$$A_{m-1} = (-)^{pm} (1 - s)_{2pm} \, \zeta(1 - s + 2pm) \qquad (m \geq 1).$$

As an illustration, we take $p = 1$ to find

$$\zeta(s) = \sum_{n=1}^{\infty} n^{-s} e^{-(n/K)^2} - \tfrac{1}{2} K^{1-s} \Gamma(\tfrac{1}{2} - \tfrac{1}{2}s)$$

$$- \chi(s) \sum_{m=1}^{M} \frac{(-)^m}{m!} (2\pi K)^{-2m} A_{m-1} + R_M. \qquad (8.2.15)$$

This formula exhibits a simple exponential smoothing in which the terms in the main sum effectively 'switch off' for values of n given by $n^* \sim K$, where the parameter K, as yet, has not been specified. The choice of K, however, is unfortunately not entirely at our disposal. This results from the large-m behaviour of the terms in the finite sum in (8.2.15) which, since $\zeta(1 - s + 2pm) \to 1$ as $m \to \infty$, is controlled essentially by the behaviour of

$$\left(\frac{t}{2\pi K}\right)^{2m} \frac{|(1-s)_{2m}|}{m! \, t^{2m}} \qquad (m \gg 1).$$

It is easily shown that, for the terms in this sequence to possess an asymptotic character, it is necessary to choose $K \gtrsim t/2\pi$. With K chosen in this manner, the terms in (8.2.15) will at first decrease to a minimum value at $m = M_0$, before finally diverging in typical asymptotic fashion. The optimal truncation point M_0 is given approximately by $M_0 \simeq (t/2\lambda)^2$, when $\lambda \equiv t/2\pi K = O(1)$. Since the main sum is then smoothed after $n^* \simeq t/2\pi$ terms, the formula (8.2.15) consequently has the character of a Gram-type formula, rather than that of a more powerful Riemann-Siegel-type formula.

In (8.2.15) we now choose $K = t/2\pi$ and confine our attention to the critical line $\sigma = \tfrac{1}{2}$, where, from (8.2.10), we note that $\chi(\tfrac{1}{2} + it) = \exp(-2i\vartheta(t))$. A bound on the remainder term R_M is discussed in Paris & Cang (1997a), where

---

† Another obvious choice is $a = \tfrac{1}{2}$, since $Q(\tfrac{1}{2}, z^2) = \text{erfc}(z)$, and the terms in the main sum are then smoothed by $\text{erfc}\{(n/K)^p\}$.

Table 8.1. *Values of the coefficient*
$G_M(t)$ *when* $t = 50$ *and* $t = 100$

M	$G_M(50)$	$G_M(100)$
5	1.09122	1.02230
10	1.79844	1.16447
15	5.99574	1.62045
20	48.5599	2.98869
25	1042.88	7.82060

it is shown that on the critical line (when $a = p = 1$)

$$|R_M| < \frac{2^{\frac{3}{2}}\zeta(2M + \frac{3}{2})C_M(t)}{\pi\Gamma(M + \frac{3}{2})}. \tag{8.2.16}$$

The coefficient $C_M(t)$ is given by

$$C_M(t) = G_M(t)\{e^{-\chi}\sinh\chi + \chi^{-N}\Gamma(N + 1, \chi)\cosh\chi\},$$

where

$$G_M(t) = \left(\tfrac{1}{2}t\right)^{-N}\left|\left(\tfrac{1}{4} - \tfrac{1}{2}N + \tfrac{1}{2}it\right)_N\right|$$

and $\chi = \pi t/4$, $N = 2M + 1$. We remark that $\zeta(2M + \frac{3}{2}) \simeq 1$ for $M \gg 1$ and that for fixed M, $G_M(t) \to 1$ and $C_M(t) \to 1$ as $t \to \infty$. The function $G_M(t)$ is determined by direct computation; in Table 8.1 we show values of $G_M(t)$ for different M when $t = 50$ and $t = 100$.

In Table 8.2 we show the real function $Z(t)$ computed from (8.2.15) with M chosen to guarantee an accuracy of 25 decimals. For the lowest value $t = 10$, M was chosen to be the optimal value $M_0 = 24$, while for the other t values truncation of the series in (8.2.15) was highly sub-optimal. The bound for R_M in (8.2.16) is unfortunately not sharp enough to establish the asymptotic nature of the expansion in (8.2.15). For modest values of M (which would, however, be quite sufficient for most computational purposes) this bound turns out to be very realistic, but is too crude for values of M near optimal truncation. To illustrate this, we compare in Table 8.3 the bound (8.2.16) for different truncations M (when $K = t/2\pi$) with the actual error incurred in computing $Z(t)$ for two different values of t. Inspection of this table reveals that the bound is quite realistic until $M \simeq t/2$. For higher values of M, it is found that the bound begins to deteriorate and at optimal truncation is quite useless.

8.2.4 A Riemann-Siegel-Type Expansion

We demonstrate how the result in (8.2.6) can yield a different type of expansion for $\zeta(s)$. To see this, we set $a = \tfrac{1}{2}s$ with $p = 1$ and let K denote an arbitrary

Table 8.2. *Computation of* $Z(t)$ *from* (5.2.14)
for different values of t. *When* $t = 10$, *optimal*
truncation yields a value accurate to 5 decimals
only. For the other t *values* M *was chosen to*
yield an accuracy of 25 decimals

$a = 1$	$p = 1$	$K = t/2\pi$	
t		Z_{approx}	M
10		$-1.54918\ 98595$	24
20		$+1.14784\ 24121\ 85197\ 27763\ 50341$	50
30		$+0.59602\ 85192\ 39884\ 95531\ 85143$	33
40		$-1.30888\ 23934\ 56599\ 15901\ 61454$	29
50		$-0.34073\ 50059\ 55024\ 98275\ 33166$	27

Table 8.3. *Values of* $|R_M|$ *for* $t = 50$ *and* $t = 100$

	$a = 1$	$p = 1$	$K = t/2\pi$	
	$t = 50$		$t = 100$	
M	$\|Z - Z_{\text{approx}}\|$	$\|R_M\|$	$\|Z - Z_{\text{approx}}\|$	$\|R_M\|$
5	1.366×10^{-3}	4.047×10^{-3}	1.250×10^{-3}	3.453×10^{-3}
10	4.588×10^{-8}	2.070×10^{-7}	2.763×10^{-8}	1.038×10^{-7}
15	3.268×10^{-13}	2.464×10^{-12}	7.709×10^{-14}	3.706×10^{-13}
20	1.217×10^{-18}	2.330×10^{-17}	6.102×10^{-20}	3.689×10^{-19}
25	3.718×10^{-24}	5.198×10^{-22}	2.123×10^{-26}	1.654×10^{-25}
30	1.188×10^{-29}	6.189×10^{-26}	4.290×10^{-33}	4.605×10^{-32}

complex parameter satisfying $|\arg K| < \frac{1}{4}\pi$. In the derivation of this expansion we shall find it convenient to restrict our attention to the critical strip $0 \le \sigma \le 1$, appealing to analytic continuation to establish the result for general values of s and $|\arg K| \le \frac{1}{4}\pi$. From (8.2.12), we then obtain†

$$A_0 = -\frac{1}{2}\pi \operatorname{cosec} \frac{1}{2}\pi s / \Gamma(1 - s) \tag{8.2.17}$$

with $A_m \equiv 0\ (m \ge 1)$. The finite sum over m in (8.2.14) now reduces to the single term with $m = 0$

$$\frac{(2\pi K)^{-s} \chi(s)}{\frac{1}{2}s\Gamma(\frac{1}{2}s)} A_0 = -\frac{K^{-s}}{s\Gamma(\frac{1}{2}s)}.$$

This situation corresponds to the integrand in (8.2.8) possessing only a single simple pole at $z = -s$ on the left of the path of integration, with the remaining

---

† This follows from the result that $\zeta(1 + 2x) \sin \pi x \to \frac{1}{2}\pi$ as $x \to 0$.

poles of the gamma function being cancelled by the trivial zeros of $\zeta(s+z)$ situated at $z = -s - 2k$, $k = 1, 2, \ldots$. Then the expansion (8.2.14) in this case becomes

$$\zeta(s) = \sum_{n=1}^{\infty} n^{-s} Q(\tfrac{1}{2}s, (n/K)^2) + \frac{K^{-s}}{\Gamma(\tfrac{1}{2}s)} \left(\frac{\pi^{\frac{1}{2}} K}{s-1} - \frac{1}{s} \right) + R_M,$$

where, from (8.2.13),

$$R_M = \frac{i}{2\pi \Gamma(\tfrac{1}{2}s)} \int_{-c-\infty i}^{-c+\infty i} \Gamma(\tfrac{1}{2}s + \tfrac{1}{2}z) K^z \zeta(s+z) \frac{dz}{z}$$

with $|\arg K| < \tfrac{1}{4}\pi$. Since there are no poles of the above integrand to the left of $z = -s$, the condition on c can be relaxed to $c > \sigma$.

Use of the functional relation for $\zeta(s)$ in (8.2.9) then shows that

$$R_M = \frac{\chi(s)}{\Gamma(\tfrac{1}{2} - \tfrac{1}{2}s)} \frac{1}{2\pi i} \int_{c-\infty i}^{c+\infty i} \Gamma(\tfrac{1}{2} - \tfrac{1}{2}s + \tfrac{1}{2}z)(\pi K)^{-z} \zeta(1 - s + z) \frac{dz}{z},$$

where we have replaced the variable z by $-z$. This integral is seen to be of the same form as that defining S on the right-hand side of (8.2.6) (when $a = \tfrac{1}{2}s$ and $p = 1$), with s replaced by $1 - s$ and the parameter K by $(\pi K)^{-1}$. Hence, we deduce the result

$$R_M = \chi(s) \sum_{n=1}^{\infty} n^{s-1} Q(\tfrac{1}{2} - \tfrac{1}{2}s, (\pi n K)^2).$$

It then follows that in the critical strip

$$\zeta(s) = \frac{K^{-s}}{\Gamma(\tfrac{1}{2}s)} \left(\frac{\pi^{\frac{1}{2}} K}{s-1} - \frac{1}{s} \right) + \sum_{n=1}^{\infty} n^{-s} Q(\tfrac{1}{2}s, (n/K)^2)$$

$$+ \chi(s) \sum_{n=1}^{\infty} n^{s-1} Q(\tfrac{1}{2} - \tfrac{1}{2}s, (\pi n K)^2) \tag{8.2.18}$$

when $|\arg K| < \pi/4$. Since $Q(a, z) \sim z^{a-1} e^{-z}/\Gamma(a)$ as $|z| \to \infty$ in $|\arg z| < \tfrac{3}{2}\pi$, both the above sums converge absolutely for all s when $|\arg K| < \tfrac{1}{4}\pi$. By analytic continuation the expansion (8.2.18) therefore holds for all values of s ($\neq 1$) in $|\arg K| \leq \tfrac{1}{4}\pi$. The particular case with $K = 1/\sqrt{\pi}$ was effectively embodied in Riemann's famous 1859 paper [Riemann (1859)], though he did not explicitly identify the incomplete gamma functions. The result (8.2.18) is the expansion given by Lavrik (1968) for the Dirichlet L-function specialised to $\zeta(s)$.

If we make the choice $K = (\pi i)^{-\frac{1}{2}}$ in (8.2.18) and employ the conjugacy property $Q(\bar{a}, \bar{z}) = \overline{Q(a, z)}$ and $\chi(\tfrac{1}{2} + it) = e^{-2i\vartheta(t)}$, then we obtain the elegant result

$$Z(t) = 2\mathrm{Re}\, e^{i\vartheta(t)} \left\{ \sum_{n=1}^{\infty} n^{-s} Q(\tfrac{1}{2}s, \pi n^2 i) - \frac{\pi^{\frac{1}{2}s} e^{\frac{1}{4}\pi i s}}{s\, \Gamma(\tfrac{1}{2}s)} \right\}, \tag{8.2.19}$$

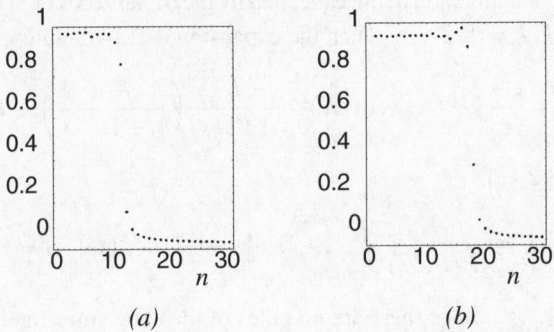

Fig. 8.7. The smoothing function $|Q(\tfrac{1}{2}s, \pi n^2 i)|$ as a function of n when (a) $t = 400$ and (b) $t = 1000$.

valid on the critical line $s = \tfrac{1}{2} + it$. For large values of t, use of the uniform asymptotics of the incomplete gamma function then enables an asymptotic formula to be constructed for $Z(t)$; for details see Paris & Cang (1997b) and Paris (2000a), where (8.2.19) is derived by an alternative approach involving a slight modification of one of the original methods used by Riemann. We observe at this point that the nature of this expansion is already apparent in (8.2.19). For large variables, the incomplete gamma function $Q(a, z)$ changes its asymptotic character in the neighbourhood of its transition point given by $z/a = 1$. Consequently, the smoothing function $Q(\tfrac{1}{2}s, \pi n^2 i)$ in (8.2.19) changes its form when $\tfrac{1}{2}s \sim \pi n^2 i$, being approximately of unit modulus for $n \lesssim n^*$ and decaying to zero for $n \gtrsim n^*$, where n^* corresponds to the 'cut-off' given by $n^* \simeq \sqrt{(t/2\pi)}$ – see Fig. 8.7. The asymptotic formula generated by (8.2.19) is thus seen to be of the computationally more powerful Riemann-Siegel type.

Finally, we mention that a similar procedure can be applied for integer values of $p > 1$. If we now set $a = s/2p$, the coefficients A_m are again given by (8.2.17) with $A_m \equiv 0\,(m \geq 1)$. The evaluation of the associated remainder term R_M is, however, not so straightforward and is found to involve a sum of 'generalised' incomplete gamma functions. We do not enter into any of the details here but merely content ourselves with a statement of the generalised expansion in the form

$$
\zeta(s) = \frac{K^{-s}}{\Gamma(s/2p)} \left(\frac{\Gamma(1/2p)K}{s-1} - \frac{p}{s} \right) + \sum_{n=1}^{\infty} n^{-s} Q\left(\frac{s}{2p}, (n/K)^{2p} \right)
$$

$$
+ \chi(s) \sum_{n=1}^{\infty} n^{s-1} \mathcal{Q}_p\left(\frac{1-s}{2p}, (\pi n K)^{2p} \right) \tag{8.2.20}
$$

valid for all $s\;(\neq 1)$ when $|\arg K| \leq \tfrac{1}{4}\pi$. The function $\mathcal{Q}_p(a, z)$ denotes the generalised incomplete gamma function defined by

$$Q_p(a, z) = \frac{\Gamma(\frac{1}{2} - ap)}{p\Gamma(ap)\,\Gamma(1/2p - a)} \int_z^\infty u^{a-1} F_{2p}(2u^{1/2p})\,du \quad (|\arg z| < \pi),$$

where $F_p(z)$ is defined in (8.1.4), and is normalised so that $Q(a, 0) = 1$. When $p = 1$, we have $F_2(2u^{\frac{1}{2}}) = e^{-u}$ so that $Q_1(a, z) \equiv Q(a, z)$, and (8.2.20) reduces to (8.2.18). Details are given in Paris (1994a) together with a discussion of the significance of this result.

8.3 The Asymptotics of Pearcey's Integral

8.3.1 Introduction

An important problem in the theoretical treatment of many short-wavelength phenomena, such as wave propagation and optical diffraction, is the uniform asymptotic evaluation of oscillatory integrals with several nearly coincident stationary phase (or saddle) points. The uniform approximation of such integrals can be expressed in terms of certain canonical integrals and their derivatives [Ursell (1972); Connor (1973)]. The role played by these canonical diffraction integrals in the analysis of caustic wavefields is analogous to that played by complex exponentials in plane wave theory.

The family of canonical oscillatory integrals is classified according to the hierarchy introduced to describe the type of singularity arising in catastrophe theory. For one-dimensional integrals, the exponents are the polynomial transformations associated with the so-called cuspoid catastrophes. The simplest case, corresponding to the fold catastrophe, involves two coalescing stationary points whose positions depend on a single real parameter. The canonical integral in this case is the familiar integral defining the Airy function. The next integral in the hierarchy corresponds to the cusp catastrophe and involves three coalescing stationary points and two real parameters X and Y. The canonical form of this integral is Pearcey's integral, given by

$$P(X, Y) = \int_{-\infty}^\infty \exp\left\{i(u^4 + Xu^2 + Yu)\right\} du. \tag{8.3.1}$$

The next two integrals in the sequence correspond to the so-called swallowtail and butterfly integrals and involve, respectively, four and five stationary points.

The integral in (8.3.1) was first evaluated numerically by Pearcey (1946) in his investigation of the electromagnetic field near a cusp, although it appears to have been first studied asymptotically by Brillouin (1916). The utility of (8.3.1) and of the higher integrals in the above hierarchy depends on detailed information concerning their numerical values and large-parameter behaviour. Extensive work concerning the numerical evaluation of the Pearcey and the swallowtail integrals has been described by Connor and Farrelly (1981a,b), Connor & Curtis

(1982, 1984) and Connor *et al.* (1983). Recent asymptotic investigations have been
undertaken by Stamnes & Spjelkavik (1983), Kaminski (1989) and Paris (1991)
for the Pearcey integral (see also §1.1.3), and Kaminski (1992) for the swallowtail integral. An asymptotic study of the zeros of $P(X, Y)$ is given in Kaminski
& Paris (1999).

The standard procedure for dealing with the asymptotics of $P(X, Y)$ for large
real values of X and Y depends on the detailed structure of the stationary points of
the phase function $f(u) = u^4 + Xu^2 + Yu$. The stationary points are determined
by $f'(u) = 0$, which yields three real distinct stationary points for (X, Y) situated
inside the cusped caustic (specified by the curve on which $f'(u) = f''(u) = 0$)

$$Y^2 + \left(\tfrac{2}{3}X\right)^3 = 0; \qquad (8.3.2)$$

on the caustic (away from the origin) two of these stationary points coalesce to form
a double stationary point. For (X, Y) situated outside the caustic, only one stationary point is real with the other two forming a complex conjugate pair. By examining
the topography of the paths of steepest descent for real X and Y, Wright (1980) has
shown that for (X, Y) lying inside the caustic all three stationary points contribute
to the asymptotics of $P(X, Y)$, while in the interior of the second caustic (the Stokes
set) $Y^2 - (\tfrac{1}{3}X)^3(5 + 3\sqrt{3}) = 0$, centred about the positive X axis, only the real
stationary point makes a contribution. In the domain containing the Y axis between
these two caustics, the asymptotics of $P(X, Y)$ are controlled by the real stationary
point and only one of the complex stationary points. The numbers of contributory
stationary points in the different regions of the XY plane are shown in Fig. 8.8.

In physical applications the cusp diffraction integral involves only real values
of X and Y. A deeper understanding of the rich structure of this integral can
be achieved, however, by considering the analytic continuation of $P(X, Y)$ for

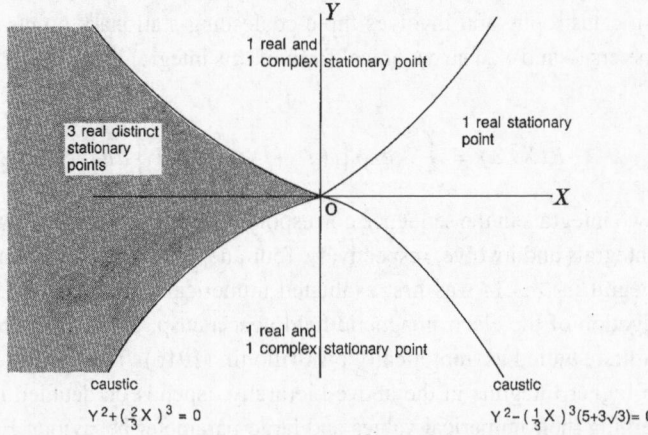

Fig. 8.8. The numbers of contributory stationary points in different domains of the XY
plane.

arbitrary complex values of X and Y. The approach we adopt here consists of expressing (8.3.1) as a Mellin-Barnes integral involving a Weber parabolic cylinder function. The asymptotics when either $|X|$ or $|Y| \to \infty$ can then be determined without reference to the above-mentioned stationary points.

8.3.2 A Mellin-Barnes Integral Representation

The integral $P(X, Y)$ is defined only for $0 \leq \arg X \leq \pi$ and real values of Y. To obtain its analytic continuation for complex values of X and Y, we rotate the path of integration in (8.3.1) through an angle of $\pi/8$, thereby removing the rapidly oscillatory term $\exp(iu^4)$, and define the new variables $t = u \exp(-\frac{1}{8}\pi i)$ and

$$x = 2^{-\frac{1}{2}}Xe^{-\frac{1}{4}\pi i}, \qquad y = 2^{-\frac{5}{4}}Ye^{\frac{1}{8}\pi i}. \tag{8.3.3}$$

Application of Jordan's lemma [Whittaker & Watson (1965, p. 115)] then shows that, for real X and Y, the contribution from the arcs from $\pm\infty$ to $\pm\infty \exp(\frac{1}{8}\pi i)$ vanishes, so that (8.3.1) may be written in the form

$$\hat{P}(x, y) = 2e^{\frac{1}{8}\pi i} \int_0^\infty \exp\left\{ -t^4 - 2^{\frac{1}{2}}xt^2 \right\} \cos(2^{\frac{5}{4}} yt)\, dt. \tag{8.3.4}$$

This last integral is absolutely convergent for all complex values of x and y, and so represents the analytic continuation of $P(X, Y)$ to arbitrary values of X and Y. It is easily seen that $\hat{P}(x, y)$ satisfies the symmetry and conjugacy relations

$$\hat{P}(x, y) = \hat{P}(x, -y), \qquad e^{-\frac{1}{8}\pi i}\hat{P}(x, y) = \overline{e^{-\frac{1}{8}\pi i}\hat{P}(\bar{x}, \bar{y})}, \tag{8.3.5}$$

where the bar denotes the complex conjugate.

From (8.1.28) and also (3.3.5), we obtain

$$\cos z = \frac{1}{2\pi i} \int_C \Gamma(s)z^{-s} \cos \tfrac{1}{2}\pi s\, ds \quad (z \neq 0), \tag{8.3.6}$$

where C denotes a loop that starts and finishes at $-\infty$ and encircles $s = 0$ in the positive sense. This representation is valid for all finite z ($\neq 0$) *without restriction*[†] on $\arg z$. Substitution of this integral into (8.3.4), where we suppose that $\mathrm{Re}(s) < 1$, then yields (when $y \neq 0$)

$$\hat{P}(x, y) = 2e^{\pi i/8} \frac{1}{2\pi i} \int_C \Gamma(s) \cos \tfrac{1}{2}\pi s \left(2^{\frac{5}{4}} y\right)^{-s}$$

$$\times \left\{ \int_0^\infty \exp\left\{ -t^4 - 2^{\frac{1}{2}}xt^2 \right\} t^{-s}\, dt \right\} ds$$

[†] Note that when the contour is swung round to be parallel to the imaginary axis, the resulting integral then defines $\cos z$ only in the sector $|\arg z| < \frac{1}{2}\pi$; see §3.3.1.

upon reversal of the order of integration. The inner integral can be evaluated in terms of the parabolic cylinder function $D_\nu(z)$, since [Erdélyi (1953, Vol. 2, p. 119)]

$$D_\nu(z) = \frac{e^{-z^2/4}}{\Gamma(-\nu)} \int_0^\infty e^{-\frac{1}{2}\tau^2 - z\tau} \tau^{-\nu-1} d\tau \quad (\text{Re}(\nu) < 0).$$

We then find [Paris (1991)]

$$e^{-\pi i/8} \hat{P}(x, y) = 2^{-\frac{1}{4}} e^{x^2/4} \frac{1}{2\pi i} \int_C \Gamma(s) \Gamma(\tfrac{1}{2} - \tfrac{1}{2}s)$$

$$\times \cos \tfrac{1}{2}\pi s \, D_{\frac{1}{2}s-\frac{1}{2}}(x)(2y)^{-s} \, ds \tag{8.3.7}$$

$$= 2^{-\frac{1}{4}} \pi^{\frac{1}{2}} e^{x^2/4} \frac{1}{2\pi i} \int_C \Gamma(s) D_{s-\frac{1}{2}}(x) y^{-2s} \, ds, \tag{8.3.8}$$

upon use of the properties of the gamma function and replacement of the variable s by $2s$. Since the parabolic cylinder function is an entire function of s and the integrand in (8.3.8) has no poles in the right half-plane, the restriction $\text{Re}(s) < 1$ imposed in the derivation can be removed by analytic continuation.

From the asymptotic behaviour of $D_\nu(z)$ for fixed z and large $|\nu|$ (see (8.3.25)), the convergence of the integral (8.3.8) is seen to be controlled by the ratio of gamma functions $\Gamma(s)/\Gamma(\tfrac{3}{4} - \tfrac{1}{2}s)$. Application of Rule 2 in §2.4 then shows that (8.3.8) is valid for all finite x and nonzero values of y. The integral (8.3.1) can also be viewed as a case of the generalised Faxén integral, discussed in §7.2, which is therefore expressible as a double Mellin-Barnes integral. The above representation can be derived from this form upon evaluation of the inner integral in terms of the parabolic cylinder function.

Before proceeding with the discussion of the asymptotics of $\hat{P}(x, y)$ for large $|x|$ or $|y|$, we note that the special cases $x = 0$ and $y = 0$ can be represented in terms of simpler functions and the corresponding asymptotics obtained. In the case $y = 0$ we have

$$\hat{P}(x, 0) = 2^{-\frac{1}{4}} e^{x^2/4 + \pi i/8} \left(\tfrac{1}{4}x^2\right)^{\frac{1}{4}} K_{\frac{1}{4}}\left(\tfrac{1}{4}x^2\right), \tag{8.3.9}$$

where K_ν denotes the modified Bessel function of the second kind. When $x = 0$, we find, using (8.3.4), (8.1.11) and (8.1.49) when $p = 4$, that

$$\hat{P}(0, y) = 2 e^{\pi i/8} \int_0^\infty e^{-t^4} \cos(2^{\frac{5}{4}} yt) \, dt$$

$$= \tfrac{1}{4} e^{\pi i/8} \left\{ U_{4,1}\left(2^{\frac{3}{4}} iy\right) + U_{4,1}\left(-2^{\frac{3}{4}} iy\right) \right\}.$$

From the asymptotics of the function $U_{4,1}(z)$ given in Paris & Wood (1986, §3.5) – see also (8.1.49) and §2.3 – we obtain the expansion in $|\arg y| \le \frac{1}{2}\pi$

$$\hat{P}(0, y) \sim \begin{cases} E_1(0, y) + E_2(0, y) & \text{in } |\arg y| < \frac{3}{8}\pi \\ E_1(0, y) & \text{in } \frac{3}{8}\pi < \arg y \le \frac{1}{2}\pi \\ E_2(0, y) & \text{in } -\frac{1}{2}\pi \le \arg y < -\frac{3}{8}\pi, \end{cases} \quad (8.3.10)$$

where the formal asymptotic sums $E_{1,2}(0, y)$ $(r = 1, 2)$ are given by

$$e^{-\pi i/8} E_r(0, y) = 2^{-\frac{1}{4}} \sqrt{\frac{\pi}{3}} y^{-\frac{1}{3}} \sum_{k=0}^{\infty} c_k \left(\frac{3}{2} y^{\frac{4}{3}}\right)^{-k}$$

$$\times \exp\left\{ -\frac{3}{2} y^{\frac{4}{3}} e^{\pm \frac{1}{3}\pi i} \pm \frac{2}{3}\pi i k \pm \frac{1}{6}\pi i \right\} \quad (8.3.11)$$

with the upper and lower signs corresponding to $r = 1$ and $r = 2$, respectively. The coefficients c_k satisfy the recurrence relation (8.1.54) and, from (8.1.57) with $p = 4$, the first few values are given by†

$$c_0 = 1, \quad c_1 = \frac{7}{48}, \quad c_2 = \frac{385}{4608}, \quad c_3 = \frac{39655}{663552},$$

$$c_4 = \frac{665665}{127401984}, \quad c_5 = -\frac{1375739365}{6115295232}, \ldots .$$

The function $\hat{P}(0, y)$ is exponentially small for large $|y|$ in $|\arg(\pm y)| < \frac{1}{8}\pi$ and exhibits a Stokes phenomenon on the rays $\arg y = \pm\frac{3}{8}\pi$ (and, by symmetry from (8.3.5), on the rays $\arg(-y) = \pm\frac{3}{8}\pi$). These rays are Stokes lines and are the asymptotes as $|y| \to \infty$ of the curve $\text{Im}(y^{\frac{4}{3}} e^{\frac{1}{3}\pi i} - y^{\frac{4}{3}} e^{-\frac{1}{3}\pi i}) = 0$ on which the leading behaviours of $E_1(0, y)$ and $E_2(0, y)$ are most unequal in magnitude. As one crosses the Stokes line (in the sense of increasing $|\arg y|$), the coefficient multiplying the exponentially subdominant expansion, namely $E_2(0, y)$ in $\arg y > 0$ and $E_1(0, y)$ in $\arg y < 0$, changes from unity to become zero in the sectors $\frac{3}{8}\pi < |\arg y| \le \frac{1}{2}\pi$, respectively.

8.3.3 Asymptotics of $\hat{P}(x, y)$ for $|x| \to \infty$

We first consider the asymptotic expansion of $\hat{P}(x, y)$ for large $|x|$ when y is finite. Proceeding formally for the moment, we substitute the expansion for large $|x|$ of the parabolic cylinder function [Whittaker & Watson (1965, p. 347); see also §5.1.2 and §5.4.3]

$$D_\nu(z) \sim z^\nu e^{-z^2/4} \sum_{k=0}^{\infty} \frac{(-\nu)_{2k}}{k!} (-2z^2)^{-k} \quad (|\arg z| < \tfrac{3}{4}\pi), \quad (8.3.12)$$

---

† A closed-form evaluation of the coefficients c_k in (8.3.11) has been given recently by Senouf (1996) as

$$c_k = \frac{\Gamma(k + \frac{1}{2})}{(12^k \sqrt{\pi})} \sum_{r=0}^{k} \binom{-k - \frac{1}{2}}{r + k} \binom{r + k}{2r} \left(\frac{8}{3}\right)^r.$$

into (8.3.8). When $|\arg x| < \frac{1}{2}\pi$, this yields

$$\hat{P}(x, y) \sim \pi^{\frac{1}{2}} e^{\pi i/8} \sum_{k=0}^{\infty} \frac{(-)^k}{k!} (2x^2)^{-k-\frac{1}{4}} A_k(\chi),$$

where the coefficients $A_k(\chi)$ are defined by

$$A_k(\chi) = \frac{1}{2\pi i} \int_C \frac{\Gamma(s)}{\Gamma(\frac{1}{2} - s)} \Gamma(\tfrac{1}{2} - s + 2k) \chi^{-s} \, ds, \qquad \chi = y^2/x$$

$$= \frac{\Gamma(2k + \frac{1}{2})}{\Gamma(\frac{1}{2})} e^{-\chi} {}_1F_1(-2k; \tfrac{1}{2}; \chi);$$

see (3.4.5) and Kummer's transformation in §6.2.3. In the above integral the loop C encloses the poles of $\Gamma(s)$. From the relation expressing the Hermite polynomials $H_{2n}(x)$ in terms of the confluent hypergeometric function ${}_1F_1(-n; \tfrac{1}{2}; x^2)$ [Abramowitz & Stegun (1965, p. 780)], it is seen that $A_k(\chi) = e^{-\chi} a_k(\chi)$, where

$$a_k(\chi) = \frac{\Gamma(2k + \frac{1}{2})}{\Gamma(\frac{1}{2})} {}_1F_1(-2k; \tfrac{1}{2}; \chi) = 2^{-4k} H_{4k}(\chi^{\frac{1}{2}}).$$

The first few coefficients $a_k(\chi)$ are consequently

$$a_0(\chi) = 1, \quad a_1(\chi) = \tfrac{3}{4} - 3\chi + \chi^2,$$

$$a_2(\chi) = \tfrac{105}{16} - \tfrac{105}{2}\chi + \tfrac{105}{2}\chi^2 - 14\chi^3 + \chi^4, \dots .$$

The expansion of $\hat{P}(x, y)$ as $|x| \to \infty$ in $|\arg x| < \frac{1}{2}\pi$ is then given by

$$\hat{P}(x, y) \sim 2^{-\frac{1}{4}} \sqrt{\frac{\pi}{x}} e^{\pi i/8} S_1(x, y), \tag{8.3.13}$$

where $S_1(x, y)$ denotes the formal asymptotic sum

$$S_1(x, y) = e^{-\chi} \sum_{k=0}^{\infty} \frac{(-)^k}{k!} a_k(\chi)(2x^2)^{-k}, \qquad \chi = y^2/x. \tag{8.3.14}$$

To derive the large-x expansion in the remainder of the complex x plane, we use the connection formula [Whittaker & Watson (1965, p. 348)]

$$D_\nu(z) = e^{\pm \pi i \nu} D_\nu(ze^{\mp \pi i}) + \frac{\sqrt{2\pi}}{\Gamma(-\nu)} e^{\pm \frac{1}{2}\pi i(\nu+1)} D_{-\nu-1}(\mp iz), \tag{8.3.15}$$

where, in order to apply (8.3.12) when we identify z with x, the signs will be chosen such that both arguments of the parabolic cylinder functions on the right-hand side of (8.3.15) lie between $\pm \frac{1}{2}\pi$. This means that we select the upper signs in (8.3.15) when $\frac{1}{2}\pi < \arg x < \pi$ and the lower signs when $-\pi < \arg x < -\frac{1}{2}\pi$. Substitution of (8.3.15) into (8.3.8) then yields

$$2^{\frac{1}{4}}\pi^{-\frac{1}{2}}e^{-\pi i/8 - x^2/4}\hat{P}(x,y) = \frac{1}{2\pi i}\int_C \Gamma(s)y^{-2s}\left\{e^{\pm\pi i(s-\frac{1}{2})}D_{s-\frac{1}{2}}(xe^{\mp\pi i})\right.$$

$$\left.+ \frac{\sqrt{2\pi}}{\Gamma(\frac{1}{2}-s)}e^{\pm\frac{1}{2}\pi i(s+\frac{1}{2})}D_{-s-\frac{1}{2}}(\mp ix)\right\}ds$$

so that, according to the above ranges of arg x,

$$\hat{P}(x,y) = \mp i\hat{P}(xe^{\mp\pi i}, \mp iy) + 2^{-\frac{1}{4}}\pi^{\frac{1}{2}}e^{x^2/4 + \pi i/8}H, \qquad (8.3.16)$$

where

$$H = \sqrt{2\pi}e^{\pm\pi i/4}\frac{1}{2\pi i}\int_C \frac{\Gamma(s)}{\Gamma(\frac{1}{2}-s)}D_{-s-\frac{1}{2}}(\mp ix)(ye^{\mp\pi i/4})^{-2s}ds.$$

The expansion of the first term on the right-hand side of (8.3.16) in $|\arg(-x)| < \frac{1}{2}\pi$ follows from (8.3.13) and (8.3.14) as

$$\mp i\hat{P}(xe^{\mp\pi i}, \mp iy) \sim 2^{-\frac{1}{4}}\sqrt{\frac{\pi}{x}}e^{\pi i/8}S_1(x,y). \qquad (8.3.17)$$

The expansion of the second term can be obtained by formal substitution of (8.3.12) into the integral for H to find

$$H \sim \pm i\sqrt{\frac{2\pi}{x}}e^{x^2/4}\sum_{k=0}^{\infty}\frac{(2x^2)^{-k}}{k!}B_k(\xi),$$

where the coefficients $B_k(\xi)$ are defined by

$$B_k(\xi) = \frac{1}{2\pi i}\int_C \frac{\Gamma(s)}{\Gamma(\frac{1}{2}-s)}\frac{\Gamma(\frac{1}{2}+s+2k)}{\Gamma(\frac{1}{2}+s)}\left(\frac{1}{2}\xi\right)^{-2s}ds, \qquad \xi = 2(xe^{\mp\pi i})^{\frac{1}{2}}y$$

$$= \frac{1}{2\pi i}\int_C \frac{\Gamma(s)}{\Gamma(\frac{1}{2}-s-2k)}\left(\frac{1}{2}\xi\right)^{-2s}ds = \left(\frac{1}{2}\xi\right)^{2k+\frac{1}{2}}J_{-2k-\frac{1}{2}}(\xi) \qquad (8.3.18)$$

upon use of the representation for the Bessel function in (3.4.22).

Then, as $|x| \to \infty$ in $|\arg(-x)| < \frac{1}{2}\pi$, we obtain

$$H \sim \pm i\sqrt{\frac{2}{x}}e^{x^2/4}S_2(x,y), \qquad (8.3.19)$$

where $S_2(x,y)$ is the formal asymptotic sum

$$S_2(x,y) = \sum_{k=0}^{\infty}\frac{(-)^k}{k!}(y^2/2x)^k\left(\frac{1}{2}\pi\xi\right)^{\frac{1}{2}}J_{-2k-\frac{1}{2}}(\xi)$$

$$= \sum_{k=0}^{\infty} \frac{(y^2/2x)^k}{k!} \{ p(2k, \xi) \cos \xi - q(2k, \xi) \sin \xi \}. \tag{8.3.20}$$

This last result follows from the properties of the spherical Bessel function [Watson (1966, p. 55)] where

$$J_{-2k-\frac{1}{2}}(\xi) = (-)^k \left(\tfrac{1}{2} \pi \xi \right)^{-\frac{1}{2}} \{ p(2k, \xi) \cos \xi - q(2k, \xi) \sin \xi \},$$

with the finite sums $p(2k, \xi)$ and $q(2k, \xi)$ being given by the polynomials in ξ^{-1}

$$p(2k, \xi) = \sum_{n=0}^{k} \frac{(-)^n (2k + 2n)!}{(2n)!(2k - 2n)!} (2\xi)^{-2n},$$

$$q(2k, \xi) = \sum_{n=0}^{k-1} \frac{(-)^n (2k + 2n + 1)!}{(2n + 1)!(2k - 2n - 1)!} (2\xi)^{-2n-1}.$$

From (8.3.16), (8.3.17) and (8.3.19) it follows that $\hat{P}(x, y)$ has the asymptotic expansion for $|x| \to \infty$ in $|\arg(-x)| < \tfrac{1}{2}\pi$ given by

$$\hat{P}(x, y) \sim 2^{-\frac{1}{4}} \sqrt{\frac{\pi}{x}} e^{\pi i/8} \left\{ S_1(x, y) \pm i \sqrt{2} e^{x^2/2} S_2(x, y) \right\}, \tag{8.3.21}$$

where the upper or lower sign is chosen according as $\tfrac{1}{2}\pi < \arg x < \pi$ or $-\pi < \arg x < -\tfrac{1}{2}\pi$, respectively.

The expansions (8.3.13) and (8.3.21) have been obtained by a formal process of appropriate substitution of the asymptotic expansion of the parabolic cylinder function into (8.3.8). To establish the asymptotic nature of these expansions, it is necessary to consider truncated versions of $S_1(x, y)$ and $S_2(x, y)$ and to estimate the order of their associated remainder terms. To see this for the expansion (8.3.13), we derive an alternative representation for $\hat{P}(x, y)$ in which the parameter group $\chi = y^2/x$ is made apparent. In the process, we shall see that this new form enables us to show that the sector of validity of (8.3.13) can be extended (in the Poincaré sense) to $|\arg x| < \tfrac{3}{4}\pi$, thereby dealing with the expansion of $\hat{P}(x, y)$ on the imaginary x-axis.

Employing the Mellin-Barnes integral representation for $D_\nu(z)$ in (3.4.14), we find from (8.3.8), upon reversal of the order of integration and use of (3.4.5),

$$e^{-\pi i/8} \hat{P}(x, y) = 2^{-\frac{1}{4}} \sqrt{\frac{\pi}{x}} \frac{1}{2\pi i} \int_{-\infty i}^{\infty i} \Gamma(s)(2x^2)^s$$

$$\times \left\{ \frac{1}{2\pi i} \int_C \frac{\Gamma(t)\Gamma\left(\frac{1}{2} - 2s - t\right)}{\Gamma\left(\frac{1}{2} - t\right)} \chi^{-t} dt \right\} ds$$

$$= 2^{-\frac{1}{4}} \sqrt{\frac{\pi}{x}} e^{-\chi} \frac{1}{2\pi i} \int_{-\infty i}^{\infty i} \frac{\Gamma(s)\Gamma\left(\frac{1}{2} - 2s\right)}{\Gamma\left(\frac{1}{2}\right)} {}_1F_1\left(2s; \tfrac{1}{2}; \chi\right)(2x^2)^s ds$$

$$\tag{8.3.22}$$

when $|\arg x| < \frac{3}{4}\pi$, where the path of integration is indented at $s = 0$ to separate the poles of $\Gamma(s)$ and $\Gamma(\frac{1}{2} - 2s)$. From Buchholz (1969, p. 97) or Slater (1960, p. 70), the behaviour of the confluent hypergeometric function in (8.3.22) for large $|s|$ on the contour is given by $\exp(\frac{1}{2}\chi)\cos[2\sqrt{\{(\frac{1}{4} - 2s)\chi\}}]$ valid for all finite complex χ. It then follows that on the path, the integral is dominated by the product of gamma functions $\Gamma(s)\Gamma(\frac{1}{2} - 2s)$, whence from Rule 1 in §2.4, (8.3.22) defines $\hat{P}(x, y)$ in the sector $|\arg x| < \frac{3}{4}\pi$. Displacement of the path of integration to the left over the poles at $s = k, k = 0, 1, \ldots, N - 1$, with residue $(-2x^2)^{-k}a_k(\chi)/k!$, then yields

$$e^{-\pi i/8}\hat{P}(x, y) = 2^{-\frac{1}{4}}\sqrt{\frac{\pi}{x}}e^{-\chi}\left\{\sum_{k=0}^{N-1}\frac{(-)^k a_k(\chi)}{k!(2x^2)^k} + R_N\right\}$$

when $|\arg x| < \frac{3}{4}\pi$, where R_N is the integral taken along the shifted path $\mathrm{Re}(s) = -N + \frac{1}{2}$. It is then readily shown that $|R_N| = O(|x|^{-2N+1})$ as $|x| \to \infty$ in $|\arg x| < \frac{3}{4}\pi$, thereby establishing the asymptotic nature of the expansion $S_1(x, y)$ in (8.3.13). The treatment of the expansion (8.3.19) in $|\arg(-x)| < \frac{1}{2}\pi$ is less straightforward and will not be given here; for details, see Paris (1991, § 4b).

In conclusion, we may state that the asymptotic expansion of $\hat{P}(x, y)$ for $|x| \to \infty$ and y finite is given by

$$\hat{P}(x, y) \sim \begin{cases} 2^{-\frac{1}{4}}\sqrt{\dfrac{\pi}{x}}e^{\pi i/8}S_1(x, y) & (|\arg x| < \frac{3}{4}\pi) \\[2ex] 2^{-\frac{1}{4}}\sqrt{\dfrac{\pi}{x}}e^{\pi i/8}\left\{S_1(x, y) \pm i\sqrt{2}e^{x^2/2}S_2(x, y)\right\} & (|\arg(-x)| < \frac{1}{2}\pi), \end{cases}$$

$$(8.3.23)$$

where $S_1(x, y)$ and $S_2(x, y)$ are the formal asymptotic sums defined in (8.3.14) and (8.3.20). The upper or lower sign is to be chosen according as $\frac{1}{2}\pi < \arg x < \pi$ or $-\pi < \arg x < -\frac{1}{2}\pi$, respectively. The expansion of $\hat{P}(x, 0)$ can be shown to agree with the large-x asymptotics of the modified Bessel function in (8.3.9), when the sum $S_2(x, 0)$ is evaluated by a limiting process as $y \to 0$. The above expansion shows that $\hat{P}(x, y)$ is exponentially large in the sector $|\arg(-x)| < \frac{1}{4}\pi$ and possesses a dominant algebraic behaviour in the remainder of the x plane. The first expansion in (8.3.23) is valid in the Poincaré sense in $|\arg x| < \frac{3}{4}\pi$, since, in the sectors $\frac{1}{2}\pi < |\arg x| < \frac{3}{4}\pi$, the exponential contribution is subdominant. However, this expansion is 'complete' only in the sector $|\arg x| < \frac{1}{2}\pi$, since it is associated with a Stokes phenomenon on $\arg x = \pm\frac{1}{2}\pi$. The ray $\arg x = \pi$ is also a Stokes line where the exponential expansion assumes maximal dominance over the algebraic expansion.

8.3.4 Asymptotics of $\hat{P}(x, y)$ for $|y| \to \infty$

From the symmetry property (8.3.5) it is sufficient to confine our attention to the sector $|\arg y| \le \frac{1}{2}\pi$. Using the connection formula (8.3.15) in the form (where we have taken the upper signs and replaced ν by $-\nu - 1$ and $-iz$ by z)

$$D_\nu(z) = \frac{\Gamma(1+\nu)}{\sqrt{2\pi}} \left\{ e^{\frac{1}{2}\pi i \nu} D_{-\nu-1}(iz) + e^{-\frac{1}{2}\pi i \nu} D_{-\nu-1}(-iz) \right\},$$

we obtain from (8.3.8)

$$\hat{P}(x, y) = 2^{\frac{1}{4}} \pi^{\frac{1}{2}} e^{x^2/4 + \pi i/8} \left\{ e^{-\frac{1}{4}\pi i} K(x, ye^{-\frac{1}{4}\pi i}) + e^{\frac{1}{4}\pi i} K(-x, ye^{\frac{1}{4}\pi i}) \right\},$$

where

$$K(x, y) = \frac{1}{2\pi i} \int_C \Gamma(2s) D_{-s-\frac{1}{2}}(ix)(2y)^{-2s} \, ds. \qquad (8.3.24)$$

As in (8.3.8), C is a loop with endpoints at $-\infty$ which encircles the poles of $\Gamma(s)$. Since the parabolic cylinder function is an entire function of its order there are no poles of the integrand in (8.3.24) in $\mathrm{Re}(s) > 0$. Accordingly, we now deform C such that $|s|$ is everywhere large on C; compare §5.6. On the expanded loop we may employ the asymptotics of the parabolic cylinder function for large order and finite argument [Paris (1991, Eq. (A.10))]

$$D_{-s-\frac{1}{2}}(z) \sim \frac{\pi^{\frac{1}{2}} \exp(-z\sqrt{s})}{2^{\frac{1}{2}s+\frac{1}{4}} \Gamma(\frac{1}{2}s + \frac{3}{4})} \sum_{r=0}^{\infty} (-)^r A_r s^{-r/2}, \qquad (8.3.25)$$

valid for $|s| \to \infty$ in $|\arg s| < \pi$, where

$$A_0 = 1, \quad A_1 = \frac{z^3}{24}, \quad A_2 = \frac{z^2}{48}\left(\frac{z^2}{24} - 3\right), \dots .$$

In addition, from Lemma 2.2, we have

$$\frac{\Gamma(2s)}{\Gamma(\frac{1}{2}s + \frac{3}{4})} = \sqrt{\frac{2}{\pi}}\left(\frac{3^{\frac{3}{4}}}{4}\right)^{1-2s} \Gamma(\frac{3}{2}s - \frac{1}{4})\{1 + O(s^{-1})\} \qquad (8.3.26)$$

for $|s| \to \infty$ in $|\arg s| < \pi$. Substitution of these expansions into (8.3.24) with the expanded loop shows that

$$K(x, y) = \frac{1}{2}\left(\frac{3}{2}\right)^{\frac{3}{4}} \frac{1}{2\pi i} \int_C \Gamma(\frac{3}{2}s - \frac{1}{4})\left(\frac{3}{2}y^{\frac{4}{3}}\right)^{-\frac{3}{2}s}$$

$$\times e^{-ix\sqrt{s}}\{1 - A_1 s^{-\frac{1}{2}} + O(s^{-1})\} ds.$$

The above integral can be evaluated asymptotically for large $|y|$ by means of the following lemma given in Paris (1991):

Lemma 8.1. *For arbitrary complex α and μ, the integral*

$$\frac{1}{2\pi i} \int_C \Gamma(s) e^{\alpha\sqrt{s}} z^{-s} s^\mu \, ds,$$

*where C denotes a loop with endpoints at $-\infty$ encircling the poles of $\Gamma(s)$, has
the expansion*

$$z^{\mu} \exp\{-z + \alpha\sqrt{z} - \tfrac{1}{8}\alpha^2\} \sum_{r=0}^{\infty} B_r z^{-r/2},$$

for $|z| \to \infty$ in $|\arg z| < \pi$, where

$$B_0 = 1, \quad B_1 = \tfrac{1}{8}\alpha\{1 - 4\mu - \tfrac{1}{12}\alpha^2\}, \dots .$$

Application of this lemma with $\mu = 0$ and $\mu = -\tfrac{1}{2}$, followed by some straightforward algebra, then yields the leading terms in the expansion of $\hat{P}(x, y)$ for
$|y| \to \infty$ (valid in the Poincaré sense) given by

$$\hat{P}(x, y) \sim \begin{cases} E_1(x, y) + E_2(x, y) & \text{in } |\arg y| \le \tfrac{1}{8}\pi \\ E_1(x, y) & \text{in } \tfrac{1}{8}\pi < \arg y < \tfrac{1}{2}\pi \\ E_2(x, y) & \text{in } -\tfrac{1}{2}\pi < \arg y < -\tfrac{1}{8}\pi, \end{cases} \tag{8.3.27}$$

where

$$E_r(x, y) = 2^{-\frac{1}{4}}\sqrt{\frac{\pi}{3}}\, y^{-\frac{1}{3}} e^{x^2/3 + \pi i/8} \exp\left\{-\tfrac{3}{2}y^{\frac{4}{3}}e^{\pm\frac{1}{3}\pi i} \pm ixy^{\frac{2}{3}}e^{\pm\frac{1}{6}\pi i} \pm \tfrac{1}{6}\pi i\right\}$$

$$\times \left\{1 + \tfrac{1}{6}xy^{-\frac{2}{3}}e^{\pm\frac{1}{3}\pi i}\left(1 - \tfrac{2}{9}x^2\right) + O\left(y^{-\frac{4}{3}}\right)\right\} \tag{8.3.28}$$

with the upper and lower signs corresponding to $r = 1$ and $r = 2$, respectively. When $x = 0$, the above expressions reduce to those given in (8.3.11). Both
exponential terms $E_1(x, y)$ and $E_2(x, y)$ vanish as $|y| \to \infty$ in $|\arg y| < \tfrac{1}{8}\pi$,
and $\hat{P}(x, y)$ is consequently exponentially small in the sectors $|\arg(\pm y)| < \tfrac{1}{8}\pi$.
Outside these sectors $\hat{P}(x, y)$ is exponentially large. The method employed here
is not sufficiently precise to determine the domains of validity of the subdominant
terms in the above expansion and their associated Stokes lines. This deficiency
results from the manner of approximation of the ratio of gamma functions in
(8.3.26), which does not correctly take into account the appearance of expo-
nentially small terms in the large-$|z|$ behaviour of $\Gamma(z)$ across its Stokes lines
$\arg z = \pm\tfrac{1}{2}\pi$; see §6.4. By means of alternative arguments using the saddle point
approach, the sectors of validity in (8.3.27) for the asymptotics of $\hat{P}(x, y)$ in the
'complete' sense of Olver (1964) can be shown to correspond to those in (8.3.10);
see Paris (1991, p. 415).

As in the case $x = 0$ given in (8.3.10), the function $\hat{P}(x, y)$ exhibits a
Stokes phenomenon (for fixed x) along curves which ultimately approach the rays
$\arg y = \pm\tfrac{3}{8}\pi$ as $|y| \to \infty$. Across the Stokes lines $\arg y = \pm\tfrac{3}{8}\pi$ the coefficients
multiplying the exponentially small terms $E_2(x, y)$ and $E_1(x, y)$, respectively,
change to become zero in the adjacent sectors $\tfrac{3}{8}\pi < |\arg y| \le \tfrac{1}{2}\pi$. The ray
$\arg y = 0$ is an anti-Stokes line where the two sums $E_1(x, y)$ and $E_2(x, y)$ are
equally significant.

8.3.5 Asymptotics of $P(X, Y)$ for Real X, Y

From the results of the preceding sections, we can now state the asymptotics of $P(X, Y)$ in terms of the original (real) variables X and Y. From (8.3.3), it is seen that $X < 0$ corresponds to the anti-Stokes line $\arg x = \frac{3}{4}\pi$, which is the upper boundary of the exponentially large sector (when $|x| \to \infty$). Thus, for $X \to -\infty$, $P(X, Y)$ will consist of an algebraic and an exponentially oscillatory expansion, while for $X \to +\infty$, $P(X, Y)$ decays algebraically like $1/\sqrt{X}$. From (8.3.23) (with the upper sign), the asymptotic expansion of $P(X, Y)$ for large real X and finite Y is found to be

$$
P(X, Y) \sim \begin{cases}
\sqrt{\pi/X}\, e^{\frac{1}{4}\pi i} S_1\big(2^{-\frac{1}{2}} X e^{-\frac{1}{4}\pi i}, 2^{-\frac{5}{4}} Y e^{\frac{1}{8}\pi i}\big) & (X \to +\infty) \\[2mm]
\sqrt{\pi/|X|}\, e^{-\frac{1}{4}\pi i} \Big\{ S_1\big(2^{-\frac{1}{2}} |X| e^{\frac{3}{4}\pi i}, 2^{-\frac{5}{4}} Y e^{\frac{1}{8}\pi i}\big) \\[2mm]
\quad + i\sqrt{2}\, e^{-\frac{1}{4}iX^2} S_2\big(2^{-\frac{1}{2}} |X| e^{\frac{3}{4}\pi i}, 2^{-\frac{5}{4}} Y e^{\frac{1}{8}\pi i}\big) \Big\} & (X \to -\infty).
\end{cases}
$$
$$(8.3.29)$$

Using (8.3.14) and (8.3.20) (with $\xi = Y(\frac{1}{2}|X|)^{\frac{1}{2}}$), we therefore obtain the leading behaviour of $P(X, Y)$ given by†

$$
P(X, Y) \sim \begin{cases}
\sqrt{\pi/X}\, e^{\frac{1}{4}\pi i} & (X \to +\infty) \\[2mm]
\sqrt{\pi/|X|}\, e^{-\frac{1}{4}\pi i} \Big\{ 1 + i\sqrt{2}\, e^{-\frac{1}{4}iX^2} \cos\big[Y(\frac{1}{2}|X|)^{\frac{1}{2}} \big] \Big\} & (X \to -\infty).
\end{cases}
$$
$$(8.3.30)$$

Large positive values of Y correspond to the upper boundary of the exponentially small sector $|\arg y| < \frac{1}{8}\pi$. From (8.3.27) and (8.3.28), we consequently find the asymptotic behaviour

$$
P(X, Y) \sim E_1\big(2^{-\frac{1}{2}} X e^{-\frac{1}{4}\pi i}, 2^{-\frac{5}{4}} Y e^{\frac{1}{8}\pi i}\big) + E_2\big(2^{-\frac{1}{2}} X e^{-\frac{1}{4}\pi i}, 2^{-\frac{5}{4}} Y e^{\frac{1}{8}\pi i}\big),
$$

so that

$$
\begin{aligned}
P(X, Y) \sim 2^{\frac{1}{6}} \sqrt{\frac{\pi}{3}} Y^{-\frac{1}{3}} e^{-\frac{1}{6}iX^2} \Big\{ &\exp\Big\{ i\Big(\tfrac{1}{4}\pi - 3(\tfrac{1}{4}Y)^{\frac{4}{3}} + X(\tfrac{1}{4}Y)^{\frac{2}{3}}\Big) \Big\} \\
&\times \Big[1 + \tfrac{1}{12} X(\tfrac{1}{4}Y)^{-\frac{2}{3}}\big(1 + \tfrac{1}{9}iX^2\big) + O\big(Y^{-\frac{4}{3}}\big) \Big] \\
&+ \exp\Big\{ -\tfrac{1}{12}\pi i - 3(\tfrac{1}{4}Y)^{\frac{4}{3}} e^{-\frac{1}{6}\pi i} - X(\tfrac{1}{4}Y)^{\frac{2}{3}} e^{\frac{1}{6}\pi i} \Big\} \\
&\times \Big[1 - \tfrac{1}{12} X(\tfrac{1}{4}Y)^{-\frac{2}{3}} e^{\frac{1}{3}\pi i}\big(1 + \tfrac{1}{9}iX^2\big) + O\big(Y^{-\frac{4}{3}}\big) \Big] \Big\}
\end{aligned}
$$
$$(8.3.31)$$

as $Y \to +\infty$, with $P(X, -Y) = P(X, Y)$.

---

† The exponential factor $\exp(-iY^2/4X)$ appearing in S_1 has been put equal to unity in the leading behaviour.

The leading behaviour for $X \to \pm\infty$ in (8.3.30) can be interpreted in terms of a single ($X \to +\infty$) and three ($X \to -\infty$) stationary point contributions when using the method of stationary phase. The first exponential term in (8.3.31) is oscillatory, while the second exponential term is subdominant for large positive Y. The behaviour of $P(X, Y)$ for $Y \to \pm\infty$ can be seen to correspond to the two stationary point contribution discussed in §8.3.1; see Fig. 8.8.

8.3.6 Generalisations of $P(X, Y)$

The same methods can also be applied to the generalisation of the Pearcey integral given by

$$P_m(X, Y) = \int_{-\infty}^{\infty} \exp\{i(u^{2m} + Xu^m + Yu)\}\, du \qquad (8.3.32)$$

for integer $m \geq 2$. Introduction of the new variables $x = 2^{-\frac{1}{2}} X e^{-\frac{1}{4}\pi i}$ and $y = 2^{-(m+\frac{1}{2})/m} Y e^{\pi i/4m}$ (compare (8.3.3)), followed by rotation of the path of integration through an angle of $\pi/4m$, yields the analytic continuation

$$\hat{P}_m(x, y) = e^{\pi i/4m} \int_{-\infty}^{\infty} \exp\left\{ -t^{2m} - 2^{\frac{1}{2}}xt^m + 2^{(m+\frac{1}{2})/m} iyt \right\} dt$$

for all complex values of x and y. The analysis now separates into two distinct cases, according as m is even or odd, and involves the integrals

$$\frac{1}{2\pi i} \int_C \Gamma(s) \Gamma\left(\frac{1-s}{m}\right) \begin{Bmatrix} \cos\frac{1}{2}\pi s \\ \sin\frac{1}{2}\pi s \end{Bmatrix} D_{(s-1)/m}(x)(2y)^{-s} ds.$$

The integral involving $\cos\frac{1}{2}\pi s$ corresponds to the generalisation of the Pearcey integral case $m = 2$ in (8.3.7). The asymptotic analysis of these integrals for large $|x|$ or $|y|$ then proceeds in the same manner as described in §§8.3.3–4; for details, see Paris (1994c).

The discussion of $P_m(X, Y)$ for large real values of X and Y is considerably more difficult. The phase function $f(u) = u^{2m} + Xu^m + Yu$ for the integrand in (8.3.32) possesses $2m - 1$ stationary points (given by the roots of $f'(u) = 0$) although, as for the Pearcey integral case, no more than three of these stationary points contribute to the (Poincaré) asymptotics of $P_m(X, Y)$. For certain values of X and Y two of the real roots can coalesce to form a double root. This occurs when $f'(u) = f''(u) = 0$ and corresponds to values of X, Y lying on the caustic

$$Y^m + m \left(\frac{m-1}{2}\right)^{m-1} \left(\frac{mX}{2m-1}\right)^{2m-1} = 0. \qquad (8.3.33)$$

For m even the caustic (8.3.33) is cusped and symmetrical about the negative X axis (and yields (8.3.2) when $m = 2$), while for m odd the caustic is asymmetrical. Inside the caustic, the asymptotics of $P_m(X, Y)$ are controlled by the sum of three terms which result from the contributory stationary points when using the method of stationary phase. Explicit formulation of these terms, however, is rendered difficult by the fact that the real roots of the equation $f'(u) = 0$ for general m do not seem to be expressible in closed form.

On the caustic (8.3.33), the situation is more tractable since the double root can then be expressed simply as $u_0 = X^{*1/m}$, where $X^* = (m-1)|X|/2(2m-1)$. In this case, we have on the caustic (when $X < 0$)

$$P_m(X, Y) = u_0 \int_{-\infty}^{\infty} \exp\{i X^{*2} F(\tau)\} \, d\tau,$$

where

$$F(\tau) = \tau^{2m} - \frac{2(2m-1)}{m-1} \tau^m + \frac{2m^2}{m-1} \tau$$

and we have put $\tau = u/u_0$. The scaled phase function $F(\tau)$ possesses a double stationary point at $\tau = 1$ and one negative stationary point at $\tau = -k$, $k > 0$. Straightforward application of the method of stationary phase [see Olver (1974, p. 96); Wong (1989, p. 79)] then shows that for large $|X|$ on the caustic

$$P_m(X, Y) = \frac{\pi^{\frac{1}{2}}}{m} \alpha(k) X^{*(1-m)/m} \exp\left\{i\left(\tfrac{1}{4}\pi - \beta(k)X^{*2}\right)\right\}\{1 + O(X^{*-1})\}$$

$$+ 3^{-\frac{1}{6}} \left(m^2(2m-1)\right)^{-\frac{1}{3}} X^{*(3-2m)/3m} \exp\{i(2m-1)X^{*2}\}$$

$$\times \left\{ \Gamma\left(\tfrac{1}{3}\right) - \frac{i(3m-5)}{\left[144 m^2\left(m - \tfrac{1}{2}\right)\right]^{\frac{1}{3}}} \Gamma\left(\tfrac{2}{3}\right) X^{*-\frac{2}{3}} + O\left(X^{*-\frac{4}{3}}\right) \right\},$$

$$(8.3.34)$$

where the coefficients $\alpha(k) \equiv (F''(-k)/2m^2)^{-\frac{1}{2}}$ and $\beta(k) \equiv -F(-k)$ are given by

$$\alpha(k) = \left(\frac{k}{1 + k^{2m-1}}\right)^{\frac{1}{2}}, \qquad \beta(k) = k(2m + k^{2m-1}).$$

Apart from the case $m = 2$ (where $k = 2$), it is not possible to express k in simple closed form. Values of k, together with the corresponding values of $\alpha(k)$ and $\beta(k)$, are tabulated in Table 8.4 for different m. The behaviour in (8.3.34) when $m = 2$ agrees with the leading form on the caustic obtained for the Pearcey integral in (1.1.21).

Finally, we mention a different generalisation given by

$$\int_0^{\infty} \exp\{i(u^4 + Xu^2)\} J_\nu(uY) u^{\nu+1} du \qquad (-1 < \nu < \tfrac{5}{2}),$$

where J_ν denotes the Bessel function of order ν. This integral,† which equals a multiple of Pearcey's integral when $\nu = -\tfrac{1}{2}$, occurs in the problem of image

† Integrals of this type are called Bessoid integrals.

Table 8.4. *Values of k, α(k) and β(k) for different values of m*

m	k	α(k)	β(k)
2	2.000000	0.471405	24.000000
3	0.722120	0.776916	4.474513
4	1.308203	0.416058	19.043922
5	0.829336	0.836366	8.447287
6	1.181562	0.403243	21.582945
7	0.876927	0.861572	12.436018
8	1.128631	0.397539	24.989666
9	0.903778	0.875506	16.429858
10	1.099584	0.394313	28.668478

formation in high resolution electron microscopes when $\nu = 0$. By means of the Mellin-Barnes integral representation for the Bessel function in (3.4.22), Janssen (1992) has shown that the above integral, multiplied by the factor $e^{-\pi i(\nu+1)/8}$, can be recast in the form

$$2^{-\frac{1}{4}\nu-\frac{3}{2}} y^{\nu} e^{x^2/4+\pi i/8} \frac{1}{2\pi i} \int_C \Gamma(s) D_{s-1-\nu}(x) y^{-2s}\, ds,$$

where x, y are as defined in (8.3.3) and C is the loop with endpoints at $-\infty$ enclosing the poles of $\Gamma(s)$. The same procedure has also been applied to this integral to yield the asymptotics for large X and Y, including the case in the neighbourhood of the caustic in (8.3.2). A numerical investigation of this integral when $\nu = 0$ has been carried out by Kirk *et al.* (2000).

Appendix

A Short Table of Mellin Transforms

We give in this appendix a short table of Mellin transforms for ease of reference. More extensive tables can be found in Erdélyi (1954), Oberhettinger (1974) and Marichev (1982).

$f(x)$		$M[f; s] = F(s) = \int_0^\infty f(x)x^{s-1}ds$
$f(ax)$	$(a > 0)$	$a^{-s}F(s)$
$x^a f(x)$		$F(s + a)$
$f(1/x)$		$F(-s)$
$f(x^a)$	$(a > 0)$	$a^{-1}F(s/a)$
$f(x^{-a})$	$(a > 0)$	$a^{-1}F(-s/a)$
$x^\alpha f(x^\mu)$	$(\mu > 0)$	$\mu^{-1}F((s + \alpha)/\mu)$
$x^\alpha f(x^{-\mu})$	$(\mu > 0)$	$\mu^{-1}F(-(s + \alpha)/\mu)$
$(\log x)^n f(x)$	$(n = 1, 2, \ldots)$	$F^{(n)}(s)$
$f'(x)$		$-(s - 1)F(s - 1)$
$f^{(n)}(x)$	(see (3.1.9))	$\dfrac{\Gamma(n + 1 - s)}{\Gamma(1 - s)}F(s - n)$
$\left(x\dfrac{d}{dx}\right)^n f(x)$	$(n = 1, 2, \ldots)$	$(-s)^n F(s)$
$\int_0^x f(t)\,dt$		$-s^{-1}F(s + 1)$
$\int_x^\infty f(t)\,dt$		$s^{-1}F(s + 1)$
$\begin{cases} 1 & (0 < x < 1) \\ 0 & (x > 1) \end{cases}$		$s^{-1} \qquad\qquad \mathrm{Re}(s) > 0$

$f(x)$	$M[f; s] = F(s) = \int_0^\infty f(x)x^{s-1}ds$
$\begin{cases} 0 & (0 < x < 1) \\ 1 & (x > 1) \end{cases}$	$-s^{-1}$ $\qquad\qquad$ $\mathrm{Re}(s) < 0$
$\begin{cases} x^a & (0 < x < 1) \\ 0 & (x > 1) \end{cases}$	$(s+a)^{-1}$ $\qquad$ $\mathrm{Re}(s) > -\mathrm{Re}(a)$
$\begin{cases} 0 & (0 < x < 1) \\ x^a & (x > 1) \end{cases}$	$-(s+a)^{-1}$ $\qquad$ $\mathrm{Re}(s) < -\mathrm{Re}(a)$
$(1+x)^{-1}$	$\pi \operatorname{cosec} \pi s$ $\qquad$ $0 < \mathrm{Re}(s) < 1$
$(1+x)^{-a}$ $\qquad$ $(\mathrm{Re}(a) > 0)$	$\dfrac{\Gamma(s)\Gamma(a-s)}{\Gamma(a)}$ $\quad 0 < \mathrm{Re}(s) < \mathrm{Re}(a)$
$(1+x^2)^{-1}$	$\frac{1}{2}\pi \operatorname{cosec} \frac{1}{2}\pi s$ $\quad$ $0 < \mathrm{Re}(s) < 2$
$\dfrac{x}{1+x}$	$-\pi \operatorname{cosec} \pi s$ $\qquad$ $-1 < \mathrm{Re}(s) < 0$
$(1-x)^{-1}$ $\qquad$ (Cauchy PV)	$\pi \cot \pi s$ $\qquad$ $0 < \mathrm{Re}(s) < 1$
$\dfrac{1+x\cos\phi}{1+2x\cos\phi+x^2}$ $\quad$ $(|\phi| < \pi)$	$\dfrac{\pi \ \cos\phi s}{\sin \pi s}$ $\qquad$ $0 < \mathrm{Re}(s) < 1$
$\dfrac{x\sin\phi}{1+2x\cos\phi+x^2}$ $\quad$ $(|\phi| < \pi)$	$\dfrac{\pi \ \sin\phi s}{\sin \pi s}$ $\qquad$ $-1 < \mathrm{Re}(s) < 1$
$(1+2x\cos\phi+x^2)^{-1}$ $\quad$ $(|\phi| < \pi)$	$\dfrac{\pi}{\sin\phi}\dfrac{\sin(1-s)\phi}{\sin \pi s}$ $\quad$ $0 < \mathrm{Re}(s) < 2$
$(1+2x\cos\phi+x^2)^{-\frac{1}{2}}$ $\quad$ $(|\phi| < \pi)$	$\dfrac{\pi}{\sin \pi s} P_{s-1}(\cos\phi)$ $\quad$ $0 < \mathrm{Re}(s) < 1$
$\begin{cases} (1-x)^a & (0 < x < 1) \\ 0 & (x > 1) \end{cases}$ $\qquad\qquad$ $(\mathrm{Re}(a) > -1)$	$\dfrac{\Gamma(a+1)\Gamma(s)}{\Gamma(s+a+1)}$ $\qquad\qquad$ $\mathrm{Re}(s) > 0$
$\begin{cases} 0 & (0 < x < 1) \\ (x-1)^a & (x > 1) \end{cases}$ $\qquad\qquad$ $(\mathrm{Re}(a) > -1)$	$\dfrac{\Gamma(a+1)\Gamma(-a-s)}{\Gamma(1-s)}$ $\qquad\qquad$ $\mathrm{Re}(s) < -\mathrm{Re}(a)$
$\begin{cases} \log x & (0 < x < 1) \\ 0 & (x > 1) \end{cases}$	$-s^{-2}$ $\qquad\qquad$ $\mathrm{Re}(s) > 0$
$\begin{cases} 0 & (0 < x < 1) \\ \log x & (x > 1) \end{cases}$	s^{-2} $\qquad\qquad$ $\mathrm{Re}(s) < 0$
$\begin{cases} \log^n x & (0 < x < 1) \\ 0 & (x > 1) \end{cases}$ $\qquad\qquad$ $(n = 1, 2, \ldots)$	$(-)^n s^{-n-1} n!$ $\qquad\qquad$ $\mathrm{Re}(s) < 0$
$\log(1+x)$	$\dfrac{\pi}{s \sin \pi s}$ $\qquad$ $-1 < \mathrm{Re}(s) < 0$
$\log|1-x|$	$\dfrac{\pi}{s \tan \pi s}$ $\qquad$ $-1 < \mathrm{Re}(s) < 0$

$f(x)$	$M[f; s] = F(s) = \int_0^\infty f(x)x^{s-1}ds$
$\log\left\|\dfrac{1+x}{1-x}\right\|$	$\dfrac{\pi}{s}\tan\frac{1}{2}\pi s \qquad -1 < \mathrm{Re}(s) < 1$
$-(1+x)^{-1}\log x$	$\left(\dfrac{\pi}{\sin\pi s}\right)^2\cos\pi s \quad 0 < \mathrm{Re}(s) < 1$
$(x-1)^{-1}\log x \qquad \text{(Cauchy PV)}$	$\left(\dfrac{\pi}{\sin\pi s}\right)^2 \qquad 0 < \mathrm{Re}(s) < 1$
$\arctan x$	$-\frac{1}{2}\pi s^{-1}\sec\frac{1}{2}\pi s \quad -1 < \mathrm{Re}(s) < 0$
$\mathrm{arccot}\, x$	$\frac{1}{2}\pi s^{-1}\sec\frac{1}{2}\pi s \quad 0 < \mathrm{Re}(s) < 1$
$\begin{cases} e^{-ax} & (0 < x < 1) \\ 0 & (x > 1) \end{cases}$	$a^{-s}\gamma(s, a) \qquad\qquad \mathrm{Re}(s) > 0$
$\begin{cases} 0 & (0 < x < 1) \\ e^{-ax} & (x > 1) \end{cases}$ $(\mathrm{Re}(a) > 0)$	$a^{-s}\Gamma(s, a)$
e^{-x}	$\Gamma(s) \qquad\qquad\qquad \mathrm{Re}(s) > 0$
$e^{-x} - 1$	$\Gamma(s) \qquad\qquad -1 < \mathrm{Re}(s) < 0$
$e^{-x} - 1 + x$	$\Gamma(s) \qquad\qquad -2 < \mathrm{Re}(s) < -1$
e^{-x^2}	$\frac{1}{2}\Gamma(\frac{1}{2}s) \qquad\qquad \mathrm{Re}(s) > 0$
$(e^x + 1)^{-1}$	$(1 - 2^{1-s})\Gamma(s)\zeta(s) \qquad \mathrm{Re}(s) > 0$
$(e^x - 1)^{-1}$	$\Gamma(s)\zeta(s) \qquad\qquad \mathrm{Re}(s) > 1$
$(e^x - 1)^{-2}$	$\Gamma(s)\{\zeta(s-1) - \zeta(s)\} \quad \mathrm{Re}(s) > 2$
$(1+x)^{-\frac{1}{2}}e^{-a\sqrt{1+x}} \qquad (\mathrm{Re}(a) > 0)$	$2\pi^{-\frac{1}{2}}(\frac{1}{2}a)^{\frac{1}{2}-s}\Gamma(s) \qquad \mathrm{Re}(s) > 0$ $\times K_{\frac{1}{2}-s}(a)$
$e^{-ax^2-bx} \qquad (\mathrm{Re}(a) > 0)$	$(2a)^{-\frac{1}{2}s}\Gamma(s)e^{b^2/8a} \qquad \mathrm{Re}(s) > 0$ $\times D_{-s}(b(2a)^{-\frac{1}{2}})$
$e^{-ax-b/x} \qquad (\mathrm{Re}(a, b) > 0)$	$2(b/a)^{\frac{1}{2}s}K_s(2(ab)^{\frac{1}{2}})$
$e^{-x}(\log x)^n \qquad (n = 1, 2, \ldots)$	$\Gamma^{(n)}(s) \qquad\qquad \mathrm{Re}(s) > 0$
$\sum_{n=1}^{\infty} e^{-\pi n^2 x}$	$\pi^{-s}\Gamma(s)\zeta(2s) \qquad \mathrm{Re}(s) > \frac{1}{2}$
e^{ix}	$e^{\frac{1}{2}\pi is}\Gamma(s) \qquad\qquad 0 < \mathrm{Re}(s) < 1$
$\sin ax \qquad (a > 0)$	$a^{-s}\Gamma(s)\sin\frac{1}{2}\pi s \quad -1 < \mathrm{Re}(s) < 1$
$\cos ax \qquad (a > 0)$	$a^{-s}\Gamma(s)\cos\frac{1}{2}\pi s \quad 0 < \mathrm{Re}(s) < 1$
$e^{-x\cos\phi}\cos(x\sin\phi) \quad (\|\phi\| < \frac{1}{2}\pi)$	$\Gamma(s)\cos\phi s \qquad\qquad \mathrm{Re}(s) > 0$
$e^{-x\cos\phi}\sin(x\sin\phi) \quad (\|\phi\| < \frac{1}{2}\pi)$	$\Gamma(s)\sin\phi s \qquad\qquad \mathrm{Re}(s) > -1$

$f(x)$	$M[f;s] = F(s) = \int_0^\infty f(x)x^{s-1}ds$
$J_\nu(x)$	$\dfrac{2^{s-1}\Gamma(\frac{1}{2}s + \frac{1}{2}\nu)}{\Gamma(1 + \frac{1}{2}\nu - \frac{1}{2}s)}$ $-\mathrm{Re}(\nu) < \mathrm{Re}(s) < \frac{3}{2}$
$x^\nu J_\nu(x)$	$\dfrac{2^{s+\nu-1}\Gamma(\frac{1}{2}s + \nu)}{\Gamma(1 - \frac{1}{2}s)}$ $-2\mathrm{Re}(\nu) < \mathrm{Re}(s) < \frac{3}{2} - \mathrm{Re}(\nu)$
$x^{-\nu} J_\nu(x)$	$\dfrac{2^{s-\nu-1}\Gamma(\frac{1}{2}s)}{\Gamma(1 + \nu - \frac{1}{2}s)}$ $0 < \mathrm{Re}(s) < \mathrm{Re}(\nu) + \frac{3}{2}$
$J_\nu^2(x)$	$\dfrac{2^{s-1}\Gamma(\frac{1}{2}s + \nu)\Gamma(1 - s)}{\Gamma^2(1 - \frac{1}{2}s)\Gamma(1 + \nu - \frac{1}{2}s)}$ $-2\mathrm{Re}(\nu) < \mathrm{Re}(s) < 1$
$Y_\nu(x)$	$-2^{s-1}\pi^{-1}\cos\left[\frac{1}{2}\pi(s - \nu)\right]$ $\times\, \Gamma(\frac{1}{2}s + \frac{1}{2}\nu)\Gamma(\frac{1}{2}s - \frac{1}{2}\nu)$ $\|\mathrm{Re}(\nu)\| < \mathrm{Re}(s) < \frac{3}{2}$
$\sin x\, J_\nu(x)$	$\dfrac{2^{\nu-1}\Gamma(\frac{1}{2}s + \frac{1}{2}\nu + \frac{1}{2})\Gamma(\frac{1}{2} - s)}{\Gamma(1 + \nu - s)\Gamma(1 - \frac{1}{2}\nu - \frac{1}{2}s)}$ $-1 - \mathrm{Re}(\nu) < \mathrm{Re}(s) < \frac{1}{2}$
$\cos x\, J_\nu(x)$	$\dfrac{2^{\nu-1}\Gamma(\frac{1}{2}s + \frac{1}{2}\nu)\Gamma(\frac{1}{2} - s)}{\Gamma(\frac{1}{2} - \frac{1}{2}\nu - \frac{1}{2}s)\Gamma(1 + \nu - s)}$ $-\mathrm{Re}(\nu) < \mathrm{Re}(s) < \frac{1}{2}$
$H_\nu^{(1)}(x)$	$2^{s-1}\pi^{-1}e^{\frac{1}{2}\pi i(s-\nu-1)}$ $\times\, \Gamma(\frac{1}{2}s + \frac{1}{2}\nu)\Gamma(\frac{1}{2}s - \frac{1}{2}\nu)$ $\|\mathrm{Re}(\nu)\| < \mathrm{Re}(s) < \frac{3}{2}$
$K_\nu(x)$	$2^{s-2}\Gamma(\frac{1}{2}s + \frac{1}{2}\nu)\Gamma(\frac{1}{2}s - \frac{1}{2}\nu)$ $\mathrm{Re}(s) > \|\mathrm{Re}(\nu)\|$
$K_\nu^2(x)$	$\dfrac{\pi^{\frac{1}{2}}\Gamma(\frac{1}{2}s + \nu)\Gamma(\frac{1}{2}s - \nu)\Gamma(\frac{1}{2}s)}{4\Gamma(\frac{1}{2}s + \frac{1}{2})}$ $\mathrm{Re}(s) > 2\|\mathrm{Re}(\nu)\|$
$e^x K_\nu(x)$	$\dfrac{\Gamma(s + \nu)\Gamma(s - \nu)\Gamma(\frac{1}{2} - s)}{2^s \pi^{\frac{1}{2}}\sec\pi\nu}$ $\|\mathrm{Re}(\nu)\| < \mathrm{Re}(s) < \frac{1}{2}$

$f(x)$	$M[f; s] = F(s) = \int_0^\infty f(x)x^{s-1}ds$
$e^{-x}I_v(x)$	$\dfrac{\Gamma(s+v)\Gamma(\frac{1}{2}-s)}{2^s\pi^{\frac{1}{2}}\Gamma(1+v-s)}$ $-\mathrm{Re}(v) < \mathrm{Re}(s) < \frac{1}{2}$
$\mathrm{Ai}(x)$	$\dfrac{2^{\frac{2}{3}s-\frac{7}{6}}}{2\pi}\Gamma(\tfrac{1}{3}s)\Gamma(\tfrac{1}{3}s+\tfrac{1}{3})\quad \mathrm{Re}(s) > 0$

References

Abramowitz, M. and Stegun, I. (ed.) (1965). *Handbook of Mathematical Functions* (Dover, New York).

Airey, J.R. (1937). The 'converging factor' in asymptotic series and the calculation of Bessel, Laguerre and other functions, *Philos. Mag.* **24**, 521–552.

Arnold, V.I., Gussein-Zade, S.M. and Varchenko, A.N. (1988). *Singularities of Differentiable Maps II* (Birkhäuser, Berlin).

Bakhoom, N.G. (1933). Asymptotic expansions of the function $F_k(x) = \int_0^\infty \exp(xu - u^k)\, du$, *Proc. Lond. Math. Soc.* **35**, 83–100.

Barnes, E.W. (1906). The asymptotic expansion of integral functions defined by Taylor's series, *Phil. Trans. Roy. Soc. London* **206A**, 249–297.

Barnes, E.W. (1908). A new development of the theory of hypergeometric functions, *Proc. London Math. Soc.* (2) **6**, 141–177.

Barnes, E.W. (1910). A transformation of the generalised hypergeometric series, *Qu. J. Math.* **41**, 136–140.

Batchelder, P.M. (1967). *An Introduction to Linear Difference Equations* (Dover, New York).

Bateman, H. (1942). Book Review of *Operational Methods in Applied Mathematics*, *Bull. Amer. Math. Soc.* **48**, 510–511.

Bender, C.M. and Orszag, S.A. (1978). *Advanced Mathematical Methods for Scientists and Engineers,* International Series in Pure and Applied Mathematics (McGraw-Hill, New York).

Berndt, B.C. (1985). *Ramanujan's Notebooks*, Part II, Ch. 15 (Springer-Verlag, New York).

Berry, M.V. (1986). Riemann's zeta function: A model for quantum chaos?, in *Quantum Chaos and Statistical Nuclear Physics*, ed. T.H. Seligman and H. Nishioka, Lecture Notes in Phys. **263**, 1–17 (Springer-Verlag, New York).

Berry, M.V. (1989). Uniform asymptotic smoothing of Stokes's discontinuities, *Proc. Roy. Soc. London* **A412**, 7–21.

Berry, M.V. (1990). Waves near Stokes lines, *Proc. Roy. Soc. London* **A427**, 265–280.

Berry, M.V. (1991a). Infinitely many Stokes smoothings in the gamma function, *Proc. Roy. Soc. London* **A434**, 465–472.

Berry, M.V. (1991b). Stokes's phenomenon for superfactorial asymptotic series, *Proc. Roy. Soc. London* **A435**, 437–444.

Berry, M.V. (1991c). Asymptotics, superasymptotics, hyperasymptotics, in *Asymptotics Beyond All Orders*, ed. H. Segur, H. Tanveer and H. Levine, 1–14 (Plenum Press, New York).

Berry, M.V. (1995). The Riemann-Siegel expansion for the zeta function: high orders and remainders, *Proc. Roy. Soc. London* **A450**, 439–462.

Berry, M.V. and Goldberg, J. (1988). Renormalisation of curlicues, *Nonlinearity* **1**, 1–26.

Berry, M.V. and Howls, C.J. (1990). Hyperasymptotics, *Proc. Roy. Soc. London* **A430**, 653–668.

Berry, M.V. and Howls, C.J. (1991). Hyperasymptotics for integrals with saddles, *Proc. Roy. Soc. London* **A434**, 657–675.

Berry, M.V. and Howls, C.J. (1993a). Infinity interpreted, *Physics World* **6**, 35–39.

Berry, M.V. and Howls, C.J. (1993b). Unfolding the high orders of asymptotic expansions with coalescing saddles: singularity theory, crossover and duality, *Proc. Roy. Soc. London* **A443**, 107–126.

Berry, M.V. and Howls, C.J. (1994). Overlapping Stokes smoothings: survivial of the error function and canonical catastrophe integrals, *Proc. Roy. Soc. London* **A444**, 201–216.

Berry, M.V. and Keating, J.P. (1992). A new asymptotic representation for $\zeta(\frac{1}{2} + it)$ and quantum spectral determinants, *Proc. Roy. Soc. London* **A437**, 151–173.

Bleistein, N. and Handelsman, R.A. (1975). *Asymptotic Expansion of Integrals* (Holt, Rinehart and Winston, New York). Reprinted (1986) (Dover, New York).

Boersma, J. (1975). Ray-optical analysis of reflection in an open-ended parallel-plane waveguide. I: TM case, *SIAM J. Appl. Math.* **29**, 164–195.

Boersma, J. (1995). Solution to Problem No. 94–12, *SIAM Rev.* **37**, 443–445.

Boersma, J. and de Doelder, P.J. (1979). On some Bessel-function integrals arising in a telecommunication problem, Technical Report 1979–13, Eindhoven University of Technology.

Boyd, W.G.C. (1990). Stieltjes transforms and the Stokes phenomenon, *Proc. Roy. Soc. London* **A429**, 227–246.

Boyd, W.G.C. (1993). Error bounds for the method of steepest descents, *Proc. Roy. Soc. London* **A440**, 493–518.

Boyd, W.G.C. (1994). Gamma function asymptotics by an extension of the method of steepest descents, *Proc. Roy. Soc. London* **A447**, 609–630.

Boyd, J.P. (1999). The Devil's invention: asymptotic, superasymptotic and hyperasymptotic series, *Acta Applicandæ Mathematicæ* **56**, 1–98.

Braaksma, B.L.J. (1963). Asymptotic expansions and analytic continuations for a class of Barnes integrals, *Compos. Math.* **15**, 239–341.

Brazel, N., Lawless, F. and Wood, A.D. (1992). Exponential asymptotics for an eigenvalue of a problem involving parabolic cylinder functions, *Proc. Amer. Math. Soc.* **114**, 1025–1032.

Brieskorn, E. and Knörrer, H. (1986). *Plane Algebraic Curves* (Birkhäuser, Basel).

Brillouin, L. (1916). Sur une méthode de calcul approchée de certaines intégrales, dite méthode de col, *Ann. Sci. École Norm. Sup.* **33**, 17–69.

Bromwich, T.J.I'A. (1926). *An Introduction to the Theory of Infinite Series*, 2nd ed. (Macmillan, London).

Buchholz, H. (1969). *The Confluent Hypergeometric Function* (Springer-Verlag, New York).

Bühring, W. (2000). An asymptotic expansion for a ratio of products of gamma functions. *Int. J. Math. and Math. Sciences* **24**, 505–510.

Burwell, W.R. (1924). Asymptotic expansions of generalised hypergeometric functions, *Proc. Lond. Math. Soc.* **53**. 599–611.

Chapman, R.J. (1995). Solution to Problem No. 94–12, *SIAM Rev.* **37**, 445–447.

Chapman, S.J. (1996). On the non-universality of the error function in the smoothing of Stokes discontinuities, *Proc. Roy. Soc. London* **A452**, 2225–2230.

Cherry, T.M. (1950). Asymptotic expansions for the hypergeometric functions occurring in gas-flow theory, *Proc. Roy. Soc. London* **A202**, 507– 522.

Chester, C., Friedman, B. and Ursell, F. (1957). An extension of the method of steepest descents, *Proc. Camb. Phil. Soc.* **53**, 599–611.

Connor, J.N.L. (1973). Semiclassical theory of molecular collisions: three nearly coincident classical trajectories, *Molec. Phys.* **26**, 1217–1231.

Connor, J.N.L. and Curtis, P.R. (1982). A method for the numerical evaluation of the oscillatory integrals associated with the cuspoid catastrophes: application to Pearcey's integral and its derivatives, *J. Phys. A* **15**, 1179–1190.

Connor, J.N.L. and Curtis, P.R. (1984). Differential equations for the cuspoid canonical integrals, *J. Math. Phys.* **25**, 2895–2902.

Connor, J.N.L. and Farrelly, D. (1981a). Theory of cusped rainbows in eleastic scattering: uniform semiclassical calculations using Pearcey's integral, *J. Chem. Phys.* **75**, 2831–2846.

Connor, J.N.L. and Farrelly, D. (1981b). Molecular collisions and cusp catastrophes: three methods for the calculation of Pearcey's integral and its derivatives, *Chem. Phys. Lett.* **81**, 306–310.

Connor, J.N.L., Curtis, P.R. and Farrelly, D. (1983). A differential equation method for the numerical evaluation of the Airy, Pearcey and swallowtail canonical integrals and their derivatives, *Molec. Phys.* **48**, 1305–1330.

Copson, E.T. (1965). *Asymptotic Expansions* (Cambridge University Press, Cambridge).

Courant, R. and Hilbert, D. (1953). *Methods of Mathematical Physics,* Vol. 1 (Interscience Publishers, New York).

Davies, B. (1978). *Integral Transforms and Their Applications* (Springer-Verlag, New York).

Dekking, F.M. and Mendès-France, M. (1981). Uniform distribution modulo one: a geometrical viewpoint. *J. Reine Angew. Math.* **329**, 143–153.

Delabaere, E. and Howls, C.J. (1999). Global asymptotics for multiple integrals with boundaries, Report No. 557, Laboratoire de Mathématiques, Université de Nice. Submitted to *Duke Math. J.*

Delavault, H. (1961). *Les Transformations Intégrales à Plusieures Variables et Leurs Applications*, Mémorial des Sciences Mathématiques Vol. 148 (Gauthier-Villars, Paris).

Denef, J. and Sargos, P. (1989). Polyèdre de Newton et Distribution f_+^s.I, *J. d'Anal. Math.* **53**, 201–218.

Dingle, R.B. (1973). *Asymptotic Expansions: Their Derivation and Interpretation* (Academic Press, London).

Dostal, M. and Gaveau, B. (1987). Développements asymptotiques explicites d'intégrales de Fourier pour certains points critiques dégénérés, *C.R. Acad. Sci. Paris* **305**, 857–859.

Dostal, M. and Gaveau, B. (1989). The stationary phase method for certain degenerate critical points. I, *Can. J. Math.* **41**, 907–931.

Dunster, T.M. and Lutz, D.A. (1991). Convergent factorial series expansions for Bessel functions, *SIAM J. Math. Anal.* **22**, 1156–1172.

Edwards, H.M. (1974). *Riemann's Zeta Function* (Academic Press, New York).

Elfving, G. (1981). *The History of Mathematics in Finland 1828–1918*, Societas Scientiarum Fennica (Frenckell, Helsinki).

Erdélyi, A. (ed.) (1953). *Higher Transcendental Functions* (McGraw-Hill, New York).

Erdélyi, A. (ed.) (1954). *Tables of Integral Transforms*, Vol.1 (McGraw-Hill, New York).

Evgrafov, M.A. (1961). *Asymptotic Estimates and Entire Functions* (Gordon & Breach, New York).

Faxén, H. (1921). Expansion in series of the integral $\int_y^\infty \exp[-x(t \pm t^{-\mu})]t^\nu \, dt$, *Ark. Mat. Astr. Fys.* **15**, 1–57.

Fields, J. (1966). The uniform asymptotic expansion of the ratio of two gamma functions, *Proc. Edinburgh Math. Soc.* **15**, 43–45.

Flajolet, P. and Sedgewick, R. (1995). Mellin transforms and asymptotics: Finite differences and Rice's integrals, *Theor. Computer Science* **144**, 101–124.

Flajolet, P., Régnier, M. and Sedgewick, R. (1985). Some uses of the Mellin integral transform in the analysis of algorithms, in *Combinatorial Algorithms on Words*, NATO Advanced Science Institute Series, Series F: Computer and Systems Sciences, ed. A. Apostolico and Z. Galil, Vol. 12, 241–254 (Springer-Verlag, New York).

Flajolet, P., Grabner, P., Kirschenhofer, P., Prodinger, H. and Tichy, R.F. (1994). Mellin transforms and asymptotics: Digital sums, *Theor. Computer Science* **123**, 291–314.

Flajolet, P., Gourdon, X. and Dumas, P. (1995). Mellin transforms and asymptotics: Harmonic sums, *Theor. Computer Science* **144**, 3–58.

Ford, W.B. (1936). *The Asymptotic Developments of Functions defined by Maclaurin Series*, University of Michigan Studies, Scientific Series, Vol. 11.

Forehand, D.I.M. and Olde Daalhuis, A.B. (2000). Private communication.

Frenzen, C. (1992). Error bounds for asymptotic expansions of the ratio of two gamma functions, *SIAM J. Math. Anal.* **18**, 890–896.

Friedrichs, K.O. (1955). Asymptotic phenomena in mathematical physics, *Bull. Amer. Math. Soc.* **61**, 485–504.

Gerenflo, R., Luchko, Y. and Mainardi, F. (1999) Analytical properties and applications of the Wright function, *Frac. Calc. and Appl. Anal.* **2**, 383–414.

Glasser, M.L. (1995). Solution to Problem No. 94–12, *SIAM Rev.* **37**, 448.

Gradshteyn, I.S. and Rhyzhik, I.M. (1980). *Table of Integrals, Series and Products* (Academic Press, New York).

Gram, J.P. (1903). Note sur les zéros de la fonction $\zeta(s)$ de Riemann, *Acta Mathematica* **27**, 289–304.

Grossman, N. (1997). Convergent expansions for two common trigonometric sums, *J. Math. Anal. Appl.* **205**, 577–585.

Hadamard, J. (1912). Sur la série de Stirling, *Fifth Internat. Congress Math.*, Cambridge. Reprinted in *Oeuvres de Jacques Hadamard*, Vol. 1, pp. 375–377, Éditions du CNRS, Paris, 1968.

Hankel, H. (1868). Die Cylinderfunktionen erste und zweiter Art, *Math. Annalen* **1**, 467–501.

Hardy, G.H. (1920). On two theorems of F. Carlson and S. Wigert, *Acta Math.* **42**, 327–339. Reprinted in *Collected Papers*, Vol. 4, 610–622 (Oxford University Press, Oxford).

Hardy, G.H. (1937). Ramanujan and the theory of Fourier transforms, *Qu. J. Math.* **8**, 245–254. Reprinted in *Collected Papers*, Vol. 7, 280–289 (Oxford University Press, Oxford).

Hardy, G.H. (1940). *Ramanujan: Twelve lectures on subjects suggested by his life and work* (Cambridge University Press, Cambridge). Reprinted by Chelsea, New York.

Hardy, G.H. and Littlewood, J.E. (1914). Some problems of Diophantine approximation, *Acta Math.* **37**, 192–239.

Hardy, G.H. and Littlewood, J.E. (1918). Contributions to the theory of the Riemann zeta-function and the theory of the distribution of primes, *Acta Math.* **41**, 119–196.

Harsoyo, B. and Temme, N.M. (1982). A functional equation for a series related to theta functions, Report TN 102/82, Mathematisch Centrum, Amsterdam.

Haselgrove, C.B. (1963). *Tables of the Riemann Zeta Function*, Royal Society Mathematical Tables, Vol. 6 (Cambridge University Press, Cambridge).

Heading, J. and Whipple, R.T.P. (1952). On the oblique reflection of long wireless waves from the ionosphere at places where the Earth's magnetic field is regarded as vertical, *Phil. Trans. Roy. Soc. London* **A244**, 469–503.

Howls, C.J. (1992). Hyperasymptotics for integrals with finite endpoints, *Proc. Roy. Soc. London* **A439**, 373–396.

Howls, C.J. (1997). Hyperasymptotics for multidimensional integrals, exact remainder terms and the global connection problem. *Proc. Roy. Soc. London* **A453**, 2271–2294.

Ivić, A. (1985). *The Riemann Zeta-Function* (Wiley, New York).

Janssen, A.J.E.M. (1992). On the asymptotics of some Pearcey-type integrals, *J. Phys. A* **25**, 823–831.

Jones, D.S. (1990). Uniform asymptotic remainders, *Proc. Internat. Conf. on Asymptotic and Computational Analysis*, Winnipeg, Canada, 5–7 June 1989 (ed. R. Wong), 241–264 (Marcel Dekker, New York).

Jones, D.S. (1994). Asymptotic remainders, *SIAM J. Math. Anal.* **25**, 474–490.

Jones, D.S. (1997). *Introduction to Asymptotics: A Treatment using Nonstandard Analysis* (World Scientific, Singapore).

Jones, D.S. and Kline, M. (1958). Asymptotic expansions of multiple integrals and the method of stationary phase, *J. Math. Phys.* **37**, 1–28.

Kaminski, D. (1989). Asymptotic expansion of the Pearcey integral near the caustic, *SIAM J. Math. Anal.* **20**, 987–1005.

Kaminski, D. (1992). Asymptotics of the swallowtail integral near the cusp of the caustic, *SIAM J. Math. Anal.* **23**, 262–285.

Kaminski, D. and Paris, R.B. (1996). Exponential asymptotic expansions and the Mellin-Barnes integral, Technical Report MACS 96:02, University of Abertay Dundee.

Kaminski, D. and Paris, R.B. (1997). Asymptotics via iterated Mellin-Barnes integrals: application to the generalised Faxén integral, *Methods and Applic. of Analysis* **4**, 311–325.

Kaminski, D. and Paris, R.B. (1998a). Asymptotics of a class of multidimensional Laplace-type integrals. I: Double integrals, *Phil. Trans. Roy. Soc. London* **A356**, 583–623.

Kaminski, D. and Paris, R.B. (1998b). Asymptotics of a class of multidimensional Laplace-type integrals. II: Treble integrals, *Phil. Trans. Roy. Soc. London* **A356**, 625–667.

Kaminski, D. and Paris, R.B. (1999). On the zeroes of the Pearcey integral, *J. Comp. and Appl. Math.* **107**, 31–52.

Kath, W.I. and Kriegsmann, G.A. (1989). Optical tunnelling: radiation losses in bent fibre optic waveguides, *IMA J. Appl. Math.* **41**, 85–103.

Keating, J.P. and Reade, J.B. (2000). Summability of alternating gap series, *Proc. Edinburgh Math. Soc.* **43**, 95–101.

King, A.C. and Needham, D.J. (1994). The initial development of a jet caused by fluid, body and free-surface interaction. Part I: A uniformly accelerated plate, *J. Fluid Mech.*, **268**, 89–101.

Kirk, N.P., Connor, J.N.L., Curtis, P.R. and Hobbs, C.A. (2000). Theory of axially symmetric cusped caustics: numerical evaluation of a Bessoid integral by an adaptive contour algorithm. *J. Phys. A: Math. Gen.*, **33**, 4797–4808.

Kloosterman, H. D. (1922). Een integraal voor die ζ functie van Riemann, *Christiaan Huyghens Math. Tijdschrift* **2**, 172–177.

Knopp, K. (1956). *Infinite Sequences and Series* (Dover, New York).

Kowalenko, V., Frankel, N.E., Glasser, M.L. and Taucher, T. (1995). *Generalised Euler-Jacobi Inversion Formula and Asymptotics Beyond All Orders*, London Math. Soc. Lecture Notes Series **214** (Cambridge University Press, Cambridge).

Lavrik, A.F. (1968). An approximate functional equation for the Dirichlet *L*-function, *Trans. Moscow Math. Soc.* **18**, 101–115.

Lawless, F. (1995). $\Gamma(z)$: a difference approach, Technical Report MACS 95:01, University of Abertay Dundee.

Ledermann, W. (ed.) (1990). *Handbook of Applicable Mathematics: Supplement* (Wiley, Chichester).

Liakhovetski, G.V. and Paris, R.B. (1998). Uniformity problem for asymptotics of multidimensional Laplace-type integrals, Technical Report MS 98:02, University of Abertay Dundee.

Lighthill, M.J. (1947). The hodograph transformation in trans-sonic flow. II Auxiliary theorems on the hypergeometric functions $\psi_n(\tau)$, *Proc. Roy. Soc. London* **A191**, 341–351.

Lindelöf, E. (1933). Robert Hjalmar Mellin, *Acta Mathematica* **61**, i–vi.

Liu, J. and Wood, A.D. (1991). Matched asymptotics for a generalisation of a model equation for optical tunnelling, *Euro. J. Appl. Math.* **2**, 223–231.

Luke, Y.L. (1969). *The Special Functions and Their Approximations*, Vol. 1 (Academic Press, New York).

Macfarlane, G.G. (1949). The application of the Mellin transform to the summation of slowly convergent series, *Phil. Mag.* **40**, 188–197.

Malgrange, B. (1974). Intégrales asymptotiques et monodromie, *Ann. Sci. École Norm. Sup. IV* **7**, 405–430.

Marichev, O.I. (1982). *Handbook of Integral Transforms of Higher Transcendental Functions: Theory and Algorithmic Tables* (Ellis Horwood, Chichester).

McCabe, J.H. (1983). On an asymptotic series and corresponding continued fraction for a gamma function ratio, *J. Comp. Appl. Math.* **9**, 125–130.

McLachlan, N.W. (1963). *Complex Variable Theory and Transform Calculus* (Cambridge University Press, Cambridge).

McLeod, J.B. (1992). Smoothing of Stokes discontinuities, *Proc. Roy. Soc. London* **437 A**, 343–354.

Mellin, R.H. (1896). Über die fundamentale Wichtigkeit des Satzes von Cauchy für die Theorien der Gamma- und der hypergeometrischen Funktionen, *Acta Soc. Fennicae* **21**, 1–115.

Mellin, R.H. (1902). Über den Zusammenhang zwischen den linearen Differential- und Differenzengleichungen, *Acta Math.* **25**, 139–164.

Meyer, R.E. (1989). A simple explanation of the Stokes phenomenon, *SIAM Rev.* **31**, 435–445.

Middleton, R.C. (1994). A numerical investigation of the smoothing of the Stokes phenomenon, B.Sc. Dissertation, University of Abertay Dundee.

Miller, J.C.P. (1952). A method for the determination of converging factors, applied to the asymptotic expansions for the parabolic cylinder functions, *Proc. Cambridge Phil. Soc.* **48**, 243–254.

Milne-Thomson, L.M. (1933). *The Calculus of Finite Differences* (MacMillan, London).

Moiseiwitsch, B.L (1977). *Integral Equations*, Longman Mathematical Texts (Longman, London).

Morse, P.M. and Feshbach, H. (1953). *Methods of Theoretical Physics*, Vol. 1 (McGraw-Hill, New York).

Murphy, B.T.M. and Wood, A.D. (1997). Hyperasymptotic solutions of second-order ordinary differential equations with a singularity of arbitrary integer rank, *Methods and Applic. of Analysis* **3**, 250–260.

Nielsen, N. (1906). *Handbuch der Theorie der Gammafunktion* (Teubner, Leipzig).

Nikishov, A.I. and Ritus, V.I. (1992). Stokes line width, *Teor. i Matem. Fizika* **92**, 24–40.

Ninham, B.W., Hughes, B.D., Frankel, N.E. and Glasser, M.L. (1992). Möbius, Mellin and mathematical physics, *Physica A* **186**, 441–481.

Oberhettinger, F. (1974). *Table of Mellin Transforms* (Springer-Verlag, New York).

Olde Daalhuis, A.B. (1992). Hyperasymptotic expansions of confluent hypergeometric functions, *IMA J. Appl. Math.* **49**, 203–216.

Olde Daalhuis, A.B. (1998). Hyperasymptotic solutions of higher order linear differential equations with a singularity of rank one, *Proc. Roy. Soc. London* **A454**, 1–29.

Olde Daalhuis, A.B. and Olver, F.W.J. (1994). Asymptotic solutions of ordinary differential equations. II: Irregular singularities of rank one, *Proc. Roy. Soc. London* **A445**, 39–56.

Olde Daalhuis, A.B. and Olver, F.W.J. (1995a). Hyperasymptotic solutions of second-order linear differential equations I, *Methods and Applic. of Analysis* **2**, 173–197.

Olde Daalhuis, A.B. and Olver, F.W.J. (1995b). On the calculation of Stokes multipliers for linear differential equations of the second order, *Methods and Applic. of Analysis* **2**, 348–367.

Olde Daalhuis, A.B., Chapman, S.J., King, J.R., Ockendon, J.R. and Tew, R.H. (1995). Stokes phenomenon and matched asymptotic expansions, *SIAM J. Appl. Math.* **55**, 1469–1483.

Olver, F.W.J. (1964). Error bounds for asymptotic expansions with an application to cylinder functions of large argument, in *Asymptotic Solutions of Differential Equations and their Applications*, ed. C.H. Wilcox, 163–183 (Wiley, New York).

Olver, F.W.J. (1974). *Asymptotics and Special Functions* (Academic Press, New York). Reprinted (1997) (A.K. Peters, Massachussets).

Olver, F.W.J. (1980). Asymptotic approximations and error bounds, *SIAM Rev.* **22**, 188–203.

Olver, F.W.J. (1990). On Stokes' phenomenon and converging factors, *Proc. Internat. Conf. on Asymptotic and Computational Analysis*, Winnipeg, Canada, 5–7 June 1989 (ed. R. Wong), 329–355 (Marcel Dekker, New York).

Olver, F.W.J. (1991a). Uniform, exponentially improved, asymptotic expansions for the generalized exponential integral, *SIAM J. Math. Anal.* **22**, 1460–1474.

Olver, F.W.J. (1991b). Uniform, exponentially improved, asymptotic expansions for the confluent hypergeometric function and other integral transforms, *SIAM J. Math. Anal.* **22**, 1475–1489.

Olver, F.W.J. (1993). Exponentially-improved asymptotic solutions of ordinary differential equations I: The confluent hypergeometric function, *SIAM J. Math. Anal.* **24**, 756–767.

Olver, F.W.J. (1994). Asymptotic expansions of the coefficients in asymptotic series solutions of linear differential equations, *Methods and Applic. of Analysis* **1**, 1–13.

Olver, F.W.J. (1995). On an asymptotic expansion of a ratio of gamma functions, *Proc. Roy. Irish Academy* **95A**, 5–9.

Paris, R.B. (1980). On the asymptotic expansion of solutions of an nth order linear differential equation, *Proc. Roy. Soc. Edinburgh* **85A**, 15–57.

Paris, R.B. (1991). The asymptotic behaviour of Pearcey's integral for complex variables, *Proc. Roy. Soc. London* **A432**, 391–426.

Paris, R.B. (1992a). Smoothing of the Stokes phenomenon using Mellin-Barnes integrals, *J. Comp. Appl. Math.* **41**, 117–133.

Paris, R.B. (1992b). Smoothing of the Stokes phenomenon for high-order differential equations, *Proc. Roy. Soc. London* **A436**, 165–186.

Paris, R.B. (1993). Application of the refined asymptotics of the gamma function to the Riemann zeta function, Technical Report MACS 93:11, University of Abertay Dundee.

Paris, R.B. (1994a). A generalisation of Lavrik's expansion for the Riemann zeta function, Technical Report MACS 94:01, University of Abertay Dundee.

Paris, R.B. (1994b). An asymptotic representation for the Riemann zeta function on the critical line, *Proc. Roy. Soc. London* **A446**, 565–587.

Paris, R.B. (1994c). A generalisation of Pearcey's integral, *SIAM J. Math. Anal.* **25**, 630–645.

Paris, R.B. (1996). The mathematical work of G.G. Stokes, *Mathematics Today* **32**, 43–46.

Paris, R.B. (2000a). New asymptotic formulas for the Riemann zeta function on the critical line, in *Proc. Hong Kong Workshop on Asymptotics and Special Functions*, (ed. R. Wong), 242–261 (World Scientific, Singapore).

Paris, R.B. (2000b). On the use of Hadamard expansions in hyperasymptotic evaluation. *Proc. Roy. Soc. London*, in press.

Paris, R.B. (2000c). The Hadamard expansion for $\log \Gamma(z)$, Technical Report MS 00:03, University of Abertay Dundee.

Paris, R.B. and Cang, S. (1997a). An exponentially-smoothed Gram-type formula for the Riemann zeta function, *Methods and Applic. of Analysis* **4**, 326–338.

Paris, R.B. and Cang, S. (1997b). An asymptotic representation for $\zeta(\frac{1}{2} + it)$, *Methods and Applic. of Analysis* **4**, 449–470.

Paris, R.B. and Liakhovetski, G.V. (2000a). Asymptotics of the multidimensional Faxén integral, *Frac. Calc. and Appl. Anal.* **3**, 63–73.

Paris, R.B. and Liakhovetski, G.V. (2000b). Asymptotic expansions of Laplace-type integrals. III. *J. Comp. Appl. Math.* **135**, in press.

Paris, R.B. and Wood, A.D. (1985). On the asymptotic expansion of solutions of an nth order linear differential equation with power coefficients, *Proc. Roy. Irish Acad.* **85A**(2), 201–220.

Paris, R.B. and Wood, A.D. (1986). *Asymptotics of High Order Differential Equations*, Pitman Research Notes in Mathematics, **129** (Longman Scientific and Technical, Harlow).

Paris, R.B. and Wood, A.D. (1989). A model equation for optical tunnelling, *IMA J. Appl. Math.* **43**, 272–284.

Paris, R.B. and Wood, A.D. (1992). Exponentially-improved asymptotics for the gamma function, *J. Comp. Appl. Math.* **41**, 135–143.

Paris, R.B. and Wood, A.D. (1995). Stokes phenomenon demystified, *IMA Bulletin* **31**, 21–28.

Pearcey, T. (1946). The structure of an electromagnetic field in the neighbourhood of a cusp of a caustic, *Phil. Mag.* **37**, 311–317.

Pogorzelski, W. (1966). *Integral Equations and their Applications,* International Series of Monographs in Pure and Applied Science (Pergamon Press, Oxford).

Ramanujan, S. (1920). A class of definite integrals, *Qu. J. Math.* **48**, 294–310. Reprinted in *Collected Papers* 1962, 216–229 (Chelsea, New York).

Rawlinson, A.E.J. (1954). In *Directory of National Biographies 1951–1960*, 65–67 (Oxford University Press, Oxford).

Reed, I. S. (1944). The Mellin type of double integral, *Duke Math. J.* **2**, 565–572.

Riekstiņš, E. Ya. (1977). *Asymptotic Expansions of Integrals*, Vol. 2 [in Russian] (Zinatne, Riga).

Riekstiņš, E. Ya. (1983). Application of integrals of Mellin-Barnes type for a sharp estimate of the remainder in asymptotic expansions, [in Russian] *Latv. Mat. Ezhegodnik* **27**, 172–182.

Riekstiņš, E. Ya. (1986). *Estimates of Remainders in Asymptotic Expansions* [in Russian] (Zinatne, Riga).

Riemann, B. (1859). Über die Anzahl der Primzahlen unter einer gegebenen Grösse. In *Collected Works of Bernhard Riemann* (ed. H. Weber) 1953, 145–153 (Dover, New York).

Riemann, B. (1863). Sullo svolgimento del quoziente di due serie ipergeometriche in frazione continua infinita. In *Collected Works of Bernhard Riemann* (ed. H. Weber) 1953, 424–430 (Dover, New York).

Riney, T.D. (1956). On the coefficients in asymptotic factorial expansions, *Proc. Amer. Math. Soc.* **7**, 245–249.

Riney, T.D. (1958). A finite recursion formula for the coefficients in asymptotic expansions, *Trans. Amer. Math. Soc.* **88**, 214–226.

Sedgewick, R. and Flajolet, P. (1996). *An Introduction to the Analysis of Algorithms* (Addison-Wesley, Massachusetts).

Senouf, D. (1996). Asymptotic and numerical approximations of the zeros of Fourier integrals, *SIAM J. Math. Anal.* **27**, 1102–1128.

Slater, L.J. (1960). *Confluent Hypergeometric Functions* (Cambridge University Press, Cambridge).

Slater, L.J. (1966). *Generalised Hypergeometric Functions* (Cambridge University Press, Cambridge).

Sneddon, I.N. (1972). *The Use of Integral Transforms* (McGraw-Hill, New York).

Spira, R. (1971). Calculation of the gamma function by Stirling's formula, *Math. Comp.* **25**, 317–322.

Stamnes, J.J. and Spjelkavik, B. (1983). Evaluation of the field near a cusp of a caustic, *Optica Acta.* **30**, 1331–1358.

Stieltjes, T.J. (1886). Recherches sur quelques séries semi-convergentes, *Ann. Sci. École Norm. Sup.* **3**, 201–258. Reprinted in *Complete Works*, Vol. 2, 2–58 (Noordhoff, Groningen, 1918).

Stokes, G.G. (1850). On the numerical calculation of a class of definite integrals and infinite series. In *Collected Mathematical and Physical Papers*, Vol. 2, 329–357 (Cambridge University Press, Cambridge).

Stokes, G.G. (1857). On the discontinuity of arbitrary constants which appear in divergent developments. In *Collected Mathematical and Physical Papers*, Vol. 4, 79–109 (Cambridge University Press, Cambridge).

Temme, N.M. (1979). The asymptotic expansion of the incomplete gamma functions, *SIAM J. Math. Anal.* **10**, 757–766.

Temme, N.M. (1996). *Special Functions: An Introduction to the Classical Functions of Mathematical Physics* (Wiley, New York).

Titchmarsh, E.C. (1939). *The Theory of Functions* (Oxford University Press, Oxford).

Titchmarsh, E.C. (1975). *Introduction to the Theory of Fourier Integrals* (Oxford University Press, Oxford).

Titchmarsh, E.C. (1986). *The Theory of the Riemann Zeta-Function*. Revised by D.R. Heath-Brown (Oxford University Press, Oxford).

Tricomi, F.G. (1957). *Integral Equations* (Interscience Publishers, New York).

Ursell, F. (1972). Integrals with a large parameter: several nearly coincident saddle points, *Proc. Camb. Phil. Soc.* **72**, 49–65.

Ursell, F. (1983). Integrals with a large parameter: Hilbert transforms, *Proc. Camb. Phil. Soc.* **93**, 141–149.

Ursell, F. (1990). Integrals with a large parameter. A strong form of Watson's lemma, in *Elasticity, Mathematical Methods and Applications*, ed. G. Eason and R.W. Ogden, 391–395 (Ellis-Horwood, Chichester).

Vasil'ev, V.A. (1977). Asymptotic exponential integrals, Newton's diagram, and the classification of minimal points, *Functional Analyt. Appl.* **11**, 163–172.

Watson, G.N. (1918). Asymptotic expansions of hypergeometric functions, *Camb. Phil. Trans.* **22**, 277–308.

Watson, G.N. (1966). *A Treatise on the Theory of Bessel Functions* (Cambridge University Press, Cambridge).

Whittaker, E.T. (1954). In *Obituary Notices of Fellows of the Royal Society,* Vol. IX, 15–25. (The Royal Society, London).

Whittaker, E.T. and Watson, G.N. (1965). *Modern Analysis* (Cambridge University Press, Cambridge).

Wright, E.M. (1940). The asymptotic expansion of integral functions defined by Taylor series, *Phil. Trans. Roy. Soc. London* **238A**, 423–451.

Wright, E.M. (1958). A recursion formula for the coefficients in an asymptotic expansion, *Proc. Glasgow Math. Assoc.* **4**, 38–41.

Wright, F.J. (1980). The Stokes set of the cusp diffraction catastrophe, *J. Phys.* A **13**, 2913–2928.

Wong, R. (1989). *Asymptotic Expansion of Integrals* (Academic Press, London).

Wong, R. and McClure, J.P. (1981). On a method of asymptotic evaluation of multiple integrals, *Math. Comp.* **37**, 509–521.

Wong, R. and Zhao, Y.-Q. (1999a). Smoothing of Stokes's discontinuity for the generalized Bessel function, *Proc. Roy. Soc. London* **A455**, 1381–1400.

Wong, R. and Zhao, Y.-Q. (1999b). Smoothing of Stokes's discontinuity for the generalized Bessel function. II, *Proc. Roy. Soc. London* **A455**, 3065–3084.

Wood, A.D (1971). Deficiency indices of some fourth order differential operators, *J. London Math. Soc.* **3**, 96–100.

Wood, A.D. (1991). Exponential asymptotics and spectral theory for curved optical waveguides, in *Asymptotics Beyond All Orders*, ed. H. Segur, H. Tanveer and H. Levine, 317–326 (Plenum Press, New York).

Wrench, J.W. (1968). Concerning two series for the gamma function, *SIAM J. Math. Anal.* **22**, 617–626.

Index